D1330856

# Emissions from Combustion Processes: Origin, Measurement, Control

Edited by
Raymond Clement
Ron Kagel

**Library of Congress Cataloging-in-Publication Data**

Emissions from combustion processes: origin, measurement, control/edited by Raymond Clement, Ron Kagel.

p. cm.

Includes bibliographical references.

ISBN 0-87371-172-6

1. Combustion gases—Environmental aspects. 2. Incineration—Environmental aspects. 3. Factory and trade waste—Environmental aspects. I. Clement, Raymond. II. Kagel, Ronald O.

TD885.5.C66E45 1990

628.5′32—dc20 90-6001

CIP



LEWIS PUBLISHERS, INC.
121 South Main Street, P.O. Drawer 519, Chelsea, Michigan 48118

PRINTED IN THE UNITED STATES OF AMERICA

# Preface

Environmental pollution from combustion processes is an important area of concern. Attention has recently been focused on the emissions from municipal waste incineration, although the issue is much more complex. Other sources such as rotary kilns, steel mills, pulp and paper plants, and natural sources must also be assessed. Great concern over the emission of trace levels of the chlorinated dibenzo-*p*-dioxins and chlorinated dibenzofurans has been expressed by some, but other pollutants such as acid rain precursors, polycyclic aromatic hydrocarbons, and others should not be ignored.

Fundamental questions concerning the emissions from combustion sources must be investigated before their full environmental impact can be assessed. What are the mechanisms of formation of toxic organics? How are they distributed in the environment? What analytical methods are needed to perform these investigations including sampling, analysis, and quality control? How do we assess the risk to humans and the environment of selected combustion-generated contaminants? Finally, how do we control these emissions to minimize risks, and what remedial action can be taken where significant environmental degradation has already occurred?

Solutions to the above and other related questions will be developed through careful scientific study and by the application of advanced technologies. This book has brought together leaders in many of the areas of combustion-related research who present their recent work regarding the science and technology of emissions from combustion processes. Although it would be impractical to attempt a comprehensive coverage of such a large and complex area as combustion processes in a single volume, we believe that many of the most important scientific issues have been addressed. Solutions to the environmental problems caused by emissions from combustion processes will be based, in part, on the science and technology that is described here.

**R. E. Clement, Ph.D.**, was awarded the Ph.D. degree in 1981 under the supervision of Professor Emeritus F. W. Karasek at the University of Waterloo. He joined the Ontario Ministry of Environment in 1982 and is currently a senior research scientist in the Laboratory Services Branch where he directs the research of the Dioxin and Mass Spectrometry Units. Dr. Clement is co-author of the book *Basic Gas Chromatography–Mass Spectrometry: Principles and Techniques*, and has authored more than 70 publications in the field of trace organic analysis. His principal research interest is the analytical chemistry of the chlorinated dibenzo-p-dioxins and dibenzofurans, and much of his work deals with the emissions of these compounds from combustion sources. Dr. Clement was co-chair of the 9th International Symposium on Chlorinated Dioxins and Related Compounds held September 17–22, 1989 in Toronto, Canada. Dr. Clement serves on the Editorial Board of the environmental journal *Chemosphere* and is a member of ACS, CIC, ASMS, and AOAC.

**Ronald O. Kagel, B.S., Ph.D., FAIC** B.S. University of Wisconsin 1958, Ph.D., University of Minnesota 1964, Environmental Consultant, Environmental Quality Department, The Dow Chemical Company, Midland, Michigan. Elected, Fellow, American Institute of Chemists, 1986. Chairman-elect of the Board of Directors, Coalition for Responsible Waste Incineration (CRWI). American Chemical Society, Association for the Advancement of Science. Coblentz Society, New York Academy of Sciences. Joint Committee on Atomic and Molecular Physical Data. Industrial Advisory Council, Illinois Institute of Technology. Industrial Advisory Committee, UCLA, NSF Center for Hazardous Waste. American Men and Women of Science. Who's Who in Technology Today, Who's Who in the Midwest. Alpha Chi Sigma, Phi Lambda Upsilon, Sigma XI. Past: Chairman, of the Board of Directors, CRWI; Chairman, Fourier Transform Spectroscopy Technical Groups, Optical Society of America and Society for Applied Spectroscopy. Associate editor, *Applied Spectroscopy*; Coeditor, *Transform*; Editorial Advisory Boards, *The Raman Newsletter* and *Chemecology*; Liaison Coordinator ASTM D-19; ANSI ISO/TC; National Research Council (Sub Coms.); Task Group Leader, Chemical Manufacturers Association; Environmental Quality Committee, Synthetic Organic Chemical Manufacturers Association. Author or coauthor of over 80 publications, reports, presentations, and book chapters; two books (one in press).

# Contents

## CHAPTER 1

# An Overview of Combustion Emissions in the United States

**Dale A. Pahl, David Zimmerman, and Ronald Ryan**

## INTRODUCTION

Combustion emissions are specific to individual site and device parameters, including combustor type and design, operating conditions, fuel quality, control device operation, and maintenance practices. An overview of U.S. combustion emissions cannot examine each combustion source, but must aggregate combustion processes to permit a meaningful summary and analysis of the data on a national scale.

The U.S. Environmental Protection Agency (EPA) employs several tools to analyze emissions and their transport through the atmosphere. These tools include trend analyses, detailed emissions inventories, individual source tests, and emissions species profiles developed from source tests. The latter two methods are too detailed for an overview because they look at specific emissions test data to examine particular species from specific combustors or combustor types. The first two methods can be used to gain an overview of combustion sources. Trend analyses allow emissions magnitudes, transport parameters, and major source categories to be examined over time without reference to individual sources. Detailed inventories include individual plants and combustors above a specified emissions level and allow the linkage of particular source emissions with transport parameters for each source. The standard format of the detailed data base allows the data to be aggregated to less detailed levels. These two "broadbrush" techniques—trend analyses and detailed emissions inventories—are used to summarize combustion emissions for this paper.

Because the eventual purpose of this research is to assess the impact of emissions, parameters needed to calculate the transport of emissions are as critical as the magnitude of emissions. The detailed inventory is useful because it allows one to link emissions and transport parameters. With this information, the fate of pollutants can be explored through atmospheric transport modeling and ambient monitoring. Information on the chemical constituents of emissions allows analysis of potential health and ecological effects.

Emissions sources are divided into two categories: stationary (e.g., boilers) and area/mobile sources (e.g., automobiles). These categories are treated differently by trends and detailed data bases. The trends methodologies distinguish only between stationary and mobile sources. The detailed methodology includes those stationary sources which are too small or too numerous for an individual inventory (at the national level) as area sources, in addition to mobile sources. Stationary sources emitting less than 25 tons* per year (TPY) of $NO_x$ (nitrogen oxides), $SO_2$ (sulfur dioxide), and VOC (volatile organic compounds) are therefore included in the area source category.

Stationary sources have been grouped into categories for analysis based on their contributions to total anthropogenic U.S. $SO_2$ and $NO_x$ emissions. The major categories include utilities, industrial combustion, and other combustion sources (e.g., commercial combustion and waste disposal). Data are also presented by fuel type and, within the industrial category, by industry sector. Area source categories presented here include transportation, residential combustion, and industrial combustion.

In reviewing trends data, it is sometimes not possible to separate combustion and noncombustion emissions, particularly in the industrial and miscellaneous categories, because process-level detail is not available at this level of aggregation. Such a separation is possible in the detailed data base and is explained in the following sections.

This paper will summarize combustion emissions using the results of past and ongoing research programs in emissions trends and detailed inventories. Due to the availability of data and the distribution of total U.S. emissions, $SO_2$ and $NO_x$ will be emphasized. The discussion will include both magnitude and transport emissions parameters. The paper will initially review the general methodologies employed by the referenced research, then summarize the results regarding aggregated, combustion emissions categories. Finally, brief conclusions will be drawn from the presented information.

## METHODOLOGY

Combustion sources contribute significantly to the total $SO_2$ and $NO_x$ emitted by anthropogenic sources. Major combustion sources include electric utilities, industrial boilers and internal combustion engines, smelters, natural gas engines and turbines, industrial process heaters, iron and steel furnaces, kilns, incinerators, residential fuel combustion, and transportation sources. This paper presents an overview of long-term historic trends in combustion emissions and a detailed breakdown of current (1985) emissions from combustion sources. Methodology and data sources for these two analyses are discussed here.

*For readers more familiar with metric units, a table of conversion factors follows the references.

## Historic Inventory

Historic inventories permit the identification of aggregated emissions over many years. In order to provide consistency to the estimates and trends, the methodology is limited to the use of data that are available over the entire time frame. The focus of the overview is on major categories of emissions to gain perspective on their emissions trends over time.

Three major historic data bases have been developed to examine trends for $SO_2$ and $NO_x$ emissions.[1,2,3] The primary data set used here, "Historic Emissions of Sulfur and Nitrogen Oxides in the United States 1900 to 1980 Volume 1. Results," was developed to provide long-term estimates of emissions to allow comparison with existing, long-term effects research. These data are resolved to the state level for 5-year intervals and are divided among approximately 30 sectors encompassing all combustion emissions. In order to present long-term trends, this paper will rely principally on this historic inventory from 1900 to 1980. A brief review of the methodology is provided here. This work was extended to encompass VOC emissions from 1900 to 1985 and those data are also used here.[4] All data are resolved to the state level and include total anthropogenic emissions by major sector.

Combustion-derived emissions in the historic $SO_2$ and $NO_x$ database are broken down into major fuels. Annual quantities of emissions of $SO_2$ and $NO_x$ were calculated for each of the contiguous 48 states and the District of Columbia. Emissions of each pollutant were estimated for every fifth year from 1900 to 1980. The time span from 1900 through 1980 was selected to allow a comparative study with historic environmental measurements. The 1985 data from similar trends studies were integrated to extend the analysis to 1985 for $SO_2$ and $NO_x$.[6] Five-year intervals were selected to indicate the emission trends sufficient for most effects studies and to develop a methodology that could be applied to all other years. The state level was selected because it provides the most complete and consistent body of information on an historic basis and collectively covers all geographic regions of the country. The basic steps involved in calculating state emissions of $SO_2$ and $NO_x$ are listed here as an example of the methodology employed:[1]

- Obtain national/state level information on fuel use.
- Allocate fuel quantity used by each source category.
- Develop/apply source category emission factors.
- Determine fuel sulfur content by state for each category.
- Apportion/calculate emissions, after emissions controls.

For each state, the estimates are based on the apparent annual consumption rate of fuels. The fuels include bituminous and anthracite coals, lignite, residual and distillate oil, natural gas, wood, gasoline, diesel fuel, and kerosene. The consumers of these fuels, which are also the emitters of $SO_2$ and $NO_x$, are categorized according to whether they are electric utilities, industrial boilers, commercial and residential furnaces, natural gas pipelines, highway vehicles, railroads, coke plants, smelters, vessels, or other major sources. Emissions were also estimated for wildfires, industrial processes (including in-process fuel combustion) based on production rates, and a miscellaneous source category. Collectively, these source categories account for all anthropogenic emissions in each state.

The state-level consumption and process data were multiplied by specially derived emission factors to yield estimates of uncontrolled emissions. The actual procedure varied somewhat depending on the usefulness and availability of information. In general, state-specific data were available for 1950 to 1980. Prior to 1950, these data were not always available and national data were apportioned to the states.

Another historic data base, *Estimated Monthly Emissions of Sulfur Dioxide and Oxides of Nitrogen for the 48 Contiguous States, 1975–1984,*[2] has been developed to track emissions for major source categories monthly and annually. These data are based on the *National Air Pollutant Emission Estimates, 1940–1984.*[3] Emissions are allocated to each state based on state shares in the 1980 National Acid Precipitation Assessment Program (NAPAP) emissions inventory[5] and further disaggregated using monthly activity indices for various sectors. The three historic databases differ chiefly in the temporal resolution, temporal extent, and category resolution.

## Detailed Emissions Inventory

Detailed emissions data inventories contain highly resolved process, emissions, temporal, and spatial data not available from historic inventories. The detailed inventory permits analysis of the relationships between emissions at given sources, the transport and transformation of emissions, and their impact on the local or distant environment. Detailed inventories identify the relative importance of both emissions categories and individual sources.

The most comprehensive U.S. $SO_2$ and $NO_x$ emissions inventories are the 1980 and 1985 NAPAP emissions databases, which the states and the EPA compiled for the NAPAP.[5] Both contain source-specific data and emissions estimates for their respective base years. These inventories are designed to permit the analysis of relationships between emissions sources and regional air pollution problems (e.g., acid precipitation, ozone). The 1985 database will be discussed here because it represents the most accurate and comprehensive national inventory of combustion sources in the United States.

The 1985 emissions inventory is designed to provide anthropogenic emissions data for all significant stationary and area sources of $SO_2$, $NO_x$, and VOC in the United States, as well as engineering parameters which affect the atmospheric transport of emissions. The inventory is divided into two major categories: point and area sources. Point sources are defined as sources that have precise location data and emit at least 25 TPY of a NAPAP priority pollutant ($NO_x$, $SO_2$, or VOC). Area sources comprise both mobile sources and point sources too small to list individually because collecting point source data

**Table 1. Specifications for the Historic Emissions Trends and 1985 Detailed Emissions Databases**

| Specification | Historic | 1985 Detailed |
|---|---|---|
| Geographic domain | Contiguous United States | Contiguous United States |
| Species | $SO_2$, $No_x$, VOC | $SO_2$, $NO_x$, VOC, PM, CO<br>Primary $SO_4$, HF, HCL, $NH_3$ |
| Spatial resolution | State | Emission point (point source)<br>county (area source) |
| Temporal resolution | Annual/five-year | Seasonal/annual |
| Source resolution | Major sectors | 3300 point source classes<br>100 area source classes |
| Uses | Trend analysis | Asessment research<br>Effects research<br>Projections |

for every stationary source in the United States is impractical. Combustion sources reside in both categories.

The inventory contains estimates of emissions from approximately 100,000 stationary sources in the United States. During the last decade, emissions from important process categories have been measured using standardized sampling techniques. The measurements have been compiled and analyzed for each category to develop emission factors for each of 3300 industrial processes currently reported in the United States. Many of these factors are contained in the EPA publication AP-42.[7] The emission factor is multiplied by the activity level of a stationary or area source over a given period of time to estimate emissions.

Area source emissions estimates are calculated by the EPA using a series of computer programs in a "top-down strategy" which allocates national emissions estimates to the state and county levels. Approximately 100 area source categories are used to estimate emissions in each U.S. county.

The 1985 inventory focused on the following items: annual emissions estimates for $SO_2$, $NO_x$, and VOC; the maximum design rate and operating data; the Source and Standard Industrial Classification Codes (SCC and SIC); emissions control equipment and efficiencies; fuel characteristics; stack parameters; and location data. The 1985 database contains estimates of $SO_2$, $NO_x$, and VOC; as well as two other criteria pollutants, CO (carbon monoxide) and PM (particulate matter); and four chemical species believed to play an important role in acid deposition, primary sulfate, hydrogen chloride, hydrogen fluoride, and ammonia.

Table 1 compares the specifications for the 1900 to 1980 historic emissions database and the 1985 NAPAP emissions database.

## RESULTS

Combustion-related emissions comprise more than 90% of total anthropogenic $SO_2$ and $NO_x$, and about 50% of total anthropogenic VOC in the 1985 emissions inventory. Due to the structure of both the historic and current inventories, it is not possible to entirely separate combustion and process emissions for all industrial processes. Certain processes (such as smelting, electric arc furnaces, and nitric acid production) produce $SO_2$ and $NO_x$ emissions which are not directly combustion products. If emissions from these and other industrial processes are considered combustion-related, nearly 100% of anthropogenic $SO_2$ and $NO_x$ emissions arise from combustion.

In deciding which processes were sources of combustion emissions, the following criteria were used. Only those sources where a material is oxidized to provide heat or to destroy the bulk of the material were included. Chemical processes which use an oxidation reaction to produce an intermediate stream were not included. Thus, the $SO_2$ emissions of Claus sulfur recovery units and sulfuric acid production plants are not included, even though burning is used in these processes to produce an $SO_2$ feedstream. $NO_x$ emissions from the catalytic oxidation of ammonia in the production of nitric acid are not included, nor are emissions from carbon black furnaces or charcoal kilns. Some major processes which are included in combustion emissions are refinery catalytic cracker regeneration, hydrogen sulfide gas incineration or flaring, and smelting furnaces. Smelting furnaces include emissions from fuels which are burned to produce heat as well as emissions of oxidized impurities from the ores. These emissions are not separable by the inventory. As a result, smelter $SO_2$ emissions are inordinately larger than their $NO_x$ emissions due to the oxidation of sulfides.

Some VOC emissions occur as a result of incomplete combustion, although most are from noncombustion (evaporative) sources. VOC emissions are included here due to their importance in acid deposition chemistry. It is important to note that total emissions of $SO_2$ and $NO_x$ are essentially the same as combustion emissions, but that the same is not true for VOC emissions. The major focus of this paper is therefore on $SO_2$ and $NO_x$ emissions.

The results consider atmospheric transport parameters as well as emissions magnitudes. Both data types are important to assess the fate of emissions. The critical atmospheric transport parameters include source location, stack height, exhaust gas flow rate, and temperature. In this paper, the historic and detailed inventories are analyzed in terms of stack height. The spatial distribution of major sources is presented for the detailed inventory.

### Historic Trends

Total annual emissions of $SO_2$, $NO_x$, and VOC at 5-year intervals are plotted in Figure 1, along with Gross National Product (GNP) and population totals. The lines connecting the 5-year data points are shown only to indicate the

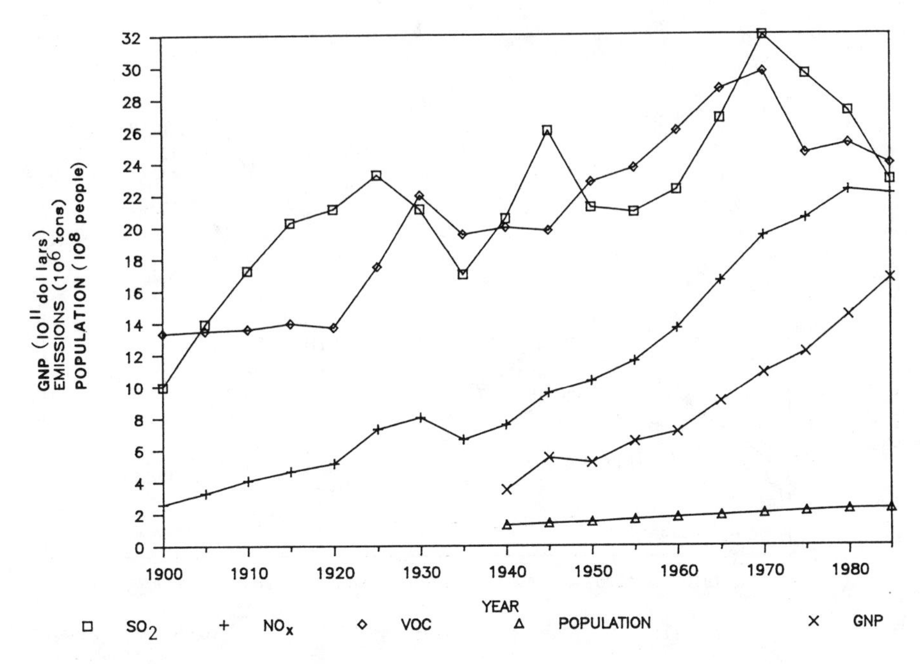


**Figure 1.** Historic U.S. emissions of $SO_2$, $NO_x$, and VOC.

trend and do not represent actual data. $NO_x$ emissions have increased more rapidly and more steadily than $SO_2$ or VOC emissions, rising from $3 \times 10^6$ tons in 1900 to a high of about $22 \times 10^6$ tons in 1980. $SO_2$ emissions have increased erratically from $10 \times 10^6$ tons (1900) to a high of $32 \times 10^6$ tons in 1970. Major decreases in $SO_2$ emissions occurred as a result of the Depression, the end of World War II, and the passage of the Clean Air Act. VOC emissions trends are similar to those of $NO_x$, with an additional major downturn in 1975. GNP and population are plotted as examples of social and economic trends which link and affect emissions trends in the United States. Many additional factors, such as the growth of the electric utility and automobile industries and the use of natural gas fuels, need to be considered to fully understand or project these trends.

Figures 2, 3, and 4 present the historic data for $SO_2$, $NO_x$, and VOC, respectively, broken down by major categories. Industrial sources include industrial boilers, space heaters, and industrial process (noncombustion) emissions. Commercial/residential includes all fuels used in these sectors. Highway vehicles reflect both gasoline- and diesel-powered highway motor vehicles. The other category includes other transportation sources (vessels, railroads, off-highway), smelters, wildfires, and miscellaneous sources.

Industrial emissions dominated total $SO_2$ until 1950 (Figure 2). Since that time, utilities have accounted for a rapidly growing proportion, reaching about 75% in 1980. The other three major sectors have all shown declining $SO_2$ emissions since 1945. Figure 3 shows that $NO_x$ emissions have increased stead-

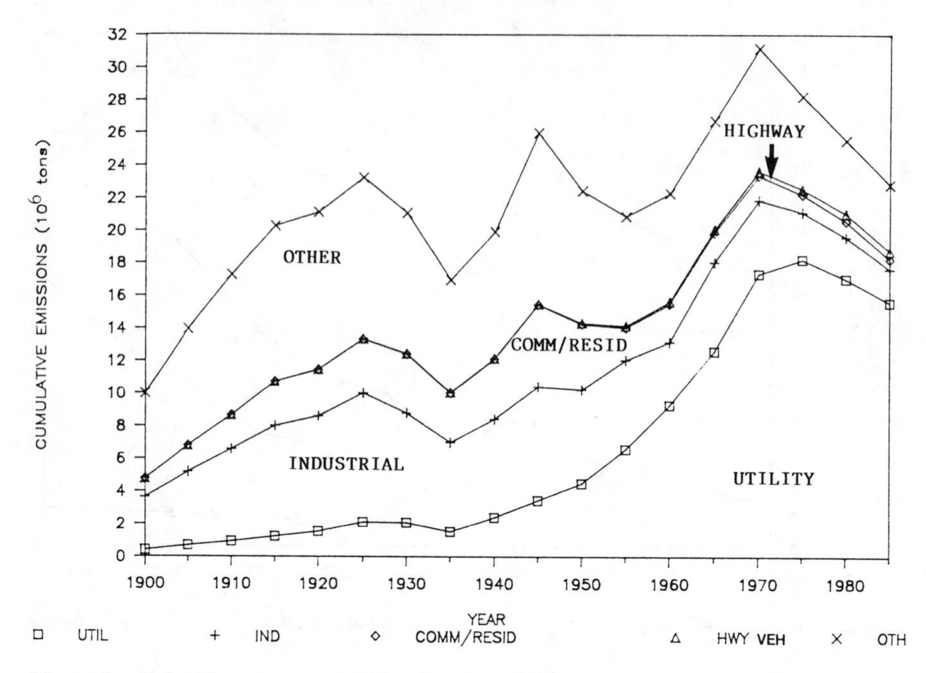


**Figure 2.** U.S. $SO_2$ emissions by major category: 1900 to 1985.

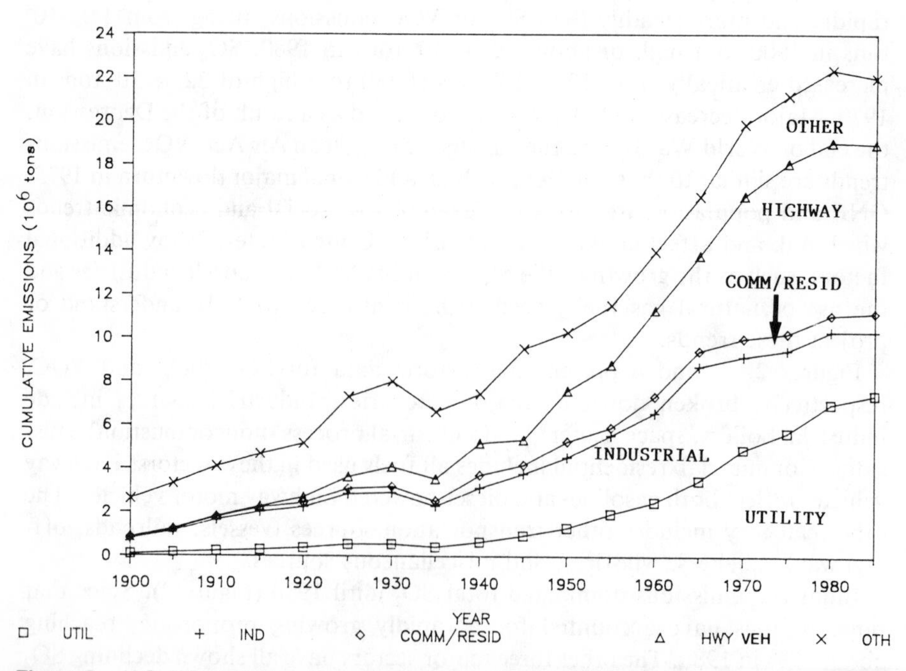


**Figure 3.** U.S. $NO_x$ emissions by major category: 1900 to 1985.

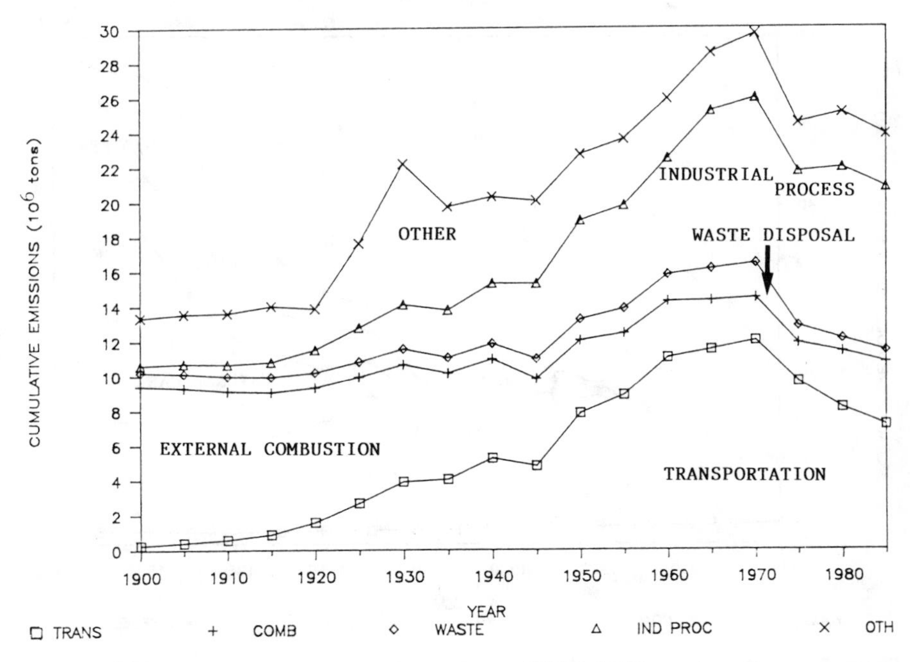


**Figure 4.** U.S. VOC emissions by major category: 1900 to 1985.

ily since 1900, dipping only during the Depression. The increase has principally resulted from the growth of utility and highway vehicle emissions. Figure 4 shows that the VOC emissions trend is the result of three major sectors varying independently.

## *Fuel Use*

Both increased total fuel use and changes in the types of fuels used greatly affect emissions. Figure 5 shows that total coal consumption has varied little since 1910, but has shown a steady, slow increase since 1960. Natural gas consumption has risen dramatically since 1935 and has contributed significantly to the similar trend in $NO_x$ emissions. Wood combustion has decreased in both absolute and relative terms since 1900. The oil category includes residual and distillate oil, but does not include gasoline and diesel fuels. Gasoline and diesel fuels are shown as a separate category in Figure 5.

## *Stack Height*

Emissions exit height is one of the important parameters affecting the transport of pollutants in the atmosphere. These data are available from the historic inventory (Figures 6 and 7) in four broad categories: less than 120 ft, 120 to 240 ft, 240 to 480 ft, and more than 480 ft. The first class includes ground-level emissions and the last, the tallest stacks. Exit velocities and temperatures are not in the historic inventory.

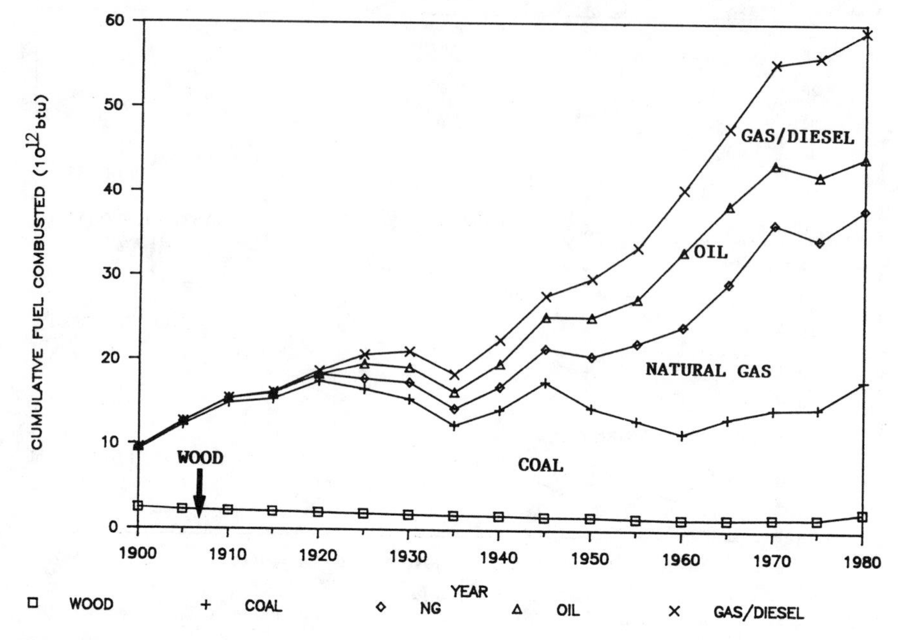


**Figure 5.** Historic fuel use by fuel type.

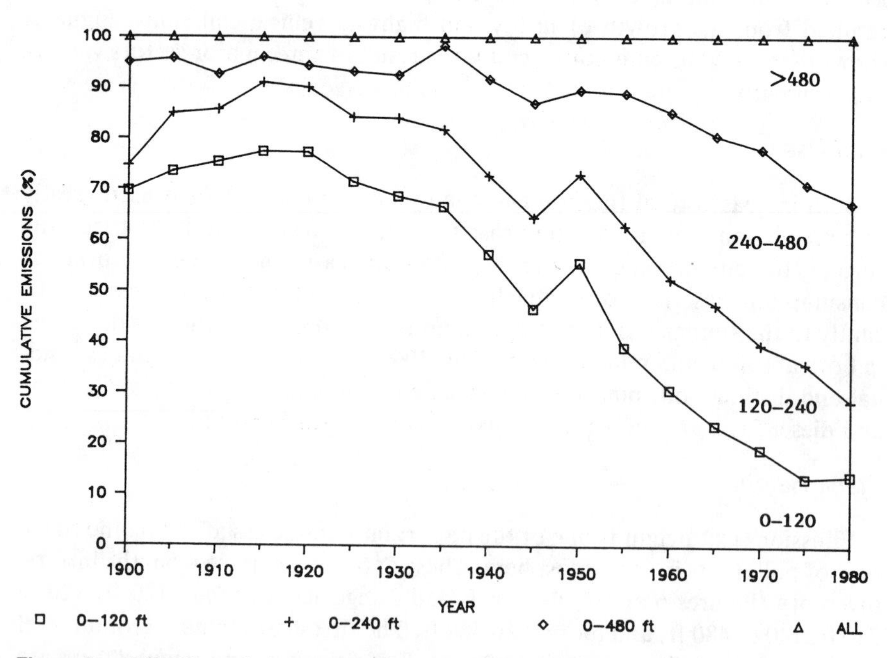


**Figure 6.** U.S. $SO_2$ emissions by stack height: 1900 to 1980.

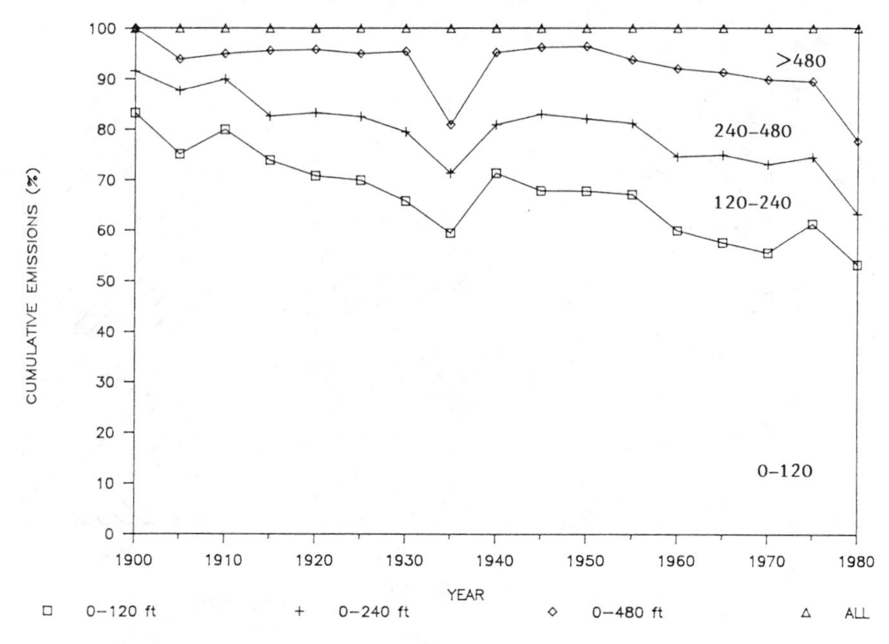


**Figure 7.** U.S. $NO_x$ emissions by stack height: 1900 to 1980.

$SO_2$ and $NO_x$ emissions distributions by stack height show the trend toward taller stacks. For $SO_2$, emissions prior to 1920 primarily came from stacks less than 120 ft. Beginning in 1920, $SO_2$ emissions from the two tallest categories increased in relation to those from the smaller stacks. As of 1980, more than 30% of $SO_2$ emissions are from stacks of 480 ft or taller, and over 70% of $SO_2$ emissions are from stacks taller than 240 ft. $NO_x$ emissions continue to be dominated by ground-level sources, although the proportion has dropped from 83% in 1900 to 53% in 1985. Since 1950, the growth in proportional distribution has been in the tallest stacks.

## 1985 Detailed Emissions Inventory

This section presents an overview of combustion emissions reported in the 1985 NAPAP anthropogenic emissions inventory. The detailed inventory provides much greater source resolution than the historic inventory, including data for discrete point sources (e.g., stacks) and for 100 categories of area sources for each U.S. county. Emissions data from the 1985 detailed emissions inventory are based on a preliminary summary of the database. These data have since been reviewed by the states as a final quality assurance step. Revisions based on this review will be reflected in the final 1985 anthropogenic emissions inventory.

The differentiation between point and area sources is fundamental to the structure of the detailed data base. Figure 8 shows how each of the three

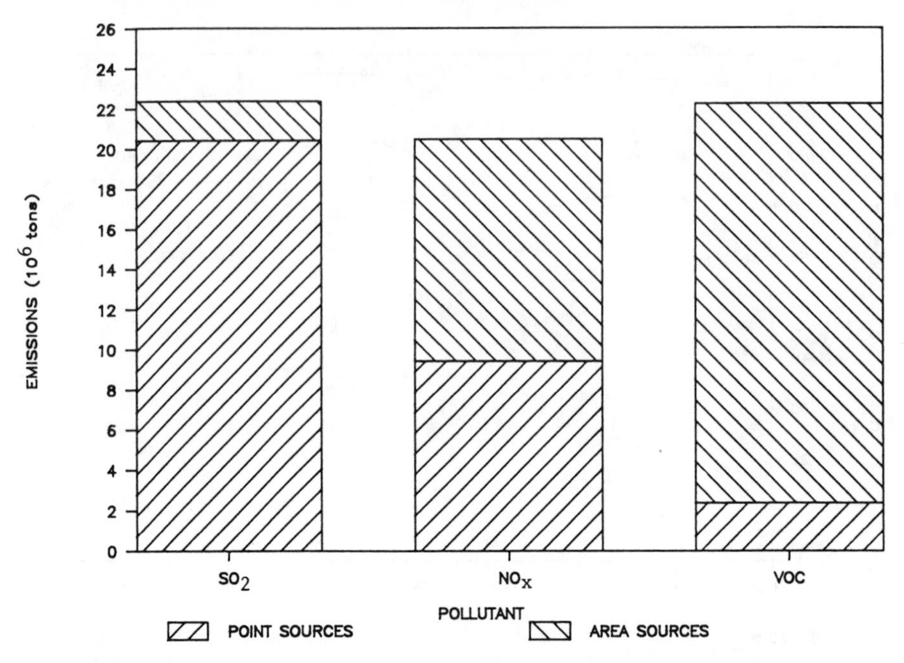


**Figure 8.** Total 1985 anthropogenic emissions from point and area sources.

pollutants is divided between point and area sources. This figure includes noncombustion emissions, which are large contributors only for VOC. The figure shows that point sources account for 90% of $SO_2$, the $NO_x$ emissions are divided nearly equally between point and area sources, and over 90% of VOC emissions are from area sources. The largest major sectors for combustion emissions are utilities, industrial (i.e., smelters, iron and steel, refineries), and transportation (highway) emissions (Table 2).

The largest single $SO_2$ emissions sector is utilities. These units contribute $15.6 \times 10^6$ tons of $SO_2$. The distribution of $SO_2$ emissions among all major sectors is shown in Figure 9. The transportation sector accounts for over 43% of total $NO_x$ emissions ($8.8 \times 10^6$ tons). Utilities contribute about 32 percent ($6.7 \times 10^6$ tons). Figure 10 shows the distribution of $NO_x$ emissions among

**Table 2. 1985 Anthropogenic Emissions by Major Category ($10^3$ ton)**

| Major Category | $SO_2$ | $NO_x$ | VOC | PM | CO |
|---|---|---|---|---|---|
| Utility combustion | 15,590 | 6,659 | 58 | 490 | 367 |
| Transportation | 864 | 8,834 | 7,350 | 4,428 | 42,697 |
| Industrial combustion | 3,729 | 2,358 | 375 | 635 | 2,359 |
| Commercial/ residential combustion | 571 | 689 | 3,037 | 1,529 | 8,899 |
| Other combustion | 572 | 712 | 591 | 715 | 4,360 |
| Total combustion (all sources) | 21,326 | 19,252 | 11,411 | 7,797 | 58,682 |
| Total anthropogenic emissions | 22,404 | 20,505 | 22,387 | 36,913 | 63,375 |

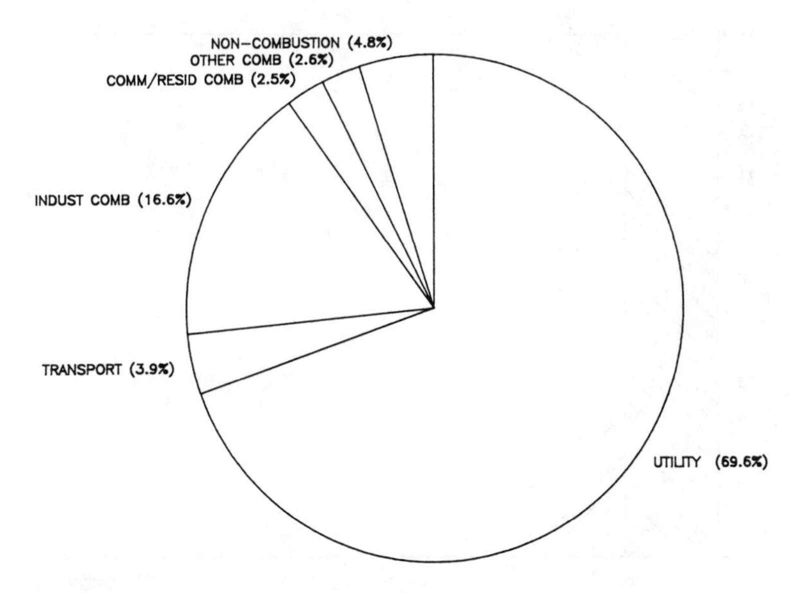


**Figure 9.** Distribution of 1985 U.S. $SO_2$ emissions by major category.

major categories. While combustion sources totally dominate $SO_2$ and $NO_x$ emissions, they account for only 59% of VOC emissions. Nearly 65% of combustion VOC emissions are from transportation sources, and more than 25% are from commercial and residential combustion (these are principally from residential wood combustion).

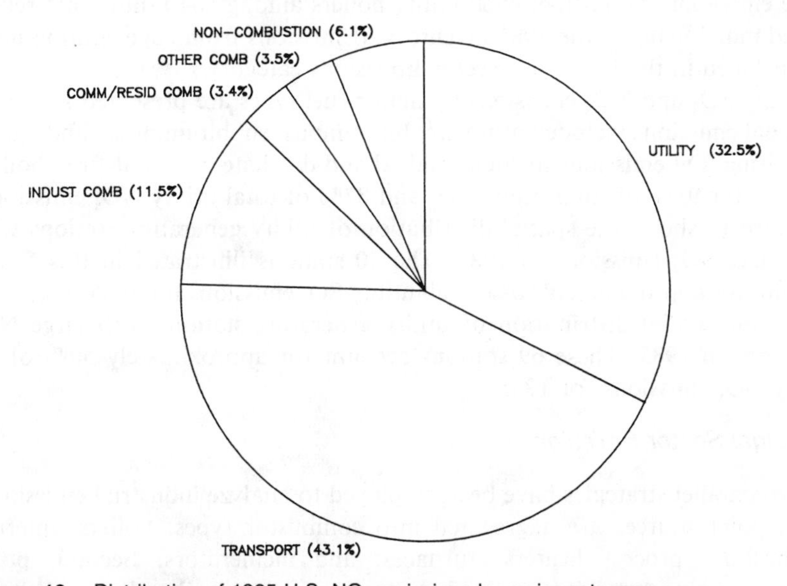


**Figure 10.** Distribution of 1985 U.S. $NO_x$ emissions by major category.

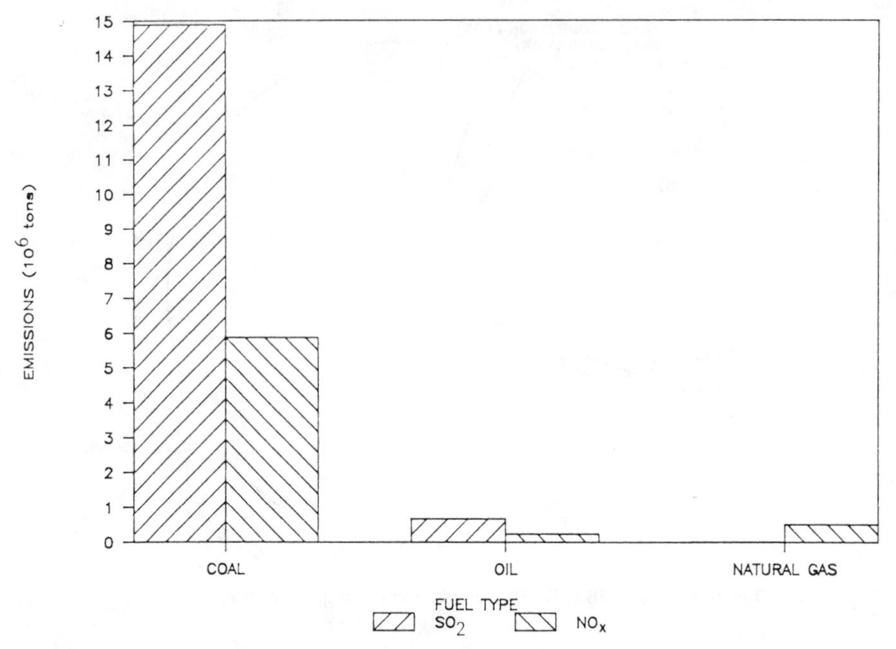


**Figure 11.** 1985 utility $SO_2$ and $NO_x$ emissions by fuel type.

## *Utility Sector Emissions*

Utilities emit $SO_2$ and $NO_x$ principally from coal-fired external combustion boilers, with oil-fired and gas-fired combustion turbines also contributing some emissions. More than 4600 utility boilers among 864 utilities are represented individually in the 1985 inventory. (Emissions from cogeneration units are included in the Industrial Sector Emissions category.)

Utility $SO_2$ and $NO_x$ emissions by major fuel types are presented in Figure 11. Coal emissions include anthracite, bituminous, subbituminous, and lignite coal firing. Oil emissions include residual and distillate oil. Coal-fired boilers account for 95% of total utility $SO_2$ and 87% of total utility $NO_x$ emissions.

Figure 12 shows the spatial distribution of utility generating stations with very large $SO_2$ emissions in 1985. The 50 stations illustrated in this figure account for approximately 50% of all utility $SO_2$ emissions for 1985. Figure 13 shows the spatial distribution of utility generating stations with large $NO_x$ emissions in 1985. These 69 stations account for approximately 50% of all utility $NO_x$ emissions for 1985.

## *Industrial Sector Emissions*

Two parallel strategies have been employed to analyze industrial emissions. First, point sources are aggregated into combustor types: boilers, internal combustion, process heaters, furnaces, and incinerators. Second, point sources are aggregated into major industry types: paper, oil and gas, smelters,

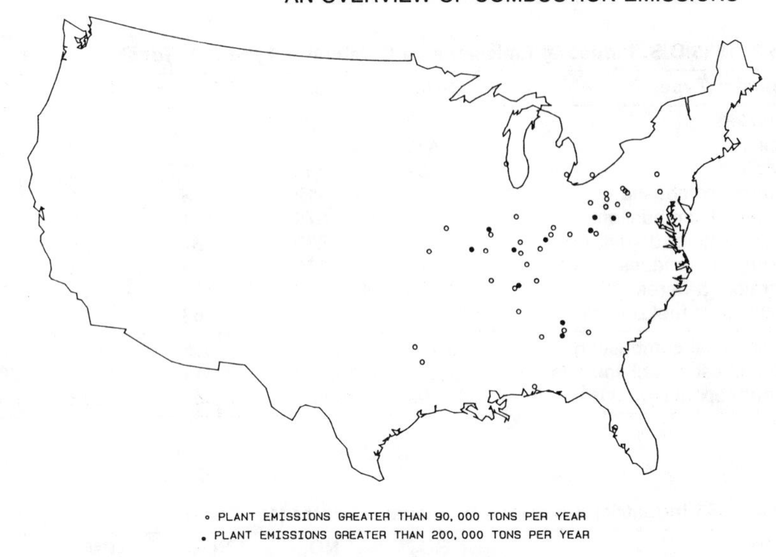


**Figure 12.** Spatial distribution of selected electric utilities with large $SO_2$ emissions.

iron and steel, cement, chemicals, and other. The detailed inventory contains both a SCC and a SIC for each point source to allow this analysis.

Table 3 shows a breakdown of industrial combustion emissions by combustor type, and Table 4 shows a breakdown of the same emissions by industry. Figure 14 shows that oil and gas extraction and refining, ore smelters and iron and steel mills, and pulp and paper mills are the largest industrial sources of $SO_2$ emissions. These industries burn fuel, by-products, or process intermediates containing sulfur contaminants. Figure 15 shows the distribution of indus-

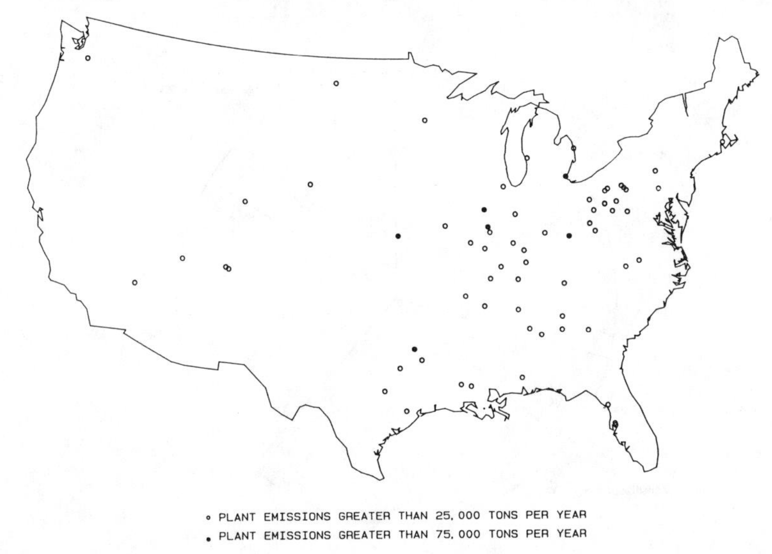


**Figure 13.** Spatial distribution of selected electric utilities with large $NO_x$ emissions.

**Table 3. 1985 U.S. Industrial Emissions by Combustor Type ($10^3$ Tons)**

| Combustor Type | $SO_2$ | $NO_x$ | VOC | PM | CO |
|---|---|---|---|---|---|
| Coal boilers | 1147 | 398 | 5 | 108 | 74 |
| Oil boilers | 476 | 167 | 6 | 63 | 38 |
| Gas boilers | 137 | 392 | 27 | 17 | 228 |
| Oil internal combustion | 2 | 12 | 5 | 1 | 52 |
| Gas internal combustion | 5 | 576 | 71 | 5 | 107 |
| Process heaters, dryers, kilns | 914 | 563 | 87 | 298 | 1103 |
| Metallurgical furnaces & ovens | 642 | 125 | 74 | 18 | 527 |
| Incinerators & flares | 376 | 23 | 27 | 11 | 58 |
| Miscellaneous fuel boilers | 31 | 101 | 63 | 113 | 173 |
| Total industrial combustion | 3,730 | 2,357 | 365 | 635 | 2,360 |
| Total combustion (all sources) | 21,326 | 19,252 | 11,411 | 7,797 | 58,682 |
| Total anthropogenic emissions | 22,404 | 20,505 | 22,387 | 36,913 | 63,375 |

**Table 4. 1985 Industrial Emissions by Industry Sector ($10^3$ Tons)**

| Industry | $SO_2$ | $NO_x$ | VOC | PM | CO |
|---|---|---|---|---|---|
| Pulp & paper mills | 603 | 278 | 91 | 144 | 767 |
| Refineries | 576 | 301 | 87 | 59 | 459 |
| Oil & gas extraction | 259 | 296 | | | |
| Copper, lead, & aluminum smelt | 459 | 9 | 138 | 218 | 369 |
| Iron & steel mills | 208 | 105 | | | |
| Cement manufacturing | 274 | 130 | 1 | 52 | 6 |
| Industrial organic chemicals | 162 | 161 | 10 | 29 | 33 |
| Industrial inorganic chemicals | 94 | 39 | | | |
| Others | 1095 | 1038 | 38 | 133 | 726 |
| Total industrial combustion | 3,730 | 2,357 | 365 | 635 | 2,360 |
| Total combustion (all sources) | 21,326 | 19,252 | 11,411 | 7,797 | 58,682 |
| Total anthropogenic emissions | 22,404 | 20,505 | 22,387 | 36,913 | 63,375 |

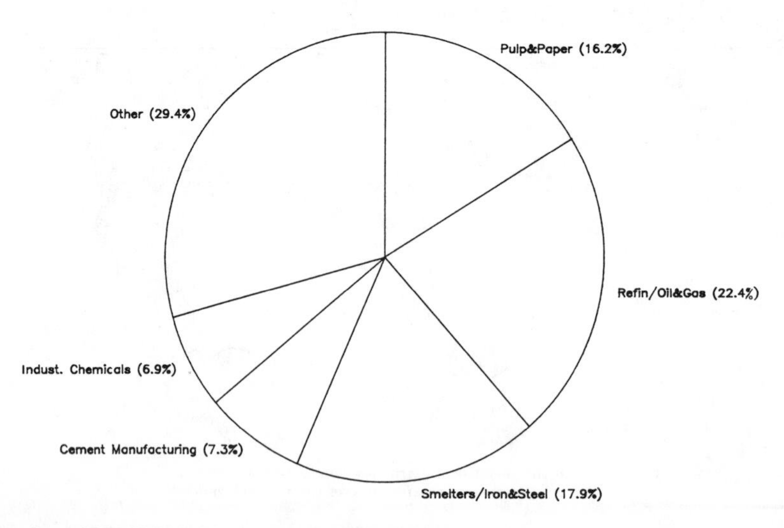


**Figure 14.** 1985 U.S. industrial combustion $SO_2$ emissions by six sectors.

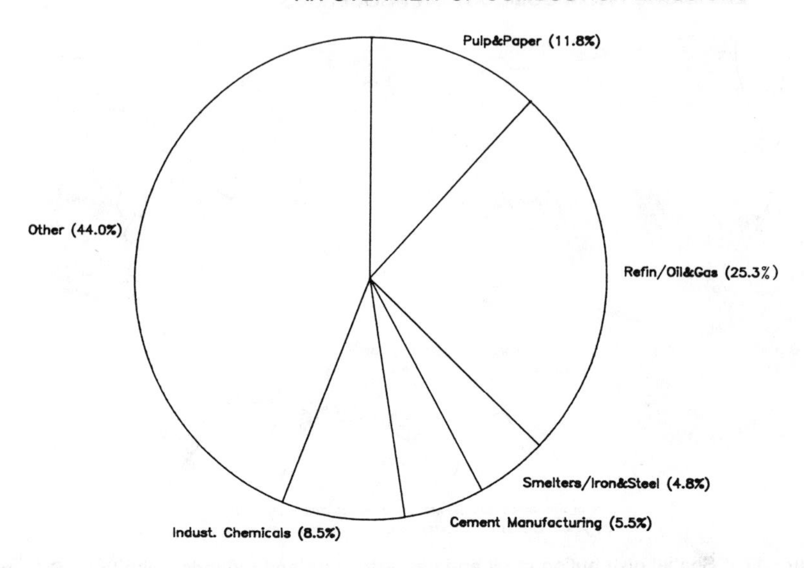


**Figure 15.** 1985 U.S. industrial combustion $NO_x$ emissions by six sectors.

trial $NO_x$ combustion emissions among the industries. Note that the smelting segment accounts for a smaller portion of $NO_x$ than it did for $SO_2$. This is because the smelter $SO_2$ emissions in Figure 14 reflect the oxidation of sulfide in the ores as well as fuel combustion emissions. Figures 16 through 20 show the locations of the largest $SO_2$-emitting plants for the major industries identified. Figures 21 through 25 show the locations of the largest $NO_x$-emitting plants in those industries.

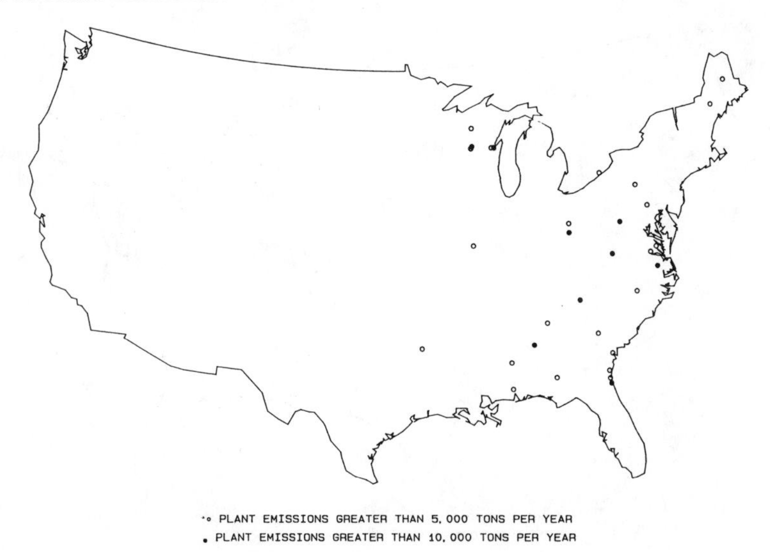


**Figure 16.** Spatial distribution of pulp and paper mills with large $SO_2$ emissions.

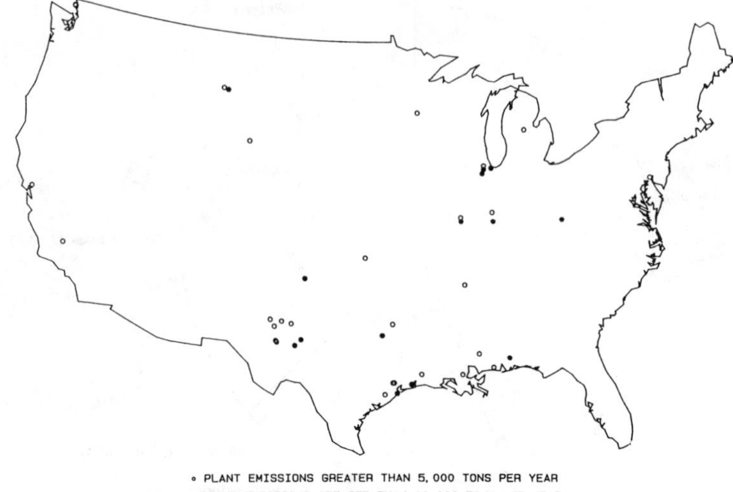


**Figure 17.** Spatial distribution of oil and gas extractors and refineries with large $SO_2$ emissions.

## *Overall Fuel Use Comparison*

Overall U.S. fuel consumption for 1985 reveals that coal is the largest energy source, with natural gas nearly as large. Coal combustion emits principally $SO_2$, while natural gas emits primarily $NO_x$. The use of these two fuels impacts $SO_2$ and $NO_x$ emissions to the greatest extent.

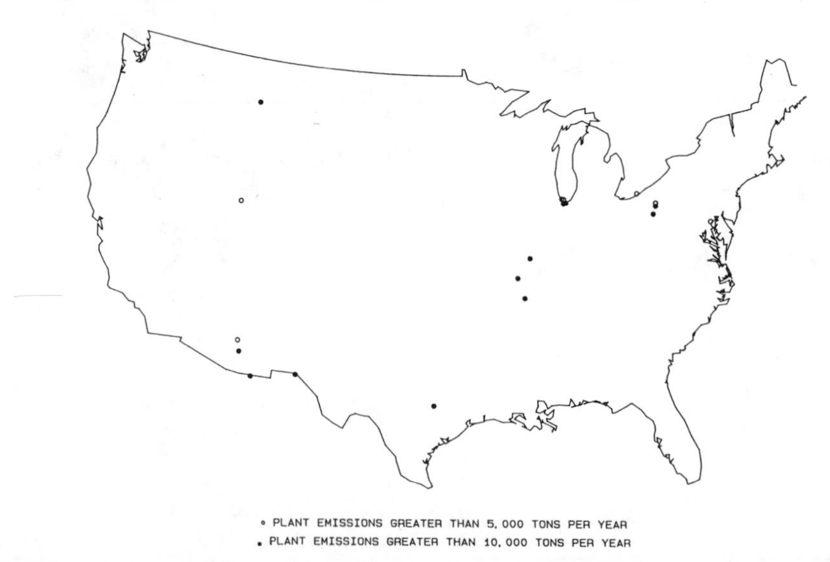


**Figure 18.** Spatial distribution of smelters and iron and steel mills with large $SO_2$ emissions.

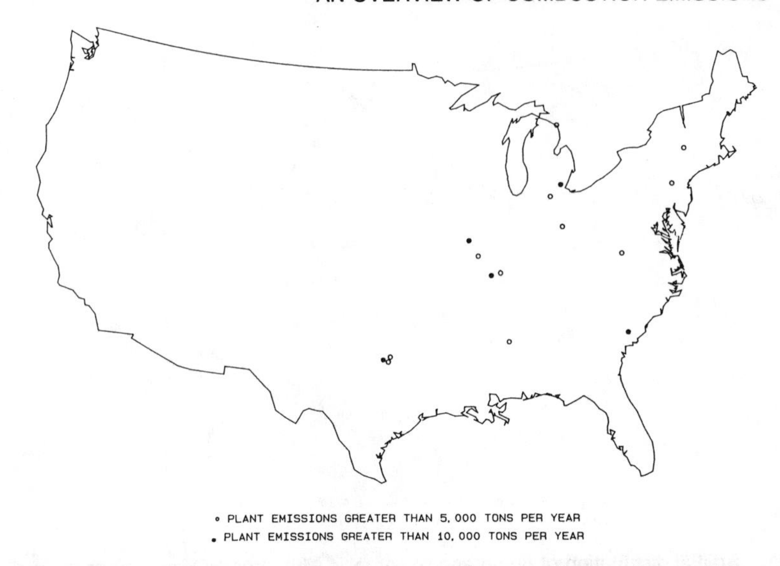


**Figure 19.** Spatial distribution of cement manufacturers with large $SO_2$ emissions.

## *Emissions By Stack Height*

The importance of release height to acid deposition models made collection of stack parameters a high priority for the 1985 inventory. Figure 26 shows the magnitude of $SO_2$ and $NO_x$ for four stack height ranges. It should be noted that this figure is for point sources only and is, therefore, not comparable to Figures 6 and 7. Area sources account for about half of all $NO_x$ emissions, and

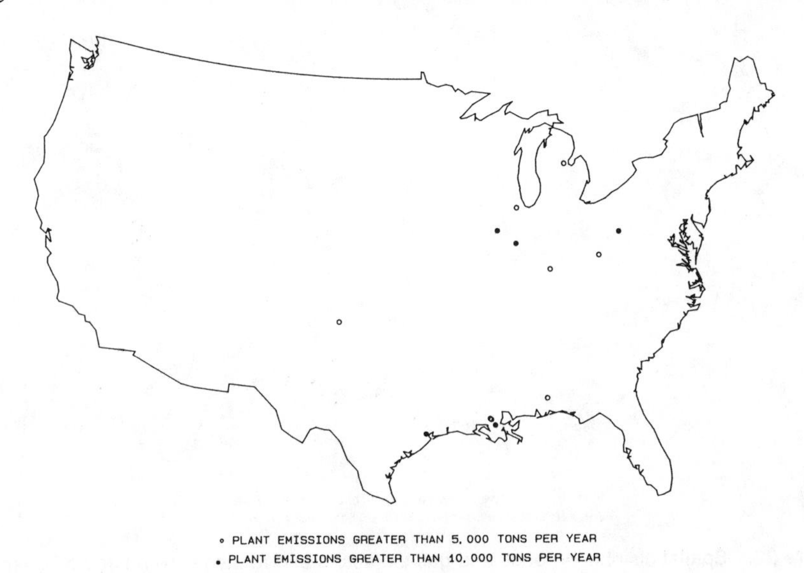


**Figure 20.** Spatial distribution of chemical manufacturers with large $SO_2$ emissions.

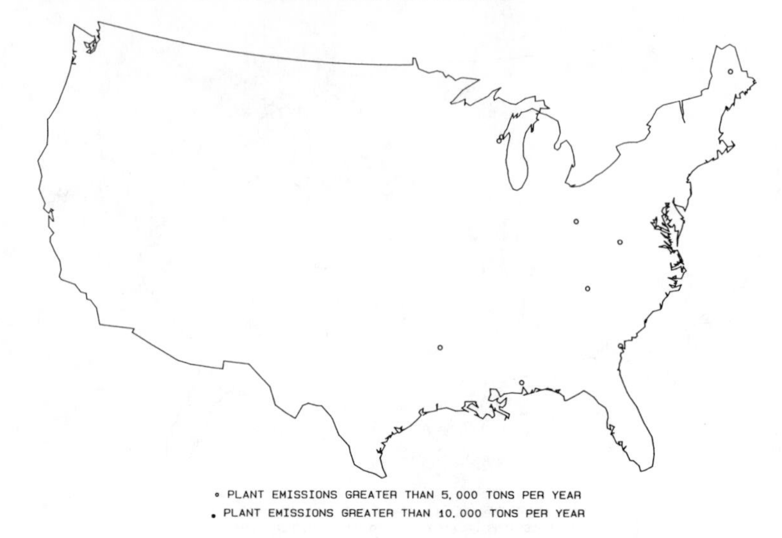


**Figure 21.** Spatial distribution of pump and paper mills with large $NO_x$ emissions.

almost all of these emissions would be at the ground level (less than 120 ft). This figure does show that the bulk of point source $SO_2$ emissions are emitted above 240 ft, while point source $NO_x$ emissions are more evenly distributed vertically.

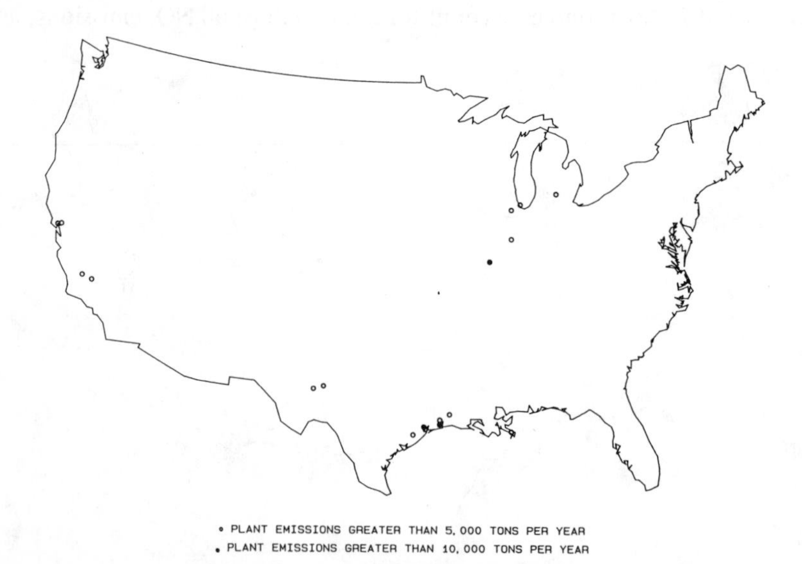


**Figure 22.** Spatial distribution of oil and gas extractors and refineries with large $NO_x$ emissions.

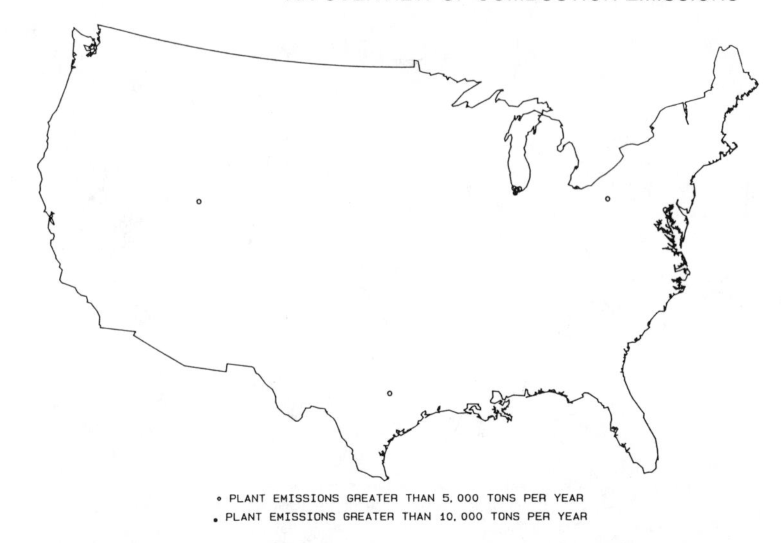


**Figure 23.** Spatial distribution of smelters and iron and steel mills with large $NO_x$ emissions.

## CONCLUSIONS

Combustion categories are the most significant sources of anthropogenic $SO_2$ and $NO_x$ emissions. Utilities contributed nearly 70% of the 1985 anthropogenic emissions of $SO_2$. The two major sources of $NO_x$ in 1985 were utilities and the transportation sector. On the other hand, VOC emissions in 1985 came from combustion and noncombustion sources, principally evaporative fugitive sources, in approximately equal amounts.

Detailed conclusions on specific emissions sources cannot be drawn from

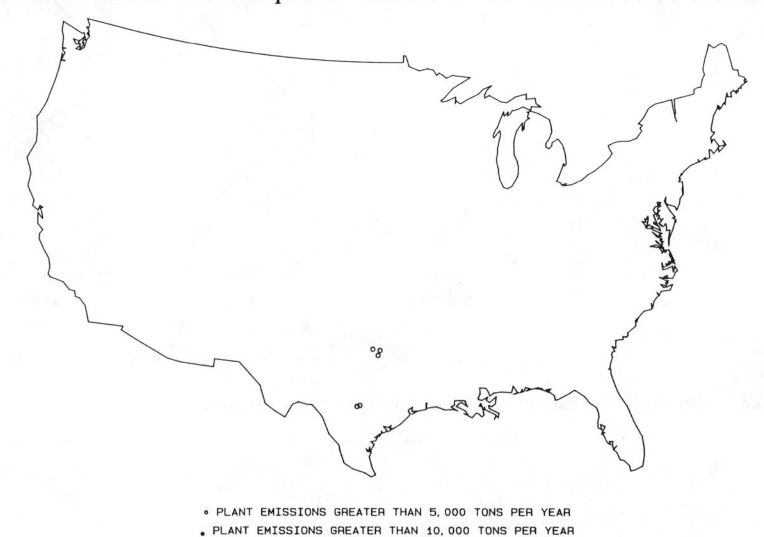


**Figure 24.** Spatial distribution of cement manufacturers with large $NO_x$ emissions.

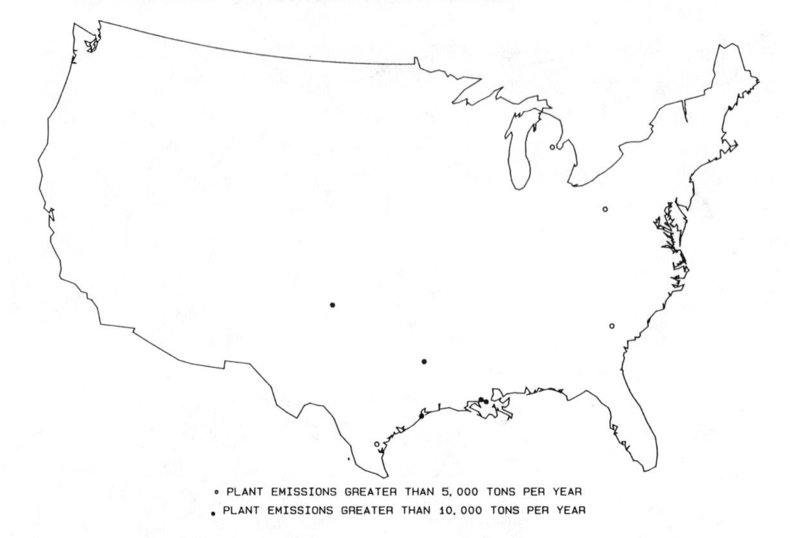


**Figure 25.** Spatial distribution of chemical manufacturers with large $NO_x$ emissions.

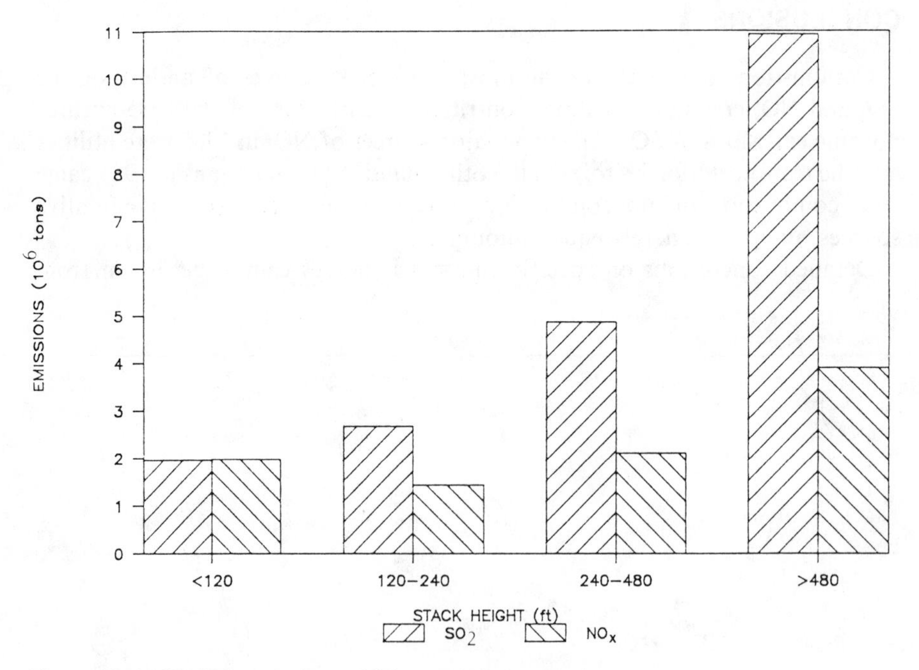


**Figure 26.** 1985 U.S. emissions of $SO_2$ and $NO_x$ by stack height.

the trends data or the single year of detailed emissions presented in this paper, but two important points can be noted. First, emissions from combustion sources must be interpreted in conjunction with information about the transport of those emissions. Long-range transport significantly affects how emissions potentially impact the environment. Second, the data indicate that the Clean Air Act of 1970 coincides with a dramatic shift in the historic trend of $SO_2$ emissions.

Emissions trend analyses and detailed emissions inventories provide the tools necessary to begin the assessment of the fate and impacts of emissions in the environment. The data presented here are representative of a much larger volume of detailed emissions data for all of the source categories and industrial processes in the detailed emissions inventories. There is a substantial body of emissions test data for most of these categories that indicates the individual chemical constituents to be expected from those processes and unit operations.

## REFERENCES


1. "Historic Emissions of Sulfur and Nitrogen Oxides in the United States from 1900 to 1980, Vol. 1," Results, EPA-600/7-85-009a (NTIS PB85-191195), U.S. Environmental Protection Agency, Research Triangle Park, NC (April 1985).
2. "Estimated Monthly Emissions of Sulfur Dioxide and Oxides of Nitrogen for the 48 Contiguous States, 1975-1984," Argonne National Laboratory Report ANL/EES-318 (1986).
3. "National Air Pollutant Emission Estimates, 1940-1984," EPA-450/4-85-014, U.S. Environmental Protection Agency, Research Triangle Park, NC (1986).
4. "Historic Emissions of Volatile Organic Compounds in the United States from 1900 to 1985," EPA-600/7-88-008a (NTIS PB88-208723), U.S. Environmental Protection Agency, Research Triangle Park, NC (May 1988).
5. "Development of the 1980 NAPAP Emissions Inventory." EPA-600/7-86-057 (NTIS PB88-132121), U.S. Environmental Protection Agency, Research Triangle Park, NC (December 1986).
6. Personal communication between Dave Zimmerman, Alliance Technologies Corporation, and Janice Wagner, E.H. Pechan & Associates (May 1988).
7. "Compilation of Air Pollutant Emission Factors, Volume I: Stationary Point and Area Sources," AP-42, 4th ed. (GPO 055-000-00251-7) U.S. Environmental Protection Agency, Research Triangle Park, NC (September 1985).


## CONVERSION FACTORS

Readers more familiar with the metric system may use the following to convert to that system:

| English Units | Multiplied by | Equals Metric |
|---|---|---|
| BTU | 1055 | Joules |
| Foot | 0.3048 | Meter |
| Ton | 0.9072 | Metric ton |

CHAPTER 2

# Formation of Polychlorodibenzo-p-Dioxins and Polychlorodibenzo Furans during Heterogeneous Combustion

**Elmar R. Altwicker, Jeffrey A. Schonberg, and Ravi Kumar N. V. Konduri**

Potentially toxic emissions from MSW- and RDF-incineration requires the continued investigation of the relationship of combustion parameters to concentrations emitted. Of much concern are the formation and emission of polychlorinated dibenzo-p-dioxins (PCDD) and dibenzofurans (PCDF) from municipal solid waste (MSW) combustion. According to recent measurements,[1] average stack gas concentrations of PCDD/PCDFs may be 1000 times those observed during hazardous (industrial) waste combustion. However, these compounds have also been found elsewhere (including the furnace), particularly on flyash in colder portions of the postcombustion regions of incinerators. The possibility exists, therefore, that there are several mechanisms of PCDD/PCDF formation. Moreover, MSW incineration may be viewed as an inherently heterogeneous process, i.e., combustion in the presence of noncombustible matter, so that a purely homogeneous description of PCDD/PCDF formation is insufficient.

Rather than study the reactions of specific compounds under homogeneous or heterogeneous conditions, it was decided to simulate MSW incineration in the laboratory using a system approach. A heterogeneous spouted bed combuster was chosen and coupled to two postcombustion stages.

This paper consists of three major sections. First, a description of the combustion facility is given including performance parameters such as temperature, pressure, and residence time distribution. Second, results of baseline studies of the combustor as well as of the combustion of chlorinated hydrocarbons leading to PCDD/PCDF formation is presented. Finally, some initial calculations relevant to the time-temperature history in MSW incinerators and their relationship to laboratory studies of PCDD/PCDF formation are given.

## LABORATORY COMBUSTION FACILITY

The approach taken in choosing a combustor was to view MSW incineration as a heterogeneous process; one in which gaseous, liquid, and solid fuels would be present. In addition, some of the waste would always be noncombustible. Thus, several types of heterogeneous processes could be present: combustion of a liquid or solid fuel leading to noncombustible residues, flyash formation, and gas-to-particle conversions in the combustor. In a typical incinerator, a distribution of residence times can be expected from the time that fuel is preheated to the departure of the combustion gases from the furnace.

The consideration of heterogeneity and residence time distribution that could (potentially) be described weighed considerably in our choice of a spouted bed combustor. In this modification of a fluidized bed, a single entry spout (fuel plus air) penetrates a cylindrical bed of inert (sand) particles. The particles displaced by the spout create a particle fountain, but gravity returns them to the annulus. Thus, a circular motion of particles (downward in the annulus, upward in the spout, Figure 1) is set up. Gas flow is through the spout, but there is leakage to the annulus. In spouted bed combustion, these spout particles move continuously through the flame, positioned on top of the bed.

Important design parameters have been summarized in Table 1a. A number of investigators have reported on spouted bed combustors. A comparison of our design with earlier designs in shown in Table 1b; the overall laboratory system is shown schematically in Figure 2. This system consists of vertically positioned Quartz combustor, 50 mm in diameter and 250 mm high; the inlet port for the air/fuel mixture is 2.89 mm in diameter. The postcombustion zones are 55 mm I.D. (Quartz) and 850 and 550 mm in length, respectively. The first is surrounded by a three-zone Lindberg furnace, the second by a single-zone furnace. The bed portion of the combustor is jacketed for air cooling. Liquid fuel is pressure fed using helium through a SS-filter and two SS-needle valves; fuel and air are mixed in a heated glass bead-filled mixing chamber. Only reagent grade fuels and fuel mixtures have been used. No experiments with solid fuels have been performed to date. The bed material is Ottawa sand (0.84 to 1 mm), but larger particles have also been used. The bed is preheated using a $H_2$ pilot flame, which also serves to ignite the fuel.

Starting at approximately 20 mm above the inlet of the reactor (Figure 2), six sampling ports are located at 25-mm intervals. Both sampling probes and thermocouples can be inserted. Approximately 40 mm above the top sampling port are two large openings (ground glass ball joints) at a 90°C angle which enable multiple probe insertions. Sampling has been conducted using both quartz and stainless steel probes in quenched and unquenched modes. In addition, samples can be taken from the postcombustion zones at each entry and exit, and transported via filters, cold traps, etc. to the measurement instrumentation. For PIC analysis, samples are collected in a cold trap, then "distilled" into a small volume trap (made of capillary tubing) from which—via heating—the sample is injected

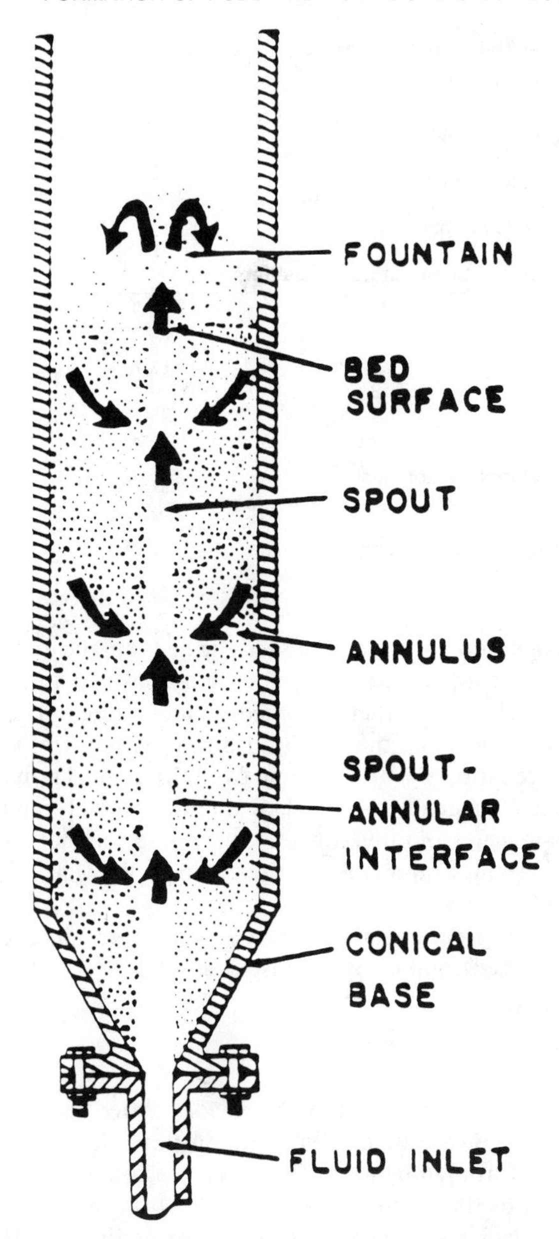


**Figure 1.** Schematic diagram of a spouted bed with a conical bottom.[3]

directly onto the capillary column in a FID-gas chromatograph (see further). Alternately, the sample is extracted with benzene or toulene for higher molecular weight compounds. The total exhaust gases can be taken through a SS cyclone followed by a SS after filter. The complete system is freestanding (accessible from three sides and topped by a powerful hood).

**Table 1a. Spouted Bed Design Parameters**

$D$ = reactor (bed) diameter
$D_i$ = inlet diameter
$D_p$ = packing (particle) diameter
$\rho_p$ = particle density
$H_s$ = packed height (static)
$U_i$, $U_s$, $U_{ms}$ = inlet, superficial, and minimum spouting velocity
$\rho_g$ = density at bed temperature

**Table 1b. Spouted Bed Combustors & Gasifiers**

| D | $D_I$ | $D/D_I$ | Reference |
|---|---|---|---|
| 40 | 3 | 13.33 | 4 |
| 53 | 3 | 17.7 | 2 |
| 300 | 31.8 | 9.4 | 5[a] |
| 12.7 × 100 | 1.6 | 25.1 | 6[b] |
| 50 | 2.89 | 17.3 | 7 |

*Notes*: D = Diameter of reactor, mm.
$D_I$ = Diameter of inlet, mm.
[a]Gasifier.
[b]Rectangular design; equivalent circular diameter = 40.2 mm.

In a typical operation of the combustor, the bed is preheated to 200 to 300°C under a warm airflow using a $H_2$ pilot flame inserted through one of the sampling ports. The fuel is then switched on and ignited. The flame is stabilized on top of the bed, and the $H_2$ pilot flame is withdrawn. Figure 3 shows a lean (equivalence ratio, $\phi = 0.63$) flame from the combustion of toluene. The flame is conical and spreads over the bed annulus. However, unlike implied by the claims of some previous investigators, the flame does not cover the annulus completely; it becomes submerged in the bed near the wall, where apparently some quenching occurs.

Here, a number of parameters relevant to the performance and understanding of a spouted bed combustor are discussed.

## Advantages of Spouted Beds

Countercurrent flow in the annulus provides better heat transfer and heat recirculation. In the case of catalytic combustion, this provides higher exit conversions. Uniform temperatures in the bed region are obtainable (see further). Operation of the reactor is possible below the lean flammability limit. Bubbling and slugging can be avoided if coarse particles ($\geq 1$ mm) are used. Any solid fuel oxidized in the reactor could reach the surface of the bed and be transported towards the inflowing oxidant by the annulus flow, unlike in fluidized beds. A complicated distributor is not required. $\Delta P_{ms}/\Delta P_{mf}$ is always less than 1.

In this research, some experiments have been conducted to understand the thermal and flow characteristics and concentration profiles in the reactor during combustion, a subject which has received little attention.

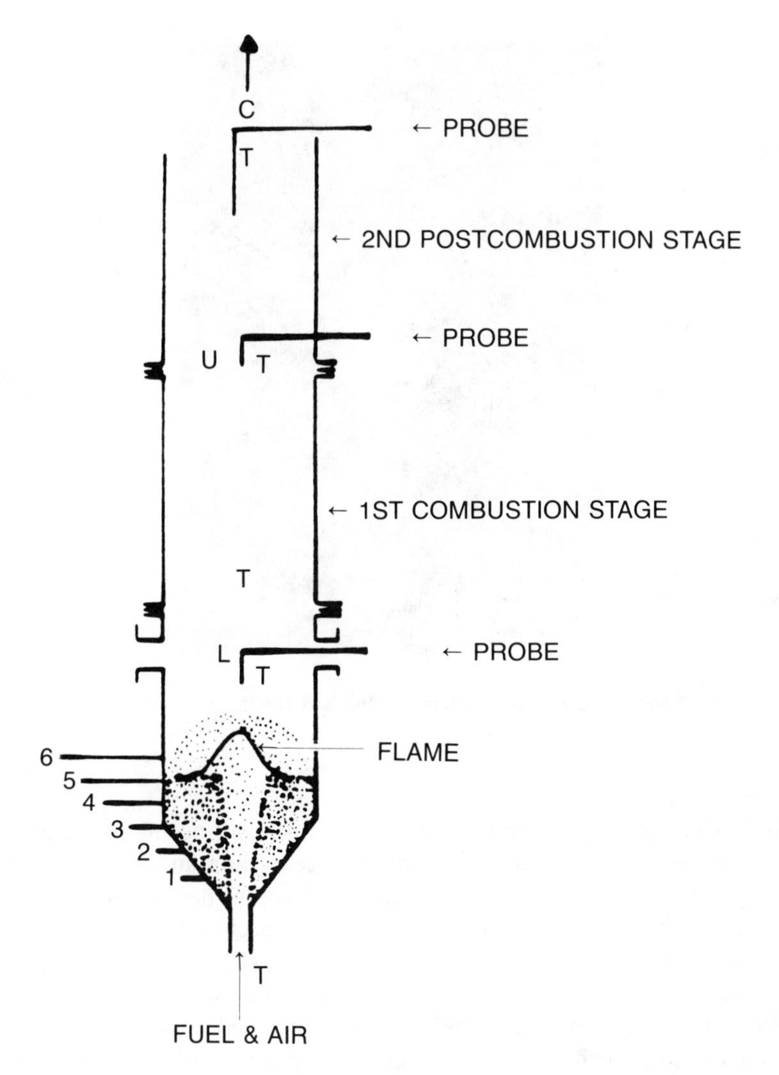


**Figure 2.** Laboratory system: spouted bed, flame, postcombustion stages.

## Temperature Profiles

Figure 4 shows the temperature* isotherms under combustion conditions. The reactor is not insulated in order to facilitate external cooling flow around the reactor. The reactor is divided into two regions, the bed and the flame or fountain region. In the bed region, the temperature remains quite uniform except in the vicinity of the flame and the walls. Figure 5 shows the radial temperature profile for an axial position close to the flame, i.e., at the top of

*All temperatures have been measured using Pt-Pt/Rh thermocouples and all temperatures are uncorrected for heat losses.

**Figure 3.** Spouted bed combustor, toluene flame, $\phi$ = 0.63.

the bed. The radial maximum—at any axial position—occurs in the annulus. Hence, the temperature increases from the centerline and then falls in the vicinity of the wall. In the flame region, i.e., above the bed, maximum temperatures are occurring on the centerline, which is expected. In contrast, Arhib et al.[2] observed that the radial temperature maxima, even in the bed region, occur on the centerline for an adiabatic reactor. Their analysis was based on several assumptions: plug flow of the gas phase, uniform bed temperature and insignificant reaction in the bed region, spherical particles of uniform diameter, and hydrodynamic parameters such as spout voidage independent of temperature.

The first assumption may be questionable because in the annulus, Darcy's flow is expected to exist for small particles resulting in a parabolic velocity profile. The first part of the second assumption is thought to be true in an order magnitude sense; the second part is not tenable. We observed that 40 to 80% of the conversion takes place in the bed region (see the DREs section). As a consequence of their fourth assumption, their numerical calculations used the hydrodynamic parameters determined under cold conditions. At high temperatures, these parameters are not well established.

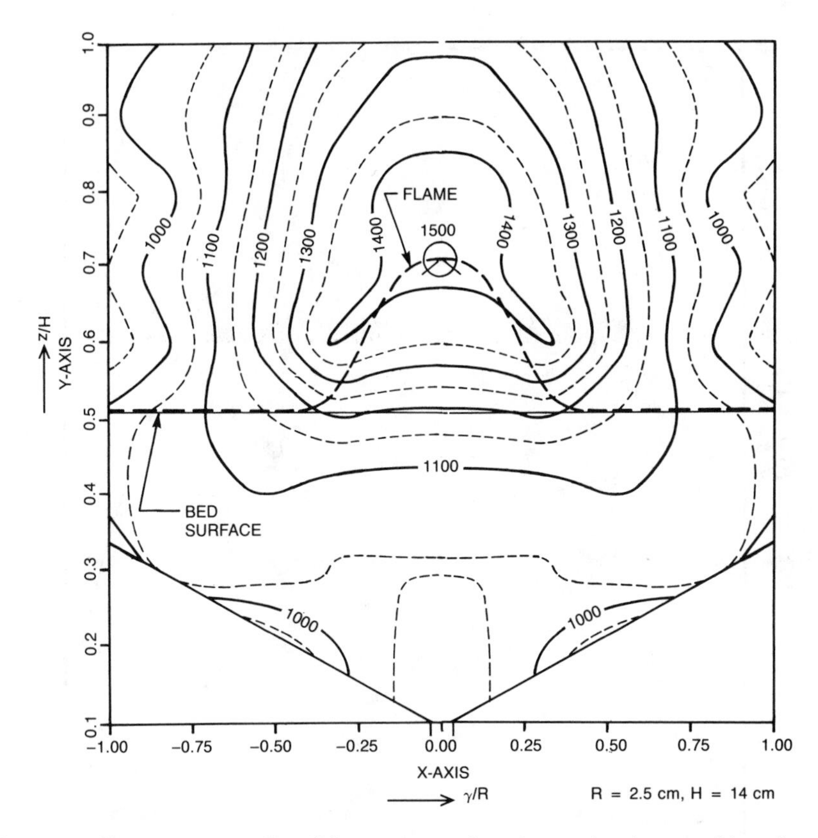


**Figure 4.** Temperature profiles (°C, uncorrected) under combustion conditions in cone-bottom spouted bed combustor.

## Minimum Spouting Velocity

The minimum spouting velocity ($U_{ms}$) is defined as the minimum fluid superficial velocity that can sustain the external spout. This parameter is an important hydrodynamic scaling parameter. Parameters like spout diameter, fountain height, and particle circulation rate are correlated or theoretically derived at minimum spouting. Especially under combustion conditions, it is not possible to operate at $U_{op}/U_{ms} = 1$ due to pulsations which will be discussed in the next section. Hence, all these parameters have to be estimated at the actual operating velocity which, in turn, depends upon the $U_{op}/U_{ms}$ ratio. For sand of 0.84 to 1 mm in diameter, this ratio is approximately 1.75.

Arhib et al.[2] also reported that the minimum spouting velocity decreased with increased temperature. Similar results have been obtained by Wu et al.[8] in the temperature range of 32 to 420°C without combustion. We observed no significant effect of temperature on this parameter.

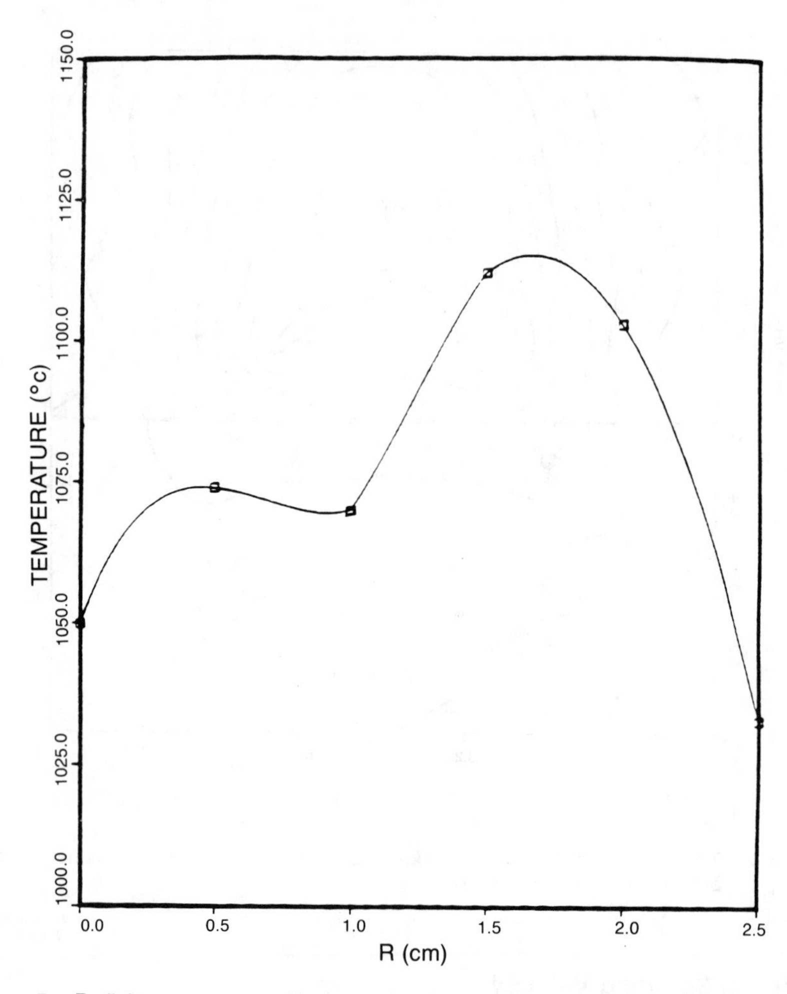


**Figure 5.** Radial temperature profile at top of bed (z = 5.9 cm) under combustion conditions.

## Stability

A key aspect in the operation of the reactor is its stability. Stable conditions mean a fountain and flame positioned symmetrically with respect to the cross section of the reactor and the absence of pulsations. The stability of the reactor depends upon many parameters. The most important are the superficial gas velocity ($U_s$), the bed temperature, and the particle size. In the case of combustion, the equivalence ratio implicitly involves temperature. The relationship between these two parameters has been shown as a regime map.[2] This regime map indicates that when the superficial velocity, $U_s$, is increased from zero, the bed essentially remains static. When the superficial velocity is slightly below the minimum spouting velocity and a flat flame is located on top of the bed, local heating causes fluidization at the top. When the superficial velocity reaches the threshold velocity required for spouting, the spout breaks through the bed and an external spout can be seen. But, under combustion conditions, the superficial velocity probably does not impart sufficient momentum to avert pinching in the spout. Therefore, the external spout collapses. Due to pressure recovery, the spout breaks again. This is periodic process. This regime is called the pulsating regime. When the superficial velocity is increased further, a stage is reached where the fluid momentum is sufficient to avert spout pinching, resulting in a stable spouting regime. Arhib et al.[2] did not observe any upper limit for the superficial velocity for stable spouting. However, at very high velocities ($U/U_{ms} > 4$) a new regime is observed where the external spout is highly turbulent and not well defined. This regime is designated as the turbulent regime in Figure 6. In addition to these regimes, a nonflammable regime is possible. There the equivalence ratio is below the threshold amount of fuel required to sustain a flame. Our operating regime is indicated by the shaded area in Figure 6.

### Pressure Profiles

Pressure distributions in spouted beds are important to an understanding of their fluid mechanics. Knowledge of the pressure profiles can be used in conjunction with particle shape to quantify the flow distribution (spout versus annulus). The nature of pressure distributions in spouted beds has been discussed by Morgan and Littman for flat bottom and Rovero et al. for cone-bottom spouted beds.[9,10] It was found that combustion temperatures have an effect on the pressure fields in the bed portion of the system. The axial pressure profile through the bed is shown in Figure 7 for cold, preheat, and combustion conditions, i.e. approximately 25°C, 225°C, and 775°C, respectively.

Evaluation of the experimental pressures using the Ergun equation indicates that approximately 85% of the total flow occurs through the spout.[11]

## EXPERIMENTAL RESULTS

Our first objective was the acquisition of reproducible results with simple, nonchlorinated fuels. Pentane and toluene were chosen. The former because it is readily combusted. The latter because in addition to ease of combustion, it is a convenient solvent for highly chlorinated compounds and a common constituent of many waste streams. Chlorinated compounds were combusted neat or, more commonly, in solutions of one of these two hydrocarbons.

This section describes the analytical techniques employed and the results obtained from the combustion of chlorinated compounds expressed principally in terms of CE (combustion efficiency), DRE (destruction and removal efficiency), and PCDD/PCDF formation.

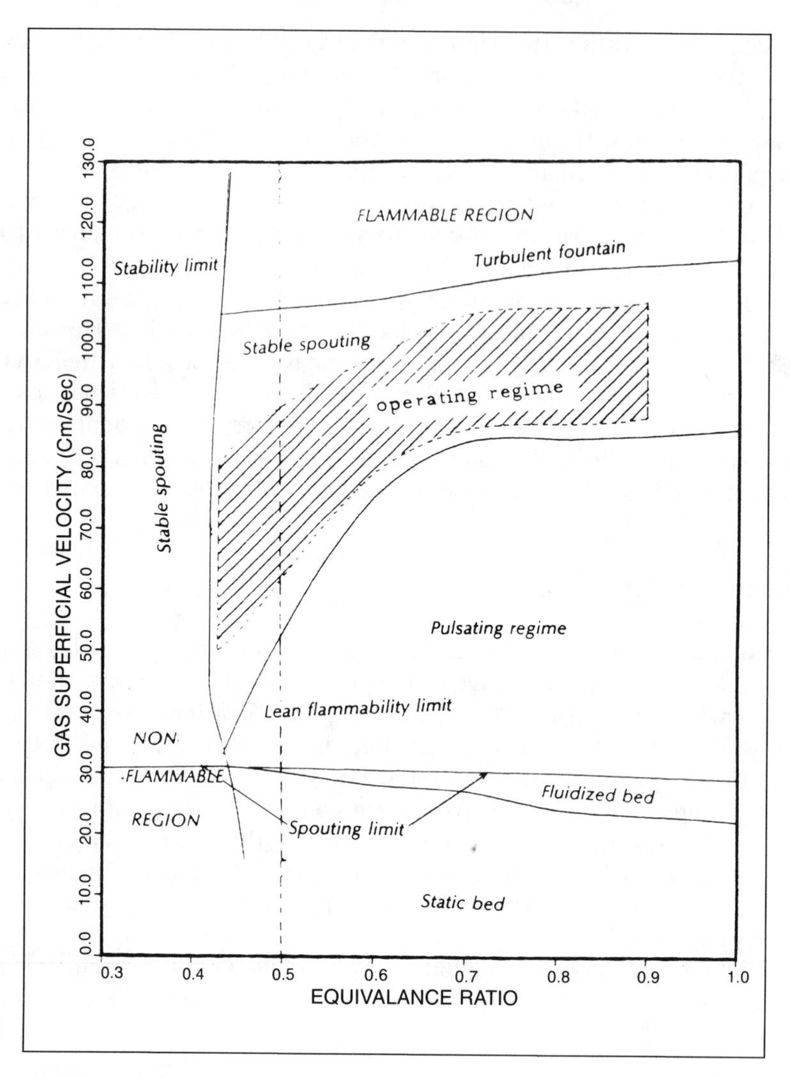


**Figure 6.** Regime diagram for spouted bed under combustion conditions; superficial velocity, $U_s$, versus equivalence ratio, $\phi$; lower dotted line indicates $U_{ms}$, minimum spouting velocity.

## Analytical Techniques

Samples are drawn through Quartz probes of open, i.e., standard bore (or quenching type) design, depending on where the probe is positioned. A Quartz wool plug is inserted in the probe tip to reduce solids carryover. The Quartz probe(s) is connected through glass-lined SS tubing to an SS filter holder at position U. At position L (see Figure 2), samples are quenched by drawing them through a cold (dry ice) trap. Most results here however were obtained

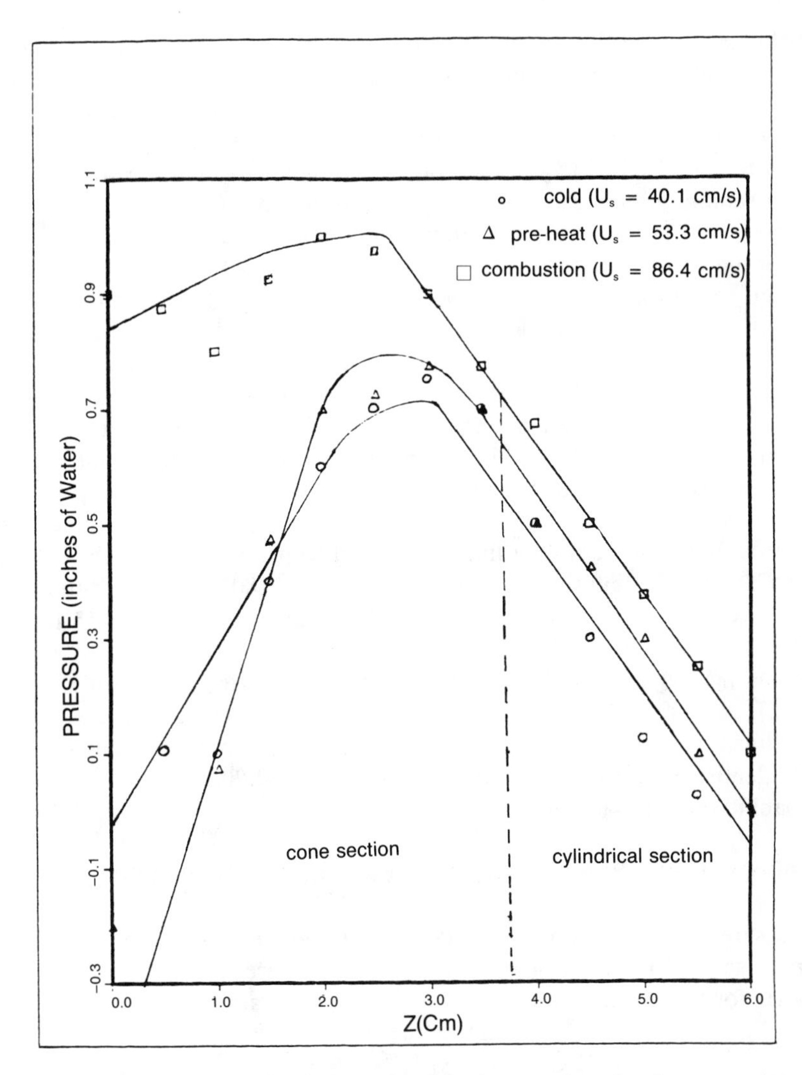


**Figure 7.** Axial pressure profiles under ambient, preheat, and combustion temperatures.

without a filter at U; where used and analyzed, the results from filter samples will be noted specifically. High temperature filter types used (which have been benzene extracted) include Gelman Type A/E 47-mm glass fiber and Fiberfrax 440 and 970. Direct flame sampling has not been used. The gas stream is analyzed for CO (continuous Beckman NDIR), $CO_2/O_2/N_2$ (1-cc sampling loop, every 6 min during an experiment, Carle 8500 gas TC-chromatograph), $H_2O$ (mass), and total acid (HCl; titration).

PIC analyses are performed on either a Perkin Elmer 900 Dual FID or a Shimadzu 9A Dual FID using wide bore tubular columns such as 10 m × 0.53 mm ID Non-Pakd fused silica, 1.2 μm RSL-200 film of nonpolar polydiphenyl-dimenthylsiloxane. Samples for PIC-analyses from the combustor are obtained using SS capillary tubing traps immersed in dry ice or $LN_2$. These U-shaped traps are equipped with two SS Swagelok Quick Connect Fittings and are wound with Nichrome wire. They are inserted into a bypass sampling line where flue gases are pulled through a switching valve equipped with a sampling loop via a small Teflon pump. The sample loop size can be changed. One or more samples from the loop are condensed in the trap. Exit flow at the pump is monitored to guard against any significant decrease (plugging due to ice formation). The trap is disconnected and inserted directly into the carrier gas flow of the gas chromatograph. The $LN_2$ is removed and the trap heated by connecting the Nichrome wire to a voltage input. The trap contents are immediately recondensed into a smaller trap just ahead of the GC column. This trap is heated and the samples injected onto the column, which is carried through a suitable time/temperature program. Individual response factors are determined using gas (or liquid) standards diluted in passive SS containers to known pressures. Currently, we are implementing a direct transfer through a short sampling line to the cold trap just ahead of the column.

PCCD/PCDF analysis is performed on a HP-5880 capillary column GC/ECD to check the elution window for these compounds (37 to 55 min). If peaks are observed, samples are analyzed for a HP-5970 GC/mass selective detector (MSD) in the selected ion monitoring mode (SIM) using initially [M+2], [M], and [M-COCl] to monitor PCDD/PCDF congeners. A GC/MSD linear mass scan on a HP5987 GC/MS/computer system is used for GC peak identification by computer matching with a 70,000 spectra library using probability-based matching (PBM).

## Combustion and Pentane and Toluene, $CO_2$ and $O_2$ Results

Measurements based on syringe samples or, more commonly, via the sample loop attached to the TC-gas chromatography were used to establish a C-balance for the system using the stoichiometric equations:

$$C_5H_{12} + X\,O_2 + (3.78)\,X\,N_2 \rightarrow 5CO_2 + 6H_2O + (X - 8)O_2 + 3.78\,X\,N_2$$

$$C_7H_8 + x\,O_2 + (3.78)\,X\,N_2 \rightarrow 7CO_2 + 4H_2O + (x - 9)O_2 + 3.78\,X\,N_2$$

All runs to date have employed $\phi \leq 1$ so that the need to consider values of X $< 8$ was eliminated. For runs where additional $N_2$ was added to achieve a higher $\phi$ value (and air reduced correspondingly to maintain stable conditions in the combustor), theoretical values for $CO_2$ and $O_2$ had to be calculated for each experimental run. After some initial difficulties due to leaks in the sampling system, observed values (at sampling position L) were typically within 3% of the calculated, expected value.

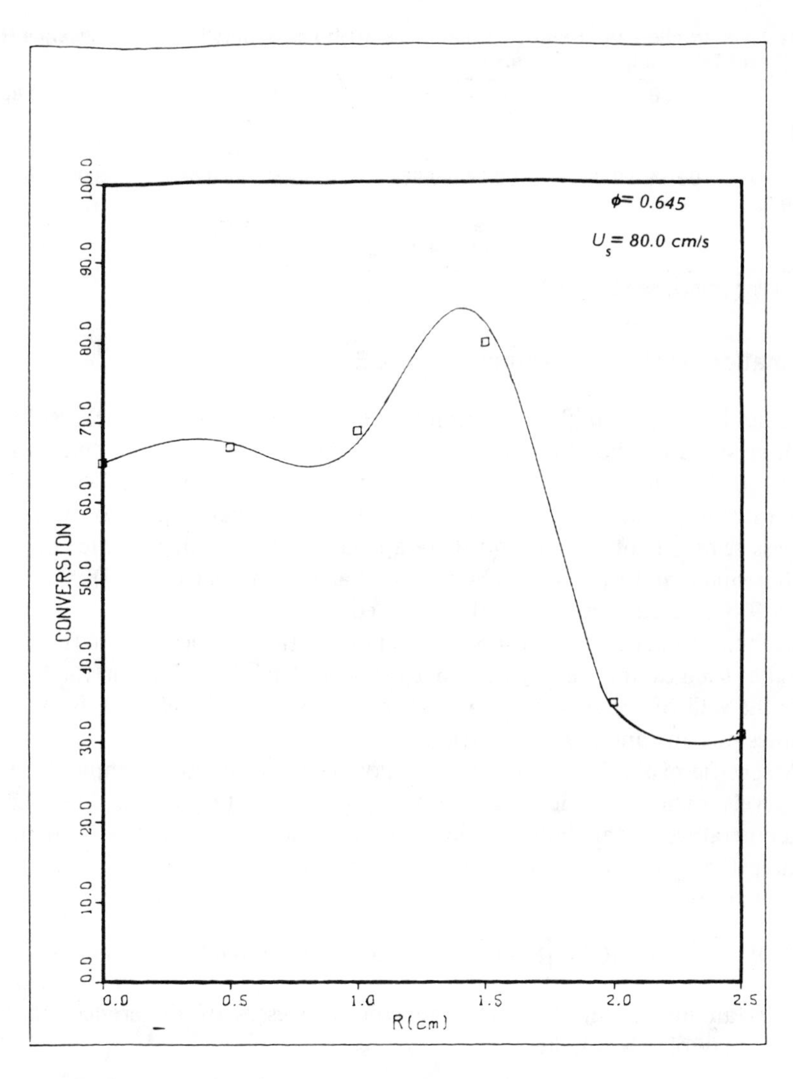


**Figure 8.** Radial conversion of toluene in bed at $\phi = 0.73$.

## DREs

Some conversion occurs in the bed upstream from the flame (Figure 8).

Additional destruction takes place in the postcombustion zone. Table 2 compares DREs at positions L and U (see Figure 2) from the combustion of pentane and toluene over a range of equivalence ratios, $\phi$. The typical residence time of the hot gases in the bed is estimated to be 0 ($10^{-2}$) sec (in the spout, through which ~85% of the gas flow passes) and 0 ($10^{-1}$) sec in the annulus, and $<0$ ($10^{-2}$) sec in the flame, and about 1 sec in the postcombustion region between L and U.

**Table 2. Destruction and Removal Efficiencies (DRE) as a Function of Equivalence Ratio at Two Sampling Points.**

| $\phi$ | 0.50–0.57 | 0.62–0.67 | 0.70–0.78 | 0.80–0.86 |
|---|---|---|---|---|
| Pentane | | | | |
| U[a] | – | 99.986–99.9996 | 99.996–99.9989 | 99.95–99.9999 |
| L[a] | 99.14–99.96 | 98.29–99.989 | 99.58–99.998 | 99.928–99.988 |
| Toluene | | | | |
| U | – | 99.998 | 99.9992 | 99. |
| L | – | 99.88–99.91 | 99.935 | 99. |

[a]Sampling points, see Figure 2.

## Formation of CO, Determination of CE

A plot based on sampling position L of CE versus $\phi$ (for pentane) results in a positive slope of about 1 in the range of CEs from 97 to 100% and $\phi$s from 0.50 to 0.87. This result is a direct consequence of the increase in bed temperature with increasing $\phi$. Combustion efficiency is higher at position U (Figure 2), less scatter is observed, and there appears to be less dependence on $\phi$ for both pentane and toluene. At higher equivalence ratios ($\phi > 0.7$), CE values of 99.9% or greater are typically observed.

Carbon monoxide formation commences in the hot bed below the flame. This has been established by extensive radial and axial sampling in the bed and near the wall of the reactor; a typical result is shown in Figure 9, in terms of isolines of constant CO concentration.

A specific example of the effect of postcombustion zone temperatures on CO levels is shown in Figure 10 for an equivalence ratio of 0.76. The increase in temperature in this laminar flow regime indicates a rate of CO oxidation greater than is calculated using the global rate expression of Dryer and Glassman:[12]

$$R_{CO} = 10^{14.6}\,[CO]\,[H_2O]^{1/2}\,[O_2]^{1/4}\exp(-40{,}000/RT),\ \text{mol/cm}^3\text{/sec}$$

This is an indication that heterogeneous processes or differences in flow regime may play a role here.

## Formation of PICs (Products of Incomplete Combustion)

PICs include any organic carbon compound, i.e., soot (free or as C on flyash) and dioxins could be considered PICs. In this research, some identification of PICs and quantification has been carried out. However, the number of PICs is very large and no attempt has been made to make specific assignments to all observed gas chromatographic peaks.

Measured PICs (but not for all runs) include many $C_2$–$C_6$ hydrocarbons, benzene, methylene chloride, chloroform, and carbon tetrachloride. Benzene was observed in every instance. A number of oxygenated and chlorinated products were identified by mass spectrometry; details can be found elsewhere.[13,14]

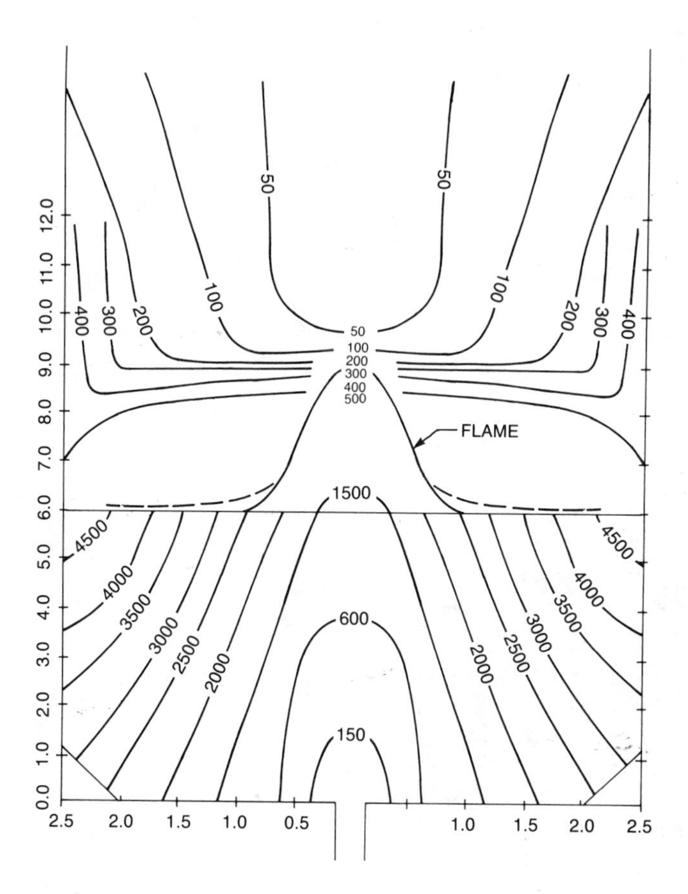


**Figure 9.** Carbon monoxide ratios, ppmv, above and below the flame under combustion conditions.

## PCDDs/PCDFs

Many of the experiments performed have been of an exploratory nature, both with respect to the types of compounds combusted as well as the sampling strategies employed. For this reason, direct quantitative comparison between different compounds reported here will only be possible in some instances. To minimize potential hazards, very low Cl/C molar ratios were employed initially until some feel had been developed for the quantities of PCDD/PCDFs present. Sampling and cost limitations precluded analyses for PCDD/PCDFs at every sampling location for some runs. Initial sampling for PCDD/PCDFs was based on filter samples taken at position U (i.e., after the first postcombustion stage, Figure 2). Later, combustion effluent was condensed directly into a trap (total trap, TT) at both positions L and U, and subsequently worked up. No separate analyses for the 2,3,7,8-T4CDD congener were performed.

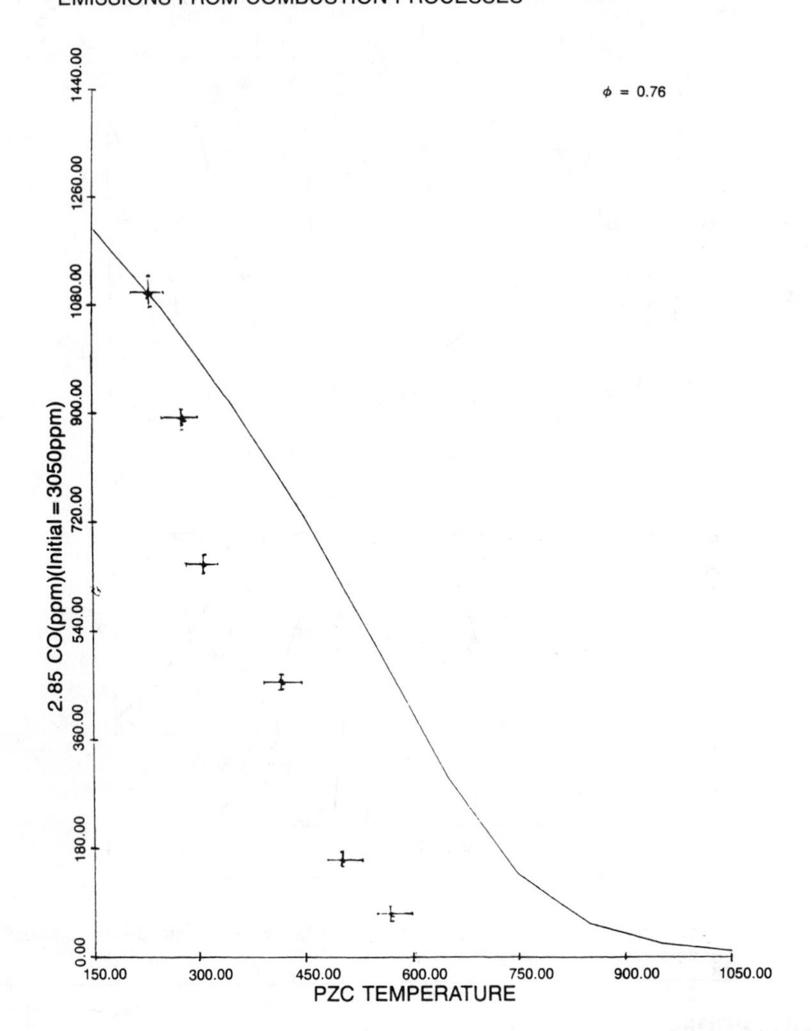


**Figure 10.** Effect of postcombustion zone temperature on CO levels at sampling point U.

### *Combustion of Pentane/1-Chloropentane and Hydrogen Chloride with Pentane, Respectively*

At Cl/C ratios of $\sim 10^{-5}$, either none or only trace quantities of PCDDs (no PCDFs) were observed. When the Cl/C ratio was raised to $5.5 \times 10^{-3}$ (for the combustion of ~4% 1-chloropentane in pentane), quantities were observed that are comparable to those from some incinerators: for a flow rate of 0.028 mol/min of 1-chloropentane and a nominal gas phase concentration of $1.15 \times 10^9$ ng/m³, 433 ng/m³ OCDD, and 650 ng/m³ H(7)CDD were obtained after a 30-min sampling at position U (volumes are corrected to STP). In addition, trace quantities (<50 ng/m³) of T(4)CDD, as well as OCDF and H(7)CDF, were found. It seems clear, therefore, that simple precursors can lead to PCDD/PCDFs (*de novo* synthesis). However, this result is based on only one experiment.

Results with HCl and pentane so far have been largely negative, although some indications for the formation of trace quantities were obtained. Further studies require a longer sampling time as well as higher Cl/C ratios.

It will also be important in future experiments to check for the presence of hexachlorobenzene (HCB) and pentachlorophenol (PCP) as these are PICs observed in other combustion reactions and may be precursors.

### *Combustion of Chlorobenzenes in Pentane and Toluene*

Chlorobenzene, 1,2-dichlorobenzene, 1,2,4-trichlorobenzene, and hexachlorobenzene have been combusted using pentane or toluene as solvents. Typical DREs for pentane and toluene under these conditions ranged from 99.9 to 99.998 at L and 99.9 to 99.9999% at U (Figure 2) with corresponding CE values of 97.8 to 99.0 and 99.89 to 99.99, respectively.

Using filter sampling (at U or C) and a Cl/C ratio of $8.2 \times 10^{-3}$ (for chlorobenzene in pentane), results were generally negative. At higher Cl/C ratios (0.037 to 0.077) and sampling through a total trap (TT, Figure 11), the major products of interest were OCDD, $10^3$ to $3.3 \times 10^3$ ng/m$^3$, PCP, $1.6 \times 10^4$ to $10^5$ ng/m$^3$; and HCB, $1.5 \times 10^4$ to $1.4 \times 10^5$ ng/m$^3$. The observed PCP/HCB ratio was approximately 1. For individual experiments, this ratio can be smaller or greater than 1; the reasons for this have not been determined. However, OCDD was the only dioxin present in measurable (ng) quantities (60-min sampling time) and no evidence for PCDFs was obtained.

Some of the most interesting results to date have been obtained with 1,2-dichlorobenzene (10% in toluene). These are summarized in Figure 11. Following Scheme 1, the only dioxin observed was OCDD and the PCP/HCB ratio was approximately 3. However, when the combustion gases were passed through a flyash trap (FAT, a short glass tube packed with flyash and glass wool at both ends) held at approximately 300°C (flyash furnished by Karasek and used in his experiments on PCP conversion over flyash)[15,16,17] the yield of PCDDs increased (in addition to OCDD, some lower congeners were formed) and PCDFs were found (Scheme 2). The nominal residence time of the gases in the flyash trap was ~0.1 sec, the sampling time was 60 min. These results have to be understood as ±30%. Although in these experiments the PCP/HCB ratio was typically $\geq 1$, in at least one experiment, this ratio was $<<1$. Average formation rates of 0.019 ng/g flyash per min PCDD, corrected for inlet OCDD (to FAT), and 0.187 ng/g/min PCDF can be calculated; the PCP consumption rate was 0.26 ng/g/min.

Combustions with HCB have been carried out using both pentane and toluene and Cl/C ratios of $2 \times 10^{-3}$ to $8.4 \times 10^{-3}$. Results with the former solvent were largely erratic. This was attributed to the precipitation of HCB in the feed line, although a concentration of less than 50% of the solubility value

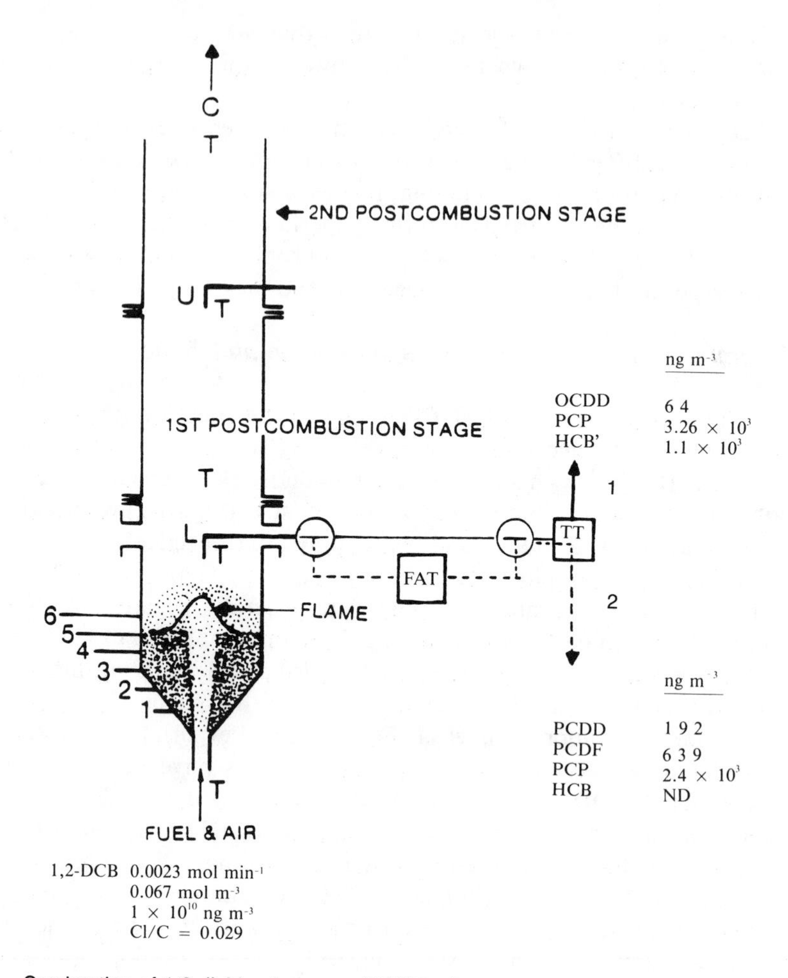


**Figure 11.** Combustion of 1,2-dichlorobenzene (DCB)/toulene.

at the boiling point of pentane was used. Although HCB solubility in toluene is much higher, the amount of HCB in toluene combustions was kept in the range of 0.25 to 0.5%. DREs of HCB ranged from 99.95 to 99.996% and PCP/HCB ratios from 13 to 49 (mostly at position L, but also at U and C). In effect, PCP was a major PIC. However, no PCDDs and PCDFs have been observed from these experiments (total of four) at any of the sampling positions. A number of other chlorinated compounds were identified by mass spectrometry, including 1,3-cyclopentadiene, 1,2,3,4-tetrachloro, 5-dichloromethylene.

A potentially interesting observation was made on the sand used as spouting medium in the reactor. Analysis, after the earlier-mentioned set of experiments was completed, indicated some OCDD >> H(7)CDD > H(6)CDD, as well as traces of OCDF and H(7)CDF. This sand was extracted and used in Karasek's

**Table 3. Pentachlorophenol (0.5% in Toluene), $\phi$ = 0.85**

| T, °C, Sampling point[a] | DRE | | CO (ppm) | OCDD[b] | PCDF | % Yield[c] × $10^5$ |
|---|---|---|---|---|---|---|
| | Toluene | PCP | | (ng/m³) | | |
| C 333 ± 3 | – | 99.999 | – | 3[d] | nd | 0.054 |
| U 593 ± 10 | 99.998 | 99.998 | 45 | 100 | nd | 0.026 |
| L 681 ± 20 | 99.94 | 99.86 | – | 230 | nd | 0.12 |

nd = not detected (at the 0.5 ng/g level).
Bed temperature = 1017 ± 50°C.
CE = 99.93 (U), 97.09 (L).
[a]See Figure 2 for location of sampling points.
[b]2 ± 30%
[c]yield of OCDD, based on PCP remaining.
[d]± 3.

apparatus[16,17] with $^{13}$C-PCP. Only OCDD was found; there was no evidence for lower cogeners.

### *Combustion of Chlorinated Phenols in Toluene*

Preliminary experiments with 2,4-dichlorophenol (in toluene, Cl/C ratio $1.4 \times 10^{-2}$) have given PCP and HCB as the principal chlorinated PICs; no PCDD/PCDFs were found (one analysis, over flyash) in these experiments. This could be attributed to (1) compounds below the detection limit over the reaction time scales employed or (2) sampling error. It has been shown that the formation of PCDDs is possible under related conditions, but with much longer gas residence times, over flyash.[17]

The work of Karasek and Dickson[15,17] provided the impetus for the study of the combustion of pentachlorophenol. PCP was effectively destroyed under typical spouted bed combustion conditions (Table 3); only OCDD appeared to be formed under these conditions. Some HCB was also formed, indicating that the substitution of OH by Cl can occur. The observations reported in Table 3 were made at an inlet mole ratio of PCP per air of $2.5 \times 10^{-5}$ and Cl/C ratios in the range of $10^{-3}$ to $10^{-2}$. The yield of PCDFs was essentially zero. In one experiment, detectable quantities were reported at sampling position L. When the combustion gases were passed over flyash (inserted between L and TT at ~300°C) and the total trap contents plus FAT contents were analyzed, PCDFs were found. The principal dioxin observed was H(7)CDD, together with OCDD, and only traces of H(6)CDD; several PCDF congeners were found.

### *Qualitative Mechanistic Considerations*

The experimental observations reported in the two earlier sections can be summarized in a qualitative PCDD/PCDF formation scheme (Figure 12) for combustion reactions carried out at $\phi < 1$; it provides somewhat greater specifics than similar schemes.[18]

Specific reactions that occur are dechlorination, chlorination, hydroxylation, OH substitution by chlorine, and condensation. Evidence for the thermal dechlorination of chlorobenzenes is only indirect. The reason for this is the following: combustion of the baseline fuels pentane and toluene results in the formation of some benzene. Since most of the combustion studies reported earlier used toluene, the formation of benzene is expected. It was also observed, however, that the yield of benzene increased when chlorobenzenes were combusted. Though this does not prove that the increase in benzene occurred due to dechlorination, it may be inferred. Experiments with chlorobenzenes (neat) or $^{13}C$-labeled chlorobenzenes would be necessary to confirm this pathway. This presumably is a gas phase declorination process; the evidence for heterogeneous dechlorination is discussed later.

It appears that chlorination and dechlorination may be part of a quasi-equilibrium system that also leads to hexachlorobenzene. Since the only chlorine source in these reactions is the single chlorobenzene chosen as a fuel, formation of hexachlorobenzene must occur at the expense of the starting material. Observations on hydrogen chloride are only qualitative at this time, however, so it is not possible to perform a chlorine balance on these reactions.

The thermal reactions of chlorine substitution by hydroxyl and hydroxyl substitution by chlorine occur. That they may occur more or less simultaneously is suggested by the formation of hexachlorobenzene from 2,4-dichlorophenol. Although for most of these reactions, regardless of the starting compound, the observed PCP/HCB ratio has been greater than 1, there have been a few exceptions to that, in a couple of instances even to the extent that the presence of PCP could not be confirmed in the presence of a substantial yield of HCB. This suggests that the relative quantities of both are very sensitive to experimental and sampling conditions. In the combustion of 2,4-dichlorophenol, the observed PCP/HCB ratio was about 3 to 4, suggesting that chlorination was favored over OH substitution by chlorine.

In the cases where the combustion effluents containing PCP and HCB were passed over flyash at ~300°C, HCB was not observed in the flyash itself or the downstream collection trap upon extraction. The quantitative aspects of these reactions are such, however, that formation of PCP (and PCDDs?) can only be inferred. This is another experiment that could benefit greatly from the use of $^{13}C$-labeled compounds (does labeled HCB lead to labeled PCP?). The hydroxylation of chlorobenzenes over a flyash-like matrix (silica) is known to occur.[19]

Pentachlorophenol appears as the principal PIC from the combustion of chlorobenzenes and 2,4-dichlorophenol. Although others (lower chlorinated) phenols have not been quantified, experimental observations suggest that they are present in lesser quantities. Since PCP, when used as a fuel, is largely destroyed, it seems evident that PCP formation and destruction represent competing processes and that the observed quantities reflect the result of that competition. At sampling point L, the DRE for PCP ranges from 99.81 to 99.94% and goes as high as 99.998 and 99.999% at U and C, respectively. Yields of PCP from chlorobenzenes are typically two orders of magnitude lower than quantities measured at L when PCP is the starting material.

During the combustion of PCP, the thermal condensation produces measurable quantities of OCDD (230 ± 120 ng/m$^3$) reproducibly. No other congeners have been observed in these reactions. Comparable quantities are formed from the combustion of chlorobenzene and 1,2-dichlorobenzene (formation from 1,2,4-trichlorobenzene has not been confirmed). Since PCP is produced from mono- and disubstituted chlorobenzenes, both chlorination and hydroxylation reactions must occur. Since the thermal condensation of PCP to OCDD is well known, the formation of OCDD in these reactions probably occurs via that route. This corresponds essentially to what is shown in Figure 12 and assigns a central role to PCP in these reactions. This, of course, does not rule out a different pathway for the chlorobenzenes. However, in all reactions studied to date, the yield of PCP has always exceeded the yield of OCDD. The formation of OCDD might be limited by some other intermediate; it certainly would not appear to be limited by the quantities of PCP. Support for a different pathway would have to come from the identification of other probable intermediates. No such compounds can be identified at present.

The observations on PCDDs formed over flyash are qualitatively in agreement with those of Karasek and Dickson.[15,16,17] Since our analyses were based on samples which combined the total trap sample (TT) and the flyash extract, no information can be gleaned from these experiments on the nature and

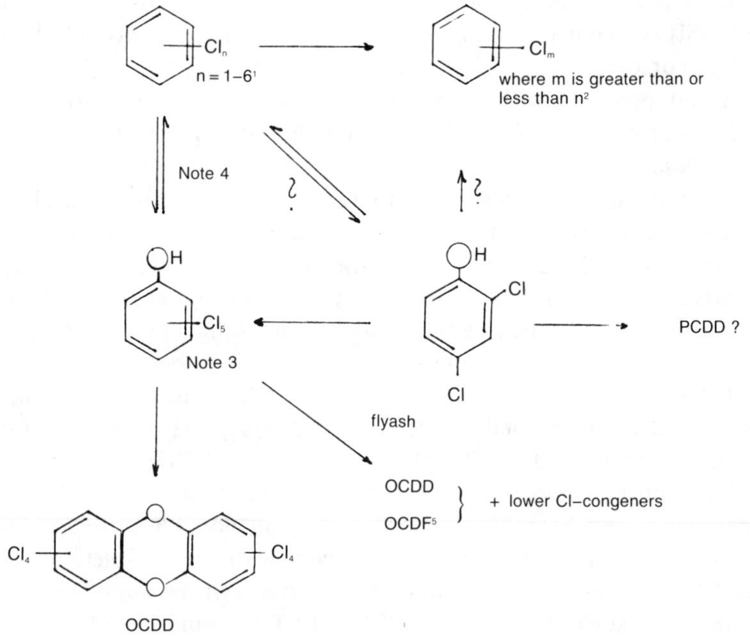

**Figure 12.** Qualitative mechanistic scheme of PCDD formation from chlorobenzenes and chlorophenols.

extent of any desorption step. Presumably, the dechlorination reactions occur on the flyash. This requires a source of hydrogen. Small quantities of hydrogen gas could be present in the combustion gases and act as a reducing agent in spite of an excess of oxygen. It should be noted, however, that in the $^{13}C$-PCP experiments over flyash[17] no hydrogen is present in the gas stream. A source of hydrogen must exist, therefore, in the flyash (it is difficult to conceive of a mechanism whereby water or the hydrogen from the phenolic hydroxyl group could be that source). If the flyash contains carbon (as it invariably does), surface H-C bonds may play a role in this reaction.

Perhaps the most surprising observation has been the formation of PCDFs over flyash in quantities greater than those of the PCDDs, both from PCP and from 1,2-dichlorobenzene as starting materials. Since $^{13}C$-labeled PCP produces only PCDDs,[15,16,17] it must follow that the combustion (or high temperatures) of PCP and chlorobenzenes can lead to precursors that can form PCDFs. These have not been identified, but might be phenoxyphenols. There is one possible caveat to the earlier-mentioned explanation. Although the flyash used was extensively extracted (and the final extracts showed no PCDD/PCDFs), the matrix undoubtedly contained nonextractable organics, including carbon. In view of the findings of Vogg and Stieglitz,[20] *de novo* synthesis from carbon is one route to PCDD/PCDFs. Although the time scales appear to be very different between the two types of experiments and the quantities of flyash used here are much smaller (this will be discussed further in a following section), this possibility could be ruled out rigorously only through the use of labeled precursors.

The result obtained with 1-chloropentane needs to be confirmed with a series of experiments under a variety of conditions. This is an example of *de novo* synthesis.

One of the important questions that has not been answered about the spouted bed reactor concerns the extent (if any) of the formation of PCDD/PCDFs in the hot bed (and before the gases enter the flame). The observation cited earlier (in the combustion of chlorobenzenes in Pentane and Toluene section) with respect to the presence of these compounds in the bed sand after a series of experiments is therefore of interest. The bed temperatures are certainly high enough ($>1000°C$) for PCDD/PCDF formation to occur. If the reaction is purely a thermal condensation, one expects principally OCDD. However, H(7)CDD and H(6)CDD, as well as some PCDFs, are also found; the reaction must be more complex. Since this observation was made after a set of experiments with hexachlorobenzenes, it is not possible to relate it to any given set of conditions. It conflicts, however, with the complete absence of PCDD/PCDFs from the combustion of HCB at any of the sampling positions, L, U, and C; it suggests that some PCDD/PCDFs could have formed in the bed (preflame region) and subsequently completely been destroyed in the flame. Only sampling for PCDD/PCDFs in the bed during an experiment will clarify this aspect.

## TRANSPORT RATES, THE ROLE OF THE TIME/TEMPERATURE HISTORY IN HETEROGENEOUS CATALYSIS BY FLYASH

Experimental studies cited earlier[15] have demonstrated the heterogeneous formation of PCDDs from precursors such as pentachlorophenol. However, heterogeneous catalysis is not a possible formation mechanism in municipal waste incinerators if the mass transfer time scales are longer than the relevant process time scales (for example, the residence time of flyash in a duct leading from the secondary combustion chamber to an air pollution control device and then a stack). Therefore, it is useful to estimate the time required for sufficient quantities of precursor to be transported to the surface of a flyash particle to potentially form the observed quantities of PCDDs. This transport time was estimated in previous work;[21] this paper discusses the reaction kinetics as well. Shaub and Tsang[21] appear to have taken the average interparticle distance for a given particle concentration and to have calculated a diffusion time based on that distance. However our work shows that interparticle mass transfer interactions are probably insignificant, i.e., the particles function independently. Therefore, the characteristic transport length is not the particle separation, but is actually the particle radius. An estimate of the time coverage of the flyash during transport with suitable dioxin precursor molecules was $5.5 \times 10^{-4}$ sec.[21]

In a municipal waste incinerator, flyash is found in at least two different situations: it is carried along by turbulent flow in ducts and it is held on the walls of heat exchangers, ducts, and air pollution control equipment. Consider the former case first. The flyash is suspended in a turbulent flow of a cooling gas. Heat is removed by conduction through the duct walls and so the gas temperature is certainly nonuniform. The nature of the flyash itself is complicated. Photomicrographs of RDF/coal flyash particles show basic morphologies.[22] The most prevalent ash particle is described as the "shredded sponge" type and is porous. Furthermore, flyash particles are polydisperse. The hydrodynamic interaction of the flyash with the turbulent flow must be considered as well as the potential influence of the turbulent flow on mass transfer rates to the flyash surface. The chemistry itself is uncertain. The concentration of precursors in the gas must be known as well as the yield of PCDDs if the mass transfer time is to be estimated. Precursors could form other compounds besides PCDDs; in fact, pentachlorophenol apparently does.[17] The chemistry of PCP is temperature sensitive and so, presumably, is the chemistry of other potential precursors.

Some of these complications may be removed (at least for an approximate calculation) because the flyash particles are so small (of micron size). Consequently, their sedimentation velocities in a stagnant gas are very low. Furthermore, the relative velocities of gas moving past the particle, due to the shearing of the carrying flow, at distances of 10 or 20 particle radii from the particle center are probably very small. Therefore, the influence of forced convection in the mass transfer process is neglected in our estimate; the characteristic mass transfer length is assumed to be less than 10 or 20 particle radii. (The characteristic mass transfer length is the distance, over which the concentration of precursor changes from the value in the bulk gas to the (lower) value(s) at the flyash surface.) Therefore, the nature of the turbulent flow of gas in the duct need not be known and only a typical particle residence time is needed. Inherent in this discussion is the notion that particles are well separated. Neglecting hydrodynamic clustering in this first approximation, this implicit assumption is valid because the particle loadings are so low, for example, 30 to 100 $mg/m^3$.[23]

As noted, the temperature of the gas surrounding the particle changes as the gas cools. If the mass transfer time is sufficiently small, the density and diffusivity will be essentially constant if the gas temperature changes gradually from the flame to the stack over the residence time of the gas. This latter assumption has two obvious flaws. First, some particles pass near the wall where the average gas temperature may change sharply over a small "boundary layer." However, we expect that most particles spend most of their time in the bulk flow away from the walls. The second flaw is that gas temperature might exhibit local fluctuations owing to the turbulent nature of the cooling process. For this calculation, the physical properties were taken to be constant and were evaluated at 300°C, chosen because laboratory experiments (and incinerator observations) suggest it to be a temperature favorable for heterogeneous dioxin formation.

The complexities of flyash itself may be reduced because of the assumption that the rate limiting step in the heterogeneous catalysis is the transport of precursor to the flyash surface. It follows that the pores in the ash may be neglected because precursor molecules diffusing inside the pores (in the gas phase) are not likely to wander much deeper than the width of the pore because of the available surface with its high reactivity. The reactivity of the surface in practice may be quite low (see References and 16 for results reported over different types of flyash on the formation of PCDDs from PCP); however, the assumption that diffusion limits the process rate is a best case scenario and so "best case" kinetics are assumed. Therefore, the outer surface of the flyash particle is taken as the boundary and this surface is assumed to be spherical.

With these assumptions, the precursor concentration field is described by:

$$\frac{\partial C}{\partial t} = D \frac{1}{r} \frac{\partial}{\partial r}\left(r \frac{\partial C}{\partial r}\right)$$

where $C = C_o$ r approaching infinity

$C = 0$ $r = 0$

The solution of this equation is given in Carslaw and Jaeger[24] as:

$$\frac{C}{C_o} = 1 - \frac{a}{r}\,\mathrm{erfc}\left(\frac{r - a}{2\sqrt{Dt}}\right)$$

If the mass transfer time is large enough, the value of the complimentary error function is equal to one for moderate and small values of r. The simpler form of the solution shows that the assumption about the range of the concentration field seems reasonable. The corresponding expression for the amount (Q) of precursor which has been destroyed at the surface of a flyash particle is

$$Q = 4\pi aDtC_o$$

This expression could be written for each precursor and used to estimate the time required to produce the observed levels of PCDDs on freely suspended flyash if the chemical kinetics are instantaneous.

This expression was used with field data to calculate the mass transfer time in an operating incinerator. Using reported emissions data from the Charlettetown PEI plant,[23] i.e., 100 $mg/m^3$ fly ash, 4000 ng $CP/m^3$, 100 ng $PCDD/m^3$, assuming all chlorophenols (CP) to have the same diffusivity as pentachlorophenol (taken to be 0.53 $cm^2/sec$ at 300°C) and assuming that they all are irreversibly converted to PCDDs, the mass transfer time was found to be 0.075 sec. If we assume that flyash particles spend roughly 2 to 3 sec going from 1200 K to 600 K, there is ample time for mass transfer, in spite of the fact that only during a portion of this time is the temperature of the flyash low enough to permit PCDD formation.[17] If the temperature window is taken as 250 to 300°C and the gas and the ash particles are assumed to cool uniformly, then there is $\sim$0.25 sec available for the mass transfer process, probably still sufficient.

The assumption regarding conversion of chlorophenols to PCDDs only is dubious, especially in light of the findings of Dickson,[17] who showed poor selectivities. In other words, side reactions consume much of the PCP; the amount of unreacted PCP is not the issue. Using a conversion of only 1% to PCDD and neglecting other possible precursors implies that the transport time would be 7.5 sec, i.e., transport would be a problem. Hence the issue of the chemistry is very important.

Suspended flyash particles should have adequate exposure to precursors in the gas stream. Flyash particles are also found on walls; on the collection plates of electrostatic precipitators, for example. In this case, the turbulent flow sweeps past the collected particles; however the particles are now packed together and exposure of individual particles to precursor may be lower. We focus, therefore, not on the individual particle (as earlier), but on the gas flow over a wall with zero precursor concentration. Again, the temperature is taken to be constant at 300°C. The average velocity of the gas between parallel plates (taken to be 0.25 m apart) is assumed to be 2.5 m/sec. A correlation for turbulent mass transfer to a tube is used, equating hydraulic diameters. It turns out that the stagnant film diffusion thickness (a fictitious concept in this case, but probably larger than the real average "boundary" layer thickness) is very small compared to the hydraulic diameter; thus, there is some basis for using the tube correlation for the actual duct. This correlation is

$$\frac{k_c d}{D} = 0.023\ Rd^{0.83}\ Sc^{0.44}$$

$$2000 < Rd < 35{,}000$$

$$0.6 < Sc < 2.5$$

The Reynolds number was calculated to be $2.65 \times 10^4$. The mass flux of precursors, N, is $N = k_c \triangle C$ where $\triangle C$ is the concentration difference between the bulk gas and the wall. The gas residence time required to form 100 $ng/m^3$ PCDDs from a bulk precursor concentration of 4000 ng $CP/m^3$ may be calculated from a material balance if the mass flux is known. The gas residence time in the electrostatic precipitator under these assumed conditions is only about 1 sec. If the yield of PCDDs from CP is only 1%, the required gas residence time would be 1.7 min.

The major point of this discussion is that more information is needed on the fate of other substances found in flue gas, which might be precursors to PCDDs, and their concentrations. What then does the best case kinetics calculation imply? Is there enough contacting time in a municipal waste incinerator to permit enough precursor to reach the idealized catalyst? The answer apparently depends on three critical factors. First of all, chlorophenols appear to generate other compounds besides PCDDs. In fact, the experiments of Karasek and Dickson[15] imply that little PCDD is actually formed. Second, other compounds in the flue gas may form PCDDs. These factors are distinct from the rates of reactions which are assumed to be infinite in this base case scenario. Third, the concentration of chlorophenols selected for the calculation may not be universally representative.

## QUANTITATIVE MECHANISTIC ASPECTS AND THEIR RELATIONSHIP TO FORMATION OF PCDD/PCDFS IN INCINERATORS

In a seminal paper in 1983, Shaub and Tsang[26] discussed dioxin formation in incinerators and presented calculations assuming a gas phase mechanism. They concluded that gas phase reactions were unlikely to explain the observed concentrations of dioxins. As one of their research objectives, they proposed that "much more effort should be directed toward understanding the nature of condensed phase reactions (catalyzed and uncatalyzed)," i.e., surface reactions, reactions in solids, etc. Unfortunately, the picture has not changed drastically since 1983 in a quantitative sense. Intrinsic kinetic rate information on relevant gas phase and especially surface reactions is still absent.

The present work so far has not provided any intrinsic rate information;

however, a better picture of the magnitude of the time scales involved has emerged. These time scales are the bed (preflame) region, the flame, the post-combustion zones, the residence time when combustion gases pass over flyash, and the relationship of these time scales to incinerators.

An estimate of the gas phase formation rate of dioxins can be obtained through the use of the Shaub and Tsang[26] rate constant for the step: two precursors → phenoxylphenol intermediate (which would cyclize to dioxin), $k = 10^9 \exp(-26{,}000/RT)$, L/mol/sec. For the lowest bed temperature during combustion (~800°C), a 10-ms gas phase residence time (reactor inlet to flame), and a precursors concentration of $10^{-7}$ mol/L the calculated PCDD concentrations are typically lower than the observed ones (100s ng/m$^3$) by a factor of 100. When the bed temperature is 1000°C, the calculated concentrations still are approximately a factor of 10 or lower. Since these calculations assume a maximum precursor concentration and since virtually all of the gases pass through the flame, the gas phase dimerization in the bed cannot explain the observed concentrations downstream from the flame. However, it is conceivable that it could explain the cumulative formation in the bed over several experimental runs. Any conclusions on this aspect of the work would require considerably more experimentation on this portion of the system (sampling from the bed during a run, variation in bed particle size and type, and changes in equivalence ratio). Most of the experiments to date have been performed under conditions where the spout flow exceeded 85% of the total flow. A more drastic shift in the ratio of spout-to-annular flow, i.e., 1:1, would change the residence time distribution in the bed.

It is unlikely that surface pyrolysis leading to soot formation played an important role in PCDD/PCDF formation since the experiments reported here have been conducted at nominal equivalence ratios well below 1 and, in most cases, fuel and air were premixed. There remains, therefore, the possibility that partial oxidation on surfaces leads to PCDD/PCDF formation. Only sampling in the bed itself and nonflame experiments at comparable bed temperatures (such experiments are planned) can provide quantitative information on this pathway. As shown in Figure 9 by the CO levels, partial oxidation is extensive. This result has not been modeled. It indicates clearly that CO formation in the annulus (longer residence time) is more extensive and that the CO concentration approaching the flame in the annulus may be 2 to 3 times the spout concentration. The residence time in the flame itself is on the order of milliseconds (based on the conical section). The nature of the flame itself has not been completely delineated. Others[2,6] have described the spouted flame as conical extending into the annulus which is covered with flamelets. The conical shape of the flame is evident (Figure 3); however, the annulus is not completely covered by a flat flame (flamelets). The flame is said to penetrate below the bed.[2] Is it conceivable, however, that the flame is quenched at the wall and that a portion of the annular flow bypasses the flame. This effect is likely to be extremely small for the following reason: consider the isoline of 500 ppm CO, above the bed surface. In the spout flow, this measurement is below the flame tip, but in the annulus it is above the flat portion of the flame. As one proceeds downstream, however, heat is rapidly lost from the walls of the reactor. This (and presumably a lack of OH radicals near the wall) leads to a slower—but nevertheless large—decrease in CO concentrations at the wall beyond the bed.

The change in CO concentration through the flame can be used to calculate an OH radical concentration; this value, in turn, can be applied to an estimate of the rate of PCDD formation through the bimolecular reaction of OH with a suitable precursor. However, this precursor can only be a phenoxy phenol (to make PCDD) or a biphenyl phenol (to make PCDF). The rate constant estimate for the dimerization to this precursor is considerably less (see earlier) than that for the OH radical abstraction step leading to cyclization ($10^9$ L/mol/sec).[25] An estimate should therefore be made of the (assumed) first step in the sequence, i.e., OH attack on a chlorophenol. A maximum estimate of this value may be obtained by using the inlet PCP concentrations and those measured at L; the decrease is on the order of $10^3$. For a ~0.1-sec reaction time (between inlet and L), this leads to an estimate of $6.9 \times 10^{-8}$ mol/L for OH. This is within an order of magnitude of the local equilibrium $(OH)_{eqn} = 10^{-1.7}$ exp(–35,000/RT) mol/L.[26] Thus, the phenoxy radical species could form rapidly at flame temperatures (>1500 K). It is still unlikely, however, that this route will produce significant quantities of PCDD/PCDFs since their destruction will proceed at comparable rates under these conditions. It is probable, therefore, that even quantities of OCDD observed at L will require some sort of heterogeneous formation mechanism.

At the temperatures between L and U ($T < 600°C$), little further destruction of PCDDs takes place; this is what would be expected on the basis of gas phase reactions. It is not clear whether the observations on the decrease in CO in the postcombustion region represent evidence of heterogeneous decomposition due to the presence of fine particles (bed attrition) with very high surface areas. The results must be tested against additional global rate expressions for wet CO oxidation available in the literature.

The present work has to some extent verified the conversion of suitable precursors over flyash. However, the quantitative aspects of this process and their relevance to dioxin levels in incinerators are far from clear. Three data sets, the Karasek and Dickson,[15,16,17] the Stieglitz and Vogg,[20] and ours, can be used to calculate a formation rate of PCDDs in units of $\mu$g PCDD per gram of flyash per min. This rate ranges from $0.7 \times 10^{-5}$ to $0.28 \times 10^{-2}$. These experiments are comparable only in a limited sense. In ours, 1 to 2 g flyash (Ontario flyash furnished by F. W. Karasek) and gas contact times of 0.1 to 2 sec were employed; Dickson[17] used 20 g flyash and contact times of 9 to 90 sec; Stieglitz and Vogg "annealed" flyash for 30 to 60 min, but did not add any potential precursors or other reactants.[19] These range of formation rates has been calculated on the basis of total sampling time, not contact time, since the yield of PCDDs represents that found on the flyash as well as downstream in a trap.

The Dickson experiment[17] indicates a transport limitation: at a flow rate that resulted in a 9-sec contact time, the formation rate was 28 times greater than at a contact time of 90 sec. In both experiments, the nominal gas phase PCP concentration was 0.167 $\mu$g/ml. In our experiments this concentration was much lower, (0) 1000 ng/m$^3$, i.e., comparable to measurements in incinerators.[23] The Stieglitz/Vogg approach does not permit an estimate of a precursor concentration.[20] None of these rate values can be said to represent the intrinsic rate of the heterogeneous surface reaction. The application of these values to incinerator measurements is therefore rather dubious since, as was shown in the transport rate section, transport limitations (i.e., getting the precursor to the flyash surface) are probably absent. However, these findings can be put in perspective. Typical MSW incinerator flyash amounts of PCDD/PCDFs are (0) 100 ng/g. Using the highest rate calculated from reported data,[20] $0.13 \times 10^{-2}$, would require on the order of 1 h to form such quantities; the highest rate reported by Dickson,[17] $0.28 \times 10^{-2}$, would require approximately 0.5 hr. Our highest observed rate is approximately a factor of 10 lower than these rates. These rates would conceivably explain the formation in sections of an incinerator where the flyash has a long residence time (on the walls of economizer heat exchanges and 3rd and 4th field sections of electrostatic precipitators), but would not appear to be applicable to situations where rapping cycles are seconds or minutes.

When applied to stack particle concentrations of 20 to 100 mg/m$^3$ and a maximum gas residence time (5 sec) in an electrostatic precipitator, the laboratory rates can account for only about 1% of the typically observed concentrations, 0(100) ng/m$^3$.[22,27] It is, of course, not necessary to account for the formation of all of the observed PCDD via reactions in the electrostatic precipitator since these compounds are present at the inlet to electrostatic precipitators. It is probable that the true rate of formation via heterogeneous reactions is much greater; in part perhaps due to much more "active" flyash than has been used in the laboratory experiments, because fresh flyash surfaces are continuously generated in the combustion process, the presence of many possible precursors, and the probability that the interaction of the precursor with the flyash can be described by the strong adsorption case, i.e., $r = k$, in the units used earlier, $\mu$g PCDD/g flyash per min. These cases need to be studied separately in the laboratory.

## CONCLUSIONS

The formation of PCDD/PCDFs from possible precursors has been investigated under conditions of heterogeneous combustion. These precursors, chlorobenzenes and chlorophenols, all form pentachlorophenol, which is obviously a plausible intermediate, but not necessarily a required one. The central role of pentachlorophenol is suggested through qualitative and quantitative considerations. Rising dechlorination, chlorination, and hydroxylation reactions are observed. The quantities of PCDD/PCDFs produced within milliseconds under conditions of $\phi < 1$ are comparable, in some cases, to those observed in incinerators. Under the sampling conditions employed (up to 60 min at 120 ml/min) Cl/C ratios $10^{-3}$ or greater were typically required to form such quantities. These observations are accommodated in a qualitative reaction scheme; however, known or suggested gas phase rate constants apparently cannot account for these observed concentrations. A clarification of the relationship between pentachlorophenol and hexachlorobenzene under a variety of conditions is also needed.

There are two heterogeneous reaction aspects to this work. In the combustor itself, oxidation (and possibly precombustion thermal oxidation) occurs in the presence of sand. Surface reactions may be promoted; however, specific quantitative evidence is still lacking. In a separate experiment, the sand used in the combustor was employed in the investigation of the dimerization of $^{13}C$-pentachlorophenol, where it exhibited no catalytic activity. In general, the lack of lower congeners in the absence of flyash suggests that catalyzed dechlorination reactions are not favored over sand in our system. However, when the combustion gases from the combustion of 1,2-dichlorobenzene and pentachlorophenol were contacted with flyash at 300°C, lower PCDD congeners were observed, in agreement with the Karasek/Dickson[15] findings. PCDFs were also formed under these conditions. The apparent rates of these and other laboratory investigations are not consistent with formation rates in incinerators, however.

The spouted bed combustor selected for the study of incineration processes encompasses many desirable features that can be controlled. Future investigations will be aimed at a different residence time distribution in the bed, the study of catalytic bed material, and solid fuels, as well as nonflame combustion. Sampling within the bed and below the flame, and under nonflame conditions (but at comparable temperatures) can address the extent of PCDD/PCDF formation in that region.

## ACKNOWLEDGMENT

The authors would like to thank Professor F. W. Karasek and his group for the GC/MS analyses of PCP, HCB, PCDD/PCDFs and many others, Professors H. Littman and M. H. Morgan from this department for assistance with the spouted bed design and many helpful discussions regarding spouted bed hydrodynamics, the members of the technical advisory council for their comments, and the New York State Energy Research and Development Authority for financial support.

## REFERENCES


1. Oppelt, T. E. *J. Air Poll. Control. Assoc.* 37:558–586 (1987).
2. Arhib, H. A., F. R. Sawyer, and F. J. Weinberg. "The Combustion Characteristics of Spouted Beds," 18th Symposium (Int.) on Combustion, The Combustion Institute, 1981 (1982).
3. Mathur, K. B., and N. Epstein. *Spouted Beds* (New York: Academic Press, 1974).
4. Khoshnoodi, M., and F. J. Weinberg. *Combust. Flame* 33:11–21 (1978).
5. Haji-Sulaiman, Z., et al. *Can. J. Chem. Eng.* 64:125–132 (1986).
6. Ohtake, J., et al. "Combustion and Flow in a Two Dimensional Spouted Bed," paper presented at the 20th symposium (Int.) on combustion, The Combustion Institute, Ann Arbor, MI, 1984.
7. Altwicker, E. R., P. Nugent, K. Golas, J. Farinas, and F. W. Karasek. "Simulation of Municipal Waste Combustion Phenomena in a Spouted Bed Combustion," paper 113f, presented at AIChE annual meeting, New York, NY, November 19, 1987.
8. Stanley, W., M. Wu, C. J. Lim, and N. Epstein. *Chem. Eng. Commun.* 62:251–268.
9. Morgan, M. H., III, and H. Littman. "General Relationships for the Minimum Spouting Pressure Drop Ratio, $\triangle p_{ms}/\triangle p_{mf}$, and the Spout-Annulus Interfacial Condition in a Spouted Bed," in *Fluidization*, J. R. Grace, and J. M. Malsen, Eds. (New York, NY: Plenum Publishing Corporation, 1980), p. 287.
10. Rovero, G., C. M. H. Brereton, N. Epstein, J. R. Grace, L. Casalegno, and N. Piccini. *Can. J. Chem. Eng.* 61(3):289–296 (1983).
11. Konduri, R. Unpublished research, Rensselaer Polytechnic Institute (1988).
12. Dryer, F. L., and I. Glassman. "High Temperature Oxidation of CO and $CH_4$" 14th Symposium (Int.) on Combustion, The Combustion Institute (1973), 987.
13. Nugent, P. MS Thesis, Department of Chemical Engineering, Rensselaer Polytechnic Institute, Troy, NY (1988).
14. Karasek, F. W. Personal communication; information supplied on the basis of GC/MS-analysis (1987, 1988).
15. Karasek, F. W., and L. C. Dickson. *Science* 23714816:754–756 (1987).
16. Dickson, L. C., and F. W. Karasek. *J. Chromatog.* 389:127–137 (1987).
17. Dickson, L. C. "Mechanism of Formation of Polychlorinated Dibenzo-P-dioxins During Municipal Refuse Incineration, PhD Thesis, University of Waterloo, Waterloo, Ontario (1987).
18. Lustenhouwer, J. W. A., K. Olie, and O. Hutzinger. *Chemosphere* 9:501–522 (1980).
19. Chaltykyan, O. A. *Copper Catalytic Reactions*, (NY: Consultants Bureau, 1966).
20. Vogg, H., and L. Stieglitz. *Chemosphere* 15:1373 (1986).
21. Shaub, W. M., and W. Tsang. "Overview of Dioxin Formation in Gas and Solid Phases under Municipal Incinerator Conditions," in *Chlorinated Dibenzodioxins and Dibenzofurans in the Total Environment II*, G. Choudhary, L. H. Keith, and C. Rappe, Eds. (Stoneham, MA: Butterworth Publishers, Inc., 1985), p. 469.
22. Taylor, R., M. A. Tompkins, S. E. Kirton, T. Mauney, D. F. S. Natusch, and P. K. Hopke. *Environ. Sci. Technol.* 16:148–154 (1982).
23. Environment Canada. "The National Incinerator Testing and Evaluation Program: Two-Stage Combustion (Prince Edward Island) Summary Report," Report EPS 3UP/1 (September 1985).

24. Carslaw, H. S., and J. C. Jaeger. *Conduction of Heat in Solids*, 2nd ed. Oxford (1959).
25. Welty, J. R., C. E. Wicks, and R. E. Wilson. *Fundamentals of Momentum, Heat, and Mass Transfer*, 2nd ed. (New York: John Wiley & Sons, Inc., 1976).
26. Shaub, W. M., and W. Tsang. *Environ. Sci. Technol.* 17:121–128 (1983).
27. Beychock, M. R. *Atmos. Environ.* 21:29–36 (1987).

## NOMENCLATURE

a = radius of flyash particle.
C = concentration of precursor in gas phase; cyclone in Figure 2.
$C_o$ = bulk concentration of precursor in the gas phase.
d = tube diameter, i.e., duct width.
D = diffusivity of the precursor in the gas phase; diameter and reactor (Table 1b).
H = bed height.
$k_c$ = mass transfer coefficient.
L = lower sampling position.
$P_{mf}$ = minimum fluidization pressure drop.
$P_{ms}$ = minimum spouting pressure drop.
Q = amount of precursor destroyed at the flyash surface.
r = radial position.
Re = Reynolds number.
Sc = Schmidt number; a value of 9 was used.[25]
t = time required to produce observed levels of PCDDs.
U = upper sampling probe position.
$U_{ms}$ = minimum spouting velocity.
$U_{op}$ = operating superficial gas velocity.
$U_s$ = superficial gas velocity.
z = axial position.

### Greek Symbols

$\phi$ = equivalence ratio.

## CHAPTER 3

# Study of the Relationship Between Trace Elements and the Formation of Chlorinated Dioxins on Flyash

**R.E. Clement, D. Boomer, K. P. Naikwadi, and F. W. Karasek**

## INTRODUCTION

In 1977, the presence of chlorinated dibenzo-p-dioxins (CDDs) in municipal incinerator flyash was first reported.[1] Soon after, the presence of CDDs in flyash from a Canadian municipal incinerator was also demonstrated.[2] Subsequent work has shown that CDDs and a series of related compounds, chlorinated dibenzofurans (CDFs), are present to some extent in the flyash of virtually every municipal waste incinerator.[3-10] These substances are of great interest because of the extremely high toxicity and other effects exhibited by some CDD/CDF congeners in laboratory animal studies.[11] Scientists at the Dow Chemical Corporation found that CDDs were in fact present in the particulate matter from many types of combustion sources and postulated that these substances are formed from trace chemical reactions occurring in fire.[12]

Many studies have been conducted in an effort to elucidate the reaction mechanisms by which CDDs and CDFs are formed during the incineration of municipal waste. Eiceman and Rghei demonstrated that under conditions of high temperature and chlorine gas, flyash from municipal incinerators may undergo gas-phase particulate reactions to produce trace concentrations of chlorinated organic compounds.[13] They also showed that chlorine substitution reactions of 1,2,3,4-TCDD to higher chlorinated CDDs occur on flyash surfaces between 50° to 250°C with an atmosphere of HCl in air.[14] Mechanistic aspects of the thermal formation of chloroorganics including the CDDs/CDFs were discussed in detail by Choudhry and co-workers.[15-18] They concluded that there are three principal explanations for the presence of CDDs/CDFs in municipal incinerator flyash:

1. They are trace components of municipal refuse that do not undergo effective thermal destruction.

2. They are produced during the combustion and pyrolysis of chlorinated precursors.
3. They are formed by complex pyrolytic processes of chemically unrelated organic compounds (*de novo* synthesis).

It has been shown that CDDs/CDFs are present in municipal refuse.[19,20] However, mass balance calculations demonstrated that the quantity of CDDs/CDFs detected in the stack emissions and flyash of two municipal waste incinerators were generally much greater than the total CDDs/CDFs initially present in the raw refuse.[21] Also, the congener distributions of input compared to output CDDs/CDFs were completely different. It is clearly possible to form CDDs/CDFs from pyrolytic reactions of chlorinated precursors. Choudhry and Hutzinger have summarized the results of many studies where CDDs and CDFs were produced from chlorinated precursor substances such as PCBs, chlorophenols, chlorobenzenes, and others.[18] They also showed the complex synthetic pathways whereby CDDs/CDFs can be formed from chemically unrelated compounds (*de novo* synthesis). Although it is now known that CDD/CDF formation by *de novo* synthesis and/or from precursors must occur in municipal incinerators, the exact mechanisms have yet to be determined.

Several studies have been presented recently to elucidate the mechanisms of CDD/CDF formation in municipal incinerators. By performing pilot-scale incineration of selected waste, Olie and co-workers concluded that burning waste with a lignine-like structure in the presence of a chlorine donor contributes significantly to CDD/CDF formation.[22] Karasek et al. added three times the usual level of PVC to a municipal incinerator and did not observe any increase in CDD/CDF concentrations on flyash.[23] Shaub and Tsang reported that the gas-phase formation of CDDs is likely to be very low at high temperatures if efficient fuel/air mixing is obtained.[24] In addition, they determined from theoretical considerations that the production of CDDs from chlorophenols depends upon the square of the chlorophenol concentration. Rghei and Eiceman studied gas-solid phase reactions on a variety of surfaces and found that the favored mechanism for CDD formation under their conditions was electrophilic substitution.[25]

The recent studies of Karasek,[25-27] Stieglitz,[28,29] Hagenmaier,[30,31] and their co-workers have confirmed that CDDs/CDFs can be formed on flyash through surface-catalyzed reactions. It is therefore suspected that the trace metal concentration in flyash may be an important variable in CDD/CDF formation. Hagenmaier observed CDD/CDF dechlorination/hydrogenation reactions on flyash surfaces under oxygen-deficient conditions, whereas CDD/CDF formation occurred with surplus oxygen conditions.[31] Copper was found to catalyze the dechlorination/hydrogenation of the octachloro CDD/CDF at low temperature.[30] Stieglitz reported that CDDs/CDFs are formed in a *de novo* synthesis from carbon and inorganic chlorides, and that the chlorination step is induced by the catalytic activity of copper.[28] Karasek and co-workers formed

CDDs on flyash surfaces by heating preextracted flyash after it had been spiked with $^{13}C$-labeled pentachlorophenol.[25-27] A range of $^{13}C$-labeled CDD congeners was formed that had the same relative distribution as the original native CDDs extracted from the flyash. The quantity of $^{13}C$-CDDs formed can be used as a general indication of relative activity of various materials tested in the $^{13}C$-pentachlorophenol combustion experiment. By determining the activity of a variety of flyash types and determining the trace metal concentrations on these materials, it may be possible to correlate activity with specific metals. Such correlations could lead to means of controlling CDD/CDF levels in incinerator emissions, either by removing the catalytic metals from the raw refuse incinerated or by inhibiting their catalytic effects. In this study, the relationship between flyash activity for CDD/CDF formation and trace metal concentrations is examined.

## EXPERIMENTAL PROCEDURES

Flyash from six different incinerators was obtained. These samples were assigned codes as follows: M1, municipal incinerator flyash from Germany; M2, municipal incinerator flyash from an Ontario, Canada, incinerator; M3, flyash from a second municipal incinerator in Germany; M4, municipal incinerator flyash from Japan; IN, industrial waste incinerator ash; and CS, ash from a copper smelter located in Ontario. Details on the operation of the incinerators are not available. However, the technical design and operation as well as composition of the feed to these incinerators are significantly different for at least four out of the six listed.

The concentrations of native CDD/CDF in the six ash samples were determined by methods described in detail elsewhere.[34] Two- to five-g ash samples were spiked with 10 ng $^{13}C$-labeled 1,2,3,4-tetrachlorodibenzo-p-dioxin (1,2,3,4-TCDD) and 20 ng octachlorodibenzo-p-dioxin ($^{13}C$-OCDD) before digesting with concentrated HCl. Filtered particulates were air dried, then extracted overnight by using a Soxhlet apparatus with a toluene solvent. Concentrated extracts were analyzed for the tetra-to-octachloro CDDs/CDFs by gas chromatography-mass spectrometry (GC-MS) operated in the selected ion monitoring (SIM) mode. The activity for CDD/CDF formation was determined after exhaustive toluene Soxhlet extraction of untreated ash to remove organics. Extracted flyash was then packed in a flow tube and treated to 350°C for 8 hr under air flow. After cooling to room temperature, glass beads downstream of the flyash were spiked with 500 g of $^{13}C$-pentachlorophenol. The apparatus containing no glass beads and flyash was heated to 300°C under air flow and the $^{13}C$-chlorinated dioxins formed from the reaction of $^{13}C$-pentachlorophenol on flyash. The chlorodioxins were extracted from flyash and analyzed by GC-MS as described earlier. Additional details of this procedure are described elsewhere in this volume and in other references.[26,27]

Flyash samples were analyzed for trace elements by using inductively coup-

**Table 1. Native Concentrations of Dibenzo-p-dioxins (CDD) in Flyash (ppb)**

| | Sample Code | | | | | |
|---|---|---|---|---|---|---|
| | M2 | M4 | M1 | M3 | IN | CS |
| Dioxin congener | | | | | | |
| tetra-CDD | 260 | 0.2 | 2.1 | 0.3 | 57 | 0.2 |
| penta-CDD | 560 | 1.3 | 5.7 | 0.2 | 160 | 1.0 |
| hexa-CDD | 660 | 14 | 25 | 0.4 | 480 | 2.8 |
| hepta-CDD | 660 | 110 | 71 | 0.7 | 680 | 5.5 |
| octa-CDD | 440 | 280 | 230 | 1.0 | 410 | 5.8 |
| Total | 2600 | 410 | 330 | 2.6 | 1800 | 15 |
| Total $^{13}$C-CDD[a] | 3300 | 460 | 13,000[b] | 1,500[b] | 2,200[b] | 29[b] |

[a]From reaction of $^{13}$C-pentachlorophenol.
[b]Only hepta- and octachloro-CDDs were observed.

led argon plasma-mass spectrometry (ICP-MS). A SCIEX Elan 250 system was employed. The 25 elements determined were Cu, Ni, Zr, Co, Fe, Mn, Al, Ca, Mg, Na, K, Na, Ba, Be, Sr, Ti, Ag, S, Pb, V, As, B, and P.

## RESULTS AND DISCUSSION

### CDD Concentration

Table 1 gives the concentrations of native dioxins in the six samples of flyash. The total concentrations of $^{13}C_{12}$-labeled chlorodibenzo-p-dioxins as determined after the reaction of $^{13}C_6$-pentachlorophenol on organics-removed flyash are also shown. For two of the samples (M1, M3) the original native CDD concentrations are about 50 times lower than those of $^{13}$C-CDDs found after the combustion experiments. For the four other samples, however, the average concentrations of $^{13}$C-CDDs and native CDDs on the flyash only differs by 20%. The $^{13}$C-PCP experiment, therefore, seems to be a good indicator of flyash activity with respect to CDD formation. The reason for large differences in $^{13}$C- and native CDDs in two of the samples is unknown. It is notable that in four of the samples, only the heptachloro-and octachloro-CDD congeners were observed after the combustion experiment, whereas some representative members of all CDD congener groups investigated (tetra to octachloro) were originally present in all six samples.

### Elemental Concentration

Concentrations of many of the elements studied were at similar low levels in all flyash samples. This group includes Ni, Cd, Co, Cr, Mn, V, Mo, Be, Sr, Ti, Ag, As, and B. Concentrations of these elements ranged from not detected to about 0.5 ppt and did not exhibit significant sample-to-sample variability.

Concentrations of the remaining elements that were determined are given in Table 2. They have been divided into two groups. The first group, consisting

**Table 2. Concentrations of Principal Elemental Components of Flyash (ppt)**

| | Incinerator | | | | | |
|---|---|---|---|---|---|---|
| **Element** | **M2** | **M4** | **M1** | **M3** | **IN** | **CS** |
| **Group A** | | | | | | |
| Cu | 0.8 | 6.5 | 1.4 | 1 | 0.4 | 89 |
| Fe | 6.7 | 6.2 | 13 | 15 | 8.3 | 15 |
| Ca | 61 | 220 | 140 | 75 | 70 | 18 |
| Mg | 9.1 | 10 | 11 | 8.4 | 9.7 | 0.6 |
| P | 7 | 5.1 | 10 | 4 | 5.2 | 17 |
| **Group B** | | | | | | |
| S | 67 | 20 | 19 | 3.9 | 18 | 110 |
| Pb | 22 | 5.5 | 3.0 | 0.4 | 2.1 | 120 |
| Al | 22 | 31 | 40 | 110 | 33 | 0.6 |
| Na | 44 | 20 | 8.9 | 13 | 19 | 2.6 |
| K | 65 | 31 | 25 | 16 | 16 | 19 |
| Zn | 31 | 6.2 | 8.7 | 1.2 | 4.8 | 69 |
| Total of Group B (Except Al) | 229 | 83 | 65 | 35 | 60 | 321 |
| Total Native CDDs (ppb) | 2600 | 410 | 330 | 3 | 1800 | 15 |

of elements Cu, Fe, Ca, Mg, and P, contains those elements for which their concentrations in the flyash samples showed no discernable trend with respect to the trend of total CDD concentrations. In the group of elements which appear to have no correlation with CDD levels is Cu, which has been shown in a previous study to catalyze dechlorination/ hydrogenation reactions of the CDDs.[31] If Cu is acting in such a manner in these samples, it might be expected to affect the CDD congener distribution. The OCDD/TCDD ratios for the four samples of municipal incinerator flyash are 1400, 110, 3.3, and 1.7 for samples M4, M1, M3, and M2, respectively. The Cu concentrations in these samples are 6.5, 1.4, 1.0 and 0.8 parts per thousand for M4, M1, M3 and M2, respectively. A linear correlation of OCDD/TCDD ratios versus Cu concentrations gives a correlation coefficient of 0.9995. With a sample size of only four, however, this cannot be considered as more than a trend. Additional study is needed to verify any relationship that may exist. Also, the observed trend is opposite to that which was expected (i.e., high Cu concentrations should catalyze OCDD dechlorination and lead to lower OCDD/TCDD ratios. The CDD/trace element data for samples of the industrial incinerator (IN) and copper smelter (CS) do not appear to exhibit any similar trends as compared to the municipal incinerator flyash samples, although it may be significant that ash from the copper smelter produces lower CDD concentrations than all the other samples, with one exception.

## Correlation of Trace Element versus CDD Concentrations

Concentrations of the elements listed under Group B in Table 2 all show some trend with respect to CDD concentrations. However, only the municipal incinerator flyash samples follow these trends. CDD concentrations on flyash taken from IN and CS do not exhibit any simple trends with respect to elemental concentrations when considered in a group with samples of municipal incinerator flyash. Considering that the feedstock of the four municipal incinerators sampled is much more similar in composition compared to that for the industrial incinerator or copper smelter, the above observation is not surprising.

By comparing CDD and elemental concentrations of only the four municipal incinerator flyash samples, it is observed that the concentrations of S, Pb, Na, K, and Zn decrease as the total CDD concentration decreases. The linear regression of the total concentrations of these five elements versus the total CDD concentrations produces a correlation coefficient of 0.9959. With such a limited sample size, this relationship should only be considered as a trend. The inclusion of Na and K in this group could indicate that inorganic chlorine sources are important to the formation of CDDs on flyash. In fact, this latter observation suggests that the active chemical site on the flyash is possibly a metal chloride and not a metal or metal oxide.

Aluminum is the only element studied for which its concentration exhibited an inverse trend with respect to CDD concentrations on municipal incinerator flyash. No special significance is attributed to this trend ($r = -0.66$) although it is worth further investigation by examination of additional flyash samples.

## Conclusions and Future Work

The formation of CDDs/CDFs on the surface of flyash particles is a complex phenomenon that may involve many different reaction mechanisms. Evidence is presented in this study to suggest that the concentrations of specific elements in flyash are related to the quantity and congener distributions of CDDs formed on municipal incinerator flyash. Ash from an industrial incinerator and a copper smelter contained CDDs, but there was no apparent relationship between trace element concentrations and CDD concentration or congener distribution. It is possible that such relationships may only exist for incinerators or smelters that have a similar feedstock in common. The specific incinerator design would seem to be a less important factor, since three of the four municipal incinerators studied here were of different designs and were from different countries.

There are limitations in this study apart from the small number of samples available. Elemental determinations were made on bulk flyash samples, whereas only the surface composition is relevant to the proposed CDD formation mechanism (i.e., surface-catalyzed reactions). Also, the specific form in which elements are present is expected to be important. By studying flyash from a single incinerator over an extended period of time, the trends shown in this investigation may be verified. If a few specific elements can be shown to catalyse CDD/CDF formation, it should be possible to substantially reduce CDD/CDF levels by removal of selected components of the incinerator feedstock and by designing a specific catalyst poison.

## REFERENCES

1. Olie, K., P. L. Vermeuler, and O. Hutzinger. "Chlorodibenzo-p-dioxins and Chlorodibenzofurans are Trace Components of Fly Ash and Flue Gas of Some Municipal Incinerators in the Netherlands," *Chemosphere* 8:445–459 (1977).
2. Eiceman, G. A., R. E. Clement, and F. W. Karasek. "Analysis of Fly-Ash from Municipal Incinerators for Trace Organic Compounds," *Anal. Chem.* 51:2343–2350 (1979).
3. Lustenhouwer, J. W. A., K. Olie, and O. Hutzinger. "Chlorinated Dibenzo-p-dioxins and Related Compounds in Incinerator Effluents: A Review of Measurements and Mechanisms of Formation," *Chemosphere* 9:501–522 (1980).
4. Karasek, F. W., and A. C. Viau. "Gas Chromatographic-Mass Spectrometric Analysis of Polychlorinated Dibenzo-p-dioxins and Organic Compounds in High-Temperature Fly Ash From Municipal Waste Incineration," *J. Chromatog.* 265:79–88 (1983).
5. Cavallaro, A., G. Bandi, G. Invernizzi, L. Luciani, E. Mongins, and A. Gorri, "Sampling, Occurrence and Evaluation of PCDDs and PCDFs from Incinerated Solid Urban Waste," *Chemosphere* 9:611–621 (1980).
6. Rappe, C., S. Marklund, L. O. Kjeller, and M. Tysklind. "PCDDs and PCDFs in Emissions from Various Incinerators," *Chemosphere* 15:1213–1217 (1988).
7. Tong, H. Y. and F. W. Karasek. "Comparison of PCDD and PCDF in Flyash Collected from Municipal Incinerators of Different Countries," *Chemosphere* 15:1219 (1986).
8. DeFre, R. "Dioxin Levels in the Emissions of Belgian Municipal Incinerators," *Chemosphere* 15:1255–1260 (1986).
9. Tanaka, M. and R. Takeshita. "Evaluation of 2,3,7,8-TCDD and PCDDs in Fly Ash from Refuse Incinerators," *Chemosphere* 16:1865–1868 (1987).
10. Morita, M., A. Yasuhara, and H. Ito. "Isomer Specific Determination of Polychlorinated Dibenzo-p-dioxins and Dibenzofurans in Incinerator Related Samples in Japan," *Chemosphere* 16:1959–1964 (1987).
11. "Scientific Criteria Document for Standard Development No. 4–84: Polychlorinated Dibenzo-p-dioxins (PCDDs) and Polychlorinated Dibenzofurans (PCDFs)," Ontario Ministry of the Environment Report (September 1985).
12. Bumb, R. R., W. B. Crummett, S. S. Cutie, J. R. Gledhill, R. H. Hummel, R. O. Kagel, L. L. Lamparski, E. V. Luoma, D. L. Miller, T. J. Nestrick, L. A. Shadoff, R. H. Stehl, and J. S. Woods. "Trace Chemistries of Fire: A Source of Chlorinated Dioxins," *Science* 210(4468):385–390 (1980).
13. Eiceman, G. A., and H. O. Rghei. "Products from Laboratory Chlorination of Fly Ash from a Municipal Incinerator," *Environ. Sci. Technol.* 16:53–56 (1982).
14. Eiceman, G. A., and H. O. Rghei. "Chlorination Reactions of 1,2,3,4-Tetrachlorodibenzo-p-dioxin on Fly Ash with HCl in Air," *Chemosphere* 11(9):833–839 (1982).
15. Choudhry, G. G., and O. Hutzinger. "Mechanisms in the Thermal Formation of Chlorinated Compounds Including Polychlorinated Dibenzo-p-dioxins," *Pergamon Ser. Environ. Sci.* 5:275–301 (1982).
16. Choudhry, G. G., and O. Hutzinger. "Mechanistic Aspects of the Thermal Formation of Halogenated Organic Compounds Including Polychlorinated Dibenzo-p-dioxins. Part I: Theoretical Background and Thermochemical Decompositions of Monomeric Aliphatics and Aromatics," *Toxical Environ. Chem* 5:1–65 (1982).

17. Choudhry, G. G., and O. Hutzinger. "Mechanistic Aspects of the Thermal Formation of Halogenated Organic Compounds Including Polychlorinated Dibenzo-p-dioxins. Part II: Thermochemical Generation and Destruction of Dibenzofurans and Dibenzo-p-dioxins," *Toxicol. Environ. Chem.* 5:67–93 (1982).
18. Choudhry, G. G., and O. Hutzinger. "Mechanistic Aspects of the Thermal Formation of Halogenated Organic Compounds Including Polychlorinated Dibenzo-p-dioxins. Part V: Hypotheses and Summary," *Toxicol. Environ. Chem.* 5:295–309 (1982).
19. Tosine, H. M., R. E. Clement, V. Ozvacic, and G. Wong. "Levels of PCDD/PCDF and Other Chlorinated Organics in Municipal Refuse," *Chemosphere* 14:821–827 (1985).
20. Ozvacic, V., G. Wong, H. Tosine, R. E. Clement, J. Osborne, and S. Thorndyke. "Determinations of Chlorinated Dibenzo-p-dioxins, Chlorinated Dibenzofurans, Chlorinated Biphenyls, Chlorobenzenes and Chlorophenols in Air Emissions and other Process Streams at SWARU in Hamilton," Ontario Ministry of the Environment Report ARB-02-84 FIRD (July 1984).
21. Ozvacic, V., G. Wong, H. Tosine, R. E. Clement, and J. Osborne. "Emissions of Chlorinated Organics from Two Municipal Incinerators in Ontario," *J. Air. Poll. Control Assoc.* 35:849–855 (1985).
22. Olie, K., M. V. D. Berg, and O. Hutzinger. "Formation and Fate of PCDD and PCDF from Combustion Processes," *Chemosphere* 12:627–636 (1983).
23. Karasek, F. W., A. C. Viau, G. Guiochon, and M. F. Gonnord. "Gas Chromatographic-Mass Spectrometric Study on the Formation of Polychlorinated Dibenzo-p-dioxins and Polychlorobenzenes from Polyvinyl Chloride in a Municipal Incinerator," *J. Chromatog.* 270:227–234 (1983).
24. Shaub, W. M. and W. Tsang, "Dioxin Formation in Incinerators," *Environ. Sci. Technol.* 17:721–730 (1983).
25. Rghei, H. O., and G. A. Eiceman. "Effect of Matrix on Heterogeneous Phase Chlorine Substitution Reactions for Dibenzo-p-dioxin and HCl in Air," *Chemosphere* 14:167–171 (1985).
26. Dickson, L. C., and F. W. Karasek, "Mechanism of Formation of Polychlorinated Dibenzo-p-dioxins Produced on Municipal Incinerator Flyash from Reactions of Chlorinated Phenols," *J. Chromatog.* 389:127–137 (1987).
27. Karasek, F. W., and L. C. Dickson. "Model Studies of Polychlorinated Dibenzo-p-Dioxin Formation During Municipal Refuse Incineration," *Science* 237:754–756 (1987).
28. Stieglitz, L., G. Zwick, H. Beck, W. Roth, and H. Vogg, "On the De-Novo Synthesis of PCDD/PCDF on Fly Ash of Municipal Waste Incinerators," *Chemosphere* 18:1219–1226 (1989).
29. Stieglitz, L., and H. Vogg, "On Formation of PCDD/PCDF in Fly Ash from Municipal Waste Incinerators," *Chemosphere* 16:1917–1911 (1987).
30. Hagenmaier, H., M. Kraft, H. Brunner, and R. Haag, "Catalytic Effects of Fly Ash from Waste Incineration Facilities on the Formation and Decomposition of Polychlorinated Dibenzo-p-dioxins and Polychlorinated Dibenzofurans," *Environ. Sci. Technol.* 21:1080–1084 (1987).
31. Hagenmaier, H., H. Brunner, R. Haag, and M. Kraft, "Copper-Catalyzed Dechlorination/Hydrogenation of Polychlorinated Dibenzo-p-dioxins, Polychlorinated Dibenzofurans and other Chlorinated Aromatic compounds," *Environ. Sci. Technol.* 21:1085–1088 (1987).

# CHAPTER 4

# Formation of Highly Toxic Polychlorinated Dibenzo-p-Dioxins by Catalytic Activity of Metallic Compounds in Flyash

**K. P. Naikwadi and F. W. Karasek**

## ABSTRACT

The formation of polychlorinated dibenzo-p-dioxins found on the flyash in municipal solid waste (MSW) incinerators is a universal phenomenon. However, the amount of dioxins detected varies considerably from incinerator to incinerator. Flyash is a very complex matrix that consists of metallic and organic compounds. It has been discovered that dioxins can form by catalytic activity of flyash. However, correlation of flyash constituents for the formation of dioxins is extremly difficult due to the complexity of the flyash. Laboratory experiments have shown that dioxins can form at temperatures between 200 to 300°C by catalytic activity of metallic compounds. A definite correlation between reaction temperature and the amount of tetra to octachlorodibenzo dioxins formed has been predicted and observed. Separation, identification, and quantitation of dioxins and other chlorinated compounds formed in various reaction were carried out using GC, HPLC, and GC-MS/EISIM techniques.

## INTRODUCTION

Polychlorinated dibenzo-p-dioxins (PCDDs) are environmental contaminants. Formation and emission of PCDDs in the environment can occur by very complex processes. Several potential sources of PCDDs have been recognized. These include chlorophenoxy-substituted insecticide spray in fields that contain traces of PCDD and dibenzofurans (PCDFs),[1] wood burning,[2] industrial and municipal waste incineration,[3] and effluents from the paper industry.[4] In all these sources, MSW incineration contributes the major PCDD emission in the environment. MSW incineration is an attractive concept that has several advantages for garbage disposal such as reduced landfill space, reduced cost of transportation, and possible energy recovery. Efforts are being made to reduce the formation and emission of toxic compounds by MSW incinerators to take advantage of its benefits.[5]

The mechanism of PCDD formation in MSW incinerators is very complex. It has been shown in our laboratory that PCDDs can form by catalytic activity of flyash at low temperatures.[6] However, it is very difficult to accurately determine catalytic activity parameters such as constituents in flyash that promote or inhibit the formation of PCDDs and PCDFs.

This paper shows optimum conditions for the formation of PCDDs from pentachlorophenol and MSW incinerator flyash. Results of native PCDDs detected and PCDDs formed by catalytic activity of flyash will be discussed. Laboratory experiments using different metals and metal oxides in the formation of PCDDs will be described.

## EXPERIMENTAL

Machida flyash (MFA) was obtained from H. Hatano (Japan), Kraefeld flyash (KFFA) was obtained from O. Hutzinger (West Germany), Ontario flyash (OFA) was obtained from R. E. Clement (Canada), and Sun Lake flyash (SLFA) was from our laboratory stock.

Flyash samples were soxhlet extracted for 48 hr using benzene. In a particular experiment, 50 g of flyash was soxhlet extracted by 350 ml of benzene. The soxhlet extract was concentrated to 500 $\mu$L by rotary evaporation and the native PCDD and PCDF were determined using the GC-MS/EISIM technique. A catalytic activity study was carried out using the assembly shown in Figure 1. Soxhlet-extracted flyash was heated to 100°C for 1 hr and then to 350°C for 8 hr using an air flow of 50 mL/min. The assembly cooled to room temperature, and the downstream portion of the thimble and impinger was replaced by a clean tube and impinger. Then 500 $\mu$g C-13 labeled pentachlorophenol in methanol was placed on glass beads. Solvent was allowed to evaporate and then the airflow was adjusted to 10 mL/min. The flyash and precursor were heated to the desired temperature and the airflow was started (10 mL/min). After a definite period of reaction, the assembly was cooled to room temperature. Airflow was halted and the flyash and impinger were taken out. The flyash was soxhlet extracted for 24 hr, and the extract was concentrated to 200 $\mu$L. The impinger was rinsed several times by benzene and all rinses were collected and concentrated to 200 $\mu$L. Both the soxhlet extract and the impinger rinses were then analyzed separately for labeled PCDD and PCDF produced. In the study of the activity of metals and metal oxides, flyash was replaced by desired metallic materials and the activity tests were carried out.

The Hewlett Packard 5890/5970 GC-MSD instrument was used for analysis of PCDDs and PCDFs. Retention windows for tetra- to octa-chlorodibenzodioxins were determined using a C-13 labeled dioxin mixture

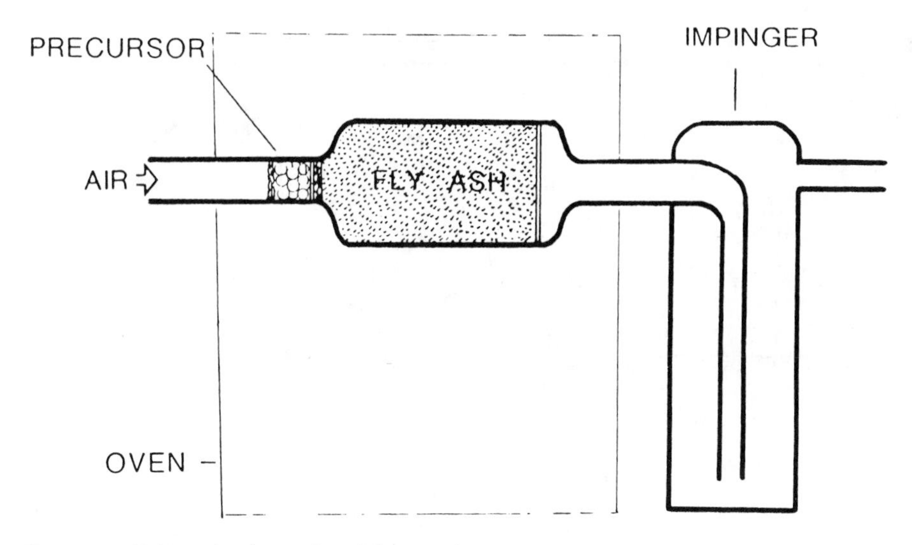


**Figure 1.** Schematic of experimental apparatus.

developed in our laboratory that contains all possible isomers in each congener group. The concentration (ng/$\mu$L) of each congener group in the C-13 labeled dioxin mixture was 0.2 [T(4)CDD], 1 [P(5)CDD], 3 [H(6)CDD], 4 [H(7)CDD], and 1.8 (OCDD).

## RESULTS AND DISCUSSION

The total ion current traces of native dioxins in OFA and dioxins produced by catalytic activity of OFA at 300°C are shown in Figure 2. Comparison of these traces shows that the number of peaks detected for each congener group for native dioxins in the flyash extract and dioxins produced by the catalytic activity of flyash are same. Although, the number of peaks in each congener group are same, the amount of each isomer with respect to other isomers is not same. The difference in peak heights of the isomers in each congener group might be due to different conditions in laboratory experiments as compared to the real incineration process. Another major difference is that in MSW incineration, PCDDs are formed from numerous precursors. However, in laboratory experiments, only pentachlorophenol was used as a precursor.

A study of catalytic activity of a SLFA sample at various temperatures shows that 300°C is the optimum temperature for formation of the maximum amount of total PCDD in the laboratory experiments (Figure 3). However, it was observed that at lower temperatures such as 180 and 240°C, only hepta- and octa-chlorodibenzo-p-dioxins were formed and an increase in the amount of P(5)CDD and T(4)CDD occurred with an increase in temperature. From this trend of formation of PCDD, it can be predicted that at lower temperatures only condensation of pentachlorophenol to OCDD is a predominant

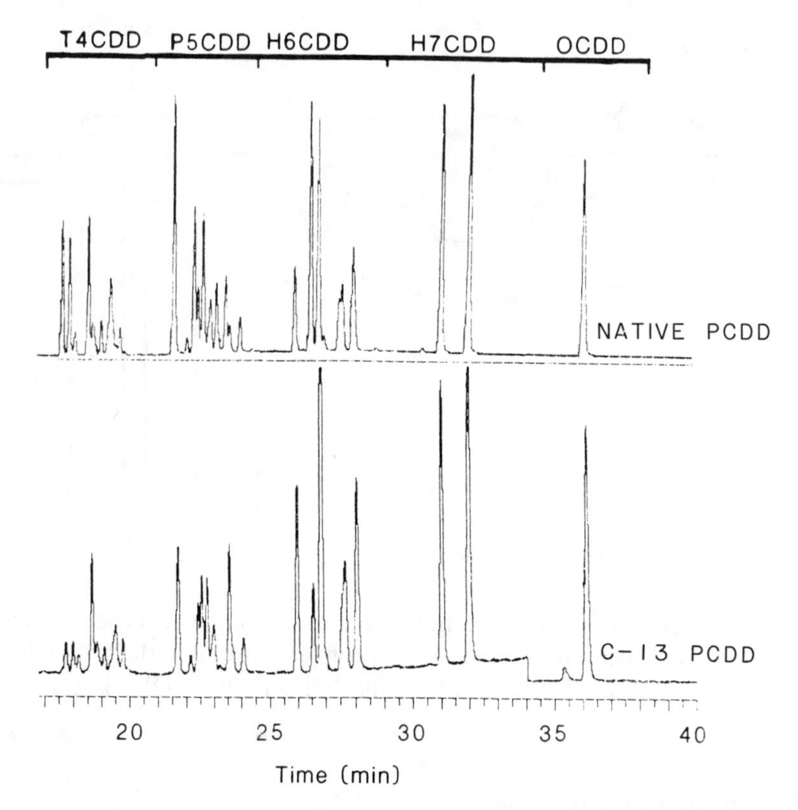


**Figure 2.** Added mass chromatograms for tetra- to octa-chlorodibenzodioxins. Mass chromatograms (upper trace) for native dioxins in Ontario flyash, ions monitored m/z = 321.9 [M + 2, T(4)CDD], m/z = 355.9 [M + 2, P(5)CDD], m/z = 389.9 [M + 2, H(6)CDD], m/z = 423.8 [M + 2, H(7)CDD], m/z = 459.7 [M + 4, OCDD]. Mass chromatograms (lower trace) draws for C-13 labeled dioxins formed on Ontario flyash by catalytic activity, ions monitored m/z = 333.9 [M + 2, T(4)CDD], m/z = 367.9 [M + 2, H(6)CDD], m/z = 401.9 [H(7)CDD], m/z = 435.8 [M + 2, H(7)CDD], m/z = 471.7 (M + 4, OCDD).

reaction and an increase in temperature results in dechlorination of higher chlorinated congeners followed by rearrangements to form all possible lower chlorinated PCDD congeners. The study of the catalytic activity of OFA and MFA shows a similar trend and optimim temperature (300°C) for the formation of PCDDs.

In all laboratory experiments using different flyash samples with pentachlorophenol as a precursor, only PCDDs were formed at various conditions. However, analysis of flyash sample extracts shows high levels of native PCDFs. This indicates that pentachlorophenol alone cannot produce PCDFs under laboratory conditions. Further study using unchlorinated phenol shows that tetra- and penta-chlorodibenzo-p-dioxins can form on flyash at 300°C. This indicates that inorganic ingredients in flyash can chlorinate organic com-

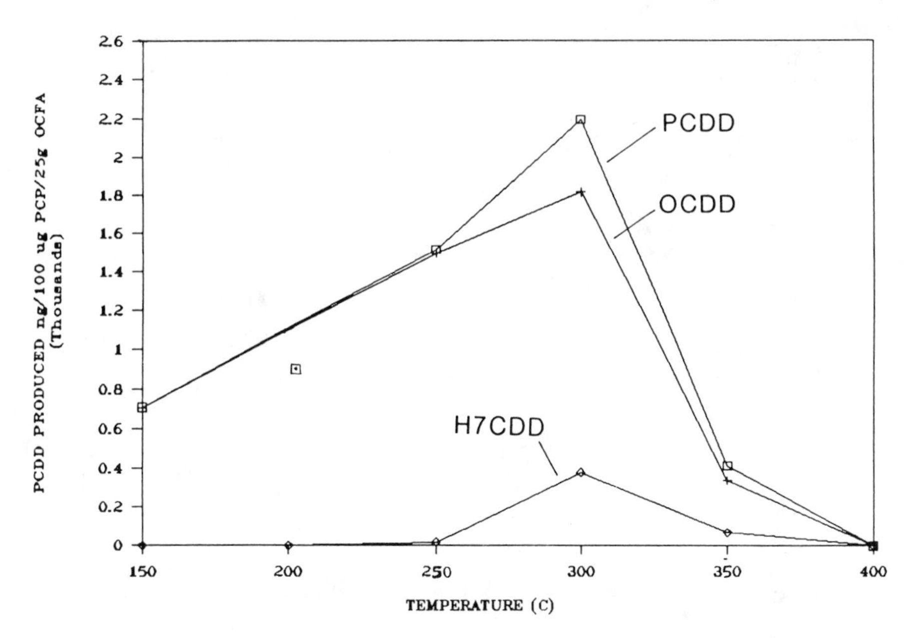


**Figure 3.** A plot showing total PCDD, OCDD, and H(7)CDD produced on flyash using a pentachlorophenol precursor at various temperatures in the laboratory tests.

pounds. An activity test using decachlorodiphenyl ether as the precursor shows formation of both PCDDs and PCDFs, which indicates that formation of PCDDs or PCDFs or both depends upon the precursor involved. Hence, it is obvious that in the formation of PCDDs and PCDFs in MSW incineration, hundreds of precursors are involved.

The MSW incinerator flyash is a very complex matrix that consists of organic and metallic compounds. Bulk elemental analysis shows that the concentration of metallic compounds in flyash samples from different incinerators varies. Analysis of flyash extracts for native dioxins shows that flyash samples having high concentrations of metals such as Zn, Cr, Mn, Al, and Cu shows more PCDDs and PCDFs as compared to flyash samples with these metals in small concentrations. In the laboratory experiments, catalytic activity of various flyash samples from different incinerators was tested using the assembly shown in Figure 1. A plot of PCDDs produced in the laboratory experiments by catalytic activity of flyash using pentachlorophenol as a precursor at 300°C, and native PCDDs, that were produced in the incineration process are shown in Figure 4. It was observed that the flyash samples that have high contents of Cr, Zn, Mn, Al, and Cu show high levels of dioxins produced both in the incineration process and in the laboratory experiments.

The study of the use of pure metals and metal oxides in our laboratory shows that metals such as Cr, Zn, Mn, Al, Cu, Ti, and corresponding metal oxides produce PCDDs and PCDFs. However, the mechanism of formation of

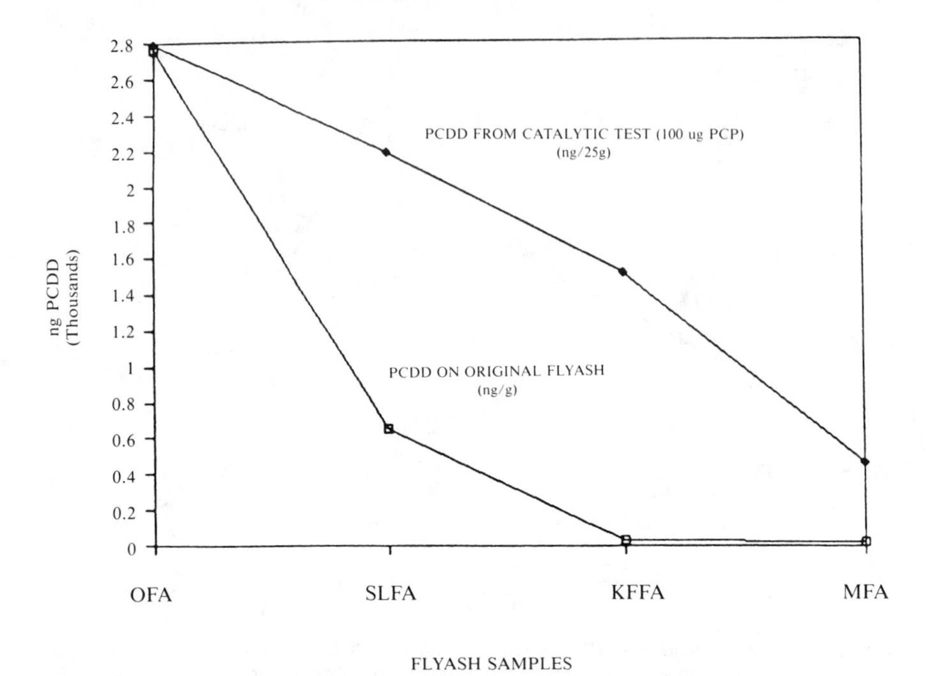


**Figure 4.** A plot showing PCDDs produced by catalytic activity of various flyash samples in laboratory experiments and native PCDDs detected on flyash samples that formed during incineration processes.

PCDDs, PCDFs, and other large numbers of compounds using metal and metal oxides is very complicated and difficult to explain.

## REFERENCES

1. Rappe, C., A. Gara, and H. R. Buser. *Chemosphere* 7:981 (1978).
2. Crummett, W. B., and D. I. Townsend. *Chemosphere* 13:777 (1984).
3. Hail, C. L., R. B. Blair, J. S. Stanley, D. P. Redford, D. Heggem, and R. M. Lucas. *Chlorinated Dioxins and Dibenzofurans in the Total Environment*, G. Choudhary, L. H. Keith, and C. Rappe, Eds. (Stoneham, MA: Butterworth Publishers, Inc., 1983) p. 439.
4. Sawyer, T., S. Dandiera, and S. Safe. *Chemosphere* 12:529 (1983).
5. Naikwdi, K. P., and F. W. Karasek. "Catalytic Activity of Flyash for Formation of Highly Toxic PCDD and Inhibitors to Prevent their Formation," 39th Pittsburgh Conference and Exposition on Analytical Chemistry and Applied Spectroscopy, New Orleans, February 22, 1988.
6. Karasek, F. W., and L. C. Dickson. *Science* 237:754 (1986).

# CHAPTER 5

# Heterogeneous Gas-Solid Reactions of Aromatic Hydrocarbons on Flyash from Municipal Incinerators

**G. A. Eiceman and R. V. Hoffman**

## ABSTRACT

Naphthalene, biphenyl, and anthracene adsorbed on flyash from municipal incinerators undergo facile chlorine substitution reactions when treated with dilute gaseous HCl at temperatures from 50 to 250°C. These polycyclic aromatic hydrocarbons also exhibited chemisorption behavior evidenced by extensive irreversible loss of compound to the flyash. Yields of chlorination products were generally ca. 10% of starting reactant mass and substitution products were consistent with chlorination via an electrophilic species rather than through free radical reactions.

## INTRODUCTION

During the mid- to late-1970s, flyash from municipal and coal-fired incinerators worldwide was found to contain organic compounds at nanogram per kilogram or microgram per kilogram levels.[1-4] Compounds of special interest in municipal incinerator flyash included polychlorinated species of dibenzo-p-dioxins (PCDDs), dibenzofurans,[1,2] phenols, and biphenyls in highly complex mixtures of as many as 150+ components.[3] Also in the late 1970s, Dow Chemical Company released findings in which levels of PCDDs were reported for various combustion processes and for which quantitative trends in PCDD content for environmental samples were observed.[5] Specifically, the ratios of PCDD congeners (e.g., hexaCDD versus heptaCDD) in soil samples downwind from an emission stack of an incinerator showed uniform changes as a function of distance downwind. Townsend proposed that the ratios reflected changes in composition of the flyash following release into ambient air and these changes were evidence of the chemical reactivity of PCDDs through oxidation and reduction reactions.[6] The fundamental premise of this hypothesis was that PCDDs (heretofore considered thermally stable and chemically unreactive) on flyash surfaces exhibited reactivity toward gases in effluent streams or ambient air at temperatures from 25 to $\geq$180°C. The presence of PCDDs on the flyash was presupposed and the original source of these PCDDs was considered a peripheral issue in Townsend's rationales.

In addition to the interpretation of these environmental trends using chemical rationales, changes in ratios could have been caused by physical effects such as prefractionation through differences in sedimentation based on particle sizes. Nonetheless, chemical reactivities were considered verifiable through simple laboratory experiments and a series of investigations[7-11] were used to ascertain the central premise of Townsend's hypothesis. In a series of investigations, major conclusions were:

1. Organic compounds on flyash could undergo chlorination to yield polychlorinated organic compounds.[7]
2. 1,2,3,4-$T_4CDD$ would undergo irreversible reactions/thermal decomposition and dechlorination when placed on flyash surfaces.[8]
3. 1,2,3,4-$T_4CDD$ will react with HCl (g) to yield higher chlorinated dioxins when the $T_4CDD$ was adsorbed on flyash surfaces at temperatures of 50 to 250°C.[9]
4. Adsorption and chlorination processes were also observed with dibenzodioxin and monochlorodioxins suggesting that the reactivities were generally exhibited by other chlorinated dioxins and possibly furans.[10]
5. Nitration reactions could also occur through heterogeneous phase gas-solid phase reactions,[11] but sulfonation[12] was not observed.
6. Finally, the nature of the solid on which the chlorinated dioxin was deposited was influential in governing reactivities.[13]

These findings were consistent with Shaub's argument that gas phase reactions alone were kinetically too slow to explain the PCDD content of flyash.[14] The veracity of these studies on the surface-mediated reactions was supported by findings of Dickson and Karasek[15] that chlorinated phenols on flyash surfaces underwent condensation to PCDDs and of Nestrick et al.[16] that benzene on $FeCl_3$ formed PCDDs and other chlorinated organic compounds at temperatures $\geq$150°C.

Despite these discoveries, several critical aspects regarding heterogeneous gas-solid phase chemistry were unclear generally for reactions of organic molecules on flyash or airborne particulate matter. These included a mechanistic model for the reactions (i.e., what molecular properties support reactivity) and a quantitative understanding of the exact role of flyash surface properties in facilitating such reactions. The objectives of our present research programs are to explore these two issues and preliminary results from the first will be discussed here.

## EXPERIMENTAL

A reaction apparatus similar to that used earlier[7-13] was constructed using all stainless steel metering valves, toggle valves, tubing, and check valves. The reaction apparatus permitted organic compounds in solution to be flash evaporated onto the flyash surface and also made possible the controlled introduction of gaseous mixtures of known composition. All procedures for sample preparation and instrumentation were identical to earlier investigations with two exceptions. In the present investigations, internal standards were added to the flyash extract before condensation and were used in quantifying results, and a chromatographic column which resolved positional isomers was used to supplement analyses with nonpolar-bonded capillary columns which separated congener groups. The column that provided resolution of positional isomers was a 3% Bentone-34 on Chromsorb W, 2 m, packed column. The flyash was obtained from the Chicago Northwest Municipal Incinerator.

## RESULTS AND DISCUSSION

Naphthalene, biphenyl, and anthracene were irreversibly bound to flyash to a large extent and only 55 to 65% of the mass transferred to the flyash could be recovered after the flyash had been heated to 150°C in inert gas flow. The balance of mass for these compounds presumably was lost through an irreversible adsorption process (i.e., chemisorption). Losses through the bed of flyash to a vapor trap showed negligible losses via vaporization (elution) thus suggesting large adsorptive activity of flyash for these molecules. Comparable behavior was previously observed with polar molecules including PCDDs[9-13] and such high activity toward nonpolar aromatic compounds was not expected. The amount of chemisorption should be mass-dependent[17] and the losses observed were consistent with the low masses used.

Naphthalene and biphenyl were highly susceptible to chlorination while adsorbed on flyash and treated with dilute gaseous HCl at ca. 150°C. However, the treatment of flyash with HCl lowered the mass balance for naphthalene suggesting that other side reactions (oxidation) were promoted upon exposure of flyash-adsorbed aromatic compounds to HCl. The yields of chlorination products were relatively low compared to dioxins and averaged ca. 15% for naphthalene with product yields as depicted in Figure 1. The trends in substituents for naphthalene were 1-chloronaphthalene > 2-chloronaphthalene as shown in Equation 1:

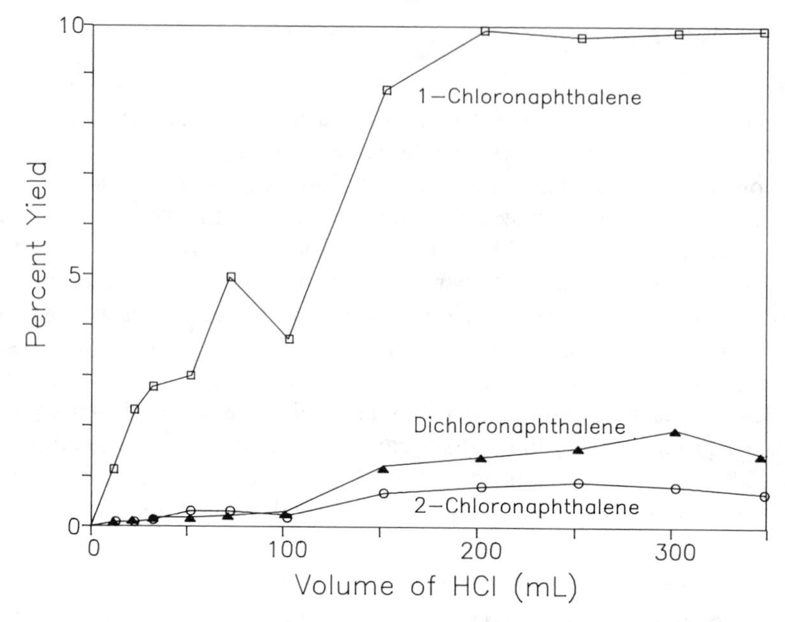


**Figure 1.** Yield of products from treatment of naphthalene with gasous HCl on flyash as a function of the amount of HCl.

HCl Cl Cl (1)

supporting a mechanistic interpretation that the reaction occurs via an electrophilic mechanism. The amounts of di- and tri-substituted naphthalenes were low at 2 to 3 and 0.1–0.3%, respectively. Studies on the amount of HCl versus the time of exposure supported a conclusion that kinetics of chlorine substitution were very fast and that stoichiometry figured more importantly than exposure time. The presence of bromochloronaphthalene was supported by scanning mass spectrometry of the appropriate GC peak and occurred presumably through exchange of natively abundant bromine in the flyash with the HCl leading to the bromine analog of the chemically active species generated with HCl treatment. This may be related to the observation of brominated aromatic compounds on flyashes from municipal incinerators,[18] but differs in that the present reactions involve only the halogenation processes.

Biphenyl also underwent chlorination, but exhibited less reactivity than naphthalene and yielded only the 2-chlorobiphenyl and 4-chlorobiphenyl as shown in Equation 2:

**Table 1. Product Yields from the Chlorination of Biphenyl on Flyash from a Municipal Incinerator. Conditions of Chlorination were 150°C for 20 Min.**

| Compound | Percent Yields as a Function of mL of HCl | | | |
|---|---|---|---|---|
| | 300 | 350 | 856 | 1713 |
| Biphenyl (recovery) | 32 | 34 | 33 | 24 |
| 2-chlorobiphenyl | 0.74 | 0.68 | 1.1 | 1.3 |
| 3-chlorobiphenyl | 0 | 0 | 0 | 0 |
| 4-chlorobiphenyl | 5.8 | 6.2 | 7.8 | 9.6 |

Precision ±20% relative standard deviation.

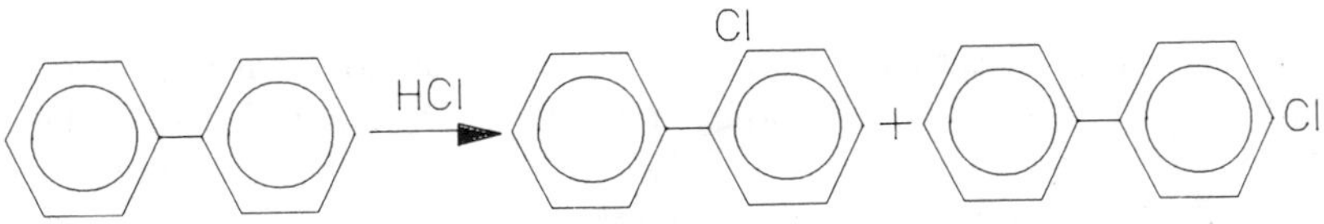

HCl → Cl + Cl (2)

The ratio of 2-/4-substitutions was 1:3 and the overall total chlorinated yield was 10%. Polychlorinated biphenyls were produced though in small amounts, but could be generated in greater yields with prolonged exposure to HCl. The relationship of yield to HCl mass is shown in Table 1 where increased HCl treatment led to more chlorinated products, but also increased the chemisorption of biphenyl that is lost in mass balances. Anthracene also underwent facile chlorination to largely 9-chloroanthracene and 9,10-dichloranthracene as shown in Equation 3:

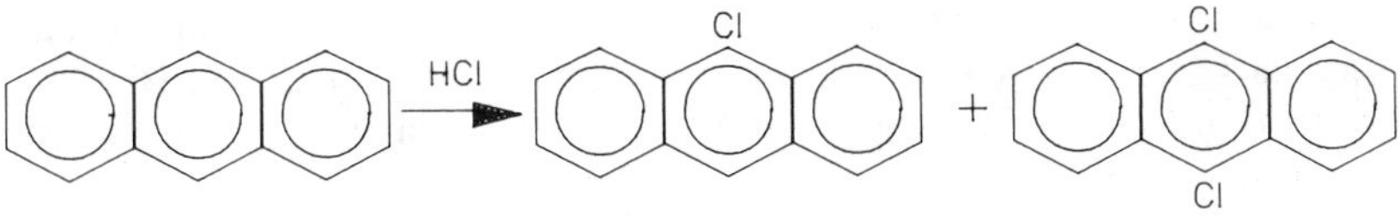

HCl → Cl + Cl Cl (3)

The ratio of these two products was 1:4 for 9-chloroanthracene:9,10-dichloroanthracene and remarkably 1-chloro- or 2-chloroanthracene were not detected in the product mixture. The percent conversion of anthracene was 25% and suggested slightly better reactivity than biphenyl and naphthalene. In each reaction, the substitutional patterns are consistent with the attack of an electrophilic species on the aromatic ring rather than a free radical reaction. Moreover, the presence of brominated aromatic hydrocarbons and bromo-, chloroPAH is consistent with an exchange of Br- with Cl- on the flyash surface, perhaps with a metal chloride, as expected with an electrophilic intermediate.

The role of the flyash surface chemistry on reactivity has been evidenced in our laboratory studies using flyashes from Toronto, Hamilton, and a coal-fired incinerator. These various flyashes gave yields for the chlorinations of naphthalene, biphenyl, and anthracene in comparable fashion to those from the Chicago flyash, but in varying yields. The causes for the differences have not been identified and the exact percent yields of chlorinated products are not reported here. Future investigations will be centered on examining the influence of activating or deactivating substituents on the reactivity of aromatic species, exploring the chemical nature of the chlorinating site on the flyash, and preparing a model to predict the reactivity of flyashes toward aromatic compounds in stack gases and ambient atmospheres.

## ACKNOWLEDGMENTS

The project was funded by the National Science Foundation through Grant no. ATM 8712088. Flyash samples were provided by Dr. Emil Nigro and Dr. R. E. Clement to whom we are grateful. Laboratory efforts of M. Collins, Y-T. Long, and M. Q. Lu are gratefully acknowledged.

## REFERENCES

1. Olie, K., P. L. Vermeulen, and O. Hutzinger. *Chemosphere* 8:455 (1977).
2. Buser, H. P., H. P. Bosshardt, and C. Rappe. *Chemosphere* 7:165 (1978).
3. Eiceman, G. A., R. E. Clement, and F. W. Karasek. *Anal. Chem.* 51:2343 (1979).
4. Ford, G. A., and C. S. Junk. *Chemosphere* 9:187 (1980).
5. Blumb, R. R., W. B. Crummett, S. S. Cutie, J. R. Glenhill, R. H. Hummel, R. O. Kagel, L. L. Lampariski, E. V. Luoma, D. L. Miller, T. J. Nestrick, L. A. Shadoff, R. H. Stehl, and J. S. Woods. *Science* 210:385 (1980).
6. Townsend, D. I. "The Use of Dioxin Isomer Group Ratios to Identify Sources and Define Background Levels of Dioxins in the Environment," Pesticide Division, Abstract #80, National Meeting of the American Chemical Society, Washington, DC, September 10, 1979.
7. Eiceman, G. A., and H. O. Rghei. *Environ. Sci. Technol.* 16:53 (1982).
8. Rghei, H. O., and G. A. Eiceman. *Chemosphere* 11:569 (1982).
9. Eiceman, G. A., and H. O. Rghei. *Chemosphere* 11:833 (1982).
10. Rghei, H. O., and G. A. Eiceman. *Chemosphere* 13:421 (1984).
11. Eiceman, G. A., and H. O. Rghei. *Chemosphere* 13:1025 (1984).
12. Rghei, H. O., and G. A. Eiceman. *Chemosphere* 14:259 (1985).
13. Rghei, H. O., and G. A. Eiceman. *Chemosphere* 14:167 (1985).
14. Shaub, W. M., and W. Tsang. *Environ. Sci. Technol.* 17:721 (1983).
15. Dickson, L. C., and F. W. Karasek. *J. Chromatogr.* 389:127 (1987).
16. Nestrick, T. J., L. L. Lamparski, and W. B. Crummett. *Chemosphere* 16:777 (1988).
17. Eiceman, G. A., and V. J. Vandiver. *Atmos. Environ.* (1982).
18. Schwind, K.-H., J. Hossenpour, and H. Thoma, *Chemosphere* 17:1875 (1988).

# CHAPTER 6

# Computer–Automated Laboratory Thermal Oxidizer

**Robert D. VanDell, Robert A. Martini, and Nels H. Mahle**

## ABSTRACT

This article describes a computer-automated laboratory thermal oxidizer (LTOX). The LTOX is a continuously fed flame oxidizer containing an afterburner. Inputs and temperatures are independently controlled and measured to produce a cylindrical diffusion flame from liquids, gases, or solids which is reproducible. The afterburner section is designed to produce laminar flow of flue gas to facilitate quantitative collection of all combustion products to be analyzed and is totally independent of the flame temperature. Within the afterburner, residence times and oxygen-to-fuel equivalence ratios can be separately varied and measured for clear process definition. Ambient operating pressures are controlled and maintained at all times during a burn.

Flue gas is continuously analyzed for $NO_x$, $CO_2$, CO, $O_2$, $N_2$, and $CH_4$. Nongaseous products and particulates are quantitatively collected via adsorption and cryogenic traps for gas chromatography and gas chromatography-mass spectrometry analysis.

## INTRODUCTION

A great deal of literature has been published on the emission of the products of incomplete combustion (PICs) from industrial, municipal, pilot, and laboratory incinerator systems. These systems include waste thermal oxidizers, fire-tube boilers, open-hearth burners, and rotary kilns, to name a few. Much of the experimental work carried out on these systems suffers from a lack of quantitation which can be used to elucidate the fundamental chemistry involved. This is due to lack of control over feedstock compositions, oxidation conditions, and complex geometry and configuration. There are, however, a few laboratory studies which have attempted to define these parameters more precisely so that data obtained from them can be sufficiently analyzed. Notable is the early work using flame-free, hot narrow tubes by Duvall and co-workers[1] which is still ongoing.[2] Essentially, their system consists of a narrow bore (>1 mm i.d.), lengthy capillary mounted in an electrically heated furnace for temperature control. Milligram-size samples of the fuel are batch loaded at the front end via a sample boat. Combustion air is controllably passed over the latter to volatilize and/or pyrolyze the sample whereupon it subsequently is oxidized while passing down the capillary. From the measured flow rate, temperature, and pressure drop, residence times may be accurately calculated. A trap at the opposite end of the capillary serves to collect the unburned fuel and/or PICs for analysis or the capillary is sometimes connected directly to a gas chromatograph or a gas chromatograph-mass spectrometer. Data on the minimum destruction temperature for a compound and/or plots of destruction efficiency versus residence time may be constructed. These data supply information on the overall destruction kinetics of the fuel. Also, data concerning the formation and destruction of PICs may be obtained.

Since this approach is flameless and batch rather than continuous, flame destruction data are not available. Also, the exact air-to-fuel ratio during early oxidation of the fuel in or near the boat is unknown. Since the technique is primarily a study of gas-phase destruction, the role of carbon or other particulates is unavailable. These limitations have made it difficult, in the opinion of some, to relate their data to full-scale incinerators.

An attempt to more closely imitate full-scale thermal oxidizers and boilers which do utilize flames without sacrificing definition has been made in this laboratory. The following is a description of a continuous, computer-controlled, laboratory thermal oxidizer.

## GENERAL DESCRIPTION

The LTOX,[3] is shown in Figure 1. It was designed to simulate (in the laboratory) the main features of a full-scale thermal oxidizer or a heat recovery thermal oxidizer. Similar to the full-scale burner, the LTOX contains two distinct burning zones. They are the flame region and the afterburner or post-flame region. The flame, produced from a gas or vaporized liquid, is cylindrical and laminar. Its geometry is closely controlled via a fixed vaporizing nozzle and controlled variable air and fuel delivery rates. The afterburner section is a 1-cm i.d. quartz tube, 40 cm long. Gas flow in this tube is laminar to facilitate the quantitative collection of particulates and related partial combustion products during a burn. The tube is mounted in a tube furnace manufactured by Applied Test Systems Inc. (Model 3320) to maintain any desired afterburner temperature ranging from 500 to 1300°C. A vacuum pumping system is provided to remove flue gas from the burner at a rate equal to its flow in the combustion tube in order to maintain atmospheric pressure on the flame and afterburner.

A trap system is attached to the exit of the combustion tube for quantitative trapping of particulates and low- and high-boiling PICs. The nature of the

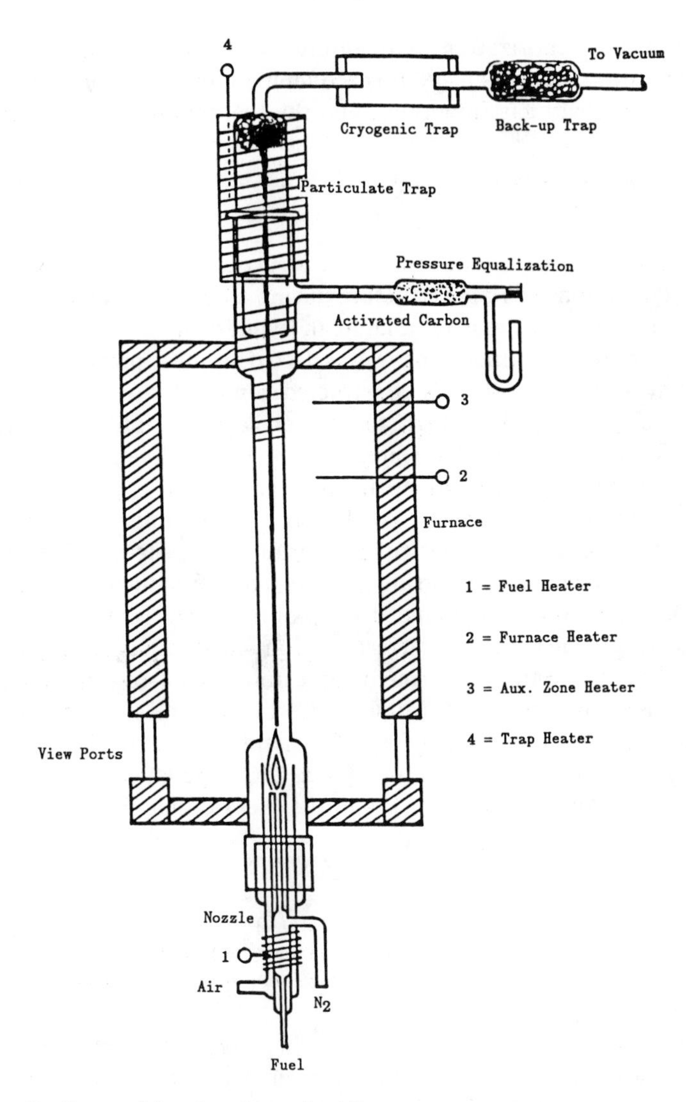


**Figure 1.** Continuous laboratory thermal oxidizer.

traps varies with the experiment. They may be removed after a burn for product analysis by gas chromatography and/or gas chromatography-mass spectrometry. Gaseous products and excess air passing through the traps are quantitatively analyzed online for NO and $NO_2$ with a chemiluminescent analyzer and for $CO_2$, $CH_4$, CO, $O_2$, and $N_2$ with a thermal conductivity detector gas chromatograph (Fisher model 1200 gas partitioner). Data from the online analyzer, along with flow rates and temperatures, are sent to a microcomputer (DEC MicroPDP-11/23) for storage and processing. The computer also sets and maintains air and fuel flow rates and temperatures. These temperatures include the fuel vaporization temperature, furnace temperature, particulate trap temperature, and auxiliary furnace temperature. The flow rates include the air and auxiliary gas flow rates, and the fuel pumping rate.

## NOZZLE

To obtain meaningful data, the flame structure must be reproducible and simple. Fluctuations in flame structure during a burn will cause scatter in the resulting data. For this reason, a laminar, cylindrical flame was chosen. The nozzle design and its dimensions were selected to meet precise specifications and is shown in Figure 1. It is constructed completely of quartz. Liquid or gaseous fuel is pumped through the center capillary into the vaporization chamber where the liquid fuel is heated to boiling and vaporized.

The vapor may be mixed with an auxiliary gas such as $N_2$, $CH_4$, or $H_2$ before passing to the tip of the nozzle. At the tip, the fuel mixture meets with air, which flows through the outer tube to produce the flame. The outer tube diameter and the inside and outside diameters of the capillary were selected to produce laminar flow in the flame. The tip of the nozzle is mounted well inside the hot region of the furnace to affect ignition. Air and auxiliary gas flows are controlled by the computer through the external control circuits of a four-channel mass-flow controller (Model 8249) manufactured by Matheson Gas Products. Air rates range from 0 to 5000 scfm ± 1 scfm. The vaporization chamber temperature range is 0 to 300° ± 2°C. The linear flow velocities of the air ($\nu_a$), the auxiliary gas ($\nu_{aux}$), and the fuel ($\nu_x$) may be calculated from:

$$\nu_a = (F_a/\pi Z^2)(T_f/T_o) \tag{1}$$

$$\nu_{aux} = (F_{aux}RT_f)/(\pi P_t L^2) \tag{2}$$

$$\nu_x = (F_x RT_f)/(\pi P_t L^2) \tag{3}$$

respectively, where $F_a$ is the flow rate or air in cubic centimeters per second and $F_x$ and $F_{aux}$ are the flow rates of fuel and auxiliary gas in moles per second, respectively. $T_f$ and $T_o$ are the flame and ambient temperatures, and R is the gas constant. L and Z are the effective inside radius of the inner and outer tube of the nozzle, and $P_t$ is the total pressure. Conditions for laminar flow are most closely approached when $\nu_{aux} + \nu_x = \nu_a$,[3] and therefore under this condition:

$$(F_{aux} + F_x)/F_a = (P_t/RT_o)(L/Z)^2 \tag{4}$$

$(L/Z)^2$ is the nozzle geometry. The introduction of an auxiliary gas allows one to independently control air and fuel flow rates making it possible to vary oxygen-to-fuel equivalence ratios without affecting the laminar structure of the flame.

**Table 1. Comparison of $dm_s$ Calculated from Flue Gas Analysis with that Determined from Soot Yield**

| $S_F$ Equivalent $O_2$ / Equivalent Fuel | $dm_s$ (mol/ft$^3$ Flue Gas) (Analysis) | $dm_s$ (mol/ft$^3$ Flue Gas) (Yield) |
|---|---|---|
| 0.5 | 0.2152 | 0.2146 |
| 0.7 | 0.1590 | 0.1620 |
| 1.0 | 0.1078 | 0.0951 |
| 1.4 | 0.0665 | 0.0633 |
| 1.7 | 0.0434 | 0.0417 |
| 2.0 | 0.0284 | 0.0219 |
| 2.5 | 0.00608 | 0.00766 |

## AFTERBURNER

The purpose of the afterburner is to further destroy particulates, fuel, and PICs emitted from the flame. In the LTOX, it is desirable to control the afterburner and the flame independently. This is accomplished by separately controlling the air and auxiliary flow rates at any selected furnace temperature. An average flame length of 1 to 2 cm guarantees that the flame temperature, usually much higher than the afterburner, does not significantly raise the latter. Because, in the electric tube furnace, the temperature is not uniform over its entire length, an auxiliary heater is needed at the upper end of the combustion tube to ensure thermal uniformity. During operation, the auxiliary temperature is set equal to that of the furnace.

Since the flue gas flow is laminar in the combustion tube, the residence time can be estimated from:[4]

$$t_r = (\pi Z^2 J/F_a)(T_o/T)(1 + (P_d/P_o)) \qquad (5)$$

within ± 10%. $P_d$ is the pressure drop across the combustion tube and J is the tube length.

Separate control of the flow rates and the furnace temperature allows variation in residence time and oxygen-to-fuel equivalence rates independently.

As a result of the laminar flow, particulates and PICs emitted from the flame rise in the afterburner tube in the form of a thin stream (approximately 1 to 2 mm thick) and enter the trap system without impinging the tube walls, facilitating quantitative product collection. Table 1 is a comparison between the amount of soot actually collected from a series of o-dichlorobenzene burns and that calculated from the $CO_2$, $N_2$, CO, and $O_2$ concentrations in the flue gas using the appropriate material balance equations.

## TRAP SYSTEM

The type of traps used in the trap system depends on the particular experiment. In general, the first trap quantitatively collects particulates. The temperature of this trap is controlled and may vary from 0 to 400°C. This allows one to study postburn quenching temperatures and reaction mechanisms taking place in downstream air pollution control equipment. Beyond the particulate trap, adsorption (i.e., activated carbon or TENAX, etc.) and/or cryogenic traps may be used for low- and high-boiling fuels and PICs not associated with the particulate. All of the traps may be independently removed and extracted for analysis by gas chromatography and/or gas chromatography-mass spectrometry.

Since the LTOX is continuous, the amount of sample collected depends on the duration of the burn and is limited only by the size of the trap allowing one to collect a sufficient amount to minimize analytical difficulty.

## PRESSURE CONTROL

Pressure fluctuations can cause changes in the flame structure. Therefore, it is important to maintain ambient pressure to obtain reasonable reproducibility.

To maintain ambient pressure on the flame, a controlled vacuum is pulled on the entire system and is balanced against the flue gas flow. A Bernoulli tube (see Figure 1), attached to the combustion tube just below the soot trap, is used to maintain the balance and is open on the opposite end to the room. The tube contains packing of activated carbon, which removes impurities from the air in case the latter passes into the combustion tube. One end of the tube contains a fine capillary facilitating the Bernoulli effect. A small manometer, hung between the packing and the capillary, is open to the room on the other end. Acetone is used in the manometer because of its low density. During operation, gases passing in or out of the tube cause an imbalance in the manometer, which is balanced by adjusting the vacuum. This arrangement insures quantitative collection of nonvolatile combustion products without pressurizing or depressurizing the flame.

Finally, a quartz viewing window is provided in the shell of the furnace to optically measure flame height and/or temperature or to generally observe the flame.

## PROCESS CONTROL

### MicroPDP Computer System

The process control computer is based on a Digital Equipment Corporation (DEC) MicroPDP 11/23 computer system. The hardware consists of an 11/23 computer with a floating point processor, 512 KB of memory, six serial communication ports, a real-time clock, an analog-to-digital converter, a parallel I/O, a watchdog timer, and a disk controller. The system has a 31-MB Winchester hard disk drive (RD52) and two 5 1/4-in. floppy drives (RX50).

734 Thermal Oxidizer

| ALARMS | | | | | |
|---|---|---|---|---|---|
| AI LABEL | FuelTemp | FurnTemp | Aux Temp | TrapTemp | Dil Air |
| SET POINT | 135.0 Deg C | 850.0 Deg C | 850.0 Deg C | 160.0 Deg C | 0.0 SCCM |
| MEAS VALUE | 135.0 Deg C | 850.0 Deg C | 850.0 Deg C | 172.4 Deg C | 0.0 SCCM |
| AO LABEL | Heater A | FurnaceB | Heater C | Heater D | Dil Air |
| AUTO/MAN | AUTO | AUTO | AUTO | AUTO | AUTO |
| PROP GAIN | 0.7 | 0.7 | 0.6 | 0.3 | 0.1 |
| INTEGRAL | 10.0 | 10.0 | 9.0 | 15.0 | 1.0 |
| DERIVATIVE | 0.0 | 2.0 | 0.0 | 0.0 | 0.0 |
| OUTPUT | 24.4 % | 48.2 % | 45.2 % | 15.0 % | 0.0 % |

**Figure 2.** Analog screen updates.

The PDP 11/23 runs under a single user, multitasking RT-11 real-time operating system. Using the multitasking feature, the computer executes the control program, screen updates, and menu simultaneously.

The burner control program, BURNER, is a Fortran IV program which uses the Data Translation's DTLIB subroutine library to handle process I/O, the VTLIB subroutine library to perform screen updates, and modified Fortran subroutines designed to emulate various process control systems. The entire control program is menu driven for ease of use and performs extensive error checking on all operator inputs.

## Analog Inputs and Outputs

The current control system software is designed for 10 analog inputs and 10 analog outputs. Each analog input signal is fed into the computer through an analog devices signal conditioning module. The input modules are magnetically isolated and provide the necessary amplification and linearization for various types of input signals. Output modules convert the 0- to 10-V analog outputs from the computer to 4- to 20-mA current signals which are used to drive the final control elements. For analog set points and a control variables CRT display, see Figures 2 and 3.

## Digital Inputs and Outputs

The computer system hardware can also accept 16 digital inputs and generate 16 digital outputs. For this system, only eight digital inputs and eight

734 Thermal Oxidizer

| ALARMS | | | | | |
|---|---|---|---|---|---|
| AI LABEL | Comb Air | Aux #1 | Aux #2 | FuelRate | C12 |
| SET POINT | 200.0 SCCM | 50.0 SCCM | 0.0 SCCM | 1.5 cc/hr | 0.0 PPM |
| MEAS VALUE | 200.0 SCCM | 49.1 SCCM | 0.1 SCCM | 0.0 cc/hr | 0.0 PPM |
| AO LABEL | Comb Air | Aux #1 | Aux #2 | FuelPump | |
| AUTO/MAN | AUTO | AUTO | AUTO | AUTO | AUTO |
| PROP GAIN | 0.0 | 0.1 | 9.7 | 8.5 | 0.1 |
| INTEGRAL | 1.0 | 1.0 | 1.0 | 1.0 | 1.0 |
| DERIVATIVE | 0.0 | 0.0 | 0.0 | 0.0 | 0.0 |
| OUTPUT | 3.4 % | 3.4 % | 0.0 % | 12.7 % | 0.0 % |

**Figure 3.** Analog screen updates.

734 Thermal Oxidizer

ALARMS

DIGITAL INPUTS

| DI LABEL | NO/NOx | Hood Fan | Spare | Spare | Spare | Spare | Spare | Start |
|---|---|---|---|---|---|---|---|---|
| VALUES | FALSE | TRUE | FALSE | FALSE | FALSE | FALSE | FALSE | TRUE |

DIGITAL OUTPUTS

| DO LABEL | Dil Air | Comb Air | Aux #1 | Aux #2 | Spare | Spare | Spare | Power |
|---|---|---|---|---|---|---|---|---|
| AUTO/MAN | AUTO | AUTO | AUTO | AUTO | AUTO | AUTO | AUTO | AUTO |
| VALUES | FALSE | TRUE | TRUE | FALSE | FALSE | FALSE | FALSE | TRUE |

**Figure 4.** Digital screen updates.

digital outputs are necessary. Each digital I/O signal is optically isolated with an Opto-22 module which will allow either AC or DC (10- to 30-V) devices to be controlled. For a digital inputs/outputs CRT display, see Figures 2, 3, and 4.

## Control Program

The computer control program fully automates the burner during an experiment from start to shutdown. It provides directions to the computer as it controls different elements of the experiment while at the same time allows the user to intervene or change parameters. A typical scenario of events is as follows: First, the program ramps the furnace and auxiliary zone heater to the setpoint and enables the fuel and trap heaters. Then it enables all four gas flows and the fuel pump, waits for the system to come to equilibrium, and then starts data collection. Finally, it provides for possible emergency shutdown if temperatures and flows are outside prescribed limits (see Table 2).

At the end of the experiment, the program will turn off the dilution air, the auxiliary No. 2 gas, and the fuel, leaving the combustion air and furnace heaters on. Any fuel remaining in the vaporization chamber is consumed. Finally, the program turns off the combustion air and furnace heaters and sets all outputs to their fail-safe position (off/closed).

**Table 2. Alarm and Shutdown Summary**

| Analog Input | Alarm | Shutdown |
|---|---|---|
| Fuel temperature | ±3°C | ±10°C |
| Furnace temperature | ±10°C | ±50°C |
| Auxiliary temperature | ±10°C | ±50°C |
| Trap temperature | ±10°C | ±50°C |
| Dilution air | ±10 sccm | |
| Combustion air | ±10 sccm | ±50 sccm |
| Auxiliary #1 | ±10 sccm | ±50 sccm |
| Auxiliary #2 | ±1 sccm | |
| **Digital Input** | **Alarm** | **Shutdown** |
| Pump | True | |
| Hood fan | | False |
| Stop | | False |

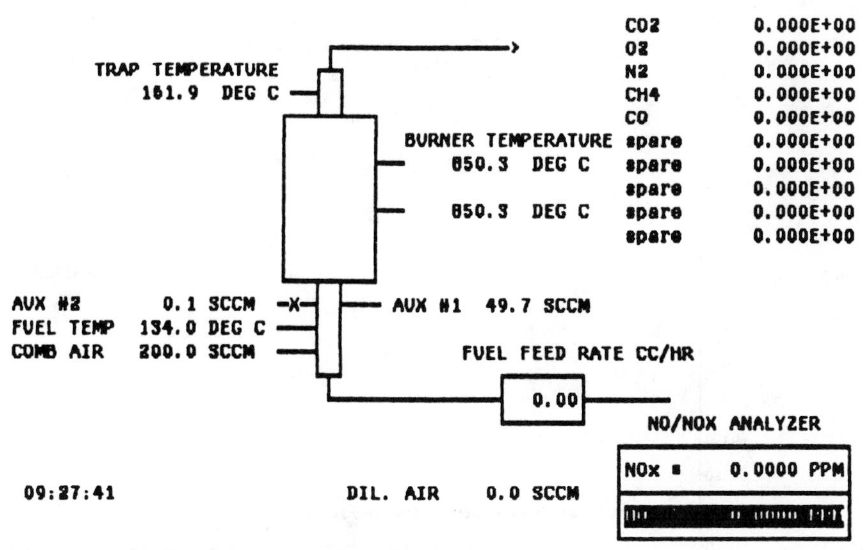


**Figure 5.** CRT screen updates.

## Alarms

Alarms are provided for each analog and digital input. When an alarm condition occurs, a message is printed indicating the control loop in alarm. The word "ALARM" is displayed on the terminal above the loop in alarm and the terminal will beep until acknowledged. These alarms are summarized in Table 2.

## Shutdowns

Shutdown statements are included in the control program to halt the experiment if a process variable should exceed a safe operating limit. Unlike alarms, when the shutdown condition becomes true, the computer will take specific actions. In this case, the program will cause the burner to shutdown and all outputs will be in a fail-safe condition. The shutdown criteria are shown in Table 2.

## Data Collection

Data collection may be activated or deactivated at the will of the operator via the menu. When the collection mode is selected, data are stored in a file every 10 sec until collection is terminated. During collection, the phrase "collection data" appears on the screen update shown in Figure 5.

At the end of the experiment, the data file may be retrieved, edited, and sent to the VAX for manipulation in RS/1. Figures 6, 7, and 8 show examples of

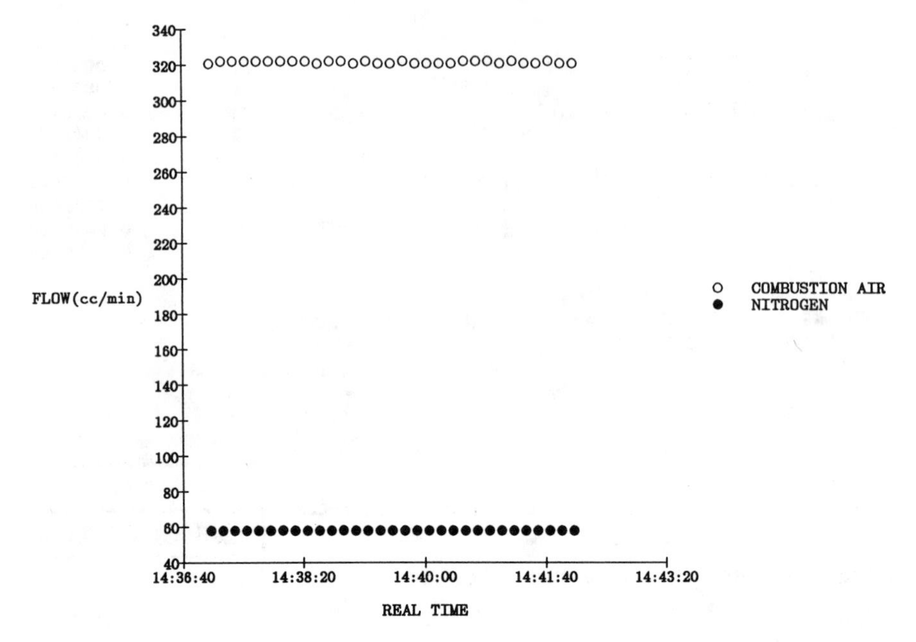


**Figure 6.** Gas flow control.

real-time plots of flow rates, temperatures, and NO emission during the burning of pyridine.

## Temperature Controllers

The main furnace was originally equipped with a single-loop proportional integral (PI) controller and an SCR power circuit to control the high-temperature heating elements. The furnace control unit was modified to accept a 4- to 20-mA signal from the computer to control the furnace temperature instead of using the PI controller. The computer can now ramp the furnace temperature to the setpoint and interlock the controller to process conditions such as alarms, etc. The computer provides tighter control than the PI controller and safer operation.

Due to the placement of the furnace heaters, the upper quarter of the furnace did not maintain the same temperature as the rest of the furnace. Therefore, an auxiliary zone heater was installed to provide a linear temperature profile throughout the furnace. The setpoint for the auxiliary zone temperature is set equal to that for the main furnace.

In order to control the fuel vaporization, trap and auxiliary heaters from the computer, current-to-time proportional circuits to switch the AC voltage on and off over a period of 1 sec were used. Zero crossover solid-state relays were used to reduce noise caused by switching the AC.

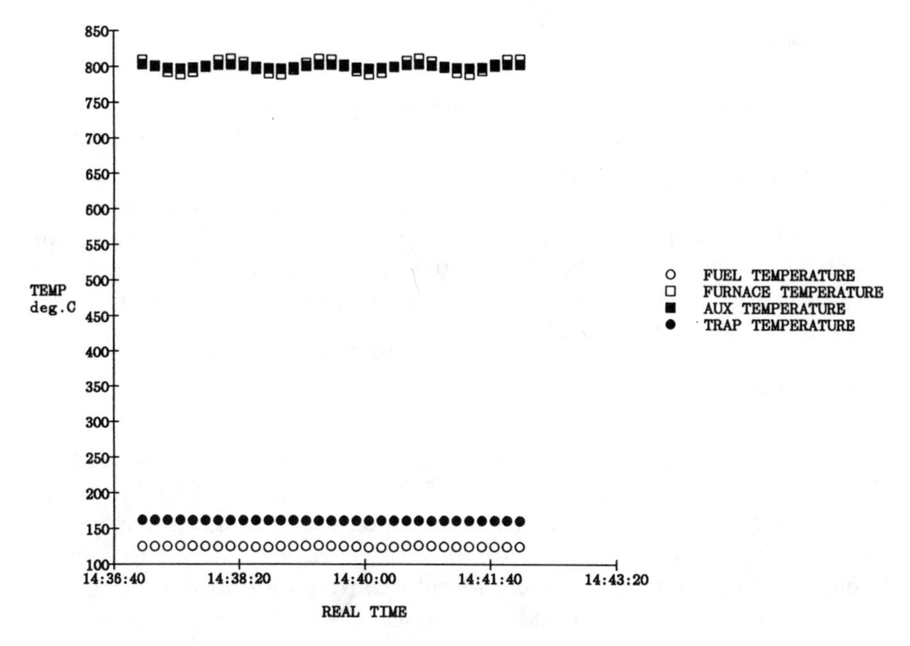


**Figure 7.** Temperature control.

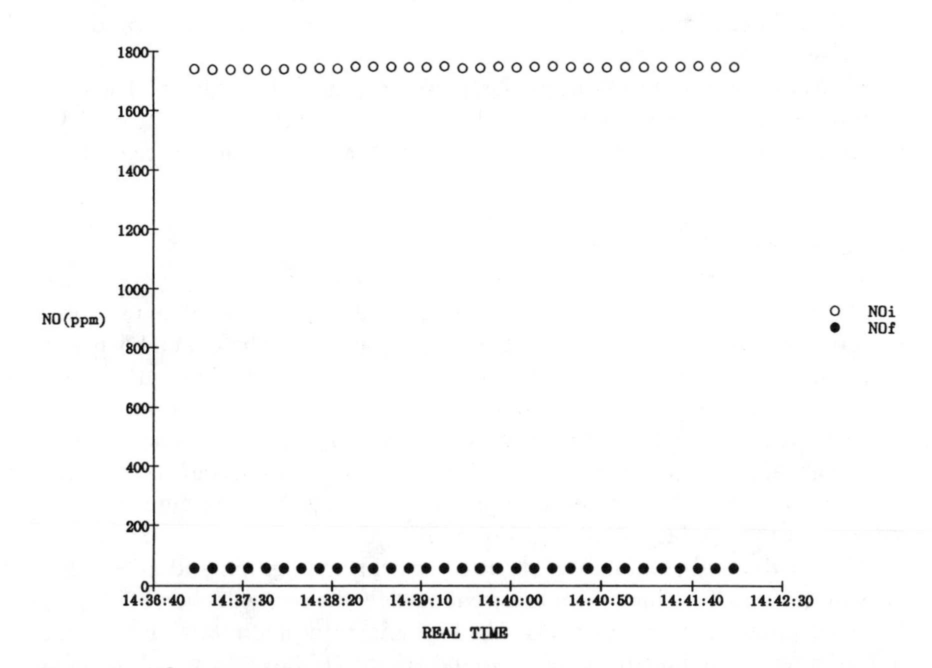


**Figure 8.** NO emissions from pyridine burn.

### Gas Flow Controllers

The dilution air (to get the $NO_x$ analyzer in range), combustion air, auxiliary #1 ($N_2/CH_4$), and auxiliary #2 gas flow rates are controlled by using four Matheson mass-flow controllers and a control/display unit. The Matheson control/display unit can accept remote (0 to 5 V) or local setpoints and sends a signal (0 to 5 V) proportional to the gas flow rate. The computer uses a feed-forward controller, to send the setpoint to the Matheson controller, which acts as a cascade control loop. All gas flows have Nupro automatic, bellows seal valves to provide positive shutoff.

### Fuel Feed Pump

Due to the very small fuel flow rates required to operate a thermal oxidizer of this scale, a Waters Associates Model 510 HPLC pump was used to pump fuel from the reservoir into the fuel vaporization chamber. The pump's standard operating range is 0.1 to 9.9 cc/min (6 to 594 cc/hr). In order to operate the pump via the computer, a current-to-pulse converter was used to bypass the manual thumbwheel switches and send pulses directly to the pump's stepper motor. In this configuration, the flow rates can be controlled from 0.1 to 12 cc/hr.

Since the flow rates are so small, a feedback measurement would be difficult. Therefore, a feed-forward controller is used based on a calibration of flow rate versus computer output.

The Waters pump has a built-in high pressure limit and shutdown. The high pressure signal is connected to a digital input so the computer can alarm if the pump shuts itself down due to excess pressure (i.e., a plug in the feed line).

## ANALYTICAL INSTRUMENTATION

A Chemiluminescent $NO/NO_x$ analyzer is used to provide online measurement of the NO and $NO_x$ concentration in the stack gas. The manual detector mode switch is connected to a digital input and the detector output is connected to an analog input. Depending on the value of the digital input, the computer can tell if the detector is in the NO or $NO_x$ mode. The analyzer also has a manual amplifier range switch. Since there is no simple way to interface the range switch to the computer, the analyzer range must be entered via a menu choice in the burner control program.

A Fisher thermal conductivity detector gas chromatograph and a Hewlett-Packard (H-P) 3392A integrator are used to detect the concentration of $O_2$, $N_2$, CO, $CO_2$, and $CH_4$ in the stack gas. The H-P integrator has a serial output and is configured to send the peak report to the MicroPDP after the gas chromatograph scale factors and retention times are entered into the MicroPDP. After each sample, the control program can scan the peak report for peaks within the retention time windows and calculate the corresponding gas concentration.

It should be noted that the integrator communicates with the computer over an RS-232 line. The control software has been modified to accept data from this type of "smart sensor." The modifications are general, but since each instrument uses a different command syntax, a separate communication subroutine was written for the H-P 3392A.

## SAFE DESIGN FEATURES

### Heaters

The single-loop PI controller is equipped with a high-limit alarm to provide additional protection from overheating the furnace. The high-limit alarm will shut off power to the SCR power circuit when the temperature is greater than 1200°C. When the high-limit alarm activates, a separate audible alarm is triggered.

In order to prevent overheating and damage to the fuel or trap heat tapes, transformers are used to reduce the line voltage to 30 to 50 VAC. All heat tapes are fuse protected and have manual On/Off switches.

### Gas Flows

All of the Matheson mass-flow controllers are fail-off devices and each gas flow has a fail-closed shutoff valve. Each regulator has special threads so that the wrong cylinder can not inadvertently be attached to a gas flow line.

### Emergency Shutdown Switch

An emergency shutdown switch is provided to rapidly stop the experiment and force the computer outputs to their fail-safe condition. The shutdown switch is connected to one of the digital inputs and will signal a shutdown condition when the stop button is pushed. In addition, pushing the shutdown switch will disconnect electrical power to the furnace, heaters, fuel pump, air controllers, and air solenoid valves. It will not disconnect power to the analog or digital signal conditioning modules.

### Watchdog Timer

The computer has a watchdog timer board to monitor the integrity of the hardware and software. The computer must signal the watchdog at the end of each pass through the control program (once a second). If the watchdog does not receive a signal for 10 sec, it will initiate a shutdown.

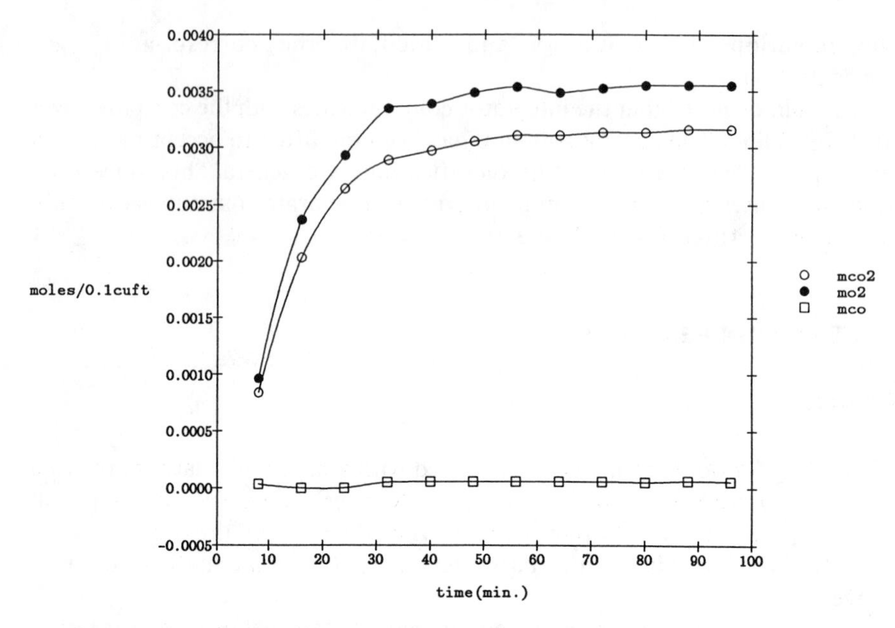


**Figure 9.** Concentration of $CO_2$, $O_2$, and CO in flue gas at 850°C.

## Utility Failures

The following utility failures have been examined:

| | |
|---|---|
| Electricity | Computer will stop, controllers will go to their fail-safe state. |
| Air pressure | Gas feed shutoff valves will close. |
| Combustion air | Flame will blow out. Computer will shutdown after 30 sec. |
| Nitrogen (aux #1) | Flame will blow out. Computer will shutdown after 30 sec. |
| Hood fan | Hood digital input will be false and computer will shutdown. |
| Computer halt | Watchdog timer will time out after 10 sec and will trip the emergency shutdown switch. |
| Computer failure | All outputs will go to their fail-safe state. |

## DISCUSSION

Control of the LTOX, especially the flame, is clearly demonstrated in Figure 6, which shows the gas flows, and Figure 7, which shows temperatures plotted against run time. Figure 8 is the NO emissions from the combustion of pyridine. Small fluctuations in the flame structure results in a wide variation in $NO_x$ emissions. The latter is an excellent measure of flame stability. Finally, Figure 9 shows the emission of $CO_2$, CO, and $O_2$ from the combustion of chlorobenzene. Again, once steady state is reached, the data indicate excellent oxidative reproducibility.

## REFERENCES

1. Duvall, D. S., and W. A. Rubey. "Laboratory Evaluation of High Temperature Destruction of Polychlorinated Biphenyls and Related Compounds," EPA Report, EPA-600/2-77-228 (December 1977).
2. Taylor, P. H., B. Dellinger. *Environ. Sci. Technol.* 22:438–447 (1988).
3. VanDell, R. D., and L. A. Shadoff. *Chemosphere*, 13:1177–1192 (1984).
4. Duvall, D. S., and W. A. Rubey. "Laboratory Evaluation of High Temperature Destruction of Kepone and Related Pesticides," EPA Report, EPA-600/2-76-299 (December 1976).

CHAPTER 7

# The Role of Carbon Particulate Surface Area on the Products of Incomplete Combustion (PICs) Emission

**Robert D. VanDell and Nels H. Mahle**


## ABSTRACT

Monochlorobenzene was burned using a computer controlled laboratory thermal oxidizer (LTOX). All conditions for the burner were constant and closely controlled with the exception of the afterburner temperature. The latter varied from 800 to 1000°C. Carbon particulates and products of incomplete combustion (PICs) were collected and analyzed by gas chromatography and gas chromatography-mass spectrometry. Also, surface areas per gram of soot were measured using $CS_2$ adsorption. Comparison of chromatograms at various afterburner temperatures indicated that some of the PICs displayed typical gas-phase destruction behavior while others exhibited destruction curves similar to that of carbon particulates, suggesting their association with soot while traversing the afterburner. With the exception of the very low boiling compounds, all of the PICs were found on the particulate. Finally, a minimum in the loss of soot surface area due to carbon oxidation appeared at 850°C, which coincided with a maximum in PIC emissions.


## INTRODUCTION

A study of the emissions of products of PICs and carbon particulates for the burning of o-dichlorobenzene has indicated that PICs are produced and/or destroyed along with carbon surface in the burning process.[1] Carbon particulates grow in size as they traverse the length of the flame, broadening their size distribution from 30 to several hundred angstroms. This process[2] is believed to be the result of the coagulation of carbon "crystalites" in the range of 20 to 30 Å. In the postcombustion zone (afterburner), particles continue to coagulate into chains or clusters, without significantly reducing the total available surface area, producing a broad or flattened size distribution. These processes are accompanied by parallel reductions in PIC concentrations in the flue gas. Concentrations of PICs not associated with carbon particulates were found to be one to two orders of magnitude lower, further indicating the importance of the role played by the particulates in the production and destruction of PICs.

A continuation of the o-dichlorobenzenes study was initiated. The purpose was to determine the effects of the postflame region temperatures in the combustion tube on PIC destruction and distribution between the carbon particulate and the gas-phase.

## EXPERIMENTAL

### Laboratory Thermal Oxidizer

The LTOX (see Figure 1) is essentially the same as that described before [1,3] with the exception of two modifications. An auxiliary heater was added to the upper end of the furnace to insure a linear temperature profile over the entire combustion tube length. A microcomputer (DEC MicroPDP-11/23) has been employed to control all temperatures and flow rates used to set experimental conditions. The latter was used in conjunction with a four-channel Matheson mass flow controller (Model 8249) to control gas flow rates and a Waters Model 510 liquid chromatography pump to deliver fuel to the nozzle. Temperatures of the fuel vaporization chamber, furnace, auxiliary furnace heater, and trap were maintained through analog feedback loops controlled by the computer. Experimental conditions were monitored by the operator on the CRT via a number of displays (see Figure 2). Any small deviations from set points were accompanied by audio and visual alarms. For safety, any sizable deviations caused the initiation of shutdown procedures. This resulted in excellent control of conditions required for reproducibility.

Because of laminar flow, soot rises in the tube in the form of a fine stream (approximately 1 to 2 mm thick) to the upper end of the combustion tube without impinging the walls. Here, it is trapped at 160°C. The filtered flue gas, containing partial combustion products not associated with the carbon, subsequently flows into a cryogenic trap, where the products are collected at −78°C. After a burn, these traps were separated and independently analyzed.

To maintain ambient pressure on the flame, a controlled vacuum is pulled on the entire system and is balanced against the flue gas flow. A Bernoulli tube, attached to the combustion tube just below the soot trap, is used to maintain the balance and is open on the opposite end to the room. The tube contains a packing of activated carbon, which removes impurities from the air in case the latter passes into the combustion tube. One end of the tube contains a fine capillary facilitating the Bernoulli effect. A small manometer, hung between the packing and the capillary, is open to the room on the other end. Acetone is used in the manometer because of its low density. During operation, gases passing in and out of the tube cause an imbalance in the manometer,

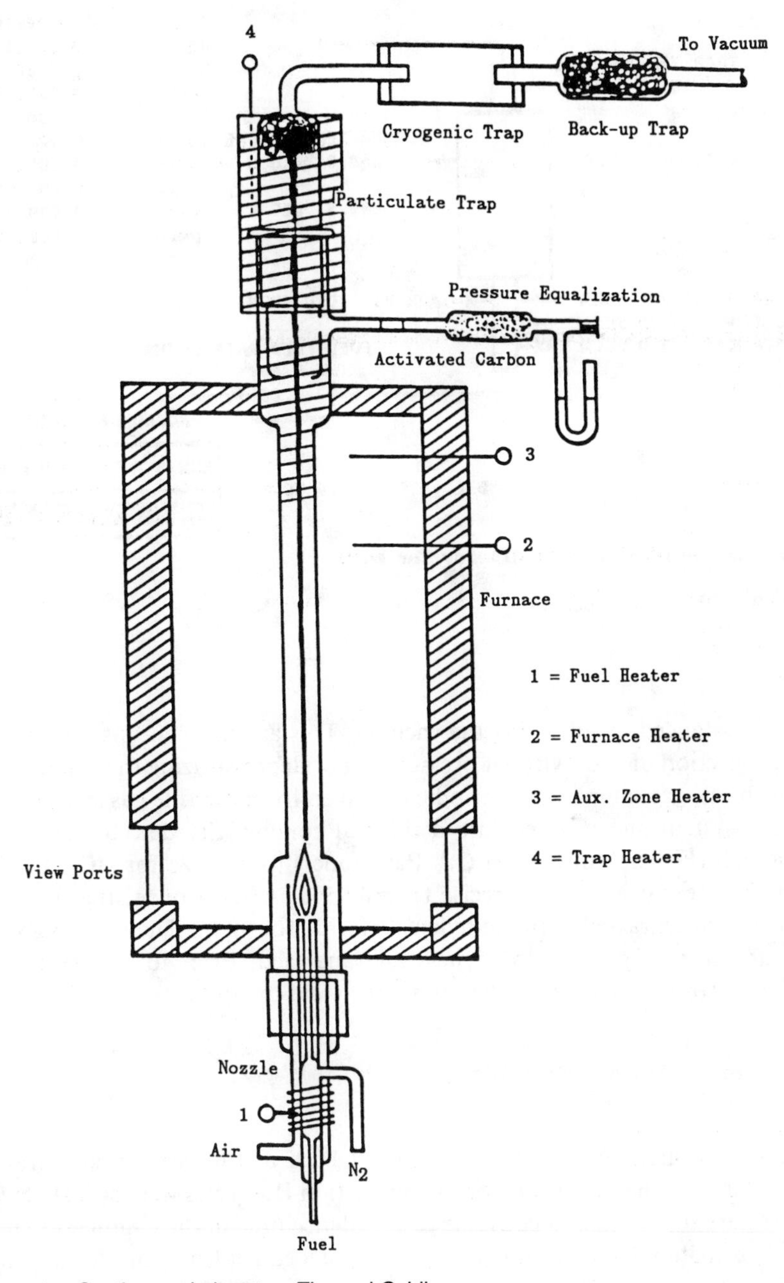


**Figure 1.** Continuous Laboratory Thermal Oxidizer.

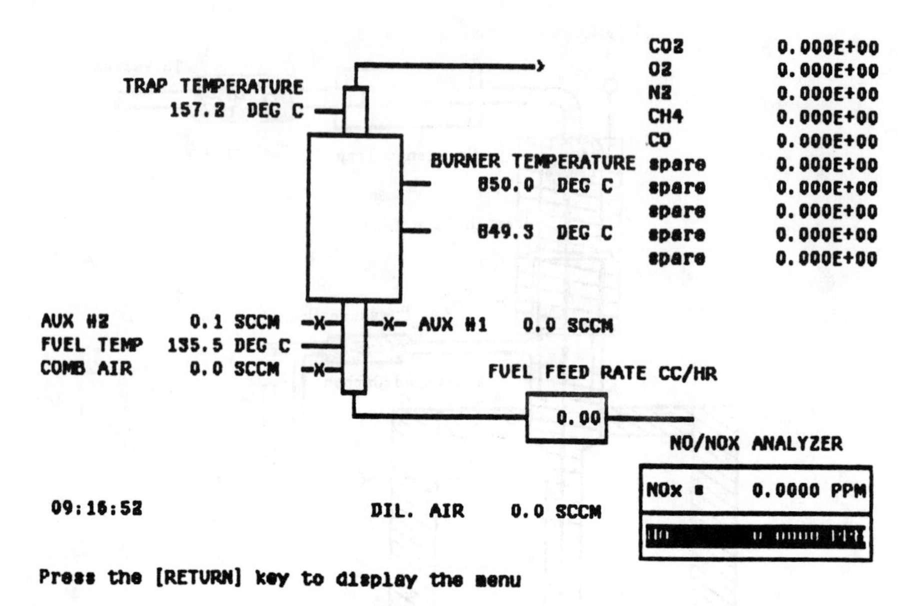


**Figure 2.** CRT screen updates.

which is balanced by adjusting the vacuum. This arrangement insures quantitative collection of soot without pressurizing or depressurizing the flame. During the burn, gas samples are periodically taken from the flue gas stream after the trap system and injected into a thermal conductivity detector gas chromatograph (Fisher Model 1200 Gas Partitioner) to analyze for $CO_2$, $O_2$, $N_2$, and CO. The partitioner is precise to $\pm 0.05\%$. After a burn, the traps are removed and extracted with methylene chloride for 48 hr. The resulting solutions are then evaporated to dryness and analyzed. Prior to extraction, the tared soot trap is weighed to determine the yield of soot.

## Experimental Burner Conditions

Monochlorobenzene was burned in the LTOX at furnace temperatures of 800 to 1000°C. The fuel delivery and combustion flow rates were adjusted with temperature to maintain a constant gas residence time in the combustion tube of approximately 1.40 sec and an oxygen to fuel equivalence ratio of approximately 1.0. Nitrogen flow rates were held constant at 50 sccm, resulting in a particulate residence time of approximately 0.05 to 0.02 sec in the flame zone. The latter is a function of the soot stream width and combustion tube temperature (see Table 1). The fuel vaporization chamber was held constant at 135°C (bp = 132°C), while the trap temperature was controlled at 160°C.

**Table 1. Experimental Conditions for Chlorobenzene Burns**

| 0 | 1 Temperature (°C) | 2 Air (sccm) | 3 Nitrogen (sccm) | 4 Fuel Rate (cc/hr) | 5 O/Fuel Ratio | 6 Tube Residence Time (sec) |
|---|---|---|---|---|---|---|
| 1 | 800 | 211 | 50 | 1.58 | 0.998 | 1.436 |
| 2 | 850 | 200 | 50 | 1.50 | 0.997 | 1.433 |
| 3 | 900 | 189 | 50 | 1.42 | 0.995 | 1.435 |
| 4 | 950 | 180 | 50 | 1.35 | 0.997 | 1.431 |
| 5 | 1000 | 200 | 52 | 1.50 | 0.997 | 1.254 |

## Gas Chromatography Analysis

Extraction samples were redissolved in 100 $\mu$L of methylene chloride and a 1 $\mu$L aliquot was injected into a high-resolution Carlo Erba Series 4160 gas chromatograph using a 30-m, 0.32-mm i.d., $d_f$ = 0.25 micrometer, DB-5 J&W capillary column. Because the detector was electron capture, a split ratio of 100:1 was maintained using an argon (5% methane) carrier. A make-up of 90 sccm argon (5% methane) was used for the detector, which was operated in the constant period mode. GC conditions were the same as those employed for the gas chromatography-mass spectrometry analysis. The initial temperature of 35°C was held for 1 min, and then ramped to 280°C at a rate of 8°C/min. The final temperature was held for 30 min before cooling. The column flow rate was approximately 1 sccm argon (5% methane). In spite of large numbers of peaks, typically obtained for combustion samples, resolution was excellent, with most peaks being base-line resolved. Chromatograms obtained from samples which were collected at various temperatures could not only easily be compared to determine the fate of the various compounds under different burning conditions, but could be compared to chromatograms obtained by gas chromatography-mass spectrometry for identification. Graphs of peak heights, normalized to carbon particulate sample weight, versus temperature were prepared. These are shown in Figures 3 and 4.

## Gas Chromatography-Mass Spectrometry Analysis

The gas chromatography-mass spectrometry was done on Hewlett-Packard Models 5985 and 5996 GC/MS instruments. A similar DB-5 capillary column as was described was used. Electron impact ionization was used at 70 eV electron energy. Mass spectra were obtained from m/z 35 to m/z 450. The mass scale was calibrated against a perfluorotributylamine reference standard. Selected Ion Monitoring (SIM) was also used to identify isomers of chloronaphthalenes and chlorobiphenyls. The proposed identifications are shown in Table 4.

## Surface Area Measurement

Before extracting, the traps containing the soot samples were dried in a vacuum oven at 50°C until a constant weight was obtained. The traps were weighed to determine soot yield and placed in a adsorption chamber. The

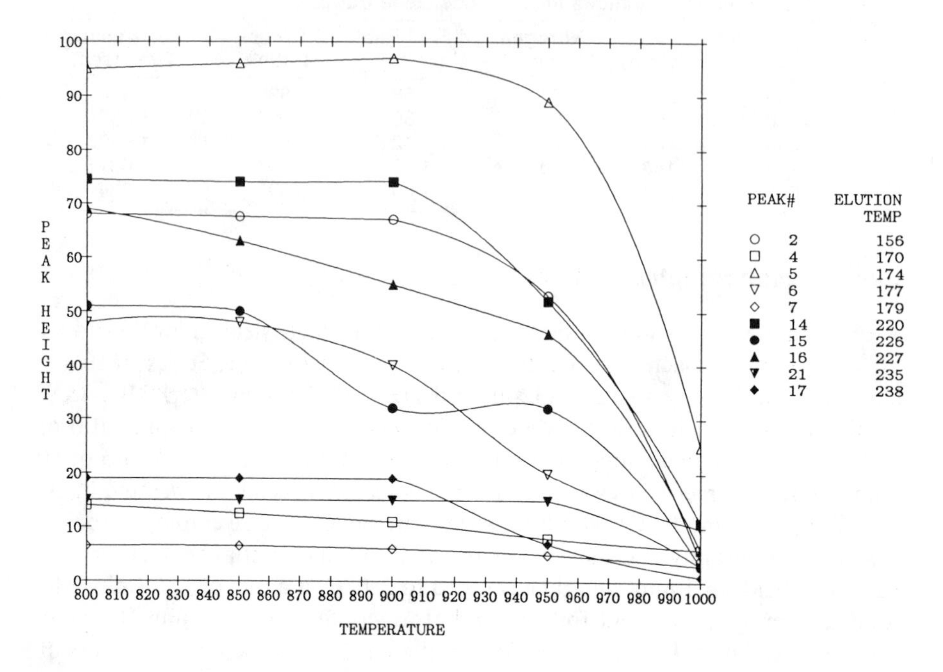


**Figure 3.** Thermal destruction of gas-phase PICs.

chamber was jacketed so that its temperature could be controlled to 50°C. After placing the trap in the chamber, the latter was flushed with argon to remove any remaining oxygen or water vapor. It was then sealed off and aliquots of $CS_2$ were injected via a septumed injection port. After equilibrium was established, 25-$\mu$L gas samples were withdrawn and injected into the GC for quantitative analysis. Peak heights were determined for each $CS_2$ aliquot injected and compared to a previously obtained calibration curve to determine the concentration of $CS_2$ remaining in the gas-phase and allow the calculation of the amount of $CS_2$ adsorbed onto the soot. The 50°C temperature is above the dew point of the $CS_2$ and prevented condensation.

Adsorption isotherms were constructed by plotting the concentration of $CS_2$ adsorbed versus that in the gas-phase. A typical example is shown in Figure 5. In all of the isotherms, an inflection point was observed where the monolayer coverage is complete. This was used along with the sample weight to calculate the total surface area per gram of soot ($\epsilon$) for the sample. Values of $\epsilon$ ranged from 166 to 619 $m^2/g$. These values could appear to be high, however, not nearly as high as those typically obtained from nitrogen adsorption, which is the reason the latter was not used. Assuming a surface area of 4 $Å^2$ per carbon atom, a maximum value of 2000 $m^2/g$ is theoretical. In view of this, $\epsilon$ values obtained by this method appear quite reasonable. The use of the inflection

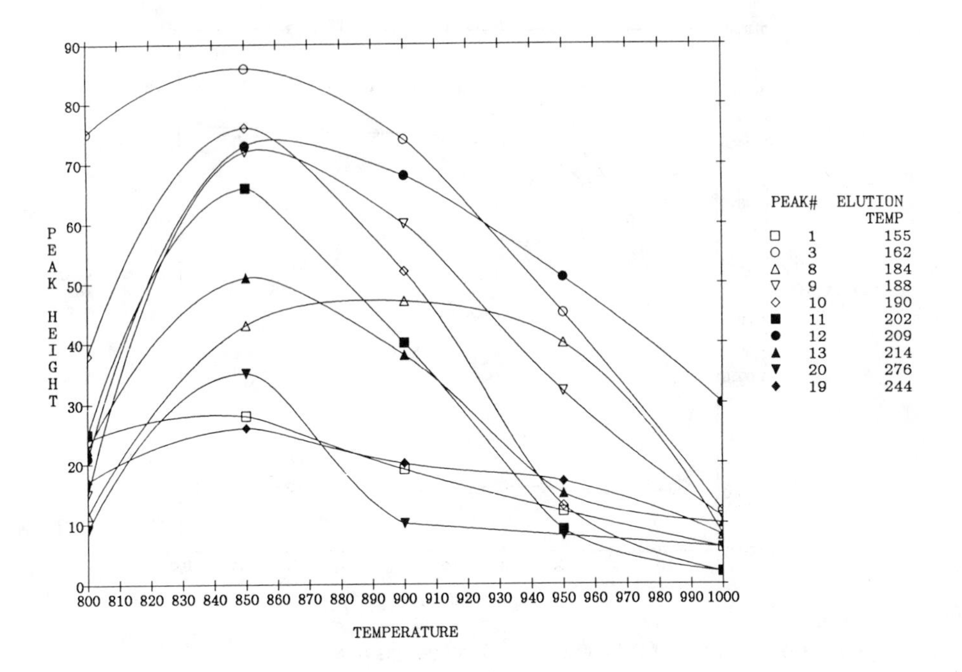


**Figure 4.** Thermal destruction of particulate-phase PICs.

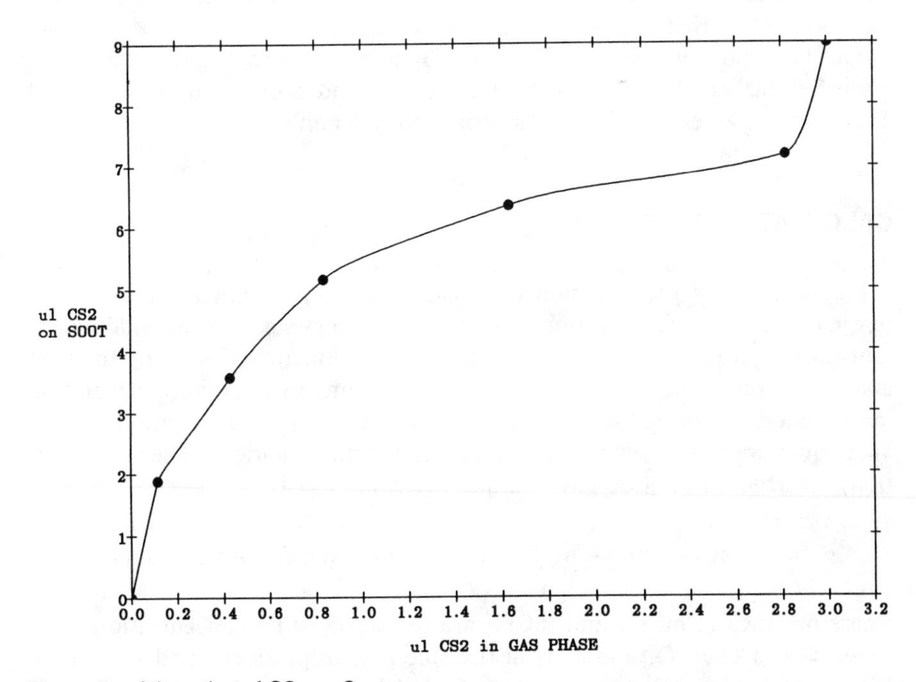


**Figure 5.** Adsorption of $CS_2$ on Soot.

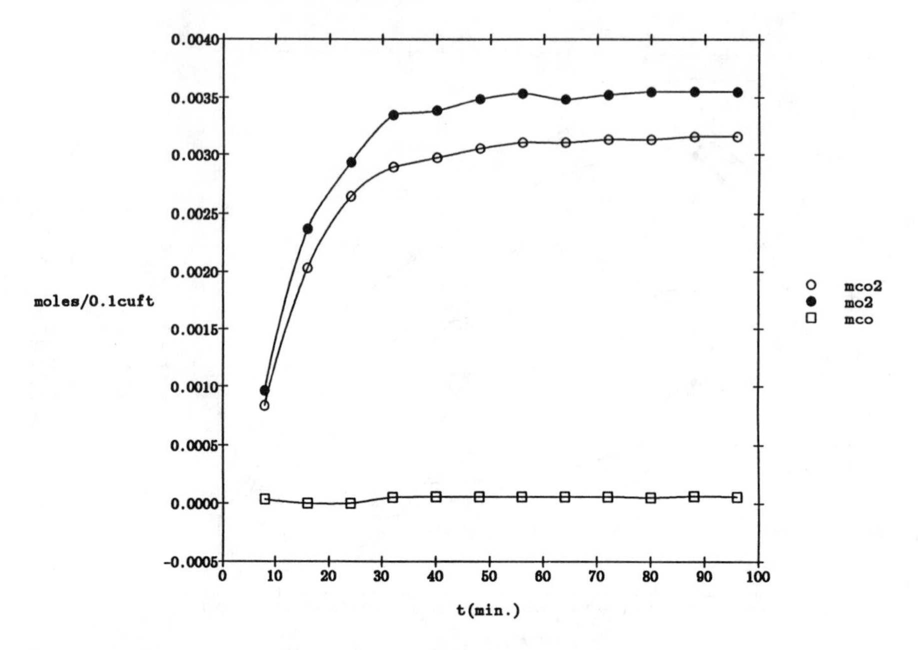


**Figure 6.** Concentration of $CO_2$, $O_2$, and CO in flue gas at 850°C.

point to determine monolayer formation seemed to be sufficient. BET-type plots were also tried and resulted in similar values for $\epsilon$.

The GC and column used for the $CS_2$ analysis were the same as those discussed earlier. The $CS_2$ peaks were sharp and base lined resolved and appeared at a retention time of approximately 1 min.

## CALCULATIONS

Figure 6 shows the emission of carbon dioxide, carbon monoxide, and oxygen consumed with run time. The burner reaches steady state in about 30 min. Because of the wide difference between the kinetics of burning the soot and the overall combustion kinetics of the fuel, any mass balance calculation must include the formation and destruction of carbon. Steady-state mass balance equations have already been developed for this situation.[4] The concentration of carbon in the flue gas in moles per cubic feet is:

$$dm_s = am_s^* - m_s = am_x - mCO_2 - mCO - mCH_4 \qquad (1)$$

where $m_x$, $mCO_2$, mCO, and $mCH_4$ are the steady-state concentration of fuel, $CO_2$, CO, and $CH_4$ in the flue gas, respectively, and a is the number of carbons per fuel molecule.

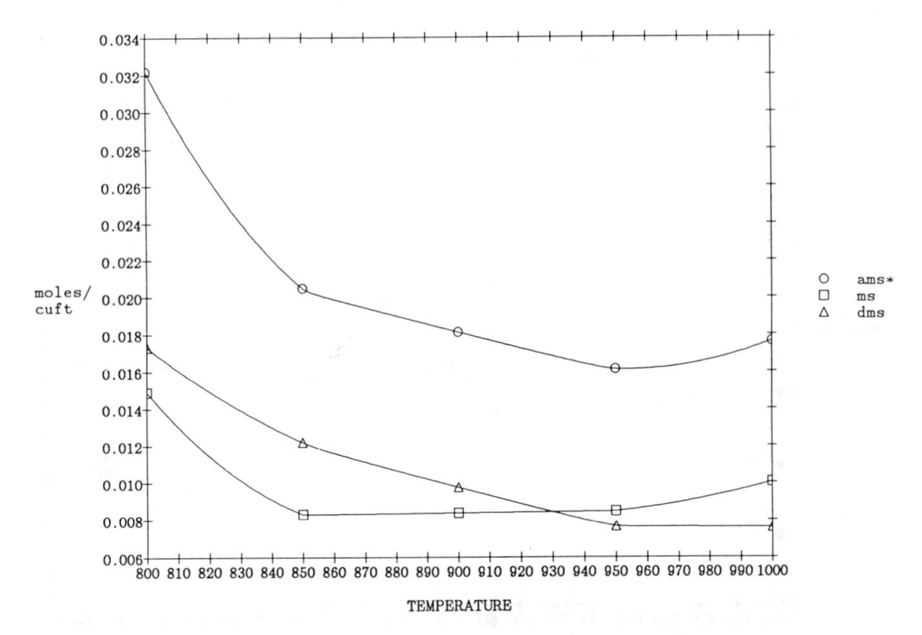


**Figure 7.** Soot made (ams*), burned (ms), and emitted (dms) at steadystate.

$m_s$* is the moles of fuel converted to carbon and may be calculated from:

$$m_s^* = (mCO_2 + Bm_x - mO_2 + {}^1/_2\, mCO + {}^1/_2\, mCH_4)/B \qquad (2)$$

where $B = [(b - d)/4 - (e/2)]$

b, d, and e are the number of hydrogen, chlorine, and oxygen atoms in the fuel molecule, respectively. For example, $B = 1$ for chlorobenzene. $am_s$* is then the moles of carbon produced per cubic foot of flue gas.

$m_s$ is the moles of carbon burned per cubic foot of flue gas and may be calculated from:

$$m_s = mCO_2 - am_x + am_s^* + mCO + mCH_4 \qquad (3)$$

The ratio $am_s^*:m_s$ has been defined as p, which is a measure of the soot point of the burner.[4] When p is less than 1, the burner is below the soot point and no soot escapes the combustion tube, while for values of p greater than 1, soot along with partial combustion products is emitted. When p equals 1, the burner is at its soot point. Values of $m_s$, $m_s$*, and $dm_s$ were calculated at various temperatures and plotted in Figure 7. The quantity p is plotted against temperature in Figure 8.

It should be noted that trace quantities of the various PICs produced are ignored in the mass balance equations simply because their concentrations are several orders of magnitude lower than those of the carbon, oxygen, carbon

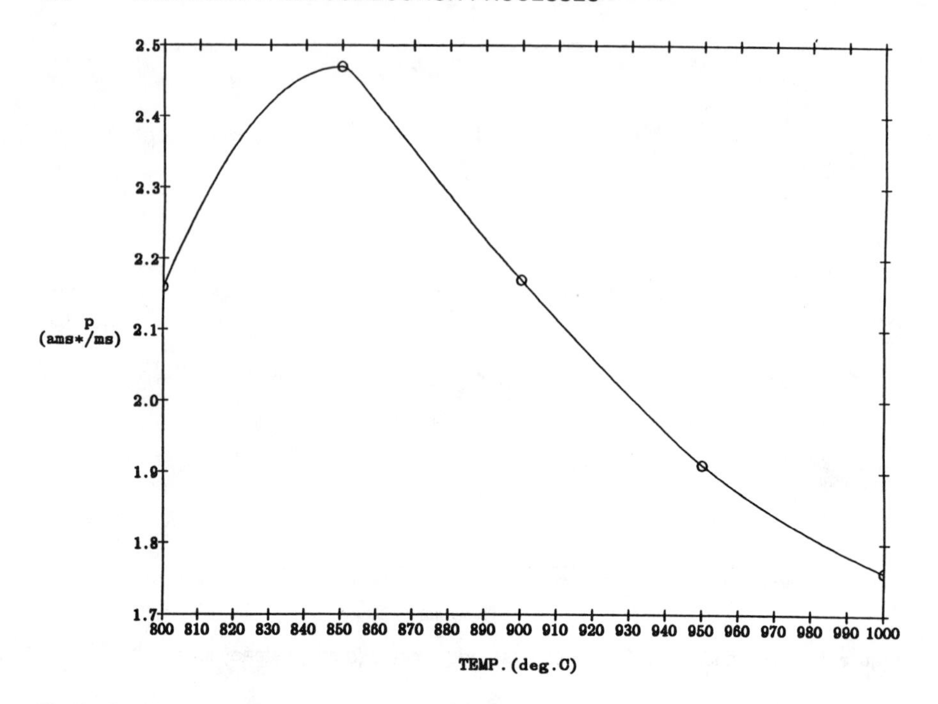


**Figure 8.** Ratio of soot made/burned.

dioxide, carbon monoxide, methane, hydrogen chloride, and water which are the major products of combustion.

Flue gas carbon concentrations ($dm_s$) calculated from the previous equations agree well with those determined from the weight of the soot collected.[1] In fact, the two are routinely compared to insure the quantitative trapping of soot and related PICs during a burn.

If one makes the assumptions that the number and density of soot particles are constantly moving through the combustion tube and their shape is spherical, a relationship between the initial surface area per gram ($\epsilon_i$) and the final experimentally observed surface area per gram ($\epsilon_f$) may be derived as follows: The surface area per gram expressed in $m^2/g$ is

$$\epsilon = (6 \times 10^{-4})/(\rho_c D) \quad (4)$$

where $\rho_c$ is the particle density and D is the statistical diameter.

Since the mass of a particle is

$$m_p = (\rho_c \pi D^3)/6 \quad (5)$$

then the number of particles per cubic foot of flue gas deposited in the particulate trap is

$$N_f = (M_c dm_s)/m_p = 8.842 \times 10^9 \times dm_s \rho_c^2 \epsilon^3 M_c \quad (6)$$

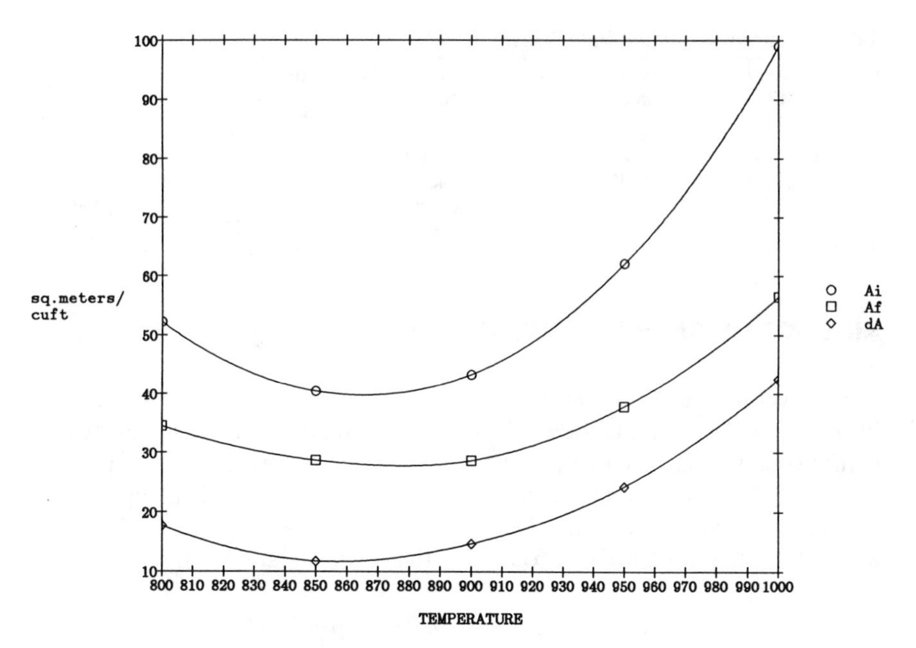


**Figure 9.** Surface area/cubic feet of flue gas.

where $M_c dm_s$ is the total mass deposited per cubic foot and $M_c$ is the atomic weight of carbon.

Similarly, the number of particles per cubic foot of flue gas synthesized by the flame is

$$N_i = 8.842 \times 10^9 \times am_s^* \rho_c^2 \epsilon_i^3 M_c \tag{7}$$

Assuming a constant particle density and number ($N_f = N_i$) as soot traverses the combustion tube:

$$\epsilon_f (dm_s/am_s^*)^{1/3} = \epsilon_i \tag{8}$$

This model also assumes no postflame coagulation other than loose clusters or chains. Using $\epsilon_f$, $\epsilon_i$, $am_s^*$, $dm_s$, and $m_s$, the surface area in square meters per cubic feet of flue gas may be calculated from

$$M_c \epsilon_f (dm_s) = A_f \tag{9}$$

$$M_c \epsilon_i (am_s^*) = A_i \tag{10}$$

The area reduction (dA) resulting from burning of the carbon is then $A_i - A_f$. dA, $A_i$, and $A_f$ are plotted in Figure 9.

**Table 2. Carbon Oxidation Rate Constants**

| 0 | 1 Temperature (°C) | 2 k (Equation 13) (sec – 1) | 3 k (Equation 12) (sec – 1) | 4 Tube Residence Time (sec) | 5 p ($am_s$*/$m_s$) |
|---|---|---|---|---|---|
| 1 | 800 | 0.1069 | 0.4329 | 1.436 | 2.16 |
| 2 | 850 | 0.2288 | 0.3622 | 1.433 | 2.47 |
| 3 | 900 | 0.3674 | 0.4305 | 1.435 | 2.17 |
| 4 | 950 | 0.4788 | 0.5181 | 1.431 | 1.91 |
| 5 | 1000 | 0.6277 | 0.6697 | 1.254 | 1.76 |

## KINETICS OF CARBON COMBUSTION

The overall kinetics of carbon combustion has been discussed in considerable detail by Weisz and Goodwin.[5] They called attention to the fact that carbon combustion is a surface phenomena and follows first-order kinetics. One conclusion of their work was that the rate of carbon oxidation is independent of the source or type of carbon. The same results were obtained on both graphite and coke. They obtained the following first-order relationship:

$$k = (1/C)(dC/dt) = 3 \times 10^7 \exp\{-36700/RT\} \quad (11)$$

assuming that a carbon atom occupies $4.1 \times 10^{-16}$ cm$^2$ and the partial oxygen pressure is 0.21 atm. Assuming this to be correct and using a notation consistent with this writing, the integrated form of this equation would be:

$$\ln(dm_s/am_s^*) = -kt \quad (12)$$

where

$$k = (1.4286 \times 10^8)(pO_2) \times \exp\{-36700/RT\} \quad (13)$$

and $pO_2$ is the partial pressure of oxygen in the combustion tube.

It was assumed that t, the residence time of soot particles traversing the combustion tube, may be calculated from:

$$t = (\pi r^2 L T_o)/(F_T T) \quad (14)$$

assuming no pressure drop across the combustion tube. L, $T_o$, $F_T$, T, and r are the combustion tube length, ambient temperature at which the flow rates are measured, total flow rate, combustion tube temperature, and tube radius, respectively. Values obtained in this manner have been recorded in Table 2 along with calculated t values from Equation 14. Values of k calculated from both Equations 12 and 13 have also been placed there for comparison. This comparison indicates that soot also follows these kinetics, which is consistent with the findings of Walsh and Green.[6]

## DISCUSSION

Figures 7 and 8 deal with the formation and burning characteristics of soot. Plots similar to Figure 8 are typically used in this laboratory to determine the point where soot begins to appear in the flue gas (extrapolation to p = 1). The shape of the curve is a function of the soot crystalite growth and coagulation in the flame.[7] These phenomena are controlled by flame temperature, oxygen partial pressure, burning velocity of the fuel, and the fuel atomic ratios.[4] In this particular case, a maximum appears at 850°C, which is partially explained by the minimum which appears in Figure 9 (dA). This minimum denotes the temperature where the least amount of surface area is destroyed as the particulate traverses the length of the combustion tube. It would be expected that PICs associated with the particulate, that is adsorbed on it or trapped interstitially in it as the latter emerges from the flame, would necessarily suffer the same fate as the carbon. The kinetics of PIC formation and destruction would be controlled by the same factors as the carbon particulate itself and kinetics associated with the chemical characteristics of the PIC itself would be overridden. On the other hand, PICs which are not associated with carbon particulates in the combustion tube would be dissolved in the gas-phase and exhibit kinetics similar to those reported by Duvall and co-workers,[8] obtained in hot-tube experiments over a temperature range at a fixed residence time. The latter indicates the minimum temperature where oxidative destruction begins and is a function of the specific chemistry of the PIC.

All of the PICs, with the exception of the highly volatile ones, end up on the particulate in the particulate trap simply because at lower temperatures, they condense there (see Table 3). Figure 3 shows a number of PICs which were destroyed in the gas-phase because the characteristic shape of the curves is identical to those obtained in hot-tube experiments. The minimum destruction temperatures displayed for the various PICs vary indicating their dependances on PIC chemistry. However, all of the curves displayed in Figure 4 characteristically display a maximum near 850°C similar to Figure 9 indicating that they are associated with the particulate and are subject to the same oxidation kinetics of carbon. These results strongly indicate that one should not apply gas-phase destruction kinetics to all species emitted from combustion processes, particularly when decisions need to be made concerning the conditions and hardware required for air pollution control.

## CONCLUSIONS

These results demonstrate the utility of the LTOX for determining the state of the products of incomplete combustion in the postflame region. A knowledge of this state is necessary to any understanding of the mechanism of PIC formation or destruction. It is also important, in some cases, if one is contemplating a postflue gas treatment for air pollution control. The data presented

**Table 3. Percent on Soot Surface**

| Temperature (°C) | Peak #1 | Peak #2 | Peak #3 | Peak #4 | Peak #5 | Peak #6 | Peak #7 |
|---|---|---|---|---|---|---|---|
| 800 | 100 | 100.0 | 100 | 97.2 | 99.4 | 100.0 | 95.6 |
| 850 | 100 | 98.3 | 100 | 95.4 | 99.2 | 90.1 | 84.4 |
| 900 | 100 | 93.4 | 100 | 92.4 | 99.2 | 92.4 | 93.8 |
| 950 | 100 | 90.6 | 100 | 88.9 | 100.0 | 98.5 | 80.7 |
| 1000 | 100 | 90.0 | 100 | 90.0 | 100.0 | 100.0 | 92.0 |

| Peak #8 | Peak #9 | Peak #10 | Peak #11 | Peak #12 | Peak #13 | Peak #14 |
|---|---|---|---|---|---|---|
| 98.3 | 100.0 | 100.0 | 100.0 | 86.8 | 88.2 | 100.0 |
| 99.1 | 99.0 | 99.2 | 98.7 | 98.7 | 96.6 | 95.9 |
| 99.6 | 95.1 | 98.1 | 98.3 | 99.0 | 97.4 | 99.5 |
| 100.0 | 97.9 | 100.0 | 96.8 | 99.4 | 75.4 | 100.0 |
| 100.0 | 100.0 | 100.0 | 100.0 | 100.0 | 93.0 | 100.0 |

| Peak #15 | Peak #16 | Peak #17 | Peak #18 | Peak #19 | Peak #20 |
|---|---|---|---|---|---|
| 99.6 | 100.0 | 89.8 | 98.2 | 82.1 | 100 |
| 100.0 | 99.1 | 91.5 | 100.0 | 96.7 | 100 |
| 96.4 | 99.6 | 94.3 | 97.9 | 97.1 | 100 |
| 100.0 | 98.9 | 94.3 | 100.0 | 100.0 | 100 |
| 100.0 | 100.0 | 98.0 | 100.0 | 100.0 | 100 |

**Table 4. Proposed Compound Identification**

| Peak Number | Compound |
|---|---|
| 1 | unknown (possible chlorinated aliphatic hydrocarbon) |
| 2 | unknown (possible chlorinated aliphatic hydrocarbon) |
| 3 | dichloronaphthalene |
| 4 | dichloronaphthalene |
| 5 | dichloronaphthalene |
| 6 | dichloronaphthalene |
| 7 | dichlorobiphenyl |
| 8 | dichlorobiphenyl |
| 9 | dichlorobiphenyl |
| 10 | dichlorobiphenyl |
| 11 | trichlorobiphenyl |
| 12 | trichlorobiphenyl |
| 13 | trichlorobiphenyl |
| 14 | trichlorobiphenyl |
| 15 | tetrachlorobiphenyl |
| 16 | tetrachlorobiphenyl |
| 17 | tetrachlorobiphenyl |
| 18 | unknown |
| 19 | unknown chlorinated aromatic hydrocarbon |
| 20 | unknown |
| 21 | unknown |

here indicate that PICs adsorbed on or interstitially trapped in carbon particles are subject to entirely different destruction kinetics than those free in the gas-phase, namely the kinetics observed for graphite and coke surfaces. PICs which remain in the gas-phase while in the postcombustion zone follow the normal gas-phase destruction kinetics which depend upon the specific PIC chemistry. All of the PICs, however, with the exception of the lowest boiling species are associated with the carbon particulate once the temperature is sufficiently reduced.

## REFERENCES

1. VanDell, R. D., and G. U. Boggs. *Chemosphere* 16: 973–982 (1987).
2. Gaydon, A. G., and H. G. Wolfhard. *Flames*, 4th ed. (London: Chapman and Hall, 1979), p. 214.
3. VanDell, R. D., and L. A. Shadoff. *Chemosphere* 13: 1177–1192 (1984).
4. VanDell, R. D., and L. A. Shadoff. *Chemosphere* 13: 763–775 (1984).
5. Weisz, P. B., and R. B. Goodwin. *J. Catalysis* 6: 227–236 (1966).
6. Walsh, D. E., and G. J. Green. *Ind. Eng. Chem. Res.* 27: 1115–1120 (1988).
7. Prado, G., J. Lahaye, and B. S. Haynes. *N.A.T.O. Conference Series, Series VI: Materials Science*, Lahaye, J., and G. Prado, Eds., (New York: Plenum Publishing Corporation, 1983), p. 145.
8. Duvall, D. S., and W. A. Rubey. "Laboratory Evaluation of High Temperature Destruction of Polychlorinated Biphenyls and Related Compounds," EPA Report, EPA-600/2-77-228 (December 1977).

# CHAPTER 8

# Products of Incomplete Combustion from the High Temperature Pyrolysis of Chlorinated Methanes

**Debra A. Tirey, Philip H. Taylor, and Barry Dellinger**

## ABSTRACT

The chlorinated methanes [methyl chloride ($CH_3Cl$) methylene chloride ($CH_2Cl_2$), chloroform ($CHCl_3$), and carbon tetrachloride ($CCl_4$)] are commonly found in hazardous and toxic wastes. Because they are the simplest chlorinated hydrocarbons (CHCs), the study of their thermal decomposition may furnish fundamental information with which to model the combustion of more complex CHCs and the incineration of complex waste streams. Fused silica flow reactor experiments in conjunction with high resolution gas chromatographic/mass spectrometric analysis have been utilized to collect quantitative product information from the gas-phase thermal decomposition of chloromethanes. Data were obtained for a fuel/oxygen equivalence ratio of 3.0, over a temperature range of 300 to 1000°C, and for a mean residence time of 2.0 sec. The parent stability (defined by the temperature required for 99% destruction for a mean residence time of 2.0 sec) was evaluated to be:

$CH_3Cl$ : 910°C, $CH_2Cl_2$ : 755°C, $CHCl_3$ : 600°C, and $CCl_4$ : 880°C.

Evaluation of parent decay curves and the formation/destruction curves for intermediate species has resulted in the following major observations: (1) different destruction mechanisms are operative for each chloromethane; (2) $CCl_4$ decay is not autocatalytic and parent compound reformation occurs; (3) $C_2Cl_6$, $C_2Cl_4$, and $C_2Cl_2$ are key intermediates from $CHCl_3$ and $CCl_4$ decomposition; (4) Cl displacement facilitates condensation of chlorinated olefins and acetylenes; and (5) $CCl_4$, $C_2Cl_4$, and $C_6Cl_6$ are stable products from $CHCl_3$ and $CCl_4$ decomposition.

## INTRODUCTION

It has been estimated that approximately 20% of the 575 million metric tons (MMT) of hazardous wastes generated yearly in the United States consists of halogenated organic compounds.[1] As traditional methods of disposal such as landfilling become increasingly expensive, incineration has become a viable disposal option. Because of their toxicity, the "incinerability" of CHCs has become an area of intense environmental and scientific interest.[1]

The chlorinated methanes are prevalent in hazardous and toxic wastes. Because they are the simplest CHCs, the study of their high-temperature (>1000 K) thermal decomposition properties may furnish fundamental information with which to model the combustion of more complex CHCs and the incineration of complex waste streams.

Early literature concerning the study of the chlorinated methanes includes reports of flow reactor studies of the pyrolysis of methyl chloride,[2] methylene chloride,[3-5] chloroform,[6-8] and carbon tetrachloride.[2] These studies provided evidence that the initiation of the thermal decomposition of methyl chloride and carbon tetrachloride involved rupture of the C-Cl bond. C-Cl bond rupture and HCl molecular elimination (involving a three-center transition state)[9] were proposed to be important initiation pathways for methylene chloride and chloroform, with the latter path dominating for chloroform.[6]

Recent studies of chloromethane thermal decomposition have included the flow reactor pyrolysis of methyl chloride[10] and shock-tube pyrolysis of methyl chloride,[11] chloroform,[12] chloromethane mixtures.[13] From a fairly complete product analysis of the high-temperature pyrolysis of methyl chloride,[10] an elementary reaction kinetic mechanism (along with thermodynamic and kinetic parameters of the species involved) was proposed. High-temperature shock-induced decomposition of methyl chloride[11] and chloroform[12] produced Arrhenius rate parameter expressions in the high pressure limit. In agreement with earlier flow reactor studies, the initiation of methyl chloride decomposition was found to be C-Cl bond fission, while that for chloroform was found to be three-center HCl elimination. In measurements of gas-phase soot formation from the shock-tube pyrolysis of chloromethane mixtures, amounts of carbon formed were larger than that of their nonchlorinated analogues.[13] Sooting behavior and product distributions were rationalized with different reaction mechanisms depending on the H:Cl atomic ratio. Chlorine catalyzed decomposition involving H abstraction by chlorine atoms was proposed as the dominant destruction pathway for methyl chloride and methylene chloride.[10,13-17]

The effect of the H:Cl atomic ratio on the inert and oxidative pyrolysis of mixtures of chlorinated methanes in fused silica flow reactors for 2.0-sec reaction times and exposure temperatures of 300 to 1100°C has also recently been reported.[18] A detailed analysis of products of incomplete combustion (PICs) indicated that the nature and yields of PICs produced were strongly dependent on several factors including oxygen concentration, temperature, and elemental composition. This investigation demonstrated that complex, perchlorinated aromatic compounds (e.g., hexachlorobenzene and octachlorostyrene) were formed at high temperatures from simple initial reactants under pyrolytic, high chlorine (H:Cl = 1:5) conditions.

In work related to chloromethane pyrolysis, global oxidation kinetic parameters for methylene chloride, chloroform, and carbon tetrachloride have been reported from flow reactor experiments obtained over a temperature range of 300 to 1100°C.[19] The stability of these compounds was also compared to other hazardous chlorinated organics. In addition to global oxidation rate parameters, elementary reaction kinetic studies of hydroxyl radical attack on the chlorinated methanes with abstractable hydrogen atoms have also been conducted at temperatures up to 527°C.[20] Rate coefficients (normalized to the number of abstractable hydrogen atoms) increased with decreasing C-H bond energy with chloroform being most reactive followed by methylene chloride and methyl chloride.

Based on kinetic calculations and correlations between laboratory and full-scale data, we have previously argued that pyrolysis reactions in incinerators are responsible for a majority of the emissions of undestroyed waste feed and toxic PICs.[21-23] This analysis suggests that although incinerators operate under nominally oxygen-rich conditions, oxygen-starved pockets exist in the system due to locally poor mixing conditions. Consequently, flow reactor pyrolysis studies appear to be particularly well suited for the study of hazardous waste incinerability.[24] For typical incinerator gas-phase residence times ($\bar{t}_r$) of 2.0 sec, the decomposition chemistry of most organics under pyrolytic conditions occurs in the temperature range of 500 to 1100°C. Our flow reactor studies were conducted at fuel-to-air equivalence ratios ($\phi$) of 3.0 which are outside of the range at which flat-flame combustion studies are feasible due to the intrinsic difficulty in maintaining a spatially stable, one-dimensional propagating flame for such high fuel-to-air equivalence ratios.

## EXPERIMENTAL

### Instrumental

Experiments were conducted using the Thermal Decomposition Analytical System (TDAS). The TDAS is a closed-continuous, in-line, fused silica flow reactor system capable of accepting a solid, liquid, or gas-phase sample, exposing this sample to a highly controlled thermal environment, and then performing an analysis of the effluents resulting from this exposure.[25,26] Specifically, chloromethane samples were introduced into the insertion region of the TDAS, which was held isothermally at 250°C. Gaseous samples were then swept with carrier gas through heated transfer lines (275°C) to a flow reactor where controlled high-temperature exposure occurred. Figure 1 presents a block diagram of the TDAS and specific details of the flow reactor and its housing.

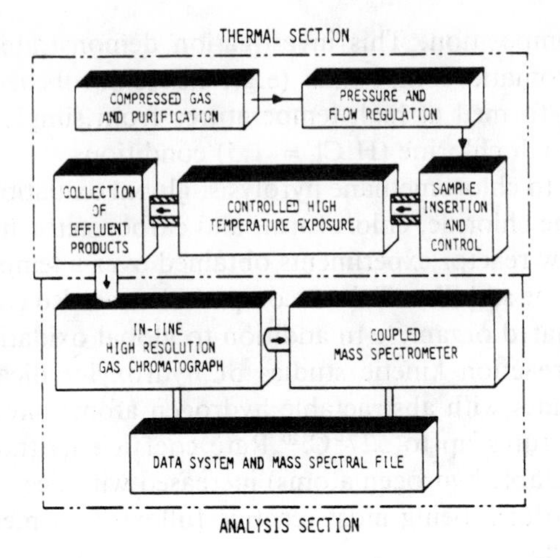


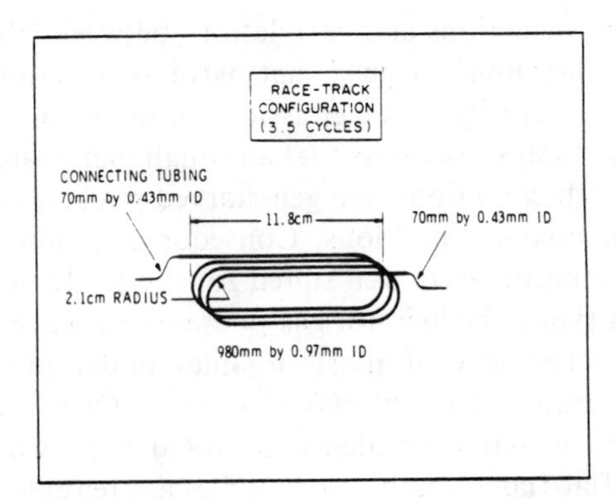


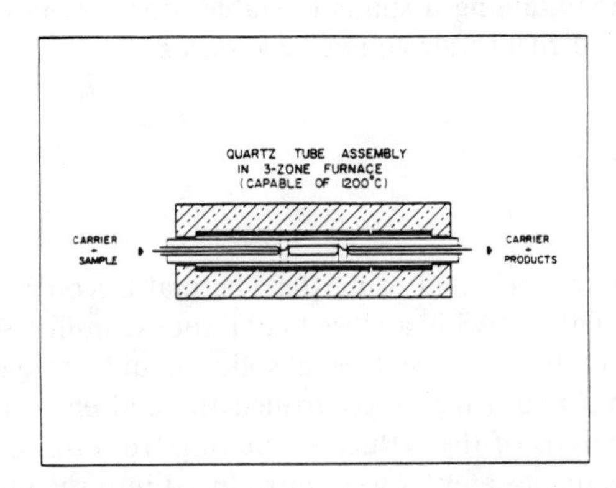


**Figure 1.** (a) Generalized schematic diagram of the Thermal Decomposition Analytical System, (b) schematic diagram of fused silica tubular reactor, (c) schematic diagram of three-zone furnace.

Following thermal exposure, the resulting effluents were swept by carrier gas through another transfer line (held at 275°C) to an HP 5890 GC where they were cryogenically focused at the head of a capillary column located inside the GC oven. Later, after a purge of the system with helium, the oven temperature was raised at a specific rate and the separated compounds eluting from the column were sent to the source of an HP 5970 mass selective detector (MSD). Data acquisition and analysis for the TDAS were accomplished with the aid of an HP 59970 ChemStation and the accompanying system software, which includes an online NIH-EPA mass spectral library.

## Procedure

Each of the chloromethane samples was prepared by adding $8 \times 10^{-4}$ mol of the pure compound to a 1-L bulb filled with nitrogen. Oxygen was then added to the bulb such that a fuel:oxygen equivalence ratio of 3.0 was attained. The necessary amount of oxygen required to satisfy this ratio was found via the stoichiometry of the combustion reaction as determined using an equilibrium code:[27]

$$CH_3Cl + 3/2\ O_2 \rightarrow HCl + CO_2 + H_2O$$

$$3/2\ CH_2CL_2 + 2\ O_2 \rightarrow HCl + 3/2\ CO_2 + H_2O + Cl_2$$

$$CHCl_3 + O_2 \rightarrow HCl + CO_2 + Cl_2$$

$$CCl_4 + O_2 \rightarrow CO_2 + 2\ Cl_2$$

These values were substituted into the fuel:oxygen equivalence ratio expression:

$$\phi = \frac{\text{actual fuel/oxygen}}{\text{stoichiometric fuel/oxygen}} = 3.0$$

to arrive at a value for the actual fuel:oxygen ratio. Since the amount of fuel added to the bulb for each of the chloromethane samples was fixed, $\phi$ was kept constant by varying the amount of oxygen added to the bulb. No other source of oxygen was made available to the samples for combustion; all experiments were run in a reactor atmosphere of pure dry nitrogen.

In addition to $\phi$, a volumetric concentration of 1000 ppm in the reactor of parent species was also kept constant throughout all four sets of experiments. This was accomplished through the use of a syringe pump which injected a volume of each sample into the insertion region of the TDAS at a rate of 10.1 $\mu$L/sec. Since the concentration of all parent species was fixed at $8 \times 10^{-4}$ mol/L and the flowrate of nitrogen carrier into the reactor was identical for all compounds, the volumetric ratio of fuel:carrier flowing through the reactor was constant in all experiments.

Decomposition data were obtained for a 250-$\mu$L gas-phase sample of $CH_2Cl_2$, $CHCl_3$, and $CCl_4$ over the temperature range 300 to 1000°C for a mean residence time of 2.0 sec. A 500-μL gas-phase sample was required for the $CH_3Cl$ experiments to provide a total mass to the detector comparable to the other three compounds. However, it is important to note that this increase in sample size did not change the value of $\phi$ or the parts-per-million concentration of parent species exposed in the reactor.

The effluents resulting from thermal exposure were focused at the head of a DB-5, 30-m, 0.25-μm, 0.25-mm i.d. capillary column at a trapping temperature of -60°C, using liquid nitrogen as coolant. Separation of the effluents was attained by programming the GC as follows: -60°C (22-min hold) to 290°C (5-min hold) @ 10°C/min.

Mass spectral data were taken from the moment of injection for each of the samples in all experiments. This was done to enable "on the fly" monitoring of compounds which could not be cryogenically focused and separated by the DB-5 column used in these experiments. Data acquisition parameters were optimized for the first 31 min of the oven program to monitor for volatile species such as HCl, $CO_2$, $Cl_2$, $COCl_2$, and $C_1$– and $C_2$–chlorinated hydrocarbons. This included running the MSD in full-scan, E1 (70 eV) mode, 5.71 scans/sec, 1700 EM, from 35 to 110 amu. From 31 to 62 min of the GC temperature program, data were taken in full-scan mode as outlined earlier, except that the mass range scanned was 35 to 400 amu. All quantifications of parent species as well as products were accomplished via the running of standard calibration curves.

## RESULTS

Figures 2 to 5 present thermal decomposition profiles for $CH_3Cl$, $CH_2Cl_2$, $CHCl_3$, and $CCl_4$, respectively. Mole fraction of chlorinated species detected relative to the initial number of moles of parent injected (on a logarithmic scale) are plotted versus exposure temperature. The analytical technique was optimized to ensure that all chlorinated species could be detected including lower-molecular-weight species such as molecular chlorine ($Cl_2$), phosgene ($COCl_2$), hydrogen chloride (HCl), and any higher molecular weight species of m/e from 50 to 400. Any chlorine-containing species with molecular weights of less than 400 were detectable for yields of ~0.1%. Data were also obtained for carbon dioxide ($CO_2$); however, carbon monoxide (CO) could not be quantified and no attempt was made to analyze for $C_1$– and $C_2$–hydrocarbons.

Table 1 summarizes maximum product yields from the decomposition of chloromethanes. Percent recovery of C, H, and Cl in the product gases, averaged over at least four different temperatures, was as follows: $CH_3Cl$, 100.9% C, 78.8% H, 89.4% Cl; $CH_2Cl_2$, 67.7% C, 84.8% H, 84% Cl; $CHCl_3$ 84.6% C, 92.2% H, 70.3% Cl; and $CCl_4$ 76.9% C, 60.3% Cl.

Chlorine mass closure for the individual chloromethanes varied inversely with increasing chlorine substitution. This may be due to the expected increase in $Cl_2$ in the product yields as the parent compound becomes more highly

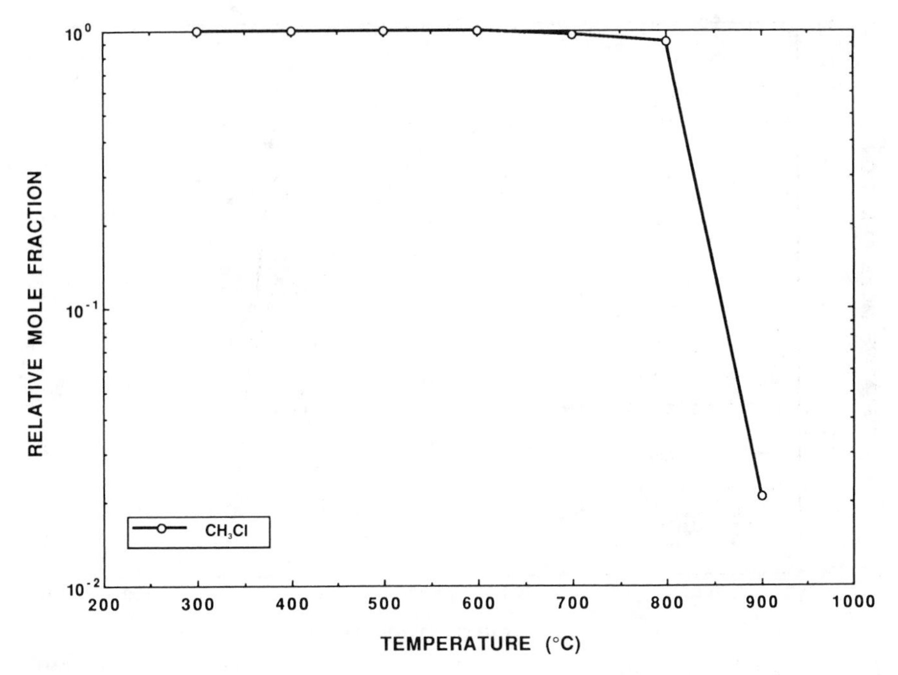


**Figure 2.** Thermal decomposition profile for $CH_3Cl$; $\phi = 3$, $[CH_3Cl] = 8 \times 10^{-4}$ mol/L, $\bar{t}_r = 2$ sec.

chlorinated. The large electronegativity of chlorine has a tendency to decrease its ionization efficiency in the ion source of the MSD, making it difficult to detect in the electron impact (EI) ionization mode. In the future, work will be done to more fully characterize the ionization behavior of chlorine in the EI mode with a greater variety of reactor concentrations so that better chlorine closures may be possible.

Table 2 summarizes the parent stability data for this study and other flow reactor studies that we have performed in pure nitrogen and air atmospheres.[18]

**Table 1. Maximum Product Yields from Chlorinated Methanes**

| | Parent Compounds | | | |
|---|---|---|---|---|
| **Products** | **$CH_3Cl$** | **$CH_2Cl_2$** | **$CHCl_3$** | **$CCl_4$** |
| 1,1-$C_2H_2Cl_2$ | | 0.016[a](750) | | |
| 1,2-$C_2H_2CL_2$ | | 0.024(750) | | |
| $C_2HCl_3$ | | 0.074(750) | 0.002(600)[b] | |
| $CCl_4$ | | | 0.101(650) | |
| $C_2Cl_4$ | | | 0.243(700) | 0.211(800) |
| $C_3Cl_4$ | | | 0.004(600) | |
| $C_2Cl_6$ | | | 0.063(600) | 0.011(600) |
| $C_4Cl_6$ | | | 0.002(700) | |
| $C_6Cl_6$ | | | 0.008(750) | 0.005(900) |

[a]Mole fraction of parent at temperature of maximum formation.
[b]Temperature of maximum formation (°C).

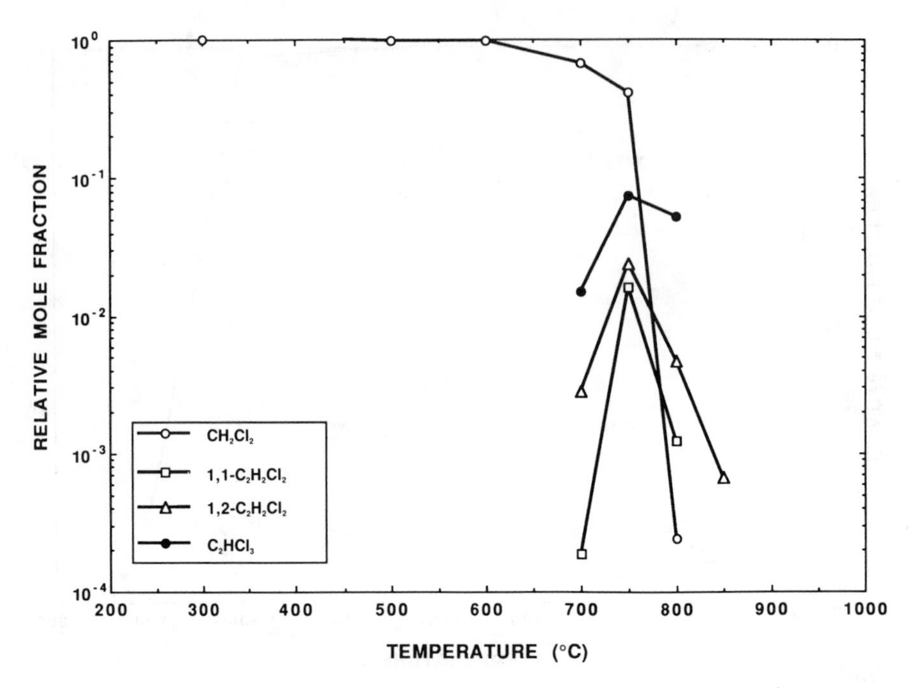


**Figure 3.** Thermal decomposition profile for $CH_2Cl_2$; $\phi$ = 3, $[CH_3Cl]$ = 8 × $10^{-4}$ mol/L, $\bar{t}_r$ = 2 sec.

The parent stability is defined by the temperature required for 99% destruction at $\bar{t}_r$ = 2 sec [$T_{99}$(2)(°C)]. $CHCl_3$ is consistently the most fragile of the series, with $CH_3Cl$ being the most stable. $CH_2Cl_2$ is the second most stable under oxidative conditions; however, for pyrolytic conditions, the curves of $CH_2Cl_2$ and $CCl_4$ cross, with methylene chloride being apparently more stable at temperatures <800°C and $CCl_4$ more stable at temperatures >800°C.[18] The crossing of these curves is unusual, occurring in less than 10% of the compounds we have studied.[24] As will be discussed later, the phenomenon is apparently related to the reformation of $CCl^4$ as a PIC under pyrolytic conditions.

It is interesting to note that no chlorinated organic PICs with molecular weights lower than that of the parent species were observed for any of the test

**Table 2. Thermal Stability Data for Chlorinated Methanes**

| | Pure Compound | | | Mixture[18] | | | |
|---|---|---|---|---|---|---|---|
| | | | | Cl/H = 1 | | Cl/H = 5 | |
| Compound | Nitrogen[18] | $\phi$ = 3[a] | Air[18] | $\phi$ = 0.05 | Nitrogen | $\phi$ = 0.05 | Nitrogen |
| $CH_3Cl$ | 1000 | 910 | 805 | 900 | 1100 | 830 | 985 |
| $CH_2Cl_2$ | 835 | 780 | 755 | 760 | 840 | 820 | 815 |
| $CHCl_3$ | 645 | 600 | 595 | 645 | 680 | 625 | 660 |
| $CCl_4$ | 940 | 880 | 750 | 790 | 705 | 780 | 930 |

[a]This study.

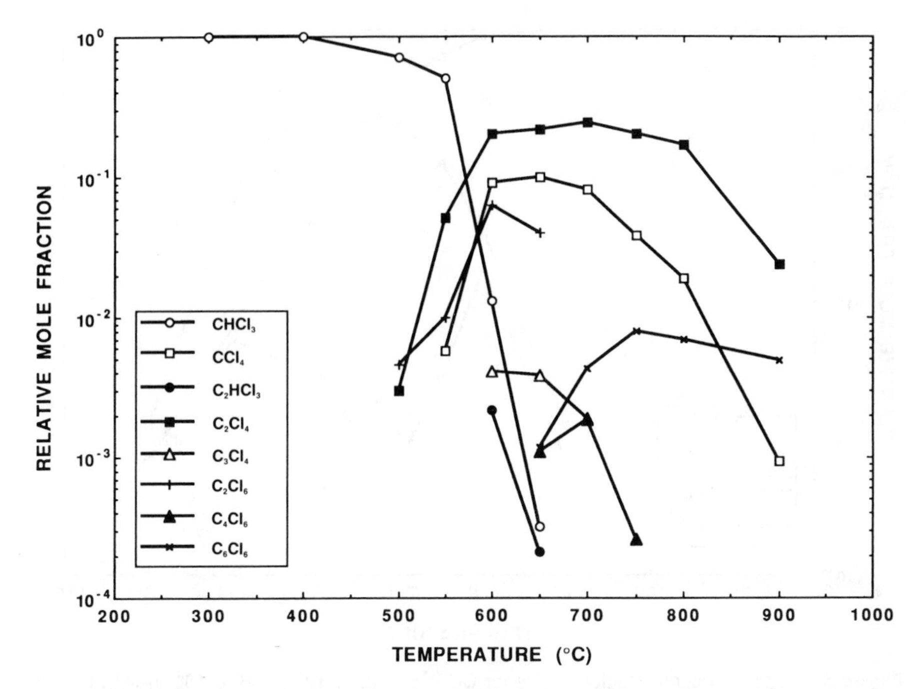


**Figure 4.** Thermal decomposition profile for $CHCl_3$; $\phi = 3$, $[CH_3Cl] = 8 \times 10^{-4}$ mol/L, $\bar{t}_r = 2$ sec.

compounds. In contrast, carbon tetrachloride and chloroform formed a number of higher molecular weight species, including significant yields of hexachlorobenzene ($C_6Cl_6$).

## DISCUSSION

### Parent Decomposition

The initiation of decomposition of the chlorinated methanes at $\phi = 3.0$ can, in principle, proceed by seven pathways: H abstraction by molecular oxygen ($O_2$); Cl abstraction by $O_2$; C-Cl bond rupture; C-H bond rupture; and concerted elimination of $H_2$, $Cl_2$, or HCl. In practice, only C-Cl bond rupture and HCl elimination (for chloroform and methylene chloride) need be considered due to the high endothermicity of the other pathways and the low $O_2$ concentration. Once decomposition has been initiated, radical-molecule reaction pathways dominate the actual destruction of the parent.

C-Cl bond rupture is the only energetically feasible pathway for initiation of destruction of carbon tetrachloride (BDE = 70.9 kcal/mol).[28] HCl elimination appears to be the preferred initiation path for chloroform (Ea = 54.5 kcal/mol, A = $2 \times 10^{14}$ $sec^{-1}$).[12] Methylene chloride initiation channels are domi-

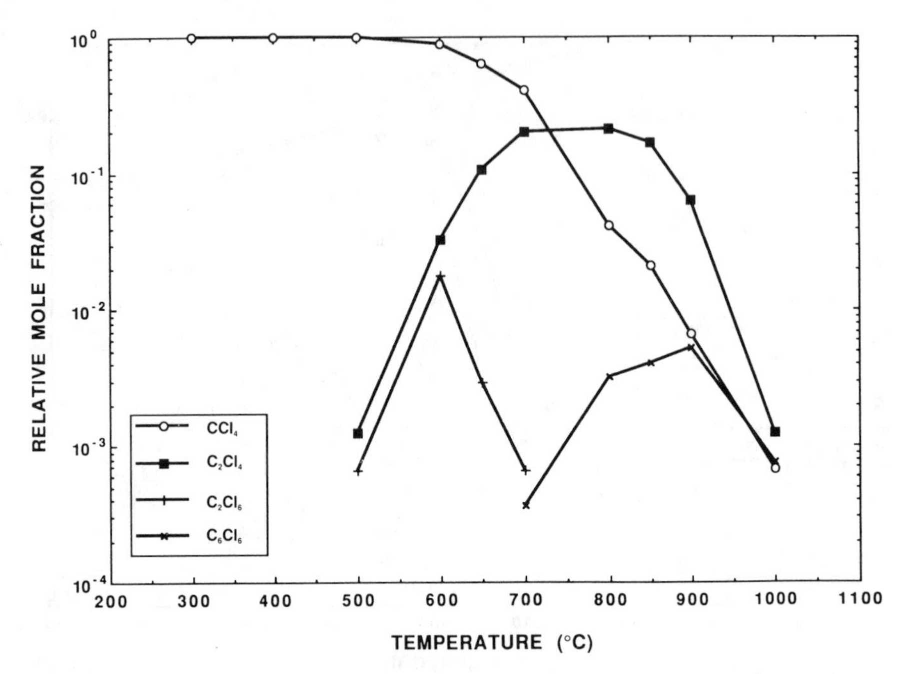


**Figure 5.** Thermal decomposition profile for $CCl_4$; $\phi = 3$, $[CH_3Cl] = 8 \times 10^{-4}$ mol/L, $\bar{t}_r = 2$ sec.

nated by HCl elimination (Ea ~70.0 kcal/mol) and C-Cl bond rupture.[5] C-Cl bond rupture is the only initiation pathway available for methyl chloride, its high relative stability being due to the strong C-Cl bond (Ea = 84.2 kcal/mol, $A = 2.5 \times 10^{15}$ $sec^{-1}$).[10,12]

Following C-Cl bond fission, H abstraction by Cl is a likely pathway for destruction of $CH_3Cl$, $CH_2Cl_2$, and $CHCl_3$.[10,13] H abstraction by various organic radicals and diradicals may also contribute to the destruction of $CH_3Cl$, $CH_2Cl_2$, and $CHCl_3$.[10,13] Hydroxyl radical (OH) may, in principle, contribute to hydrogen abstraction; however, equilibrium calculations at elevated temperatures for $\phi = 3$ show that OH concentrations are negligible for hydrogen-containing chloromethanes, and thus, reactions involving Cl atoms should dominate.[27]

Table 3 provides a comparison of the major proposed destruction channels for each of the chloromethanes. Arrhenius parameters for most of the reaction channels have not been experimentally measured. Thus, thermochemical principles and transition state theory have been employed to provide reasonable estimates.[9]

## Product Formation

As shown in Figure 2, no chlorinated organic products were observed from the decomposition of $CH_3Cl$. This observation is consistent with thermody-

**Table 3. Chloromethane Destruction Kinetics**

| Compound | log k<br>C-Cl Bond Rupture ($sec^{-1}$) | HCl Elimination ($sec^{-1}$) | H Abstraction (Cl) ($cm^3mol^{-1}sec^{-1}$) |
|---|---|---|---|
| $CH_3Cl$ | 15.40–84.2/2.3RT[10] | 14.0–102.0/2.3RT[10] | 13.5–3.1/2.3RT[10] |
| $CH_2Cl_2$ | 16.42–79.8/2.3RT[a] | 13.85–70.0/2.3RT[a] | 13.35–2.3/2.3RT[a] |
| $CHCl_3$ | 16.19–77.4/2.3RT[a] | 14.3–54.5/2.3RT[12] | 13.2–1.5/2.3RT[a] |
| $CCl_4$ | 16.30–70.6/2.3RT[a] | – | – |

[a]Estimated using thermochemical transition state theory as outlined in Reference 16.

namic predictions and the recent study of Weissman and Benson.[10] The major organic products observed in their study were ethylene ($C_2H_4$) and acetylene ($C_2H_2$), which likely formed via concerted HCl elimination from unstable intermediate products [e.g., ethyl chloride ($C_2H_5Cl$) and vinyl chloride ($C_2H_3Cl$)]. None of these latter products were observed in this study and only trace quantities were observed by Weissman and Benson.[10]

For $CH_2Cl_2$, as illustrated in Figure 3, three CHCs were observed in addition to HCl. At a temperature of 750°C, trichloroethylene ($C_2HCl_3$) was observed in approximately 7% yields with 1,1 dichloroethylene (1,1-$C_2H_2Cl_2$) and 1,2 dichloroethylene (1,2 $C_2H_2Cl_2$) observed at approximately 1.6 and 2.4% yields, respectively. All three products were relatively fragile, decomposing below detectable limits at temperatures greater than 850°C.

If HCl elimination were the dominant degradation pathway for $CH_2Cl_2$, one would expect large quantities of 1,2-$C_2H_2Cl_2$ with smaller quantities of 1,1-$C_2H_2Cl_2$, as predicted by the following sequence of reactions (observed products are underlined):

$$CH_2Cl_2 + M = CHCl + \underline{HCl} + M \tag{1}$$

$$2\ CHCl + M = \underline{1,2\text{-}C_2H_2Cl_2} + M \tag{2}$$

$$CHCl + CH_2Cl_2 + M \rightarrow 1,2,2\text{-}C_2H_3Cl_3 + M \tag{3}$$

$$1,2,2\text{-}C_2H_3Cl_3 \rightarrow \underline{1,2\text{-}C_2H_2Cl_2}(\text{major}) + \underline{1,1\text{-}C_2H_2Cl_2}(\text{minor}) + \underline{HCl} \tag{4}$$

C-Cl bond rupture followed by hydrogen abstraction can also lead to the formation of the observed dichloroethenes and $C_2HCl_3$ through the following series of reactions:

$$CH_2Cl_2 + M = CH_2Cl + Cl + M \tag{5}$$

$$CH_2Cl_2 + Cl \rightarrow CHCl_2 + \underline{HCl} \tag{6}$$

$$CH_2Cl + CHCl_2 + M = 1,2,2\text{-}C_2H_3Cl_3 + M \tag{7}$$

$$2CHCl_2 + M = 1,1,2,2\text{-}C_2HCl_4 + M \tag{8}$$

$$1,2,2\text{-}C_2H_3Cl_3 \rightarrow \underline{1,2\text{-}C_2H_2Cl_2}(\text{major}) + \underline{1,1\text{-}C_2H_2Cl_2}(\text{minor}) + \underline{HCl} \quad (4)$$

$$1,1,2,2\text{-}C_2H_2Cl_4 \rightarrow \underline{C_2HCl_3} + \underline{HCl} \quad (9)$$

The formation of $C_2HCl_3$ occurs through the recombination of $CHCl_2$ radicals and subsequent HCl elimination.

One can also include pathways for H abstraction from $CH_2Cl_2$ by $CH_2Cl$ radicals and CHCl diradicals. The $CH_2Cl$ pathway produces $CHCl_2$ radicals which subsequently recombine (and eliminate HCl) to form $C_2HCl_3$. In a similar fashion, the CHCl diradical pathway results in the formation of both $CH_2Cl$ and $CHCl_2$ radicals which subsequently recombine (and eliminate HCl) to form $C_2H_3Cl$, $C_2H_2Cl_2$, and $C_2HCl_3$. Although $C_2H_3Cl$ was not observed, we cannot eliminate the CHCl pathway because the $C_2H_3Cl$ steady-state concentration may be too low to be observable due to its apparent instability (i.e., it was not observed from $CH_3Cl$ although it should have been formed).[29]

Each of the hydrogen-containing olefins apparently decompose rapidly to an undetectable lower molecular weight specie. This is in contrast to the reactions resulting from $CHCl_3$ and $CCl_4$ decomposition which produces higher molecular weight species including $C_6Cl^6$. The increase in the number of products observed implies that a relatively complex series of reactions may be occurring for these two chlorinated methanes.

The following reactions may be responsible for $CHCl_3$ destruction (see Figure 4) which leads to the formation of key stable intermediate PICs including tetrachloroethylene ($C_2Cl_4$), $C_2HCl_3$, hexachloroethane ($C_2Cl_6$), and $CCl_4$:

$$CHCl_3 + M \rightarrow CCl_2 + \underline{HCl} + M \quad (10)$$

$$CHCl_3 + M = CHCl_2 + Cl + M \quad (11)$$

$$CHCl_3 + Cl \rightarrow CCl_3 + \underline{HCl} \quad (12)$$

$$2\ CCl_2 + M = \underline{C_2Cl_4} + M \quad (13)$$

$$CCl_2 + CHCl_3 \rightarrow \underline{C_2HCl_5} \quad (14)$$

$$\underline{C_2HCl_5} \rightarrow \underline{C_2Cl_4} + HCl \quad (15)$$

$$\underline{C_2HCl_5} + Cl \rightarrow C_2Cl_5 + \underline{HCl} \quad (16)$$

$$C_2Cl_5 \rightarrow \underline{C_2Cl_4} + Cl \quad (17)$$

$$CCl_3 + Cl + M = \underline{CCl_4} + M \quad (18)$$

$$2\ CCl_3 + M = \underline{C_2Cl_6} + M \tag{19}$$

$$2\ CHCl_2 + M = 1,1,2,2\text{-}C_2H_2Cl_4 + M \tag{8}$$

$$1,1,2,2\text{-}C_2H_2Cl_4 \rightarrow \underline{C_2HCl_3} + HCl \tag{9}$$

$$CCl_3 + CHCl_2 + M = \underline{C_2HCl_5} + M \tag{20}$$

$$\underline{C_2HCl_5} + M = C_2HCl_4 + Cl + M \tag{21}$$

$$C_2HCl_4 = \underline{C_2HCl_3} + Cl \tag{22}$$

Product analysis indicates that H abstraction is clearly a contributing pathway as recombination of trichloromethyl ($CCl_3$) radicals is the only probable pathway to $C_2Cl_6$ formation. One could also include paths for hydrogen abstraction from $CHCl_3$ involving $CHCl_2$ and $CCl_2$. These reactions would result in the formation of $CCl_3$ radicals for the former and $CCl_3$ and $CHCl_2$ radicals for the latter. Of the remaining $CHCl_3$ destruction pathways, kinetic calculations indicate that HCl elimination is several orders of magnitude faster than C-Cl bond fission at 600°C. The factor of 100 greater yields of $C_2Cl_4$ as compared to $C_2HCl_3$ at this temperature further demonstrates the dominance of HCl elimination from $CHCl_3$ (Equations 10, 13-17). Although $C_2HCl_3$ may form from decomposition of $C_2HCl_5$ at high temperatures through Equations 21 and 22, this pathway should be minor versus its formation at 600°C from Equations 8 and 9. The abstraction of H atoms from $CHCl_3$ and $CCl_2$ can lead to the formation of $C_2HCl_3$ by recombination of resulting $CHCl_2$ radicals and subsequent HCl elimination. If these types of diradical abstraction reactions are sufficiently rapid, this pathway may dominate over the C-Cl bond rupture path for the formation of the observed traces of $C_2HCl_3$. Although the recombination of $CCl_2$ via Equation 13 for $C_2Cl_4$ seems more conventional than the bond insertion mechanism via Equations 14 and 15, Shilov and Sabirova[6] have shown that the insertion reaction may dominate. Benson and Weissman[30] have proposed that C-Cl bond fission is the most important initiation step for $C_2HCl_5$. This suggests that $C_2Cl_4$ production from $C_2HCl_5$ must be a two-step process involving H abstraction by Cl atoms followed by C-Cl bond rupture (Equations 16 and 17).

Once formed, $C_2HCl_3$ may decompose via HCl elimination, which has been previously reported for the olefin, vinyl chloride:[29]

$$\underline{C_2HCl_3} \rightarrow C_2Cl_2 + \underline{HCl} \tag{23}$$

Dichloroacetylene ($C_2Cl_2$) has been previously observed from $C_2HCl_3$ pyrolysis.[18] $C_2Cl_6$ may decompose by C-C bond rupture (BDE = 71.9 kcal/mol),[28] which is the reverse of Equation 19. $CCl_3$ radicals may then react with $C_2Cl_2$ to produce tetrachloropropyne ($C_3Cl_4$):

$$CCl_3 + C_2Cl_2 \rightarrow \underline{C_3Cl_4} + Cl \tag{24}$$

This chlorine displacement reaction is significant in that it illustrates how higher molecular products may be formed. Recent studies have shown that addition reactions followed by the elimination of H atoms may be important reaction channels in hydrocarbon combustion at high temperatures.[31-33] The presence of C-Cl bonds with their lower C-Cl bond dissociation energies and more favorable equilibrium facilitates condensation-type reactions in CHC systems.

Similarly, displacement reactions involving trichlorovinyl ($C_2Cl_3$) radicals formed from:

$$\underline{C_2HCl_3} + Cl \rightarrow C_2Cl_3 + \underline{HCl} \tag{25}$$

can result in the formation of hexachlorobutadiene ($C_4Cl_6$):

$$C_2Cl_3 + \underline{C_2Cl_4} \rightarrow \underline{C_4Cl_6} + Cl \tag{26}$$

Previous studies have shown that $C_4Cl_6$ rapidly decomposes via C-Cl bond rupture to form the resonance-stabilized pentachlorobutadienyl ($C_4Cl_5$) radical:[24]

$$\underline{C_4Cl_6} = C_4Cl_5 + Cl \tag{27}$$

This radical can then react with $C_2Cl_2$ via Cl displacement to produce $C_6Cl_6$:

$$C_4Cl_5 + C_2Cl_2 \rightarrow \underline{C_6Cl_6} + Cl \tag{28}$$

We favor a concerted reaction involving cyclization and Cl displacement to form $C_6Cl_6$ although other researchers have proposed separate addition, cyclization, and H elimination reactions for the formation of benzene from butadienyl radical and acetylene.[34] The relatively high yields of $C_2Cl_4$ (24%) and $C_6Cl_6$ (0.8%) are striking and were also observed from $CCl_4$ pyrolysis.

$CCl_4$ must initially decompose (see Figure 5) via C-Cl bond rupture, which is the reverse of Equation 18. $CCl_3$ radicals then recombine to yield $C_2Cl_6$ (via Equation 19). Following $C_2Cl_6$ production, two parallel pathways apparently occur. $C_2Cl_6$ may undergo C-Cl bond rupture followed by $\beta$-bond homolysis to yield $C_2Cl_4$:

$$\underline{C_2Cl_6} + M = C_2Cl_5 + Cl + M \tag{29}$$

$$C_2Cl_5 = \underline{C_2Cl_4} + Cl \qquad (30)$$

$C_2Cl_6$ may also undergo C-C bond rupture via the reverse of Equation 19 followed by recombination of Cl (Equation 18) to reform $CCl_4$. Based on unimolecular decomposition via C-Cl bond rupture with an estimated A factor of $2 \times 10^{16}$ $sec^{-1}$ and Ea ~70 kcal/mol, $CCl_4$ should have a $T_{99}(2)$ of ~690°C. The thermal decomposition profile in Figure 5 clearly shows evidence of reformation at temperatures greater than 700°C with $CCl_4$ exhibiting a $T_{99}(2)$ of 880°C. Of the four chlorinated methanes studied, this self-inhibiting reaction was evident only for $CCl_4$.

$C_2Cl_4$ produced from $CCl_4$ pyrolysis may decompose to form $C_2Cl_2$ by successive loss of Cl:

$$\underline{C_2Cl_4} + M = C_2Cl_3 + Cl + M \qquad (31)$$

$$C_2Cl_3 = C_2Cl_2 + Cl \qquad (32)$$

$C_2Cl_3$, $C_2Cl_4$, and $C_2Cl_2$ may then react as in Equations 26 to 28 to form $C_6Cl_6$. The intermediates $C_3Cl_4$, $C_2Cl_2$, and $C_4Cl_6$ were not observed in the formation of $C_6Cl_6$ from $CCl_4$ as they were from $CHCl_3$. This is most likely due to the greater thermal stability of $CCl_4$ and its reformation through $C_2Cl_6$, which ties up the necessary species until higher temperatures (>800°C). At these higher temperatures, the reactions resulting in the disappearance of $C_3Cl_4$, $C_2Cl_4$, and $C_4Cl_6$ may sufficiently increase versus their rate of formation, such that the less stable intermediates have steady-state concentrations which are not observable. The yields of $C_6Cl_6$ were lower by ~60% from $CCl_4$ as compared to $CHCl_3$, consistent with the earlier analysis.

## SUMMARY

The most interesting aspect of the study is the high yields of formation of the chlorinated aromatic, $C_6Cl_6$, from the higher chlorinated methanes, $CHCl_3$ and $CCl_4$. We have proposed a simple mechanism involving the formation of key intermediate species $C_2HCl_3$, $C_2Cl_4$, and $C_2Cl_2$ which is consistent with the observed product distribution. The unchlorinated homologs of these species have also been proposed as key intermediates in soot formation from hydrocarbons and $CH_3Cl$.[10,34,35] However, based on the absence of production of chlorinated aromatics from $CH_3Cl$ and $CH_2Cl_2$ as compared with $CHCl_3$ and $CCl_4$, it appears that chlorine facilitates condensation reactions resulting in the formation of higher molecular weight chlorinated species. We propose that the displacement of Cl from olefinic and acetylenic species by other olefinic species is favored for CHCs as compared with the direct addition (followed by hydrogen loss) for hydrocarbons.[10,34,35] This is responsible for the high yields of chlorinated aromatics for the chlorinated CHCs. Furthermore, the highly chlorinated species are more resistant to oxidative degradation pathways and therefore more available for condensation-type reactions.

Frenklach et al.[13] have observed high yields of soot formation from $CH_2Cl_2$ at initially higher concentrations. They have proposed mechanisms involving HCl elimination producing the CHCl biradical along with C-Cl bond fission forming $CH_2Cl$. Recombination of these radicals forms vinyl chloride ($C_2H_3Cl$), which upon HCl elimination produces $C_2H_2$. $C_2H_2$ may then react with CHCl to form $C_3H_3$, which has been proposed as an active soot precursor.[36] Frenklach et al.[13] have also proposed the traditional $C_2$ mechanism[36] for soot formation from higher chlorinated $C_1$– and $C_2$–CHCs which were observed to soot at higher temperatures.

Although our work does not directly address soot formation, the formation of aromatics has been proposed as an intermediate step in the process. Thus, understanding the mechanism of their formation may be valuable in understanding soot formation. It would be most interesting to obtain values for the chlorine content of soot from CHC combustion.

The study of the key intermediate $C_2$–CHCs is clearly indicated as our theories would suggest that chlorinated aromatics can be directly formed from $C_2Cl_4$ and $C_2Cl_6$. Also of key interest is the pathway of $C_2HCl_3$ degradation (i.e., HCl elimination versus hydrogen abstraction) and the yields of $C_2Cl_2$. Both experimental and theoretical work are indicated to determine if the radical olefin (acetylene) addition reactions can occur at high temperatures or if the displacement-type reactions that we have proposed are necessary.

Finally, we comment on the apparent reformation of $CCl_4$ and its resulting stability. In a more complex system with hydrocarbon radicals present, the reformation may not be as significant. Yet, $CCl_4$ is commonly observed as an emission from the incineration of CHCs[37] although we have previously shown it to be kinetically fragile.[19] We currently believe that the $CCl_3$ radical may be a kinetically and thermodynamically stable radical which may exit high temperature zones of incinerators in relatively high concentration. In the postflame zone, it may then undergo recombination reactions with available Cl and H atoms to form $CCl_4$ and $CHCl_3$ in concentration beyond that predicted by simple decomposition kinetics.[38]

We are continuing our experimental studies of the inert and oxidative pyrolysis of CHCs and are combining these studies with detailed kinetic computations. This will hopefully produce a self-consistent chemical kinetic mechanism for each species studied over a wide range of fuel:oxygen equivalence ratios.

## ACKNOWLEDGMENTS

We gratefully acknowledge J. Pan for performing the chlorinated methane equilibrium calculations.

## CREDIT

The research was supported in part by the U.S. Environmental Protection Agency under Cooperative Agreement CR-813938.

## REFERENCES

1. Oppelt, E. T. *Environ. Sci. Technol.* 20:312 (1986).
2. Shilov, A. E., and R. D. Sabirova. *Rus. J. Phys. Chem. Abst.* 33:30 (1959).
3. Shilov, A. E. *Dokl. Acad. Nauk. SSR* 98:601 (1954).
4. Hoare, M. R., R. G. W. Norrsh, and G. Whittingham. *Proc. Roy. Soc.* 250(A):180 (1959).
5. Santacesaria, E., A. Morini, and S. Carrla. *La Chimica E. L'Industria* 56:747 (1974).
6. Shilov, A. E., and R. D. Sabirova. *Rus. J. Phys. Chem.* 34:408 (1960).
7. Semeluk, G. P., and R. B. Bernstein. *J. Am. Chem. Soc.* 75:3793 (1954).
8. Semeluk, G. P., and R. B. Bernstein. *J. Am. Chem. Soc.*, 79:46 (1957).
9. Benson, S. W. *Thermochemical Kinetics*, 2nd ed. (New York: John Wiley & Sons, Inc., 1976), p. 110.
10. Weissman, M., and S. W. Benson. *Int. J. Chem. Kinetics* 16:307 (1984).
11. Kondo, O., K. Saito, and I. Murakami. *Bull. Chem. Soc. Jpn.* 53:2133 (1980).
12. Schug, K. P., H. G. Wagner, and F. Zabel. *Ber. Bensenges Phys. Chem.* 83:167 (1979).
13. Frenklach, M., J. P. Hsu, D. L. Miller, and R. A. Matula. *Combust. Flame* 64:141 (1986).
14. Miller, D. L., D. W. Senser, V. A. Cundy, and R. A. Matula. *Haz. Waste* 1(1):1 (1984).
15. Senser, D. W., J. S. Morse, and V. A. Cundy. *Haz. Waste Haz. Mat.* 2(4):473 (1985).
16. Senser, D. W., V. A. Cundy, and J. S. Morse. *Comb. Sci. Tech.* 51:209 (1987).
17. Granada, A., S. B. Karra, and S. M. Senkan. *Ind. Chem. Res.* 26(9):1901 (1987).
18. Taylor, P. H., and B. Dellinger. *Environ. Sci. Technol.* 22:438 (1988).
19. Dellinger, B., D. L. Torres, W. A. Rubey, D. L. Hall, and J. L. Graham. *Haz. Waste* 1:137 (1984).
20. Taylor, P. H., J. J. D'Angelo, M. C. Martin, J. H. Kasner, and B. Dellinger. *Int. J. Chem. Kinetics* 21:829(1989).
21. Dellinger, B., M. D. Graham, and D. A. Tirey. *Haz. Waste Haz. Mat.* 3:293 (1986).
22. Dellinger, B., D. L. Hall, J. L. Graham, S. L. Mazer, W. A. Rubey, and P. H. Taylor. Paper 86-24.7, Proceedings of the 79th Annual Meeting of the Air Pollution Control Association, Minneapolis, MN, June 1986.
23. Graham, M. D., P. H. Taylor, and B. Dellinger. *Proceedings of the Eastern State Section-Combustion Institute Meeting* (San Juan, PR: 1986), 1–57.
24. Taylor, P. H. and B. Dellinger. Proceedings of the International Flame Research Committee Symposium on the Incineration of Hazardous, Municipal, and other Wastes, Palm Springs, CA, November 1987.
25. Rubey, W. A., and R. A. Carnes. *Rev. Sci. Ins.* 56:1795 (1986).

26. Rubey, W. A. "Design Considerations for a Thermal Decomposition Analytical System," US-EPA Report, EPA-60012-80-098 (1980).
27. Reynolds, W. C. STANJAN Equilibrium Program, Version 3.0, Department of Mechanical Engineering, Stanford University, Stanford, CA (1986).
28. McMillen, A. F., and D. M. Golden. *Ann. Res. Phys. Chem.* 33:493 (1982).
29. Zabel, F., *Int. J. Chem. Kinetics* 9:651 (1977).
30. Benson, S. W., and M. Weissman. *Int. J. Chem. Kinetics* 14:1287 (1982).
31. Dean, A. M., *J. Chem. Phys.*, 89, 4600 (1985).
32. Westmoreland, P. R., J. B. Howard, J. P. Longwell, and A. M. Dean. *Am. Inst. Chem. Eng. J.* 32:1971 (1986).
33. Dean, A. M., and P. R. Westmoreland. *Int. J. Chem. Kinetics* 19:207 (1987).
34. Cole, J. A., J. D. Bittner, J. P. Longwell, and J. B. Howard, *Combust. Flame* 56:51 (1984).
35. Homann, K. H., J. Warnatz, and C. Wellnan. "16th Symposium International on Combustion," The Combustion Institute, (1977), 853.
36. Palmer, H. B., and C. F. Cullis. *Chemistry and Physics of Carbon*, P. L. Walker, Jr., Ed. (New York: Marcel Dekker, Inc., 1965), 265.
37. Midwest Research Institute, "Performance Evaluation of Full-Scale Hazardous Waste Incinerators," US-EPA, HWERL Report, EPA-600/2-84-181 (1984).
38. Dellinger, B., P. H. Taylor, D. A. Tirey, J. C. Pan, and C. C. Lee. "14th Research Symposium on Land Disposal, Remedial Action, Incineration, and Treatment of Hazardous Waste," US-EPA, Washington, DC, 289.

# CHAPTER 9

# Importance of Data Quality for Understanding and Controlling Emissions

**John K. Taylor**

## INTRODUCTION

Measurement data provides the information used for most decisions concerned with understanding and controlling emissions from hazardous sources. Research studies of formation and destruction mechanisms and the environmental fate of emitted materials are critically dependent on reliable measurement data. Environmental regulations based on risk assessment are only credible as the database for decisions is sound. Emission inventories, based on mass balance considerations, are validated by good data and questionable when the data quality is questioned.

Abatement technology depends on data quality in several ways. Engineering design obviously requires high-quality physical and mechanical data. While not a subject of the present discussion, the principles related to acquiring quality data of this kind parallel the ones described later. Moreover, the testing of equipment during design and construction and the evaluation of its performance are only as reliable as the databases used. Monitoring of sources for compliance with regulations requires data of known and adequate quality obtained in well-designed and properly executed measurement programs. The safety of the public and equitable and just regulatory decisions are at stake whenever poor data or data of unknown quality are used.

## WHAT IS DATA QUALITY?

The concept of the need and requirement for quality data is not difficult to establish, but the definition of what constitutes quality data may not be uniformly understood. The following discussion presents the background for establishing data quality criteria.

The quality of data is described in part by its accuracy. The *qualitative identification* of what is measured must be known with confidence approaching "certainty." The *quantitative accuracy* of measured values always has some degree of uncertainty, but this can be estimated when a properly designed and operated measurement system is used (more about this later)[1] and statistically based limits can be established.

Three additional indicators of data quality are becoming generally recognized:[2]

- *Completeness*—A measure of the amount of data obtained as compared with the amount expected in a measurement program.
- *Representativeness*—The degree to which the data represent a characteristic of a population parameter, variation of a property, a process characteristic, or an operational condition.
- *Comparability*—The confidence with which one data set may be compared or used in connection with another.

## DATA QUALITY OBJECTIVES

Before any data are obtained, the requirements that it must meet to be useful should be determined. This has given rise to the concept of *data quality objectives* (DQOs). DQOs are best developed as a collaborative effort of experts in the discipline in which the data are to be used, statisticians knowledgeable in experimental design, and measurement experts. They must make cost-effective decisions on what quality is desirable and what is practically attainable. The measurement process must be designed and operated so that the attained accuracy approaches the attainable accuracy as closely as practical. A good rule of thumb is that the attainable accuracy should be better than the DQO by a factor of at least 3. If the attained accuracy is less than the DQO, no decisions on compliance can be made, based on single measurements, unless gross violations are involved. When the DQO just equals the attainable accuracy, multiple measurements are required for credible decisions. When the attained accuracy is considerably greater than the DQO, single measurements can be utilized for decision purposes.

## THE MEANING OF ACCURACY

Accuracy is based on the closeness of a measured result to the true value of the parameter. Precision is defined as the closeness of individual repetitive results. Inaccuracy results from two sources. Data will scatter due to imprecision and the results of measurement involving small data sets, and especially for single measurements, can be grossly inaccurate due to this cause. Also, the measurement process may involve systematic errors producing bias. In such a case, all results depart from the true value by the amount of bias present. The attainable accuracy obviously depends on the imprecision *and* bias of the measurement process. The precision of a measurement process is measured by its standard deviation. Bias is evaluated by measuring reference materials and/or by experimentally comparing the results of a measurement process with those of a reference method. In the absence of these, spiked samples may be used. Bias often may be eliminated or minimized by suitable experimental procedures (calibrations and blank corrections, for example) and this should be done to the extent feasible. Any uncertainty in bias correction becomes a systematic uncertainty that must be considered along with measurement imprecision. In brief, random components are added in quadrature and systematic components are added algebraically and the total uncertainty is considered to be the sum of the contributions from these two sources. This is discussed in some detail in References 1, 5, and 6, as well as in other places.

## THE IMPORTANCE OF ACCURACY

Accuracy is correctness. Inaccuracy can lead to misunderstanding, faulty control, and uncertainty of compliance. In the research area, inaccurate measurements can lead to false scientific conclusions. Accuracy is of greatest importance when questions of the attainment of specified values are of concern. Biased results could result in false confidence in apparent attainment when the true value of the parameter is actually higher or lower than the measured one due to unsuspected measurement bias. Also, false decisions on violations could be made due to bias existing unknowingly in a measurement process.

In the case of comparisons of results made by different analysts, laboratories, or by regulators and regulatees, accuracy is of utmost importance. The statistical significance of differences can be easily decided, using precision considerations, and this should always be the first step in any dispute. When significant differences exist, the question of which value, if either, is unbiased must be resolved.

To forestall questions of significant differences, the accuracy, i.e., both the precision and bias, must be known. The question of comparability of data must be anticipated in measurement programs and appropriate procedures should be adopted to minimize, if not eliminate, any questions arising from this source. This has not always been the case and many data compilations are confounded by questions as to how one set agrees with other sets and whether the data from various sources can be combined in any logical manner.

## THE IMPORTANCE OF PRECISION

Adequate precision is important from several points of view. It must be sufficient to minimize the number of replicate results required for a given decision. Also, the imprecision must be smaller than any level of bias which, if present, is considered to be of importance. It is difficult, costly, and time consuming to identify bias of magnitude equal to the precision and infeasible to look for it when it exceeds the imprecision of measurement.[1]

Adequate precision is necessary to distinguish differences such as those existing between samples, for example. The sensitivity of a measurement process depends on the precision and is given by the expression:

$$\Delta = z\, s_c \sqrt{2}$$

where $\Delta$ is the minimum distinguishable difference of two measured values at the level of significance selected, z is the standard normal variate at that level (e.g., 1.96 for a 95% level of confidence), and $s_c$ is the standard deviation of the measurement process at concentration level c.[1] Thus, for a 95% level of confidence:

$$\Delta \approx 3\, s_c$$

Adequate precision is also of concern when questions of detection are of importance. Detection may be considered as the decision on whether a measurement result is significantly different from that which would be expected from a sample containing no analyte of concern. The situation is similar to the one discussed earlier (i.e., the significance of a small measured value at or near zero level). The expression to evaluate detection is the same, except that the value for the standard deviation is that estimated for the zero level of concentration, namely $s_O$. The method detection level (MDL) for 95% confidence is then:[1,3,4]

$$\text{MDL} \approx 3\, s_O$$

Participants in a measurement program must attain peer precision if their detection limits are to be comparable. Unfortunately, estimates of standard deviations, s, have considerable uncertainty if based on a few measurements. Thus, 45 measurements are required to establish a value of s within 20% of its true value with a 95% confidence.[6] Based on 10 measurements, there is a 95% confidence that the estimate is only within 45% of its true value. The need to evaluate detection limits (and their maintenance) using well-established control charts, rather than sporadic estimations using a few (even as many as seven) measurements, should be evident.

Adequate precision is also required when unbiased data sets are combined, the only case in which it is logical to do so. It is obvious that data of differing quality does not enjoy peer status and should not be weighted equally when used to compute such values as grand averages, for example. When averaging data sets, it is recommended[1,5] that the means of sets be weighted inversely as their variances (i.e., $W_i = 1/\sigma^2_i$). In the weighting process, s, the best estimate of $\sigma$ is used.

It is obvious from what has been said earlier that participants in a measurement program must have peer status with respect to attained precision if the data of all is to be given equal weight. Anything less not only confuses its statistical analysis, but is also wasteful when the question of weighting is considered.

## THE ATTAINMENT OF ACCURACY AND PRECISION

It should be obvious that it is of utmost importance for individual scientists, laboratories, and participants in a measurement program to attain requisite precision and accuracy of their measurements. Individuals and laboratories need to be peer performers to maintain a competitive position in the scientific and technological community. Programs must demand and receive peer performance of participants if data is to achieve its full potential.

Appropriate quality assurance programs are becoming generally recognized as providing the best approach for the generation of evaluated data.[1] Individual scientists must have adequate skills, the desire to produce reliable data, and follow good laboratory practices (GLPs), good measurement practices (GMPs), and standard operations procedures (SOPs). All measurements should be made according to quality assurance plans designed for the specific analytical situation (PSPs). These will range from quality assurance program plans (QAPPs) and quality assurance project plans (QAPjPs) for a program and its projects to quality assurance task plans (QATPs) for specific tasks within a project. The individual analyses made by a laboratory, even though not necessarily a part of a project or program, require QATPs as well. It is well established that laboratories and scientists who regularly follow quality assurance programs produce better data than those who do not do so. Such data is likely to be sound and defensible. Moreover, its limits of uncertainty can be evaluated and stated in statistical terms.

Managers of measurement programs have the responsibility to design appropriate quality assurance programs and to monitor them adequately.[7] This objective is best achieved by joint involvement of participants at all levels and by peer review of proposed plans by experts in the subject area of an investigation, together with statisticians and analysts.

## MEANINGFUL DATA

As important as it is, quality assurance in itself does not assure that measurement data is meaningful. Analysis must be considered as a system requiring:[1]

- a correct model
- a faultless plan
- relevant samples
- appropriate methodology
- adequate calibrations
- quality assurance

Statistical design is an important adjunct of planning, sampling, measuring, calibration, and data evaluation. But above all, the model must be correct or else the best of data will have limited, if any, utility. Peer review is an essential element of the entire measurement process, but it is especially important to attain concurrence that the model is correct. Unless there is substantial agreement that the right data is obtained, there can be little agreement on the significance of the data.

## REFERENCES

1. Taylor, J. K. *Quality Assurance of Chemical Measurements*, (Chelsea, MI: Lewis Publishers, Inc., 1987).
2. Stanley, T. W., and S. S. Verner. "The U.S. Environmental Protection Agency's Quality Assurance Program," in *Quality Assurance for Environmental Measurements*, ASTM STP 867, J. K. Taylor and T. W. Stanley, Eds. (Philadelphia, PA: ASTM, 1985), pp. 12–19.
3. Glasser, J. A., et al. "Trace Analysis for Waste Waters-Method Detection Limit," *Environ. Sci. Technol.* 15:1426–35 (1981).
4. American Chemical Society, "Principles of Environmental Analysis," *Anal. Chem.* 55:2210–18 (1983).
5. Ku, H. H., "Statistical Concepts in Metrology," in *Precision Measurement and Calibration*, *NBS Special Publication 300, Vol. 1*, H. H. Ku, Ed. (Gaithersburg, MD: National Institute of Standards and Technology).
6. Natrella, M. G. *Experimental Statistics*, *NBS Handbook 91* (Gaithersburg, MD: National Institute of Standards and Technology, 1963).
7. Taylor, J. K. "Quality Assurance for a Measurement Program," in *Environmental Sampling for Hazardous Wastes, ACS Symposium Series 267*, G. E. Schweitzer and J. A. Santolucito, Eds. (Washington, DC: American Chemical Society, 1985), pp. 105–108.

## CHAPTER 10

# Quality Assurance Recommendations for Toxic Emission Measurement Systems

**Darryl J. von Lehmden**

## INTRODUCTION

The U.S. Environmental Protection Agency (EPA) Quality Assurance Handbook for Air Pollution Measurement Systems (Volume 1 – Principles)[1] recommends 23 elements to consider in planning quality assurance for a measurement system. Three of the elements have the greatest impact on the quality of results from toxic emission measurements systems. These elements are (1) use validated test methods, (2) use high-quality calibration standards, and (3) conduct periodic audits on the toxic measurement system.

### Test Methods

Wherever possible, the test methods selected for sampling and analysis should be validated under conditions similar to those planned for toxic emission testing. Bias and precision for the test method selected should be known. The EPA has validated many of the toxic emission measurement methods commonly used today. Examples of methods that have been validated include (1) the Volatile Organic Sampling Train (VOST)[2] for the measurement of ppb volatile compound emissions during RCRA hazardous waste incineration trial burn tests, (2) the semi-VOST method[3] for the measurement of ppb semivolatile compound emissions during RCRA trial burn tests, and (3) the toxic metals train[4] intended for testing at hazardous water incinerators and municipal solid waste combustors. The selection of a validated method means the method has been laboratory and field tested to determine optimum operating parameters and identify interferences. In addition, method bias and precision under actual testing conditions is a standard part of a validated method.

### Calibration Standards

The quality of toxic emission measurements is no better than the quality of the standards used to calibrate the sampling and analytical measurement equipment. Toxic sampling and analytical equipment should be calibrated with the highest accuracy standards currently available. This means the National Institute of Standards and Technology (NIST) standard reference materials (SRMs) should be used, where available, to establish and provide a traceable concentration or value for the calibration standards used for sampling and analysis.

In the air pollution emission measurement area, the EPA has worked with the NIST since 1972 in the development of SRMs for toxic emission measurements. In that year, research on a SRM for trace elements in coal flyash was initiated by the NIST with support from the EPA. This SRM was certified for several toxic metals (including mercury, arsenic, lead, beryllium, nickel, cadmium, chromium, zinc, manganese, and selenium) and is currently available from the NIST[5] as SRM 1633a.

In recent years, the EPA has supported the NIST in the development of toxic organic gas standards. The NIST has issued SRMs for toxic organic gas standards. The NIST has issued SRMs for aliphatic gas mixtures (carbon tetrachloride, chloroform, perchloroethylene, and vinyl chloride) and aromatic gas mixtures (benzene, toluene, chlorobenzene, and bromobenzene) at 0.25 and 10 ppm. During the fall of 1988, the NIST plans to issue SRMs for hydrogen sulfide gas at 5 and 20 ppm. Any organization conducting toxic gas emission measurements which include any of these compounds should make sure that commercially purchased calibration gas standards are analyzed using these SRMs.

## Performance Audits

By definition, a performance audit is a quantitative assessment of the bias (accuracy) of a measurement system or a portion thereof (e.g., analytical portion only). The best audit is one that provides an accurate assessment of the combined sampling and analysis. For this reason, cylinder gases have become very popular as audit materials for toxic emission measurement systems. However, not every toxic gas has sufficient vapor pressure to allow the preparation of a cylinder gas. Also, some toxic cylinder gases are not sufficiently stable to be used as audit material. In this case, a liquid or solid audit material must be prepared.

When toxic emission testing is done to demonstrate compliance to federal, state, or local agency regulations, the performance audit is a very powerful tool. In this era of gas chromatography and mass spectrometry, the onsite observation on how the sample is collected and analyzed is not sufficient evidence to assure good quality results. The performance audit is needed and serves two purposes. First, it provides an assessment of measurement accuracy during the time when actual sampling and analysis are performed. Second, the inclusion of the measurement accuracy in the test report documents the quality of the test measurement system.

Since performance audits are such an important tool in toxic emission measurements, the balance of this article will deal with this subject.

## DISCUSSION ON PERFORMANCE AUDITS

### ppm Organic Cylinder Gases

In 1978, the EPA Environmental Monitoring Systems Laboratory (EMSL) established a program of developing parts-per-million (ppm) cylinder gases and studying their stability. The purpose of this program is to provide audit materials that can be requested by EPA, state, and local agencies to assess the accuracy of source test measurements. Typically, these cylinder gases are used to audit source emission measurements made by source owners or their contractors during performance and compliance tests of EPA New Source Performance Standards.[6]

Currently, 45 gaseous compounds have been investigated as audit materials. Six compounds have been found to be unstable in cylinders and not suitable as audit materials. The other 39 gaseous compounds in compressed gas cylinders are suitable for conducting performance audits during source testing. A summary of the 45 compounds is shown in Table 1. Where possible, the audit gases have been developed in two concentration ranges. The low concentration range cylinders are useful in assessing the accuracy of source measurements made to demonstrate compliance to existing emission regulations and during testing, which may lead to future regulations. The high-concentration range cylinders are useful in assessing the accuracy of source measurements of uncontrolled pollutant emissions. Stability of these compounds is determined by periodic measurements of all audit cylinders. Stability data are available for all 39 compounds for at least 2 years and for some compounds as long as 9 years. A summary of all source test audit results and stability data for all compounds is available.[7] The balance gas in all audit cylinders is nitrogen.

### ppb Organic Cylinder Gases

Partly because of the success of the ppm audit program, but mostly because of the need to audit parts-per-billion (ppb) source test measurement systems, in 1983 the EMSL initiated a program to develop ppb cylinder gases and study their stability. The purpose of this program is to provide audit materials that can be requested by EPA regional offices and state agencies to assess the accuracy of volatile organic emission measurements during RCRA trial burn tests. During a RCRA trial burn, organics called principal organic hazardous constituents (POHCs) are selected for measurement. The source owner or a contractor must conduct source tests on the hazardous waste incineration system during a trial burn and demonstrate at least 99.99% destruction removal efficiency for each selected POHC in order to obtain an operating permit.

**Table 1. ppm Audit Materials Currently Available[a]**

| Compound | Low Concentration Range (ppm) | High Concentration Range (ppm) | Stability Data Available (years) |
|---|---|---|---|
| Benzene | 5–20 | 60–400 | 9 |
| Ethylene | 5–20 | 300–700 and 3,000–20,000 | 8 |
| Propylene | 5–20 | 300–700 | 8 |
| Methane/ethane | — | 1,000–9,000(m) and 200–800(e) | 8 |
| Propane | 5–20 | 300–700 and 1,000–20,000 | 8 |
| Toluene | 5–20 | 100–700 | 7 |
| Hydrogen sulfide | 5–50 | 100–700 | 8 |
| Meta-xylene | 5–20 | 300–700 | 7 |
| Methyl acetate | 5–20 | 300–700 | 9 |
| Chloroform | 5–20 | 300–700 | 8 |
| Carbonyl sulfide | 5–20 | 100–400 | 3 |
| Methyl mercaptan | 3–10 | — | 8 |
| Hexane | 20–90 | — | 8 |
| 1,2-dichloroethane | 5–20 | 100–600 | 8 |
| Cyclohexane | — | 80–200 | 8 |
| Methyl ethyl ketone | 5–50 | — | 6 |
| Methanol | 30–80 | — | 7 |
| 1,2-Dichloropropane | 3–20 | 300–700 | 8 |
| Trichloroethylene | 5–20 | 100–600 | 8 |
| 1,1-Dichloroethylene | 5–20 | 100–600 | 8 |
| Perchloroethylene | 5–20 | 300–700 | 8 |
| Vinyl chloride | 5–30 | — | 8 |
| 1,3-Butadiene | 5–60 | — | 4 |
| Acrylonitrile | 5–20 | 300–500 | 4 |
| Methyl isobutyl ketone | 5–20 | — | 5 |
| Methylene chloride | 1–20 | — | 4 |
| Carbon tetrachloride | 5–20 | — | 5 |
| Freon 113 | 5–20 | — | 5 |
| Methyl chloroform | 5–20 | — | 5 |
| Ethylene oxide | 5–20 | — | 5 |
| Propylene oxide | 5–20 | 75–200 | 5 |
| Allyl chloride | 5–20 | 75–200 | 3 |
| Acrolein | 5–20 | 100–300 | 4 |
| Chlorobenzene | 5–20 | — | 5 |
| Carbon disulfide | — | 75–200 | 2 |
| EPA method 25 mixture[b] | 100–200 | 750–2000 | 4 |
| Ethylene dibromide | 5–20 | 50–300 | 3 |
| 1,1,2,2-Tetrachloroethane | 5–20 | — | 3 |

[a]The following compounds were found to be unstable in the cylinders: Formaldehyde, Para-dichlorobenzene, Ethylamine, Aniline, Cyclohexanone, and 1,2-Dibromoethylene.
[b]The gas mixture contains an aliphatic hydrocarbon, an aromatic hydrocarbon, and carbon dioxide in nitrogen. Concentrations shown are in ppm.

Currently, 27 gaseous compounds have been investigated at ppb concentrations as audit materials for auditing volatile POHCs. Two compounds have been found to be unstable in cylinders and not suitable as audit materials. The other 25 gaseous compounds are suitable for conducting performance audits during RCRA trial burn tests. A summary of the 27 compounds is shown in Table 2. Multicomponent gas mixtures have been developed as the ppb audit

**Table 2. ppb Audit Materials Currently Available**

| Group 1 | Group 2 | Group 3 | Group 4 |
|---|---|---|---|
| Five organics in $N_2$: | Nine organics in $N_2$: | Seven organics in $N_2$: | Six organics in $N_2$: |
| Carbon tetrachloride | Trichloroethylene | Vinylidene chloride | Acrylonitrile |
| Chloroform | 1,2-Dichloroethane | F-113 | 1,3-Butadiene |
| Perchloroethylene | 1,2-Dibromoethane | F-114 | Ethylene oxide[a] |
| Vinyl chloride | F-12 | Acetone | Methylene chloride |
| Benzene | F-11 | 1,4-Dioxane | Propylene oxide[a] |
| | Bromomethane | Toluene | Ortho-xylene |
| | Methyl ethyl ketone | Chlorobenzene | |
| | 1,1,1-Trichloroethane | | |
| | Acetronitrile | | |
| Ranges of cylinders currently available: | Ranges of cylinders currently available: | Ranges of cylinders currently available: | Ranges of cylinders currently available: |
| 7–90 ppb | 7–70 ppb | 7–90 ppb | 7–90 ppb |
| 430–10,000 ppb | 90–430 ppb | 90–430 ppb | 430–10,000 ppb |
| Stability data available for four years | Stability data available for three years | Stability data available for one year | Stability data available for six months |

[a]This compound was found to be unstable in the cylinders.

materials. This allows the audit of more than one POHC measurement at the same time with the same audit cylinder. During 1983, a five-compound Group I audit mixture was developed. In subsequent years, a nine-compound Group II, a seven-compound Group III, and a six-compound Group IV audit mixture were developed. All audit groups are available in concentrations between 7 and 90 ppb for each compound within the group. Groups I, II, and III are also available in concentrations between 90 and 430 ppb. In addition, Groups I and IV are available in concentrations between 430 and 10,000 ppb. Stability data are available for all 27 compounds at the 10-ppb nominal concentration level. A summary of all source test audit results during RCRA trial burn tests and stability data for all compounds are available.[8] The balance gas in the audit cylinders is nitrogen.

## RCRA Audit Results

The recommended method for volatile POHC measurements during a RCRA trial burn test is the VOST.[9] As stated, in the VOST method, a performance audit must be conducted simultaneously with the actual planned test. A performance audit prior to a trial burn test to assess the proficiency of the measurement system (including the sampling and analytical personnel) is recommended but not required.

As of January 1988, 89 audits have been initiated to assess the accuracy of measurement methods during or prior to hazardous waste trial burn tests. Of the 89 RCRA audits, 74 are complete and the results are summarized in Table 3. For the 74 completed audits, the frequency of how many times each compound was selected for audits is shown. Seven of the 27 compounds have been found to be more popular for performance audits. These seven compounds are

**Table 3. Summary of Audits Performed for RCRA Trial Burn Tests (through January 1988)**

| | Based on 74 Completed Audits | | | | |
|---|---|---|---|---|---|
| | | Measurement Systems Audited | | | |
| Groups I to IV Compounds | Frequency Compounds Selected for Audits | VOST[a] | Bag[b] | Other[c] | Frequency VOST Auditee Results within ±50% of Audit Concentration[d] |
| Carbon tetrachloride | 32 | 25 | 4 | 3 | 21 out of 25 |
| Chloroform | 19 | 14 | 2 | 3 | 11 out of 14 |
| Perchloroethylene | 16 | 15 | 1 | 0 | 11 out of 15 |
| Vinyl chloride | 11 | 11 | 0 | 0 | 7 out of 11 |
| Benzene | 12 | 12 | 0 | 0 | 9 out of 12 |
| Trichloroethylene | 9 | 8 | 1 | 0 | 7 out of 8 |
| 1,2-Dichloroethane | 3 | 3 | 0 | 0 | 3 out of 3 |
| 1,2-Dibromoethane | 2 | 2 | 0 | 0 | 1 out of 2 |
| Acetonitrile | 3 | 2 | 1 | 0 | 0 out of 2 |
| Trichlorofluoromethane (F–11) | 4 | 3 | 0 | 1 | 3 out of 3 |
| Dichlorodifluoromethane (F–12) | 4 | 3 | 0 | 1 | 1 out of 3 |
| Bromomethane | 3 | 3 | 0 | 0 | 1 out of 3 |
| Methyl ethyl ketone | 2 | 2 | 0 | 0 | 1 out of 2 |
| 1,1,1-Trichloroethane | 4 | 4 | 0 | 0 | 3 out of 4 |
| Vinylidene chloride | 2 | 2 | 0 | 0 | 1 out of 2 |
| 1,1,2-Trichloro-1,2,2-trifluoroethane (F–113) | 2 | 2 | 0 | 0 | 2 out of 2 |
| 1,2-Dichloro-1,1,2,2-tetrafluoroethane (F–114) | 1 | 1 | 0 | 0 | 0 out of 1 |
| Acetone | 1 | 1 | 0 | 0 | 1 out of 1 |
| 1,4-Dioxane | 1 | 1 | 0 | 0 | 0 out of 1 |
| Toluene | 7 | 6 | 1 | 0 | 4 out of 6 |
| Chlorobenzene | 4 | 4 | 0 | 0 | 4 out of 4 |
| Acrylonitrile | 1 | 0 | 1 | 0 | 0 out of 0 |
| 1,3-Butadiene | 0 | 0 | 0 | 0 | 0 out of 0 |
| Ethylene oxide | 0 | 0 | 0 | 0 | 0 out of 0 |
| Methylene chloride | 0 | 0 | 0 | 0 | 0 out of 0 |
| Propylene chloride | 0 | 0 | 0 | 0 | 0 out of 0 |
| Ortho-xylene | 0 | 0 | 0 | 0 | 0 out of 0 |

[a]GC/MS was used for all VOST analysis.
[b]GC with a specific detector (FID/ECD) was used for bag samples.
[c]"Other" includes direct injection and glass bulbs.
[d]Frequency of auditee results within ±50% of the audit concentration are 10 out of 11 for bag sampling and 6 out of 8 for "other" methods.

carbon tetrachloride, chloroform, perchloroethylene, vinyl chloride, trichloroethylene, benzene, and toluene. Of the 89 audits initiated, 68 are on the VOST, 16 with Tedlar bags, 1 with glass bulbs, and 4 by direct injections. The majority of the audits employed GC/MS for analysis and only a few used GC with a specific detector. It is interesting to note from the table that the auditee results for the VOST method are usually within the ±50% accuracy limited stated in the VOST method.

Currently, the VOST method has been validated by the EPA for carbon tetrachloride, chloroform, perchloroethylene, benzene, and trichlorofluoromethane (F-11).[2] The high frequency of VOST audit results within the ±50%

accuracy limit suggests that this method may be an effective means of measuring a large number of POHCs in addition to the five compounds validated.

The high frequency of acceptable audit results for the VOST method suggests another important purpose of performance audits. Previously, the main purpose of audits was stated as providing measurement method accuracy and documentation of the quality of results in source test reports. As illustrated here for the VOST, the performance audit may also be used to determine the applicability of a method for the measurement of toxic organics for which method validation studies have not yet been completed.

### Availability of Audit Cylinders

The ppm and ppb audit cylinders are available at no charge to federal, state, and local agencies or their contractors to assess the quality of source emission measurement systems. To initiate an audit, contact:

Robert L. Lampe
U.S. EPA, Environmental Monitoring Systems Laboratory
Quality Assurance Division (MD-77B)
Research Triangle Park, NC 27711
Phone: Commercial 919-541-4531 or FTS 629-4531

## SUMMARY

Of the 23 elements that should be considered in planning quality assurance for toxic emission measurement systems, three elements have the greatest impact on the quality of measurement results. These elements are (1) use validated test methods, (2) use high-quality calibration standards, and (3) conduct periodic performance audits on the measurement systems. A performance audit serves several purposes. First, it provides an assessment of measurement accuracy during the time when actual sampling and analysis are performed. Second, the inclusion of the measurement accuracy in the test report documents the quality of the test measurement system. Third, audit results provide a screening of the test method for applicability for those compounds for which method validation has not yet been completed. It is for these reasons that performance audits have become recognized as a powerful tool in ensuring the quality of measurements.

## REFERENCES

1. "Quality Assurance Handbook for Air Pollution Measurement Systems, Volume 1 – Principles," EPA-600/9-76-005 (1984).
2. Fuerst, R. G., T. J. Logan, M. R. Midgett, and J. Prohaska. "Validation Studies of the Protocol for Volatile Organic Sampling Train," *J. Air Poll. Control Assoc.* 37:388–394 (1987).

3. Margeson, M. H., J. E. Knoll, M. R. Midgett, D. E. Wagoner, Wagoner, J. Rice, and J. B. Homolya. "An Evaluation of the Semi-VOST Method for Determining Emissions from Hazardous Waste Incinerators," *J. Air Poll. Control Assoc.* 37:1067–1074 (1987).
4. Ward, T. E., M. D. Midgett, G. D. Rives, N. F. Cole, and D. E. Wagoner. "Evaluation of a Sampling and Analysis Method for the Measurement of Toxic Metals in Incinerator Stack Emissions," Paper 1033, Annual Meeting of the Air Pollution Control Association, June 1988.
5. "NBS Standard Reference Materials Catalog," NBS Special Publication 260. Available from NBS, Office of Standard Reference Materials, Gaithersburg, MD, 20899.
6. "Standards of Performance for New Stationary Sources," Code of Federal Regulations, Part 60, U.S. Government Printing Office (July 1987).
7. Howe, G. B., J. R. Albritton, C. K. Sokol, R. K. M. Jayanty, and C. E. Decker. "Stability of Parts-Per-Million Organic Cylinder Gases and Results of Source Test Analysis Audits—Status Report #9," NTIS Publication PB88-158761 (January 1988).
8. Jayanty, R. K. M., C. K. Sokol, and C. E. Decker. "Stability of Parts-Per-Billion Hazardous Organic Cylinder Gases and Performance Audit Results of Source Test and Ambient Air Measurement Systems—Status Report #4," EPA Contract No. 68-02-4125 (January 1988).
9. "Volatile Organic Sampling Train, Method 0030, SW-846 Manual (Test Methods for Evaluating Solid Waste, Physical/Chemical Methods)," 3rd ed., available from the Superintendent of Documents, U.S. Government Printing Office, Washington, DC 20402, Document Number 955-001-00000-1.

# CHAPTER 11

# Development of a New Tracer Technology Using Enriched Rare-Earth Isotopes

**J. M. Ondov and W. R. Kelly**

## ABSTRACT

We are developing a new particulate tracer technology suitable for tracing power plant emissions using readily available, highly enriched, nontoxic, chemically and radiologically stable isotopes of Nd and Sm. The relative abundances of most RE isotopes are invariant in nature to within 2 parts/$10^5$, a precision readily attained by thermal ionization mass spectrometry (TIMS). This degree of invariance, coupled with the high precision attainable with TIMS, should in principle permit removal of bias from the ambient background and analytical processing which degrade precision for whole-element tracers. Herein we discuss the propagation of error from these sources and present the results of limited in-plant tests of the tracer release equipment. The error analyses suggest that the technique offers a >40,000-fold advantage over previously proposed whole-element tracer schemes. Results of the field tests suggest that S:N ratios adequate for detection at 100 km could be achieved in ambient aerosol particles $\leq 10\ \mu m$ from a 100-MW(e) coal-fired power plant with a release rate as little as 13 mg/hr. The S:N ratios of submicrometer particles are enormous and should permit detection over much larger distances. The tests further suggest that flyash particles can be effectively "tagged" by coagulation with fine residue particles made with simple two-fluid atomizers operated at elevated pressure.

## INTRODUCTION

Primary and secondary particles, and gases emitted from high-temperature combustion sources are thought to travel long distances and degrade the quality of air and depositing particles at sites located hundreds, even thousands of kilometers away. Clearly, the ability to determine the contributions of specific types of sources or source regions is of vital importance in formulating abatement strategies. To this end, an enormous effort has been devoted to the development of emission-and receptor-based source apportionment modeling schemes. The former use emission data and meteorological dispersion models. The latter rely on so-called "tracers of opportunity," i.e., distinct chemical and physical properties, usually of the emitted particles, to resolve components measured at a receptor site. Despite decades of research, rigorous application of the former is impeded by the lack of detailed emission inventories and the need for highly resolved meteorological data. Needing only composition data for source and receptor aerosol, the latter circumvent the need for information on emissions and transport processes and have increasingly become a practical alternative to emission-based modeling. In recent years, simple urban-scale receptor modeling has been extended to elucidate the impacts of distant source regions[1] and to apportion secondary sulfate.[2] Yet despite these developments, the true relationship between emissions and receptor air quality has never been experimentally determined[3] and the goal of definitive source apportionment remains elusive. This is especially true when similar sources need be resolved and long distances are concerned. In this regard, perhaps the most difficult and important problem facing apportionment science is that of resolving the contributions of near and distant coal-fired power plants.

Receptor and particulate deposition models can in principle be tested with an intentional particulate tracer. Intentional tracers including fluorescent particles,[4] $SF_6$,[5] and perfluorocarbon compounds (PFC),[6,7] have long been used in the development of meteorological dispersion models: PFCs permit air-mass tracking over 1000-km distances. However, comparable long-distance intentional particulate tracer materials suitable for application in high-temperature combustion sources have not been applied. The principal reasons for this lack of application include toxicity, chemical or radiologic instability, inadequate sensitivity, and cost. Furthermore, particulate tracers have generally been applied as discrete particles of the tracer material which form separate aerosol populations, despite rather important differences in particle size and physical properties, especially hygroscopicity. Accordingly, their environmental equivalence is questionable, especially when large distances are concerned.

A potentially more powerful set of particulate tracer materials well suited for use in high-temperature combustion sources are found in the stable isotopes of the rare-earth (RE) elements. The nonradioactive RE elements are virtually nontoxic and chemically stable, and therefore suitable for stack release. Most importantly, however, is that the relative abundances of most RE isotopes are invariable in nature and can be measured with great precision (2 parts/$10^5$) by TIMS. By virtue of this fact, separated isotopes can be administered to a source in what amounts to an isotope dilution analysis. The absence of natural or industrial sources of the enriched RE isotopes makes them excellent candidates for application as long-distance particulate tracer materials. Furthermore, several sources could be "tagged" simultaneously using different isotopes of the same element. A unique feature of the method is the ability to determine the natural background of the enriched isotope in the ambient

**Table 1. Assumptions Used in Estimating Tracer Requirements for Detection in Ambient Air 100 km from a 425–MW(e) Coal-Fired Power Plant**

| | |
|---|---|
| $^{148}$Nd, isotopic purity | 94% |
| Stack emission of natural Nd, g/J | $2.1 \times 10^{-13}$ |
| Energy input rate, J/sec | $1.1 \times 10^{9}$ |
| Stack gas flow rate, $m^3$/s | $1.0 \times 10^{3}$ |
| Ambient Nd background, ng/$m^3$ | 0.6 |
| Natural isotopic variability | $2/10^5$ |

aerosol, filter, and reagents simultaneously with the tracer isotope measurement.

Herein, we describe the proposed application of enriched RE isotopes to coal-fired power plants, including estimates of tracer needs and cost as a function of the desired confidence level. A major difficulty in applying the technique is to assure that the tracer and emitted particles behave identically in the atmosphere. Also discussed are results of a pilot study to test flyash particle tagging by coagulation with fine tracer particles made with an air atomizing nozzle.

## RELEASE REQUIREMENTS

### Tracer emission rate

Release requirements and cost of an enriched isotopic tracer depend primarily on its concentration in ambient air, the isotopic enrichment of the tracer material, extent of atmospheric dispersion, and the signal-to-noise (S:N) ratio desired at the receptor site. For the purpose of our discussion, we estimate the amount of $^{148}$Nd that must be released from a single 425-MW(e) pulverized coal-fired power plant (described as Plant A in Ondov et al.)[8] to achieve a S:N ratio of 500:1 at a distance of 100 km. The maximum instantaneous ground-level concentration of Nd accounted for by flyash at a distance of 100 km from the power plant can be estimated by the product of the emission rate and $\chi/Q$, the gaussian plume dispersion parameter. The value of this parameter depends on both engineering parameters (i.e., those that govern plume rise) and quite strongly on meteorological conditions. For plant parameters listed in Table 1, $\chi/Q$ ranges from $\approx 5 \times 10^{-9}$ under unstable conditions to $> 10^{-7}$ under stable conditions and is estimated as $10^{-8}$ sec/$m^3$ for neutral conditions.[11] The abundance of $^{148}$Nd relative to $^{142}$Nd in natural materials is known to be constant to within $\pm 2$ parts/$10^5$.[9] However, the background concentration of Nd in ambient air is highly variable. In the Washington, DC, area, Kitto[10] reports a geometric mean concentration for Nd of 0.62 ng/$m^3$ and a range of 0.13 to 1.80 ng/$m^3$ for 27 samples collected over a two-year period. For neutral conditions, these data (Table 1) lead to a source contribution of about 2.3 pg Nd/$m^3$ and are thus negligible in comparison with the background estimate of 0.6 ng Nd/$m^3$ from the other sources.

Given that the naturally occurring ratio of $^{148}$Nd:$^{142}$Nd is invariable to within

2 parts/$10^5$, we must displace this ratio from its natural value by 1000 parts/$10^5$, i.e., by 1%, to achieve a S:N ratio of 500. That is to say we must increase the amount of $^{148}$Nd atoms in ambient air at the receptor site by 1%. The natural abundance of $^{148}$Nd is 5.76% (atom%), so that accounting for the small differences in the molar masses, 0.6 ng Nd/m$^3$ corresponds to about 35 ρg $^{148}$Nd/m$^3$. Of this, 1% is 0.35 pg/m$^3$, which, dividing by $\chi/Q$ gives a stack emission rate of $3.5 \times 10^{-5}$ g/sec, or 126 mg/hr. The isotope is currently available at 94% isotopic purity for $1.50/mg, so that the materials cost is only about $200/hr. Samarium is about 3.5-fold less abundant than Nd in crustal materials, and the cost of suitable isotopes is two-fold lower. A considerable cost reduction could be achieved if $^{150}$Sm were used in place of $^{148}$Nd. However, in general, more or less tracer may be needed depending on the confidence level desired and the precision of the analysis.

## Effect of analysis precision on uncertainty and tracer requirements

In a high-precision mass-spectrometric analysis, one measures the atom ratios of two isotopes in a sample and compares them with the same ratio for some "standard" or reference material. The tracer mass (i.e., excess $^{148}$Nd, denoted $^{148}$Nd*) is given as follows:

$$^{148}\mathrm{Nd}^* = (R_s - R_{std}) \cdot C, \tag{1}$$

where $R_s$ is the ratio of $^{148}$Nd to $^{142}$Nd in the ambient aerosol sample, $R_{std}$ is the same ratio in the reference material, and C is the mass of the reference isotope, i.e., $^{142}$Nd in the sample. The relative standard deviation from the error propagation, $\sigma_{rel}$, in the determination of $^{148}$Nd* is given by Equation 2.

$$\sigma_{rel} = \sqrt{\left(\frac{dR_s}{(R_s - R_{std})}\right)^2 + \left(\frac{dR_{std}}{(R_s - R_{std})}\right)^2 + \left(\frac{dC}{C}\right)^2} \tag{2}$$

Clearly, high-precision analysis is required when $R_s$ is similar in magnitude to $R_{std}$. Note that the background concentration of the tracer isotope does not appear as a separate term in either Equation 1 or 2 and thus the tracer is determined independently of the background concentration. This arises because the ratios of the isotopes in the background are invariable, so that in effect, the background is measured in the sample by means of the reference isotope. This is not to say that the background has no effect. Rather, as the natural background increases, $R_s$ approaches $R_{std}$ and the uncertainty in the measurement of $^{148}$Nd* increases. The result is that better precision is required to achieve the same level of confidence.

The results of the general error analysis are shown graphically in Figure 1, in which we plot $\sigma_{rel}$ in a tracer isotope $^m$E* as a function of the increment (expressed in percent) in the ratio of the tracer atoms $^m$E* to $^n$E, the reference isotope, for an ambient sample 100 km away and for several values of the

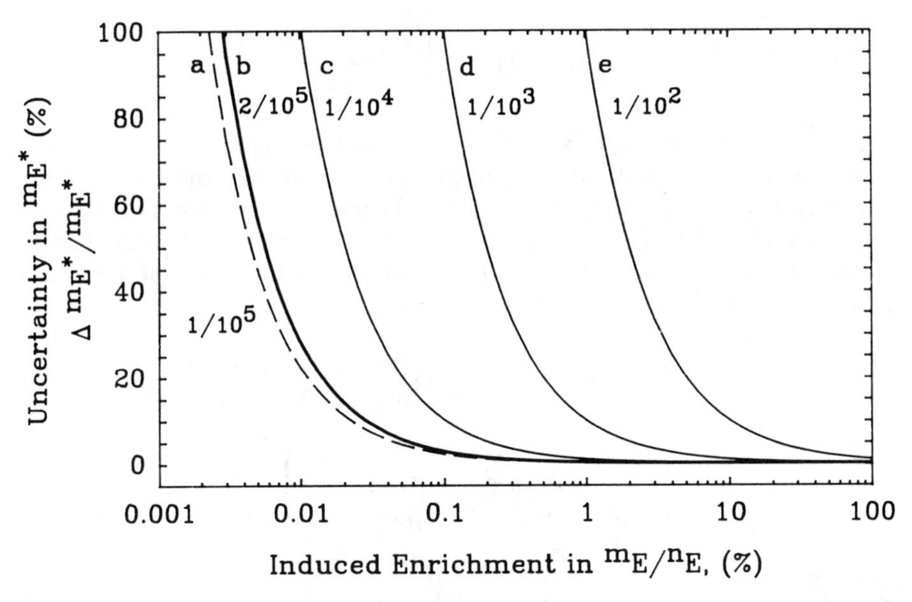


**Figure 1.** Plot of $\sigma_{rel}$ in the determination of an isotopic tracer as a function of the induced enrichment for different measurement precisions.

measurement precision. This increment is essentially equivalent to the percentage increase in tracer atoms induced at the receptor site and, as discussed earlier, the S:N is given simply by dividing the (fractional) increment by 2 parts/$10^5$. The release rate is calculated as the product of the fractional increment, background concentration, and $Q/\chi$. The results indicate that with a measurement precision of 2 parts/$10^5$, one could determine $^{148}Nd^*$ to within 15% with an enrichment of only 0.02%. This would, therefore, require 50-fold less tracer than proposed earlier for detection at the same distance. If the analytical precision were somewhat less than 2 parts/$10^5$, then more tracer would be needed. Clearly, measurement precision is gained at the cost of increased analysis time, so that trade off between analysis time and the cost of the tracer must be considered.

Giorgi et al.[12] have recently proposed the use of Dy as a tracer of particulate emissions from diesel engines. In such an experiment, the tracer concentration at the receptor site ([Dy*]) is given by the difference between the measured and background concentrations, denoted $[Dy_m]$ and $[Dy_b]$, respectively. In this case, it would be impossible to measure the tracer background simultaneously with the tracer concentration. As a result, the variance in the background concentration must be determined and the overall confidence level is degraded by this variance. Although a rigorous error analysis is beyond the scope of this paper, the uncertainty in [Dy*] primarily depends on the ratio of tracer to background, the measurement precision, and the background variance. The effect of background variance may be expressed in terms of a bias parameter, $\beta$, such that:

$$(\beta + 1) = \frac{[Dy]_{b,\,t=1}}{[Dy]_{b,\,t=2}} \quad (3)$$

where the numerator represents the background concentration as actually measured at time t = 1, and the denominator represents the background concentration that actually occurred during the tracer experiment, i.e., at time t = 2. Assuming that the emission rate of Dy in flyash is negligible with respect to the amount injected, the relative standard deviation, $\sigma'_{rel}$, of [Dy*] as adapted from Kelly and Ondov[13] is as follows:

$$\sigma'_{rel} = \left\{ \left[ \left[1 + \alpha + (1 + \beta)\right] \frac{[Dy]_{b,\,t=2}}{R_{Dy} \cdot \chi/Q} \cdot \frac{d[Dy]_m}{[Dy]_m} \right]^2 + \left[ (1 + \beta) \frac{[Dy]_{b,\,t=2}}{R_{dy} \cdot \chi/Q} \cdot \frac{d[Dy]_{b,\,t=1}}{[Dy]_{b,\,t=1}} \right]^2 \right\}^{1/2} + \left| \beta \cdot \frac{[Dy]_{b,\,t=2}}{R_{Dy}{}^* \cdot \chi/Q} \right| \quad (4)$$

where $R_{Dy}{}^*$ is the tracer injection rate in mass per unit time and $[Dy]_m$ is the concentration of tracer actually measured in the receptor sample.

In Figure 2, $\sigma'_{rel}$ for any whole element tracer [E*] is plotted versus induced enrichment in its background concentration (i.e., the ratio of the tracer (i.e., E*) to its background concentration, $E_b$) for various values of the background bias. Measurement precisions of 0.1 and 1% are, respectively, assumed in panels A and B. As in the isotopic case, the emission rate required is given by the products of the background concentration, the fractional enrichment, and $Q/\chi$. The plots clearly show that better measurement precision substantially improves $\sigma'_{rel}$ only when the background variance is quite small. If the fluctuation in the Dy background were only 100% (i.e., two-fold), one would need an induced enrichment of about 800% to assure a $\sigma'_{rel}$ of 15% in the determination of the net tracer concentration at 100 km. This translates to a 40,000-fold advantage in favor of the separated isotope tracer. In practice, the advantage could be 10-or 20-fold greater because the curves in Figure 1 assume that the isotopic abundance of the tracer isotope in the background aerosol is 100%, whereas the abundances of useful isotopes such as $^{148}Nd$ and $^{150}Sm$ are in the 5 to 15% range.

## TRACER ADMINISTRATION TESTS

Ideally, the tracer should be administered in a way such that (1) its atmospheric behavior be identical to that of power plant particles and (2) that it represent all of the particles of interest. In a typical pulverized CFPP, particu-

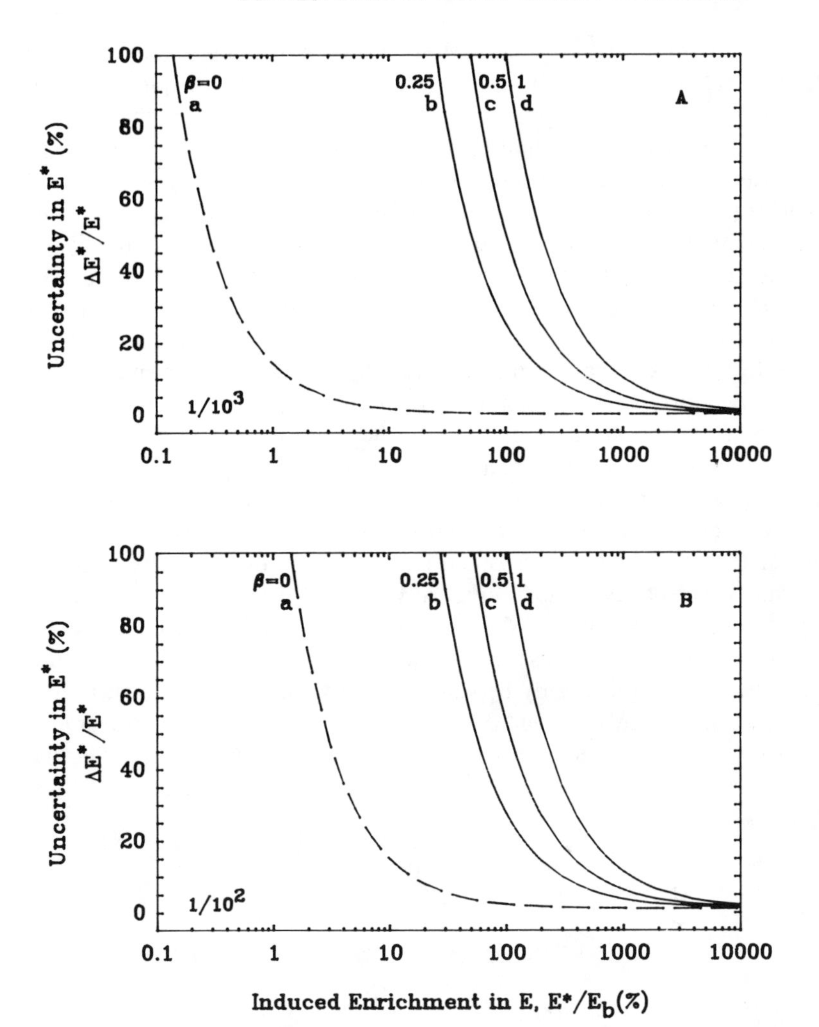


**Figure 2.** Plot of $\sigma'_{rel}$ in the determination of an elemental tracer as a function of induced enrichment for different values of the background bias ($\beta$) and measurement precisions of 0.1 (panel A) and 1% (panel B).

late emissions are bimodally distributed and the full particle size range of interest extends from about 0.05 to $\approx 10$ $\mu$m in diameter.

Direct furnace injection is precluded by cost on account of the high removal efficiencies of particulate control devices. However, computer simulations with the MAEROS[14] suggest ultrafine tracer particles would become attached to flyash particles of all sizes by coagulation during the time the flue gas travels from the outlet of the emission control devices to the mouth of the smokestack. To generate the needed particles, an air-atomizing nozzle was installed in the ductwork near the outlet of the air preheater of a 100-MW(e) coal-fired boiler and used to disperse solutions containing 3 ppm $^{148}$Nd (87%

isotopic purity) in methanol at a flow rate of 0.074 $cm^3$/sec at an atomizing pressure of 3400 kPa. Laboratory tests show that the droplets quickly dry, forming a distribution of residue particles in which 71% of the mass is contained in particles smaller than 18 nm, as measured with a microorifice impactor[15] after a nine-fold dilution with dry He.

Curve (a) of Figure 3 shows that initially, nearly all of the tracer mass was associated with particles $<0.050$ μm, but during transit to sampling ports in the stack breaching, the tracer became associated with particles of all sizes up to 10 μm or larger (curve b). Some of the growth might be accounted for by sulfuric acid condensation, but preliminary measurements suggest that there is too little sulfur associated with the particles for this to be a major contribution. Hence most of the observed growth appears to be the result of attachment to larger flyash particles.

The amount of tracer that can be delivered with a single nozzle is only about 1 mg/hr, so that many nozzles would be needed to achieve the 126-mg/hr rate discussed earlier. However, in the estimation of this value, we did not consider the particle size distribution of either the tracer or the material occurring naturally in the background. In Figure 4 (panel A), we plot the S:N expected at 100 km based on the size distribution of the tracer as measured in the in-plant test and two background distribution scenarios. In both, the background mass is assumed to be split evenly between fine and coarse particles with modes at 0.45 and 5 μm for curve a, and 0.45 and 10 μm for curve b. The tracer emission rate was scaled to 12.8 mg/hr, a rate achievable with 16 nozzles. In both cases,

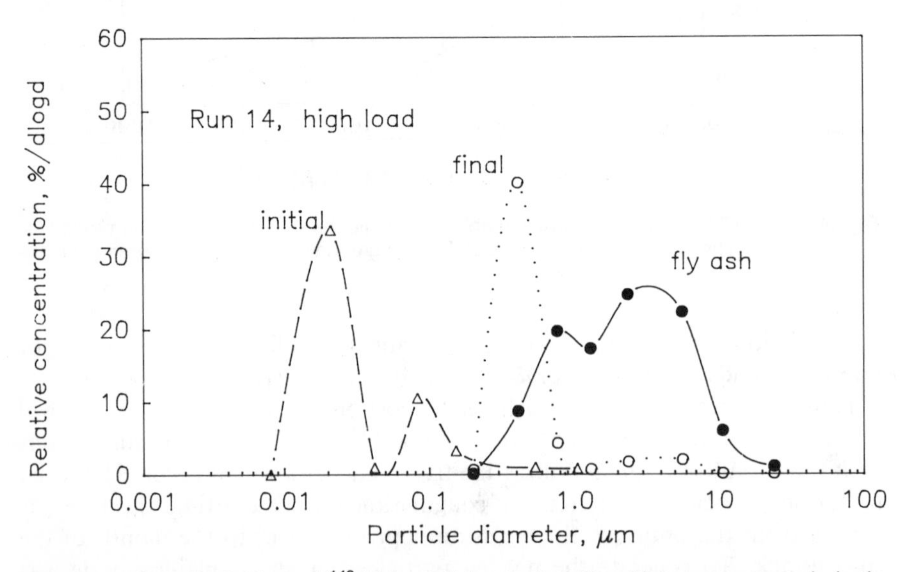


**Figure 3.** Size distributions of $^{148}Nd$ in particles made with a high-pressure nozzle in the laboratory (a) and after injection into flue gas of a 100-MW(e) coal-fired power plant (b). Curve (c) shows the distribution of naturally occurring Nd in flyash.

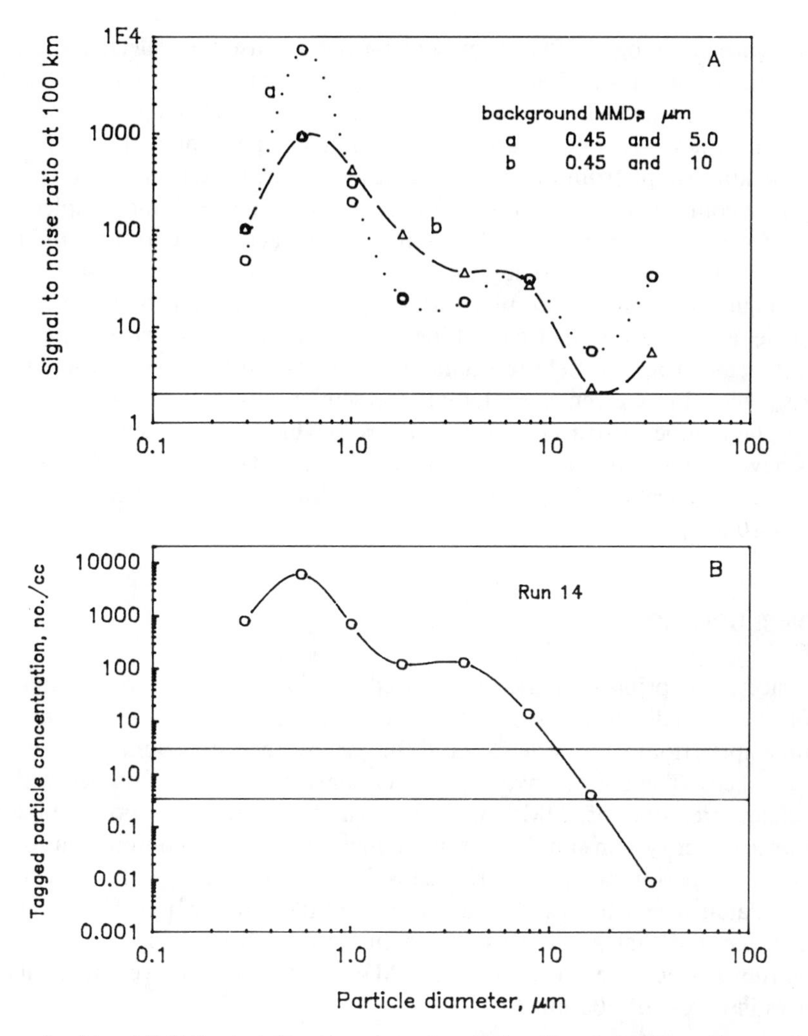


**Figure 4.** (Panel A) Estimated signal-to-noise ratio as a function of size for ambient particles 100 km distant from a 100-MW(e) coal-fired power plant. (Panel B) The number distribution of tagged flyash particles in stack. Both represent a release rate of 12.8 mg $^{148}$Nd/hr.

the S:N is estimated to be $\geq 10$ for all sizes of particles $\leq 10$ $\mu$m in diameter and enormous ratios are predicted in the submicrometer range.

## Statistical criterion

The accuracy of a particulate tracer experiment is also subject to the number of particles collected by the sampler. For example, if we assume Poisson statistics, the relative standard deviation for the sample is estimated as the square root of the number of particles counted. If we require that the 90%

confidence limit be ±10%, then 270 particles must be collected (i.e., the standard deviation is 16.5 and the 90% confidence interval is 1.64σ or ±10%). The number of particles one collects depends on the sampling rate and time, χ/Q, and stack-gas concentration. For a sampler operating at 120 Lpm for 1 hr, 100 km distant from our reference power plant, we estimate that the flue gas must contain about 3 particles/$cm^3$. A standard high-volume sampler operating for the same period would require an in-stack concentration of tracer particles of about 0.3 particles/$cm^3$.

In Figure 4, panel B, we plot the in-stack concentration of tagged flyash particles as a function of size resulting from the attachment of tracer particles, as estimated from the field test data. The calculation assumes that any tracer mass detected in a given size interval represents tracer particles of the original size. The number of tracer flyash particles to which one or more tracer particles have become attached is governed by Poisson statistics. The data suggest that the statistical criterion is easily met for either background case for particles ≤10 μm.

## CONCLUSIONS

Tracers of primary particles emitted by high-temperature combustion sources are needed to investigate the fate of aerosol particles, develop and test source apportionment models, and to apportion their contributions to receptor air sheds. The unique invariance of enriched RE isotopes coupled with the excellent precision of TIMS analysis form the basis of a powerful long-distance tracer system suitable for use in high-temperature combustion sources and far superior to previously considered, environmentally acceptable, particulate-tracer schemes. Results of in-plant tests suggest that flyash tagging is feasible and that as little as 13 mg/hr may be needed to conduct a full-spectrum tracer experiment on a 100-MW(e) coal-fired power plant at distances in excess of 100 km.

## REFERENCES

1. Rahn, K. A., and D. H. Lowenthal. "Elemental Tracers of Distant Regional Pollution Aerosols." *Science* 223:132–139 (1984).
2. Lewis, C. W. and R. K. Stevens. "Hybrid Receptor Model for Secondary Sulfate from an $SO_2$ point source." *Atmos. Environ.* 19:917–924 (1985).
3. Hidy, G. M. "Conceptual design of a massive aerometric tracer experiment (MATEX)." *J. Air Poll. Control. Assoc.* 37:1137–1157 (1987).
4. Leighton, P. A., W. A. Perkins, S. W. Grinnel, and F. X. Webster. The Fluorescent Particle Atmospheric Tracer. *J. Appl. Meterol.* 4:334–348 (1965).
5. Collins, G. F., F. E. Bartlett, A. Turk, S. M. Edmonds, and H. L. Mark. "A Preliminary Evaluation of Gas Air Tracers." *J. Air Poll. Control Asoc.* 15:109–112 (1965).

6. Jaffer, M., V. A. Dutkerwicz, and L. Husain. "Trichloro-fluoromethane as a Tracer of Urban Air Masses," *Atmos. Environ.* 15:775–779 (1981).
7. Dietz, R. N. and W. F. Dabberdt. "Gaseous Tracer Technology and Applications," Department of Applied Science, Brookhaven National Laboratory, Upton, NY (1983).
8. Ondov, J. M., R. C. Ragaini, and A. H. Biermann. "Emissions and Particle Size Distributions of Minor and Trace Elements at Two Western Coal-Fired Power Plants Equipped with Coldside Electrostatic Precipitators." *Environ. Sci. Technol.* 13:946–953 (1979).
9. Depaolo, D. J. Neodymium Isotopic Studies: Some New Perspectives on Earth Structure and Evolution. *Eos. Trans. Am. Geophys. Union* 62:137–140 (1981).
10. Kitto, M. E. "Receptor Modeling of Atmospheric Particles and Acidic Gases," PhD Thesis, University of Maryland, College Park, MD (1987), p. 89.
11. Turner, D. B. "Workbook of Atmospheric Dispersion Estimates," Public Health Service Publication No. 999-AP-26, National Air Pollution Control Administration, Cincinnati, OH (1970), pp. 6–11.
12. Georgi, B., H. Horvath, C. Norek, and I. Kreiner. "Use of Rare Earth Tracers for the Study of Diesel Emissions in the Atmosphere." *Atmos. Environ.* 21:21–27 (1987).
13. Kelly, W. R., and J. M. Ondov. "A Theoretical Evaluation of the Intentional Elemental Tracer and the Intentional Isotopic Tracer for the Study of the Fate of Emissions into the Atmosphere." *Atmos. Environ.* submitted.
14. Gelbard, F., and J. H. Seinfeld. "Simulation of Multicomponent Aerosol Dynamics." *J. Colloid Interface Sci.* 76:485–501 (1980).
15. Kuhlmey, G. A., B. Y. U. Liu, and V. A. Marple. "Micro-Orifice Impactor for Submicron Aerosol Size Classification." *Am. Ind. Hyg. Assoc. J.* 42:790–795 (1981).

12. [illegible] ... [illegible]

13. Oxburgh [illegible] ... (1985).

14. [illegible] ... (1983).

# CHAPTER 12

# $^{14}C$ Source Apportionment Technique Applied to Wintertime Urban Aerosols and Gases for the EPA Integrated Air Cancer Project

G. A. Klouda, L. A. Currie, A. E. Sheffield, B. I. Diamondstone, B. A. Benner, S. A. Wise, R. K. Stevens, and R. G. Merrill

## ABSTRACT

The $^{14}C$ source apportionment technique for tracing environmental carbon has been applied to fine (<2.5 μm diameter) atmospheric particles collected in Albuquerque, NM and Raleigh, NC, during the winter of 1984 to 85. This work was part of the EPA's Integrated Air Cancer Project (IACP). The major objective of this study was to quantify the impact of wood-burning (living carbon source) and motor-vehicle ($^{14}C = 0$) emissions on these urban airsheds through $^{14}C$ measurements. Additionally, $^{14}C$ measurements were necessary for evaluation of the EPA's single-tracer multiple-linear regression model (MLR) for source apportionment.[1] Good agreement was attained between $^{14}C$ and MLR. Future work includes applying these two techniques to samples collected in Boise, ID.

## EXPERIMENTAL

Measurements of $^{14}C$ by low-level gas counting[2] were made on the total carbon fraction of fine particles collected on quartz fiber filters at two sites in Albuquerque.[3] The first site, Zuni Park, is a residential area influenced by residential wood combustion (RWC). The second site, San Mateo, is a traffic intersection affected by motor-vehicle activities. In Raleigh, samples were collected at a residential site, Quail Hollow Swim Club, also influenced by RWC.

Ancillary $^{14}C$ measurements were made on chemical classes and individual compounds for two reasons: (1) to determine the origin of environmentally important species and (2) to potentially improve the sensitivity of the $^{14}C$ source apportionment technique. The chemical fractions of interest were elemental carbon ($C_E$),[3] total extractable organic matter (EOM), polycyclic aro-

**Table 1. Average Concentrations and Percent Residential Wood Combustion (RWC) Contribution for Total Fine Particulate Carbon and Elemental Carbon[a]**

| | Albuquerque, NM | | | | | |
|---|---|---|---|---|---|---|
| | Total carbon | | | Elemental carbon | | |
| Residential site | μg/m$^3$ | %RWC | n | μg/m$^3$ | %RWC | n |
| Night (ZP) | 45(4) | 84(2) | 8 | 5.5(0.5) | 65(4) | 8 |
| Day (ZP) | 14(4) | 57(13) | 3 | 2.8(0.7) | 43(12) | 3 |
| Traffic site | | | | | | |
| Day (SM) | 19(3) | 36(9) | 3 | 4.2(0.8) | 20(1) | 2 |
| | | Raleigh, NC | | | | |
| Residential site | | | | | | |
| Night (QH) | 55(12) | 102(7) | 4 | 3.0(0.7) | 65(9) | 2 |

[a]Sampling sites are designated as: ZP, Zuni Park; SM, San Mateo; and QH, Quail Hollow Swim Club. Numbers in parentheses represent standard errors based on measurements of n filters. Total carbon concentrations were determined by LECO carbon analyzer. Percent RWC is obtained from $^{14}C$ measurements normalized to wood with an average age of 100 years assuming rings of equal mass.

matic hydrocarbons (PAHs), and atmospheric CO and $CH_4$. These fractions, separated from the bulk fine particle and air samples, were expected to yield only microgram carbon quantities. Therefore, all these $^{14}C$ measurements were obtained by the more sensitive technique of accelerator mass spectrometry (AMS).[4]

The $C_E$ fraction was separated from the bulk fine particles by treating the filter with 70% $HNO_3$ to remove organic carbon.[5] The residue was then thermally combusted to $CO_2$ and quantified by manometry in a calibrated volume. This procedure, Wet Oxidation/Thermal (WO/T), was preferred over a straight thermal separation since charring of some organic compounds was likely to occur.[6,7] Complete separation of an individual chemical fraction from the bulk material is important for reliable $^{14}C$ results.

The $C_E$ fraction, isolated by the WO/T method, represents nonoxidizable carbon composed of graphitic and polymeric carbon. Concentrations of $C_E$ measured by this technique were reproducible at the 5 to 10% level and agreed to within ~7% of measurements by the Thermal/Optical (T/O) method[8] for aliquants taken from identical filters. The T/O method corrects for any charring which may occur during the thermal analysis by monitoring changes in light transmission through the filter with time. The $^{14}C$ results of the $C_E$ fraction show a depletion in $^{14}C$ caused by a lower contribution from RWC compared to the total carbon fraction at both residential and traffic sites, day and night (Table 1). This suggests that the $C_E$ fraction may be a better indicator of mobile activities.

The EOM was separated from three individual composite samples from Albuquerque representing contributions primarily from (1) RWC, (2) motor-vehicle activities, and (3) a mix of these two sources. $^{14}C$ measurements on these extracts were considered important since the extracts were tested for mutagenicity and were expected to contain a broad range of potentially toxic

**Table 2. Concentration and Percent Residential Wood Combustion (RWC) Contribution of Extractable Organic Matter (EOM) and Polycyclic Aromatic Hydrocarbons (PAHs)[a]**

| | Albuquerque, NM EOM | | PAH | |
|---|---|---|---|---|
| **Residential site** | **μg/m³** | **%RWC** | **ng/m³** | **%RWC** |
| Night (n = 2), (ZP) | 66(7) | 93(1) | 209(21) | 100(1) |
| Day (n = 2), (ZP) | 15(2) | – | 44(4) | 56(4) |
| Traffic site | | | | |
| Day (n = 8), (SM) | 9(1) | 44(1) | 29(3) | 53(2) |

[a]Sampling sites are designed as: ZP, Zuni Park and SM, San Mateo. Samples are composites from "n" filters to recover enough carbon for $^{14}C$ measurements. The concentrations (μg/m³) of the EOM and PAH fractions can not be directly compared to the total and elemental carbon concentrations in Table 1 since the particle filters of Table 2 were not identical to those in Table 1. The estimated error in the concentration is based on replicate extractions of NIST Urban Dust Standards. The percent RWC is obtained from $^{14}C$ measurements normalized to wood with an average age of 100 years assuming rings of equal mass. Numbers in parenthesis represent the standard error (1σ-Poisson) due to counting statistics from the $^{14}C$ measurement.

species. The EOM was obtained from the bulk fine particles using a Soxhlet extraction procedure with methylene chloride.[9] The $^{14}C$ results showed that RWC contributed 93% of the EOM during the night at the residential site (Zuni Park). Both sources contributed about equally to the EOM obtained during the day at the traffic site (San Mateo) (Table 2).

The $^{14}C$ content of the PAH fraction was of interest since some individual PAHs are known carcinogens. The apportionment of living and dead carbon associated with the PAH fraction would yield source information characteristic of the most abundant PAHs identified by GC/MS. From an aliquant of each of the earlier-mentioned EOM extracts, the total PAH fraction was chromatographically separated,[10] isolated from the solvent, and combusted to $CO_2$ for $^{14}C$ measurements. The $^{14}C$ results of the PAH fraction were in good agreement with the expected source impacts (Table 2). They also show an enrichment in $^{14}C$ compared to the EOM for both the residential (wood impacted) and traffic (road side) samples.

Nighttime integrated air samples, ~0.1 m³ volume each, were collected with Tedlar bags* in Albuquerque at the residential site, Zuni Park. Following collection, samples were transferred to evacuated 33-L cylinders for storage. The CO and $CH_4$ concentrations were determined on sample aliquants by gas chromatographic separation, conversion of CO to $CH_4$ over hot Ni on fire brick in a $H_2$ atmosphere, and measurement by a flame ionization detector. The bulk air samples were then processed under dynamic conditions to (1) cryogenically remove atmospheric $CO_2$ and hydrocarbons, (2) selectively oxidize CO and $CH_4$ to $CO_2$, and (3) cryogenically isolate each fraction individually for independent $^{14}C$ analysis.[11]

*Mention of commercial products in text does not imply endorsement by the National Institute of Standards and Technology.

The source of CO was important from a health and regulatory standpoint since the concentration during the wintertime often exceeds the Ambient Air Quality Standard of 9 ppm over an 8-hr period. In addition, both CO and $CH_4$ play important roles in the "greenhouse effect," where sources and fluxes are important input functions to models that predict future global trends of these currently increasing gases. The $^{14}CO$ results showed that 14 to 41% of the total CO at the Zuni Park site was a result of RWC. Conversely, the $CH_4$ was depleted in $^{14}C$ with respect to background $CH_4$ (largely of biogenic origin), which shows that motor-vehicle activities contribute significantly to the total urban $CH_4$ concentration.

## CONCLUSIONS

The analytical techniques developed and applied to these bulk particle and air samples yielded optimal separation and maximum recovery. $^{14}C$ measurements obtained by AMS on fractions containing only microgram quantities required special sample target preparation techniques developed by Verkouteren et al. (1987) to minimize and control the blank (contamination). These results demonstrate the importance of $^{14}C$ AMS measurements when applied to atmospheric species for source apportionment. Currently, the $^{14}C$ technique is being applied to bulk fine particles, $C_E$, and EOM collected during the winter in Boise, ID.

## ACKNOWLEDGMENTS

The authors wish to thank D. J. Donahue, T. W. Stafford, Jr., A. J. T. Jull, T. W. Linick, and L. J. Toolin for the $^{14}C$ measurements by AMS. This research was supported in part by the U.S. EPA.

## REFERENCES

1. Stevens, R. K., C. W. Lewis, T. G. Dzubay, R. E. Baumgardner, R. B. Zweidinger, R. V., Highsmith, L. T. Cupitt, J. Lewtas, L. D. Claxton, L. Currie, G. A. Klouda, and B. Zak, "Mutagenic Atmospheric Aerosol Sources Apportioned by Receptor Modeling," *Monitoring Methods for Toxics in the Atmosphere*, ASTM STP 1052, W. L. Zielinski, Jr., and W. D. Dorko, Eds., American Society for Testing and Materials, Philadelphia, 1990, pp. 187–196.
2. Currie, L. A., R. W. Gerlach, G. A. Klouda, F. C. Ruegg, and G. B. Tompkins. "Miniature Signals and Miniature Counters: Accuracy Assurance via Microprocessors and Multiparameter Control Techniques," *Radiocarbon*, 28, (2A):191 (1983).
3. Klouda, G. A., L. A. Currie, A. E. Sheffield, S. A. Wise, B. A. Benner, R. K. Stevens, and R. G. Merrill. "The Source Apportionment of Carbonaceous Com-

bustion Products by Micro-radiocarbon Measurements for the Integrated Air Cancer Project (IACP)," Proceedings of the 1987 EPA/APCA Symposium on Measurement of Toxic and Related Air Pollutants, Research Triangle Park, NC, May 1987.

4. Zable, T. H., A. J. T. Jull, D. J. Donahue, and P. E. Damon. "Quantitative Radioisotope Measurements with the NSF-Arizona Regional Accelerator Facility," *IEEE Trans. Nucl. Sci.* NS30 (April 1983).
5. Schultz, H. "Studies in Radiocarbon Dating," PhD Thesis (1962).
6. Cadle, S. H., P. J. Groblicki, D. P. Stroup. "Automated Carbon Analyzer for Particulate Samples," *Anal. Chem.*, 52:2201 (1980).
7. Klouda, G. A., and G. W. Mulholland. Personal communication (1984).
8. Cary, R. A. Personal communication (1987).
9. Lewtas, J. In: *Carcinogens and Mutagens in the Environment*, Vol. 5, H. F. Stich, Ed. (Boca Raton, FL: CRC Press, Inc., 1988), p. 59.
10. Wise, S. A., B. A. Benner, S. N. Chesler, L. R. Hilpert, C. R., Vogt, and W. E. May. "Characterization of the Polycyclic Aromatic Hydrocarbons from Two Standard Reference Material Air Particulate Samples," *Anal. Chem.* 58 (14): 3067, (1986).
11. Klouda, G. A., L. A. Currie, D. J. Donahue, A. J. T. Jull and M. H. Naylor. "Urban Atmospheric $^{14}CO$ and $^{14}CH_4$ Measurements by Accelerator Mass Spectrometry," *Radiocarbon*, 28 (2A):625 (1986).
12. Verkouteren, R. M., G. A. Klouda, L. A. Currie, D. J. Donahue, A. J. T. Jull, and T. W. Linick. "Preparation of Microgram Samples on Iron Wool for Radiocarbon Analysis via Accelerator Mass Spectrometry: a Closed System Approach," *Nucl. Inst. Meth.*, 29(B):41, (1987).

# CHAPTER 13

# Incineration—Some Things I Would Like to Know About and Why

**Walter M. Shaub**

## INTRODUCTION

Incineration of solid waste, except when energy economics are favorable, is principally a waste volume reduction technology. Air emissions are produced. These air emissions can be minimized according to various strategies, such as the employment of pollution abatement control devices. The product of the incineration process is ash residue. This ash residue is managed, either by disposal into a landfill or through limited reuse; for example, as a roadbed aggregate material.[1]

Organic compounds, such as dioxins and furans, and heavy metals, such as lead, cadmium, mercury, arsenic, and chromium, found in air emissions and in ash residue, may adversely impact human health and the environment if control of these pollutants is inadequate. The potential for harm due to environmental release depends upon exposure dynamics. For example, the intensity, duration, frequency, and pathways of the exposure and the biological response of the exposed organism must be considered.

Strategies to minimize air emissions of compounds such as dioxins and furans and ash management practices that minimize exposure to heavy metals can be established. In part, the development of management strategies to achieve desirable levels of minimization of pollutant releases requires consideration of solid waste characteristics, the science and engineering of incineration, air emissions transport, and the physical and chemical properties of ash residue.

## SCOPE OF THIS REPORT

This report identifies some technical issues which require additional examination in order to understand how chemical and physical processes relate to observed emissions and their minimization and/or control. Treatment of ash residue once generated is not addressed. The response of biological organisms to environmental exposures is not addressed. Finally, this report does not address health risks associated with exposure to gaseous or solid emissions of chemical species.

## DIOXINS AND FURANS

### Background

Dioxins and furans occur in the environment due to their release from many sources, some identified, some not identified.[2,3] Dioxins and furans are found in emissions from solid waste incineration. Much has been done to attempt to minimize incinerator emissions.

One issue of interest is whether dioxins or furans may be formed at low temperatures in the exhaust stack of a waste-to-energy facility, beyond the location of pollution control equipment.[4] It is argued that if dioxins and furans are largely formed in the stack, prospects for minimizing emissions through the use of pollution control equipment may be significantly reduced.

Low-temperature, in-stack formation of dioxins and furans appears unlikely for several reasons. Observations made during tests at a Quebec City incinerator facility are suggestive of the feasibility for dioxin and furan minimization when pollution control equipment is efficiently operated.[5] Numerical simulation studies indicate that gas-phase reactions do not promote dioxin and furan formation at low stack gas temperatures.[6] Mathematical analysis suggests that catalytic reactions "turn off" because reactive sites, e.g., on flyash, become unavailable when there is extensive condensation of material onto particulate surfaces, as is known to occur at low stack gas temperatures.[6]

However, despite reasons why it can be expected that minimization of emissions can be achieved, uncertainty remains about some choices for acceptable pollution abatement practice. There appear to be unresolved aspects of the science of dioxin and furan formation and destruction processes.

Laboratory experiments have demonstrated that dioxins and furans can be formed by catalyzed processes when certain physical and chemical conditions are met.[7-10] These experiments appear to indicate that catalysis involving flyash can promote both the formation and the destruction of dioxins and furans in certain locations in incinerators depending upon characteristic temperatures and other conditions. Some recent tests at incinerator sites also appear to support these findings.[11,12]

Other issues associated with control and minimization of dioxin and furan emissions also need to be considered, for example: The costs of periodically monitoring for dioxin and furan emissions from an incinerator facility can easily be in excess of $100,000 per full-scale stack test.[13] Apparently no continuous monitoring technologies are presently available for monitoring these species.

Sampling errors associated with waste stream inhomogeneities, process vari-

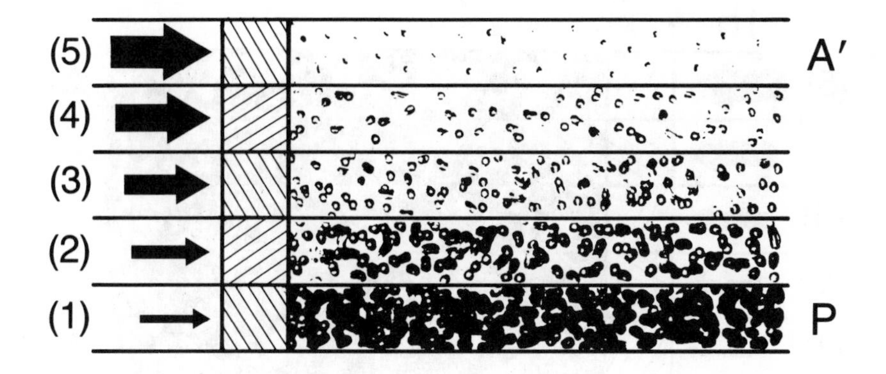


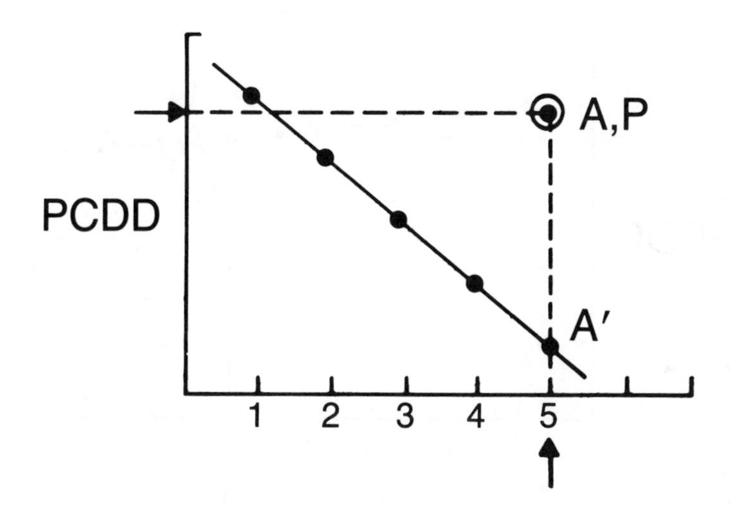


**Figure 1.** Comparison of actually observed MSW PCDD incinerator emisions (A) with PCDD emissions predicted (A′,P) from flow tube measurements if the solid-to-gas PCDD transport rate is slow due to the large desorption energy of chemisorbed PCDDs. This illustrates that in this case, flow tube experiments can fail to predict actual MSW PCDD concentrations.
v(5) is a typical MSW incinerator flow velocity.
v(1) is a typical flow tube experiment flow velocity.

ations, sample device perturbations, and analysis errors are difficult to deconvolute from one another. The practical application of statistical techniques is significantly complicated by the combination of these problems.[14]

Atmospheric lifetimes of emitted species are largely unknown or at best are reported as results of numerical estimation techniques.[15] Some work has been carried out to determine the photoreactivity of dioxins and furans,[16] but little has been done to measure reaction rates of dioxins and furans with such

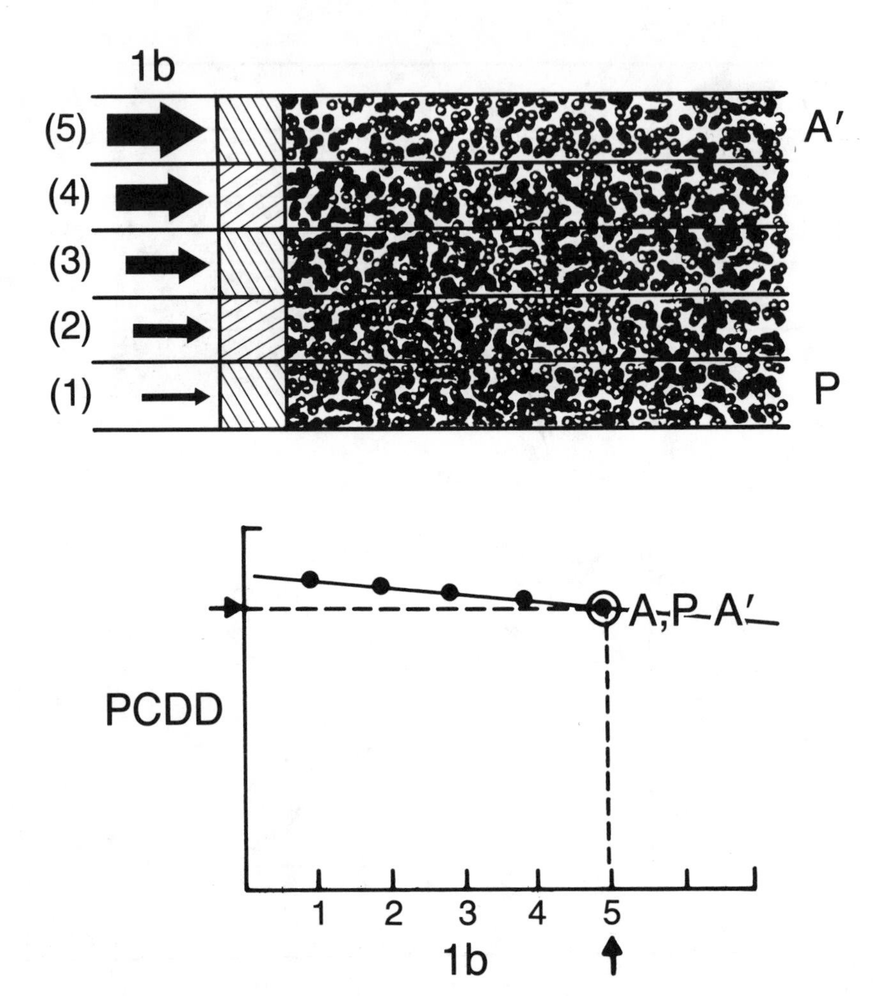


**Figure 2.** Comparison of actually observed municipal solid waste polychlorinated dibenxo-p-dioxin (MSW PCDD) incinerator emissions (A) with PCDD emissions predicted (A′,P) from flow tube measurements if solid-to-gas PCDD transport rate is fast due to the low desorption energy of chemisorbed PCDDs. This illustrates that in this case, flow tube experiments can predict actual MSW PCDD concentrations.
v(5) is a typical incinerator flow velocity.
v(1) is a typical flow tube experiment flow velocity.

reactive atmospheric species as hydroxyl radicals.[15] Much remains to be quantified about the extent of surface or volumetric distribution of these chemical species on or in particulate matter.[17] The change in isomer distribution with depth into a particle or with particle size needs to be understood.[4]

Atmospheric effects such as stagnation phenomena and particle coagulation have not been analyzed in terms of how such phenomena affect geographic distributions of emitted chemical species. Complex terrain and turbulent atmosphere models apparently require further development—there is, for example, disagreement about how to describe boundary layer effects.[18-20]

Research designed to utilize catalysts for controlling dioxin and furan emissions is currently being carried out.[7-10] Presently, little information has been reported due to the proprietary nature of much of this work. It is therefore difficult to assess issues associated with this technology.

The proportional distribution of dioxins and furans between gas and particulate phases during the incineration process and in emissions remains an area of some uncertainty.[21] The extent of buoyant entrainment or reentrainment of particulate matter containing dioxins and furans is a matter of current discussion.[18-20] The leachability of dioxins and furans from deposited particulate matter in the environment has been measured, but information is in need of compilation and careful review.[22,23] The dynamics of these phenomena are largely uncharacterized.

Additional information is needed in order to make more quantitative estimates of the role of these processes and the conditions under which they occur in incinerators.[24] Based on a review of current literature by this author, some issues which appear to require further examination are indicated here.

## Technical issues requiring additional research

### *The extent of polymerization of dioxins and furans to higher molecular weight compounds needs to be further examined*[25,26]

Polymerization is known to be sensistive, for example, to the nature of catalysts that are present and to temperature. The extent of polymerization may affect the amount of dioxins or furans that are observed to be present on flyash surfaces. This, in turn, may affect the observed distribution ratio of gas to solid-phase dioxins and furans. This may make it difficult to determine the extent of condensation of gas-phase species that may be occuring in an incinerator. The reported degree of chlorination of dioxins and furans may be shifted due to polymerization effects. The extent of chlorination and positional location of substituent chlorine atoms on dioxin and furan molecules can be expected to affect the extent of polymerization of various isomers.

### *Regarding the preparation of catalysts surfaces prior to attempting synthesis experiments*[4,6,24]

In incinerators, especially in regions near the exit point from the heat recovery section and in pollution control equipment, gas temperatures may be lower than the boiling point of high-boiling-point organic compounds known to be present in the gas stream. This means that there can be condensation of high-boiling-point organic compounds onto the surfaces of flyash particles. If there is condensation of organic compounds onto reactive catalyst sites of the flyash particles, the catalysts may be "poisoned." If catalytic sites are covered, then dioxin or furan precursors may not be able to get to or react at catalyst sites.

Coating the walls of a reaction vessel with soot is an old technique actually used by chemical kineticists to passivate the walls of reaction vessels when they are interested to isolate gas-phase reactions from solid or gas-solid reactions. Given these observations, it may be important to consider how to relate lab experiments to actual incinerators if in the laboratory experiments, the flyash has been washed or thermally treated or otherwise cleaned prior to dioxin and furan "synthesis" or "destruction" experiments.

What would be the results if "dirty" flyash is utilized or if a "dirty" gas stream that contains high-boiling-point organic compounds, e.g., soot constituents, was the carrier gas for the dioxin or furan precursors? What happens when "dirty" flyash and "dirty" gas are used?

### *Regarding competition among various chemical species for the same reactive catalyst sites on flyash*[6]

Precursors to dioxins and furans are not the only chemical species that are present in incinerators. Many other chemical species are also present. It is possible that in incinerators, dioxin and furan formation or destruction on flyash may be retarded due to competition among many chemical species for the same reactive sites. Laboratory experiments need to be carried out in order to understand the role and extent of competitive effects.

### *Effects due to apparent increases in collision cross section and collision frequency between gas-phase species and flyash surfaces*[6,24,27]

Per unit volume of an incinerator component, e.g., furnace, boiler section, or pollution control equipment, the density of particulate matter to which flowing gas is exposed changes. There is, for example, a great difference between particle-laden flue gas and gas that flows through a "packed" bed of flyash material such as flyash "cakes" that may be collected on a baghouse filter. This suggests that the collision cross section and collision frequency between gas and flyash particulate matter can be very different depending upon the nature of the gas-solid exposure scenario. This concept is important in a fundamental sense when considering the *dynamics* of the reaction process. More experiments need to be conducted in order to quantify these effects.

### *Effects due to particle morphology and composition*[6,28]

Ash that is carried in gas streams is not homogeneous. Flyash particles can be mostly carbonaceous or ferromagnetic, or mostly alumino-silicates. Some very fine particulate material may even be nearly completely metallic in nature due to condensation of volatile heavy metal compounds. Flyash can also have combinations of these characteristics.

Fractionation of physical and chemical characteristics can be particle-size dependent. This suggests, for example, that the carbon or heavy metal content of ash may vary at pollution control equipment ash collection points.

The surface area of different types of particles can vary significantly. Carbonaceous materials often have very high surface areas. The carbon content of ash, particulate surface area, and heavy metal content are believed to be related to dioxin and furan formation and destruction.

Data reported in the following tables provides representative information about these ash characteristics.

Other aspects of particle morphology and composition may also be important. Taken as a whole, these variations in particle composition and morphology may affect the nature of the surface, volumetric, or particle-size-dependent extent of chemical processes, e.g., formation, destruction, degree of chlorination, and observed isomer distribution ratios of dioxins and furans. These remarks indicate a need to more thoroughly examine compositional and morphological effects.

### *Concerning the manner in which experiments are carried out*[24,27]

It has been of interest, for example, to understand the role of various metal compounds in catalytically promoting or retarding dioxin or furan formation or destruction. In order to study these metal compound effects, metal compounds have been added to flyash prior to passing a carrier gas laden with precursor materials over the flyash.

Some questions come to mind concerning the manner in which the metal compounds have been added. For example:

1. Were they added to a clean or dirty surface?
2. Were the added metal compounds then made dirty by adding high-boiling-point organic compounds to simulate coverages which may occur during incineration?
3. Were the metal compounds deposited onto the flyash surface by use of a liquid carrier medium which was then evaporated to leave the metal compounds behind on the flyash surface *or* were the metal compounds deposited onto the fly ash surfaces by vapor deposition of these compounds, the compounds themselves having been previously introduced into the gas phase by heating?

**Table 1. Specific Surface Area and Density of Three Ash Fractions[28]**

| Ash fraction | Specific surface area ($m^2/g$) | Density (g/cc) |
|---|---|---|
| Mineral | 0.91 | 2.34 |
| Ferromagnetic | 0.28 | 3.93 |
| Carbonaceous | 63.2 | 0.87 |

**Table 2. Heavy Metal Content (mg/g) in Ash Residue[29]**

| Fraction | Lead | Cadmium | Mercury |
|---|---|---|---|
| Bottom ash | 4 | 0.06 | Low |
| ESP ash | 13 | 0.40 | 0.007 |
| Emitted | 30 | 0.80 | 0.020 |

**Table 3. Metal Content of Bottom/Flyash and Emissions[30]**

| Species | Bottom Ash[a] | Flyash[a] | Emissions[a] |
|---|---|---|---|
| Lead | 70.0 | 23.0 | 6.0 |
| Chromium | 73.7 | 23.2 | 3.2 |
| Cadmium | 73.5 | 23.0 | 3.5 |

[a]Percent by weight in fraction compared to original amount of heavy metal fraction in waste feed.

It is important to examine these questions, given the manner by which gas-phase material is deposited onto flyash during the incineration process, i.e., condensation.

*Concerning sample probe effects*[4,7–12,21,32]

When sampling hot flue gas, if the sample probe itself gets hot, what is the sample probe surface/gas contact time and what are the potential ramifications of such contacts if the surface of the sample probe is catalytic with respect to dioxin and furan formation or destruction in the environment into which the probe has been introduced?

It is known, for example, that hot sample probes can promote nitration of organic compounds. To what extent do these effects perturb observed distributions of compounds in various locations of the incinerator? Do we need to do molecular beam extraction experiments to resolve these issues unambiguously?

Recent findings about the behavior of different probes indicates that probe design and gas/particulate sampling apparatus design can affect the observed distribution of sampled gases and solids that are found in these devices. This confounds interpretation of gas- and solid-phase distributions of organic species. These findings suggest that the effects of probe designs need to be carefully quantified in order to deconvolute perturbations associated with probe effects.

**Table 4. Heavy Metal Leachate Content (Deionized Water Leach Test)[31]**

| Size fraction ($\mu$m) | Amount leached ($\mu$g/g) | | Particle surface area ($m^2$/g) |
|---|---|---|---|
| | Cadmium | Lead | |
| 6. | 0.395 | 7.71 | 15.3 |
| 2.4 | 13.4 | 12.0 | 17.9 |
| 1.5 | 53.0 | 283 | 20.8 |
| 0.65 | 252 | 2420 | 23.0 |
| 0.37 | 527 | 4720 | 30.0 |

*Why is lead apparently a good catalyst for dioxin and furan formation in automobiles, but not apparently so in incinerators, as suggested by some recently reported experiments, and why, if different metal compounds promote dioxin formation and destruction in incinerators or in automobiles, are the apparent distributions of chlorinated dioxins and furans so similar in both cases?*[37–39]

There are differences among combustion processes that need to be accounted for. For example, in automobile combustion, lead may be momentarily converted into a supersaturated "cloud" of condensing lead nuclei of very high surface area. Such an environment may not be developed in an incinerator. Note: Recently, some data have been obtained from automobile exhausts that, according to one analyst, suggests the presence of lead compounds may not be necessary in order for dioxins and furans to form and be emitted in automobile exhausts.

*What about the dynamic nature of the reaction process?*[4,24,27,28]

In most experiments that have been performed in the laboratory or in psuedo-field tests, gas flow rates past reactive surfaces such as samples of flyash have been very low—a few centimeters per second. In actual incinerators, gas flow rates can be a thousand centimeters per second and higher.

If dioxins and furans are formed by catalyzed surface reactions, the concentration of dioxins and furans in the gas phase can depend upon reaction times, desorption energies, and the rate of diffusive transport of these molecules into the gas phase. If the rate of solid-to-gas-phase transport is slow, then due to high gas flow rates in incinerators, one would expect gas-phase concentrations of these species to be low due to the dilution effect associated with passing large volumes of gas over reactive surfaces. It is therefore important to understand and quantify the rates of solid to gas transfer processes in order to determine the importance of surfaces, such as boiler tube deposits, filter cakes, or hopper ash, towards production or destruction of these compounds.

*What is the thermal desorption energy of dioxins and furans from flyash surfaces?*[6,24,40]

Depending upon the surrounding temperature, a difference of perhaps 10 kJ/mol can significantly affect estimates of the location in an incinerator in which dioxin formation or destruction may be promoted or suppressed when the reaction medium, for example, is flowing gas that contains buoyantly entrained flyash or gas flowing through a suspended filter cake of ash. This can affect observations that dioxins are either formed going into the boiler section, going out of the economizer, and so forth, depending upon the temperature profile in the incinerator.

### *What about mixing effects?* [4–6,24,27,41]

What is the role of unmixed pockets of fuel or air in affecting these processes? What about boundary layer flow and other streamlining effects—what contribution do these processes make to observed emission levels? Theoretical analyses have been reported that indicate these effects can be quite significant. More detailed studies need to be carried out to further quantify these results. Recent testing of dioxin and furan emissions from an incinerator facility in Quebec City apparently supports theoretical predictions.

There may be significant dioxin and furan and precursor formation on the grate bed of incinerators. This can be tested. Data that support this suggestion have been reported. How much material that is produced by combustion on grate beds is transported? How do such effects affect observations downstream? Can we reasonably simulate large-scale mixing numerically or in laboratory-scale environments? The Reynold's number gives some indication of turbulence, but other parameters may be desirable to express mixing or aspects of the mixing process, e.g., the role of large-scale vortex structures.

### *More about surfaces: What tests can we perform to differentiate between surface-surface and gas-surface reaction mechanisms in highly complex heterogeneous environments, and do we understand the significance of these mechanisms in terms of strategies for controlling catalytic processes? How mobile are dioxins, furans and precursor species on the surfaces of flyash particulates?* [4,6,7–10,12,17,21,24–28,31,32,37,38,40]

It is known that the translational motion of chemical species on surfaces can vary according to several different factors. Most of these findings have been derived from studies of very pristine, homogeneous environments; how do these findings carry over to predictions of molecular behavior on "dirty" surfaces? Why has it been found in some experiments that exposing flyash to ammonia does not supress catalytic formation and destruction of dioxins and furans? What does this imply about the nature of the surface? How mobile are dioxins, furans, and precursor compounds on flyash surfaces?

### *What about the behavior of dioxins and furans on particulate matter when it is released to the atmosphere?* [4,6,15,17,24]

It is *estimated* that the gas-phase atmospheric lifetime of dioxins and furans is probably a few tens of hours. Those that are on particulate surfaces at solid/gas boundaries may have shorter lifetimes. What can be determined by actual experiment? How do degradation effects alter measured distributions of the various chlorinated dioxins and furans at locations remote from an incinerator, and how is the distribution of dioxins and furans in the atmosphere quantitatively affected by these processes? (A "back-of-the-envelope" calculation suggests that the limited atmospheric lifetime of dioxins and furans is sufficient to ensure that absent sources of these compounds, the atmosphere can cleanse itself in a relatively short time.)

*What can be learned about the atmospheric lifetimes of dioxins and furans? What is the gas-to-particulate-phase distribution that describes dioxins and furans in the atmosphere? How does this distribution affect atmospheric reaction dynamics with chemical species in the atmosphere such as the hydroxyl radical?* [4,6,15,17,24]

Apparently no data is presently available that reports reaction rates of dioxins and furans with hydroxyl radicals or other reactive species in the atmosphere. This makes it difficult to estimate the role of "atmospheric sinks" in depleting such environmental emissions. More needs to be determined about how chemical species such as dioxins and furans are distributed between the gas and particulate phase in order to understand the role of atmospheric reaction dynamics. For example, chemical species may be largely unaffected by atmospheric radicals if the chemicals are volumetrically distributed in atmospheric particulates rather than being on particulate surfaces or in the gas phase. Additionally, models of volatilization dynamics of chemicals, e.g., their movement from soils to the atmosphere, also need more development.

### *Measurement errors* [3,4,6,13,14,16,21–24,42–46]

Measurement errors associated with monitoring emissions of dioxins and furans may be perturbed by such effects as feedstock inhomogeneities, process variations, probe effects, sample probes location, analysis errors, and sample aging effects. While some information is emerging that allows a partial deconvolution of such errors, much remains to be determined. At the present time, the probability for false positive and false negative error measurements appears large and prospects for developing reliable correlation techniques to look for associations between variables are difficult to evaluate.

These problems confound determinations of the role of physical and chemical processes in promoting or destroying dioxins and furans in combustion processes. Consequently, it is important to significantly reduce or quantify errors associated with measurements of dioxin and furan emissions. One current goal of facility engineers is to be able to rely upon surrogate descriptors of dioxin and furan emissions rather than upon direct measurement of these chemical species. Operationally, control of surrogate parameters is commonly accepted to be more feasible than direct control of dioxin and furan emissions. Suggestions for surrogate measurements have included monitoring process stream temperatures, carbon monoxide emissions, pollution control equipment temperatures, particulate emission rates, and other parameters. Measurement errors can confound the demonstration of associative relationships between dioxin and furan emissions and surrogate parameters. Without being able to demonstrate associative relationships, it may be difficult to justify the use of surrogate control parameters. While some data derived from both theoretical and experimental studies have been reported, more research appears to be needed to resolve uncertainties.

### *Costs of sampling and monitoring emissions* [13]

As noted in this report, continuous monitoring of dioxin and furan emissions is not feasible at the present time. Additionally, the costs of periodic monitoring for dioxin and furan emissions are high and affect the economies of scale among different-sized incineration facilities. High costs of monitoring are not solely restricted to monitoring emissions of dioxins and furans. For example, a recent report indicated that monitoring hydrogen chloroide emissions can readily cost more than $50,000.

Cheaper, more cost-effective and reliable monitoring techniques are required to ensure that desired information about species emissions can be obtained.

### *Temperature measurement* [4]

Control of incineration facility temperatures is thought by some to be one of several surrogate approaches to controlling emissions of dioxins and furans. There are several methods for measuring temperatures at locations in the process stream. Information about measurement protocol and assessment of available test results is required to ensure that measurements being made are reliable.

Temperature probes can corrode, become clogged, be improperly utilized, or be inappropriate in required applications. Situations have occurred where faulty probes have been relied upon because no easy and cost-effective means to periodically monitor probe reliability have been available. In some instances, very large measurement errors have resulted. This may affect the duty cycle of a facility.

More measurement equipment reliability test requirements could ensure timely detection of probe failures. An up-to-date assessment of equipment failure history would be useful, pertaining not only to temperature measurement, but also to other key operation parameters. It is possible that these efforts to detect failures in monitoring devices could reduce operation and maintenance costs.

Low-cost, reliable, nonintrusive optical diagnostic devices have not been practically applied for monitoring incineration facility process temperatures. Reports of optical fiber device applications are not known.

### *Environmental availability of dioxins and furans that are incorporated on or in deposits of incinerator particulates* [47]

Recently reported measurements of the leachability of dioxins and furans from environmentally deposited ash particulates from incinerators indicates that these chemical species are highly immobile. Available information needs

**Table 5. Ash Generation Statistics[48]**

| | |
|---|---|
| Operation/shakedown projects (1986) | 2,159,340 |
| Construction/advanced planning/restart (by 1989) | 7,022,235 |
| Conceptual (by 1989) | 8,527,860 |
| Total ash residue generation (by 1989) | 17,709,435 |

**Table 6. Major Uses of Lead in the United states (1984)[49]**

| Material | Tons | Percent of total |
|---|---|---|
| Storage batteries | 954,100 | 71.7 |
| Gasoline additives | 87,000 | 6.5 |
| Pigments | 84,700 | 6.4 |
| Ammunition | 52,700 | 4.0 |
| Other | 151,900 | 11.4 |
| Total: | 1,330,400 | 100.0 |

to be analyzed in composite to determine whether this information can be reported as a consensus observation. Uncertainties need to be taken into account; for example, the potential for fine particle movement in surface or other water stream flows.

## HEAVY METALS

### Background

The product of solid waste incineration is ash residue. This ash residue is managed, either by disposal into a landfill or through limited reuse, e.g., as a roadbed aggregate material. Heavy metals, e.g., lead, cadmium, mercury, arsenic, and chromium, are found in ash residue.

### Sources and distribution of heavy metals in solid waste

Heavy metals found in ash residue come from products that are disposed into the solid waste stream and incinerated. An examination of available literature indicates the following major uses of heavy metals and the distribution of these compounds in the solid waste stream.

**Table 7. Major Uses of Cadmium in the United States (1984)[49]**

| Material | Tons | Percent of total |
|---|---|---|
| Storage batteries | 1003 | 27.0 |
| Pigments | 595 | 14.0 |
| Plastics | 520 | 16.0 |
| Motor vehicles | 401 | 10.8 |
| Communications equipment | 301 | 8.1 |
| Fasteners | 234 | 6.3 |
| Other | 662 | 17.8 |
| Total: | 3716 | 100.0 |

**Table 8. Major Uses of Mercury in the United States (1986)[50]**

| Material | Tons | Percent of total |
|---|---|---|
| Batteries/electrical equip. | 1034 | 56.0 |
| Caustic chlorine | 221 | 12.0 |
| Paints | 185 | 10.0 |
| Industrial control equipment | 110 | 6.0 |
| Other | 296 | 16.0 |
| Total: | 1846 | 100.0 |

## Distribution of heavy metals: air emissions and ash residue

Solid waste incineration operations may utilize pollution abatement control equipment. Scrubber-baghouse or scrubber-electrostatic precipitator equipment combinations or their equivalent can significantly minimize air emissions of heavy metal compounds to the environment. Mercury abatement requires careful temperature control. In Table 10, representative heavy metal inlet and outlet emissions from a facility in Quebec City are reported.

## Distribution of heavy metals in ash residue fractions

A recent analysis of some heavy metals distributions at a facility located in Avesta, Sweden, is reported in Table 2. The reported data indicate enrichment in the flyash fraction due to volatilization of heavy metals, e.g., as chlorides, during incineration. Volatilization of heavy metal compounds is temperature sensitive. The percent of volatilization of some heavy metal compounds has recently been reported.[52] Results are presented in Table 11.

**Table 9. Distribution of Metals in Household Waste[51]**

| | Percentage of | | | |
|---|---|---|---|---|
| Fraction | Cd | Cr | Hg | Pb |
| Plastics | 26 | 5 | 10 | 5 |
| Paper | 4 | 7 | 13 | 3 |
| Animal matter | 1 | 1 | 2 | 1 |
| Vegetable matter | 2 | 2 | 6 | 2 |
| Textiles | 1 | 1 | 4 | 1 |
| Rubber and leather | 4 | 42 | 3 | 2 |
| Metals | 60 | 43 | 60 | 85 |
| Miscellaneous | 3 | 3 | 3 | 4 |

**Table 10. Scrubber-Baghouse Inlet/Outlet Concentrations ($\mu g/m^3$ at 8% Oxygen)[5]**

| Metal | Inlet | Outlet |
|---|---|---|
| Cadmium | 1,000—1,600 | ND—0.6 |
| Lead | 30,000—45,000 | 1—6 |
| Chromium | 1,400—3,000 | ND—1 |
| Arsenic | 80—150 | 0.02—0.07 |
| Mercury | 200—500 | 10—600 |

**Table 11. Volatilization as a Function of Heating Temperature**

| Compound | Percentage of volatilization at: | | | |
|---|---|---|---|---|
| | 700°C | 800°C | 900°C | 1000°C |
| Zinc chloride | 80 | 100 | 100 | 100 |
| Cadmium chloride | 20 | 9 | 40 | 100 |
| Lead chloride | 20 | 8.5 | 42 | 100 |
| Cadmium oxide | 0.01 | 0.08 | 0.4 | 2 |
| Lead oxide | 0.1 | 0.3 | 2 | 10 |

As reported in Table 3, heavy metals distributions in ash residue fractions have been measured.[30,53] Results can vary from facility to facility.

## Environmental availability of heavy metals in ash residue

Volatile heavy metal compounds in the incinerator that are subsequently condensed onto fine particulates are leachable due to the high surface-to-volume ratio of these particles and the expectation that these compounds are surface deposited rather than distributed throughout the particulate volume. [31] Results are reported in Table 4 which suggest leaching behavior. Anomalies occur in the distribution of heavy metal compounds and in their leaching characteristics depending upon many factors, e.g., the particle morphology, leachate medium, specific chemical composition of heavy metal compounds, specific surface area, etc. [54,55]

## Problems with ash residue heavy metal leach tests

Many regulators are in agreement that presently required laboratory tests for evaluating leaching characteristics of ash are not reliable indicators of actual leaching behavior at ash disposal sites.[56,57]

Researchers at the University of Massachusetts, Amherst have, on examination of available ash literature, reported that:[42]

> "The literature . . . provides essentially no planned experiments designed to evaluate ash chemistry and its relation to causes of its variability. Discussion of the statistical sampling design used in the research is missing from nearly every document, as is information regarding existing incinerator and environmental conditions at the time of sampling. The information which is provided often is of variable quality and lacks documentation regarding sample size, number of replicates, and incinerator parameters. Most data are simply observational. As a result, a numerical analysis can give only limited understanding of the processes of incineration. Inconsistent terminology of the various residues from the incineration process also limits and complicates data comparison. Although information was sought in this project on the effects of a variety of factors on ash properties, in many cases no coherent literature or data were found. No studies have been developed on a scientifically based sampling design: most sampling has been done by the grab method and lacks standardization . . ."

**Table 12. Actual Leachate Results Versus EP Toxicity Test Results for an Ash Monofill (mg/L)**

| Specie | Actual Leachate results | EP Toxicity test results[a] | EPA limit |
|---|---|---|---|
| Pb | 0.120, 0.160, 0.160, 0.160, 0.005, 0.65 | 19.1, 24.3, 4.46, 2.0 | 5.0 |
| Cd | 0.027, 0.027, 0.044, 0.039, <0.001 | 1.88, 2.38, 0.380, 0.70 | 1.0 |

[a]Underlined values denote that EPA limit has been exceeded.

A recent study that was conducted by researchers at the University of Massachusetts at Waltham reported problems with interpretion of required protocol:[43]

> "When residue samples which failed the EP test in our laboratory were partitioned and given to certified commercial laboratories, they usually passed for both metals (lead and cadmium). Replicate samples tested at two other State laboratories did not pass for Pb or Cd. The discrepancy seems to be due to interpretation of methodology. An added acidification step, prior to analysis, carried out in the commercial labs appears to tie up the Pb, thus reducing the available concentration. Cadmium is not affected. (Similar results were seen with flyash.) The difference in methodology leaves in doubt just how much lead is actually available."

A scientist at a recent conference [44] reported EP Toxicity test data developed by three different laboratories that examined the lead content in the *same* ash samples taken from an incinerator. Differences in the amount of lead found in the same ash samples but as measured by three different laboratories apparently varied by between one and two orders of magnitude, suggesting that measurement accuracy was questionable. Taken on face value, apparently in six out of nine cases, the three laboratories could not determine by consensus agreement that the ash samples either passed or failed the EP Toxicity test. If the interlaboratory variation in the reported data is taken as an indicator of the accuracy of the test results, then it appears that the outcome of all of the tests, relative to the true value of the leachate parameter for each ash sample, is questionable.

Data reported in Table 12 illustrate in a representative manner the extent to which currently required leachate tests fail to reproduce actual leachate analysis results.[45]

The data reported in Table 12 and data from other sources[1,42–44,56,57] suggest a need to re-evaluate the design and utilization of ash leach tests for reliably estimating the extent of heavy metals leaching in the environment.

## Technical issues requiring additional research

Information presented earlier and in referenced literature suggests that questions and associated observations posed here may be important to consider.

**Table 13. Municipal Waste Stream: Generation and Management**

| Category | Year 1970 | 1984 | 2000 |
|---|---|---|---|
| Waste generation[a] | 118.3 | 148.1 | 182.2 |
| Energy recovery | 0.3% | 4.4% | 17.5% |
| Materials recovery | 6.8% | 10.2% | 12.8% |

[a]Generation in millions of metric tons of solid waste.

### *To what extent can the ash residue be reused?*

Data reported in Table 13 suggest future trends in waste generation and management in the United States.[58]

Data reported in Table 5, indicate amounts of ash residue that may be produced from energy recovery processing. This ash residue must be either disposed of or reused. According to a recent study [58] we are nationally recycling about 10% of our disposed waste. Now let us assume a national recycling rate of 40%—an extremely optimistic assumption. At the waste stream's present growth rate, by the year 2000, we will still have to annually dispose of an amount of waste that is equivalent to about 80% of the total amount of waste that is generated today. And that is *after* the 40% is recycled. In addition, projections of incineration capacity by the year 2000 are typically below about 20 to 25% of the total volume of waste generated.

Despite these observations, we are presently closing down our nation's existing landfills at unprecedented rates. If ash residue could be reused, future landfill capacity problems may be partially mitigated. Consequently, more research is needed to determine means by which ash residue re-use may be practiced.

### *What are prospects for reducing the heavy metals content of the solid waste stream?*

Ash may contain leachable heavy metals unless steps are taken to reduce the heavy metals content of the solid waste stream. Economic impacts and technological feasibility to remove heavy metals from consumer products are unresolved. In Europe, difficulties have been reported:[59] The removal of heavy metals from products is technologically feasible in some instances, but remains economically unfavorable. A recent report considers differences between *removing* materials from a waste stream and *reutilizing* the removed materials:[60] Reutilization does *not* keep up with removal.

### *What might be done to mitigate volatilization of heavy metals if this is desired?*

The concentration, but not always the total amount, of heavy metals in flyash, can be large compared to bottom ash largely due to formation of volatile heavy metal compounds during the incineration process that are subsequently condensed onto flyash in cooler regions of the incinerator. As noted previously, heavy metals in the flyash fraction can be more leachable than those in the bottom ash fraction. This observation must be made with caution.[1,5,29–31,42–45,52–57]

One approach to reducing concentrations of heavy metals on flyash may involve removal prior to incineration of materials that may release significant amounts of volatile heavy metal compounds. It is not entirely apparent which materials ought to be or can efficiently be removed. Different solid waste stream objects may contribute to this problem to varying extents depending upon the nature of the incineration process.

Analyses of the flux of heavy metals through incinerators have been reported.[1,5,31,54,55] These studies usually do not associate emissions with specific objects in the solid waste stream. Removal of batteries [51,59] has been advocated by some. More general approaches have also been undertaken: wholesale removal of aluminum and ferrous metals and glass, grit, dirt, and other non-combustibles. Some data has been reported which indicates the effect of wholesale removals on heavy metals emissions from incinerators.[53]

Wastes in which heavy metals are present as halides or from which heavy metal halide compounds are formed during heating have sometimes been treated with silicate additives to suppress volatilization.[52] It is not known if this approach can work with municipal solid waste or in treatment of ash residues. The conversion of heavy metal halides to less volatile heavy metal oxides by reinjection of flyash back into the furnace of an incinerator has been suggested.[46] Prospects for success of this approach are unknown at present. Combinations of these approaches are apparently not reported.

### *Can a statistically valid method be developed that can be used to obtain representative samples of ash residue for analysis?*

The discussion presented in this report indicates that no statistically valid consensus method is available for obtaining representative samples of ash residue for analysis of heavy metals content. Variations occur in particle size, chemical composition of ash components, and physical morphology. There can be hourly, daily, weekly, and seasonal variations in all ash properties. Other heterogeneous characteristics occur.

Absent reliable sampling procedures, it is difficult to determine the potential for false positive and false negative measurement results. Consequently, ash residue management programs may become dependent upon application of heuristic or defined sampling procedures that may have no established scientific basis. Such approaches may be unnecessarily costly and may not afford appropriate protection of human health and the environment.

### *Can a reliable laboratory leach test be developed that can simulate actual leaching of heavy metals from ash residue as might occur when ash is disposed in the environment or managed by reuse?*

As noted in this report, leachate extraction protocol may not be clearly defined or lead to reproducible test results. The former problem can lead to procedural ambiguities. In the latter instance, current laboratory leach tests apparently do not reliably reproduce the *actual* extent of leaching that takes place when ash is disposed in landfills. Apparently, the extent of heavy metals leaching that actually occurs in the environment is *less* than that predicted by laboratory tests.

Additionally, when test results have large measurement errors associated with them and the values reported are near the regulatory limit, there can be a significant probability that an incorrect management action may be taken. The test result may be a false positive test result. The action that could be taken may, in fact, not be necessary. Conversely, when there is a false negative test result, actions that may have been necessary may not be taken. It is not possible to know exactly when this occurs.

### *About the costs of tests and analysis*

Collection and analysis of ash samples and onsite leachate testing at ashfills for heavy metals content can cost tens of thousands of dollars. One recent CORRE/U.S. EPA program of ash sample and ash leachate tests at a single incinerator facility and ash disposal site was estimated to cost $63,620 for a much scaled-down battery of tests for lead, cadmium, and a few other pollutants. This did not include any detailed plume mapping or utilization of monitoring wells, which would have added significantly to final test costs. A careful assessment of test costs and an analysis of prospects for reducing financial impacts could increase the feasibility of test programs for obtaining research data. Some effort to establish adequate surrogate indicators of leachate composition and behavior may be useful.

## CONCLUSIONS

This report has examined some information pertaining to emissions of dioxins and furans and to heavy metals in incinerator ash residue. Uncertainties that can affect abatement of dioxin and furan emissions, ash residue management, and control of heavy metals have been identified. These uncertainties can be posed as questions. Analysis of these questions may lead to indications of means to improve present management practices. Some information needs have been identified.

## REFERENCES


1. Hartlen, J., and P. Elander. "Residues From Waste Incineration," in *SGI Varia 172*, (Linkoping: Swedish Geotechnical Institute, 1986).
2. Rappe, C., R. Andersson, P.-A. Bergqvist, C. Brohede, M. Hansson, L.-O. Kjeller, G. Lindstrom, S. Marklund, M. Nygren, S. E. Swanson, M. Tysklind, and K. Wiberg. "Sources and Relative Importance of PCDD and PCDF Emissions," *Waste Manage. Res.* 5:225–237 (1987).
3. Shaub, W. M. "Sources of Emissions in Resource Recovery Plants," from transcript of oral remarks made at the N.Y. Academy of Sciences Meeting, New York, December 18, 1984.
4. Shaub, W. M. "Technical Issues Concerned With PCDD and PCDF Formation and Destruction in MSW Fired Incinerators," National Bureau of Standards Report NBSIR 84-2975 (November 1984).
5. Finkelstein, A., R. Klicius, and D. J. Hay. "The NITEP Program," presentation at the International Workshop on Municipal Waste Incineration, Montreal, Quebec, October 1 to 2, 1987.
6. Shaub, W. M. "Containment of Dioxin Emissions from Refuse Fired Thermal Processing Units – Prospects and Technical Issues," National Bureau of Standards Report NBSIR 84-2872 (February 1984).
7. Hagenmaier, H., H. Brunner, R. Hagg, and M. Kraft. "Copper-Catalyzed Dechlorination/Hydrogenation of PCDDs, PCDFs, and other Chlorinated Aromatic Compounds," *Environ. Sci. Technol.* 21(11):1085–1088 (1987).
8. Hagenmaier, H., H. Brunner, R. Haag, and M. Kraft. (Verfahren zur Zerstorung von Chloraromaten) Patent Pending DP 36 23 492.3.
9. Stieglitz, L., and H. Vogg. "On Formation Conditions of PCDD/PCDF in Fly Ash From Municipal Waste Incinerators," paper presented at the 6th International Symposium on Chlorinated Dioxins and Related Compounds, September 16 to 19, 1986.
10. Vogg, H., M. Metzger, and L. Stieglitz. Recent Findings on the Formation and Decomposition of PCDD/PCDF in Solid Municipal Waste Incineration," paper presented at a specialized seminar on Emission of Trace Organics from Municipal Solid Waste Incinerators, Copenhagen, Denmark, January 20 to 22, 1987.
11. Concord Scientific Corporation. "NITEP P.E.I. Testing Program, Volume II: A Report to Environment Canada," Downsview, Ontario (June 1985).
12. Ballschmitter, K., R. Niemczyk, I. Braunmiller, J. Ehmann, U. Duwel, and A. Nottrodt. "Chemistry of Formation and Fate of PCDD/PCDF in the Municipal Waste Incineration Process," paper SE11 presented at Dioxin '87, University of Nevada, Las Vegas, October 4 to 9, 1987.
13. Velzy, C. O. "Statement of the Research Committee on Industrial and Municipal Wastes of the American Society of Mechanical Engineers on H.R. 2787," before the Subcommittee on Health and the Environment, Committee on Energy and Commerce, U.S. House of Representatives, July 2, 1987.
14. Organization for Economic Cooperation and Development, "Energy From Waste – A Brief Assessment of the Situation in OECD Member Countries," ENV/WMP/86.6, Paris (drafted May 15, 1986).
15. Atkinson, R. "Estimation of OH Radical Rate Constants and Atmospheric Lifetimes for Polychlorobiphenyls, Dibenzo-p-dioxins, and Dibenzofurans," *Environ. Sci. Technol.* 21:305–307 (1987).

16. Rappe, C. Private communication to W. M. Shaub (October 23, 1987).
17. Stein, S. Private communication to W. M. Shaub (January 19, 1988).
18. Seinfeld, J. H. Private communication to W. M. Shaub (June 10, 1987).
19. Mackay, D. Private communication to W. M. Shaub (June 11, 1987).
20. Smith, A. H. Private communication to W. M. Shaub (June 30, 1987).
21. Hagenmaier, H., and M. Kraft. "Studies Towards Validated Sampling of PCDDs and PCDFs in Stack Gas," presentation at the International Workshop on Municipal Waste Incineration, Montreal, Quebec, October 1 to 2, 1987.
22. Shaub, W. M. "Waste-to-Energy Facilities—Compilation of Data About Lead, Lead Compounds and Lead Emissions," CORRE Technote 87-0003 (August 25, 1987).
23. Shaub, W. M. "Heavy Metals—Technical Issues Concerned With the Non-Hazardous Management of Incinerator Ash in Monofills," CORRE Technote 87-0004 (October 14, 1987).
24. Shaub, W. M. Private communication to L. Stieglitz (October 6, 1987).
25. Golden, J. H. "Polyphenylene Oxides," in *High Temperature Resistance and Thermal Degradation of Polymers*, (London: Society of the Chemical Industry, 1961).
26. Hay, A. S. "Aromatic Polyethers," *Adv. Polymer Sci.* 4:496–527 (1967).
27. Shaub, W. M. "Thermal Decontamination of Soils: A Theoretical Analysis of Soils That Contain PCDDs and PCDFs," Private document (October 1984).
28. Schure, M. R. "The Effect of Temperature Upon The Transformation of Polycyclic Organic Matter," PhD Thesis, Colorado State University (1981).
29. Carlson, K. "Heavy Metals From Energy From Waste Plants— Comparison of Gas Cleaning Systems," *Waste Manage. Res.* 4:15–20 (1986).
30. Buekins, A., and P. K. Patrick. "Incineration," in *Solid Waste Management: Selected Topics*, (M. J. Suess, Ed., (Copenhagen: 1985), World Health Organization, pp.79–150.
31. Taylor, D. R., M. A. Tompkins, S. E. Kirton, T. Mauney, P. K. Hopke, and D. F. S. Natusch. "Analysis of Fly Ash Produced from Combustion of RDF and Coal Mixtures," *Environ. Sci. Tech.* 16:148–154 (1982).
32. Shaub, W. M. Private communication to D. Barnes (September 24, 1986).
33. Risby, T. H. and S. S. Lestz. "Is the Direct Mutagenic Activity of Diesel Particulate Matter a Sampling Artifact?" *Environ. Sci. Technol.* 17:621–624 (1983).
34. Brorstrom-Lunden, E., and A. Lindskog. "Degradation of Polycyclic Aromatic Hydrocarbons During Simulated Stack Gas Sampling," *Environ. Sci. Technol.* 19:313–316 (1985).
35. Malte, P. C., and J. C. Kramlich. "Further Observations of the Effect of Sample Probes on Pollutant Gases Drawn from Flame Zones," *Combust. Sci.Technol.* 22:263–269 (1980).
36. Brorstrom, E., P. Grennfelt, A. Lindskog, A. Sjodin, and T. Nielsen. "Transformation of Polycyclic Aromatic Hydrocarbons During Sampling in Ambient Air by Exposure to Different Nitrogen Compounds and Ozone," in *PAH: Formation, Metabolism and Measurement*, M. Cooke and A. J. Dennis. Eds., (7th International Symposium), (Columbus: Battelle Press, 1983).
37. Karasek, F. "Model Studies of PCDD Formation During Incineration," presentation at the International Workshop on Municipal Waste Incineration, Montreal, Quebec, October 1 to 2, 1987.
38. Stieglitz, L. "New Aspects of PCDD/PCDF Formation in Incineration," presenta-

tion at the International Workshop on Municipal Waste Incineration, Montreal, Quebec, October 1 to 2, 1987.

39. California Air Resources Board, Engineering and Evaluation Branch Monitoring and Testing Laboratory Division, "Evaluation Test for PCDD and PCDF Emissions from Motor Vehicles," Preliminary Draft Report, (October 1987).
40. Shaub, W. M. Private communication to D. J. Hay (May 19, 1987).
41. Ghezzi, U., and M. Giugliano. "L'incenerimento ed i microinquinanti organoclorurati: effetti della post-combustione," *Ing. Ambientale* 14:175–179 (1985).
42. University of Massachusetts. "Incinerator Ash Management Project," Draft Report for NYSERDA/Mass. DEQE (October 1987).
43. Mika, J. S., and W. A. Feder. "Final Report of Incinerator Residue Research Program," University of Massachusetts, Waltham, (October 1985).
44. Sutherland, M. "Application of Statistical Analysis on Incinerator Test Data," oral presentation at the International Workshop on Municipal Waste Incineration, Montreal, Quebec, October 1 to 2, 1987.
45. Shaub, W. M., Ed. "CORRE News Release 87-0002," U.S. Conference of Mayors, Washington, DC, (December 21, 1987).
46. Organization for Economic and Development. "Control of Organic and Metal Emission from MSW Incineration", (Restricted document), Paris (October 15, 1987).
47. Modig, S. "Swedish View of the Ash Issue," paper presented at a seminar jointly sponsored by the Swedish World Trade Association and the National Resource Recovery Association, Washington, DC, March 23, 1988.
48. "Incinerator Ash: Aggregate Tonnage and Cost," NSWMA Fact Sheet, prepared 1986.
49. "The End-Use Market for 13 Non-Ferrous Metals," U.S. Department of Commerce (June 1986).
50. Carrico, L. (Bureau of Mines). Private communication to R. G. McInnes and N. H. Kohl (ERT) (June 1987).
51. National Swedish Environmental Protection Board, "Energy From Waste," (1987), p. 11.
52. Kox, W. M. A., and E. Van Der Vlist. "Thermal Treatment of Heavy -Metal-Containing Wastes," *Conserv. Recycling* 4:29–38 (1981).
53. Roos, C. E. "Is Lead a Big Problem?," *Waste Age* (February 1988), pp. 54–56.
54. Greenberg, R. R., W. H. Zoller, and G. E. Gordon. "Composition and Size Distribution of Particles Released in Refuse Incineration," Environ. Sci.Technol. 12:566–573 (1978).
55. Vogg, H., H. Braun, M. Metzger, and J. Schneider. "The Specific Role of Cadmium and Mercury in MSW Incineration," *Waste Manage. Res.* 4:65–74 (1986).
56. Private communications at Ash meeting organized by NYSERDA and Mass. DEQE; Boston (March 30 to 31, 1987).
57. Private communications at Incinerator Ash Research Seminar, University of New Hampshire (May 14 to 15, 1987).
58. Seward, R. W. NBS Standard Reference Materials Catalog 1986–87," NBS Special Publication, Department of Commerce (1987).
59. "No Drop in Landfill Volume?" *Waste Age* (November 1986), pp. 27–28.
60. Johnson, K. (Mayor, Schenectady, NY) Remarks as cited in *Soild Waste Power* (February 1988).

61. "Fate of Small Quantities of Hazardous Waste," OECD Environment Monograph No. 6, Paris (August 1986).
62. Yakowitz, H., "Waste Management Activities in Selected Industrialized Countries," paper presented at Wastech '87, sponsored by NSWMA, San Francisco, October 26 to 27, 1987.

# CHAPTER 14

# Incineration—Some Environmental Perspectives

**Walter M. Shaub**

## ABSTRACT

Minimization and control of pollutant air emissions and appropriate ash residue management practices are important objectives that must be met when incineration of solid waste is an option that is incorporated into integrated solid waste management plans. This report examines these issues and provides an outline of current information about incinerator technology and practices that, in relation to desired objectives, illustrates directions of the incinerator industry that are consistent with concerns of the public and public officials.

## INTRODUCTION

With increasing public awareness of the need to more effectively manage disposed solid waste, an interest has developed to minimize the amount of solid waste that is generated and to optimize management practices, particularly with respect to that part of the solid waste disposal stream that is perceived to pose potential problems with respect to human health and environmental impacts. Where practical, minimization efforts have concentrated upon:

1. reduction of amounts of consumer products that enter the solid waste stream
2. elimination or mitigation of potentially hazardous components or constituents of components in the solid waste stream through product modification or proscriptive covenants against hazardous waste disposal into solid waste streams
3. removal of waste stream components such as glass, paper, and ferrous fractions for reutilization (a practice which temporarily defers the time of ultimate disposal)
4. weight and volume reduction by incineration of combustible components
5. when solid wastes are incinerated or their residues are to be disposed, technology enhancements such as pollution abatement control devices, and other practices are employed

**Table 1. Recycling—Effect of Reutilization Rates Lagging Behind Removal Rates**

| Year | 1984 | 2000 | 2000 | 2000 | 2000 | 2000 |
|---|---|---|---|---|---|---|
| Waste generation (millions of metric tons) | 148.1 | 182.2 | 182.2 | 182.2 | 182.2 | 182.2 |
| Recycling goal (%) | 10.2 | 20 | 25 | 30 | 35 | 40 |
| Amount remaining after removal | — | 145.8 | 136.7 | 127.5 | 118.4 | 109.3 |
| As percentage of 1984 waste generation if rate of reutilization is: | | | | | | |
| 100% | — | 98.4 | 92.3 | 86.1 | 80.0 | 73.8 |
| 75% | — | 104.6 | 99.95 | 95.3 | 90.7 | 86.1 |
| 50% | — | 110.7 | 107.6 | 104.6 | 101.5 | 98.4 |

It is perceived by some that recycling (removal of waste components for reutilization as consumer products) offers a strong prospect to reduce the amount of solid waste ultimately to be disposed. This is a complicated issue for at least three reasons. First, material wear and degradation effects (entropy) dictate that recycling can only defer the time of, not prevent, ultimate disposal of consumed materials. Second, the processes of recycling produce their own waste streams which *add* to and change the nature and volume of waste to be disposed.

Third, there is a problem with reutilization lagging behind removal of materials from the solid waste stream. Reported in Table 1 are data about the effect of reutilization of removed solid waste upon future disposal requirements on a weight basis that has been "normalized" to 1984 solid waste generation statistics analogous to the manner by which economic activities are sometimes estimated in "constant" dollars. The waste generation statistics for 1984 and 2000 and recycling activity for 1984 reported in Table 1 were developed by Franklin Associates under contract to the U.S. EPA;[1] the other calculations have been performed by the author.

It can be seen from examining Table 1 that unless very substantial amounts of the removed waste stream components are actually reutilized, the amount of solid waste that will remain to be disposed in the year 2000 will be quite large *even if removal rates are as high as 40%*. The amount of solid waste to be disposed even after recycling apparently will remain quite substantial unless other waste management options are incorporated such as minimization of the total size of the solid waste stream that is generated in the first place or, alternatively, end-use management activities such as composting, incineration, and beneficial ash reutilization management practices are implemented.

Presently, composting activity in the United States is very minimal and incineration accounts for a means by which the volume $< 10\%$ of the solid waste stream is reduced.[1] Incineration is viewed favorably by some as, in addition to waste volume reduction, it affords the prospects of supplemental

**Table 2. Environmental Impacts of Solid Waste Management Options Some Representative Problems to Minimize or Treat**

| Process | Some potential environmental impacts |
|---|---|
| Composting | Air emissions; PCBs and heavy metals in compost product |
| Incineration | Air emissions (heavy metals, dioxins, furans, acid gases, etc.); ash residue (principally heavy metals) |
| Landfill | Air emissions (volatile organics); heavy metals, hospital waste, and hazardous wastes into ground water |
| Recycling | |
| Paper | Contaminated waste sludge; lead, cadmium and mercury in waste effluent; solvents; air emissions |
| Plastics | Solvent waste streams; air emissions |
| Ferrous | Air emissions; pickle liquor and waste sludges |
| Batteries | Lead, cadmium, mercury, etc. in sludge and air emissions |
| Glass | Respirable quartz, etc. from grinding and crushing |
| Rubber | Zinc, chromium, mercury, etc. in air emissions, etc. |
| Copper wire | Dioxin and furan emissions (when wire is plastic coated) |
| Automobiles | Organic fuels, solvents, dioxins, metals, and asbestos emissions |
| White goods | Chlorofluorocarbons |
| Construction waste | Asbestos and other respirable dusts |
| All processes | Transportation—noise and emissions; occupational exposures, e.g., when sorting waste materials |

energy generation and production of a potentially reusable ash residue for road-bed aggregate material or other purposes.

Solid waste management will require an *integrated* menu of approaches for dealing effectively with the volume of waste that is to be disposed. To employ various solid waste management practices, it is of interest to optimally minimize potential environmental impacts within the constraints of economic feasibility. Table 2 provides information which is indicative of the nature of environmental issues that require consideration in developing impact minimization strategies. It is apparent that virtually any solid waste management practice requires careful attention to avoid unnecessary risk to human health and the environment. More comprehensive risk assessments which include evaluation of *all* waste management options—recycling, composting, incineration, landfilling—need to be carried out. To date, attention has focused principally upon risk assessments of incineration practices. This is a societal disservice. Less obvious, but equally important, is a very *serious* need when an estimate of risk is constructed to also provide a reasonable estimate of the *probability* that the risk estimate will in fact be realized. There also has been a significant failure to *compare* risk estimates associated with various societal activities and this has led to a general failure to provide the public with information that could allow risks to be put into a perspective that has a comparative frame of reference.

In view of the objectives of this author to address incineration practices, it is beyond the scope of this paper to address all of these issues. Instead, this

**Table 3. A Short Inventory of Metals in Consumer Products (metric tons)[2,3]**

| | |
|---|---|
| Lead (1984) | |
| Storage batteries | 954,100 |
| Gasoline additives | 87,000 |
| Pigments | 84,000 |
| Ammunition | 52,000 |
| Other | 151,900 |
| Cadmium (1984) | |
| Storage batteries | 1,003 |
| Pigments | 595 |
| Plastics | 520 |
| Motor vehicles | 401 |
| Communications equipment | 301 |
| Fasteners | 234 |
| Other | 662 |
| Mercury (1986) | |
| Batteries/electrical equipment | 1,034 |
| Caustic chlorine production | 221 |
| Paints | 185 |
| Industrial control equipment | 110 |
| Other | 296 |

report will focus upon some aspects of potential health and environmental impacts of incineration practices.

## METALS IN CONSUMER PRODUCTS AND INCINERATOR ASH RESIDUE

Analysis of ash residue that is produced by incineration of solid waste reveals that varying amounts of heavy metal compounds are present in the residue. This indicates that the solid waste stream that has been incinerated includes disposed consumer products which when manufactured contain heavy metals that have been incorporated into them. Table 3 reports some of the uses of heavy metals such as cadmium, lead, and mercury in consumer products.

Another sort of analysis for heavy metals that has been carried out is an examination of the amounts of heavy metals that are found in the various fractions (paper, plastic, metals, etc.) of disposed solid waste. Table 4 indicates the results of one recent study of this nature that was carried out in Sweden. It can be seen that most of the heavy metals can be found in the metals fraction of the solid waste stream, although significant amounts of these compounds can be found in other fractions.

This latter observation has important ramifications for assessing the origins of heavy metals in incinerator flyash fractions. It may be that heavy metals in flyash are derived in significant quantities from the more combustible fractions of the solid waste stream, such as paper and plastic fractions. Therefore, unless levels of heavy metals in consumer products are reduced or the uses of heavy metals in consumer products are eliminated, it will be necessary to

**Table 4. Data from Swedish Survey of Some Heavy Metals in the Solid Waste Stream[4]**

| | Fraction[a] | | |
|---|---|---|---|
| | Cadmium | Mercury | Lead |
| Plastics | 26 | 10 | 5 |
| Paper | 4 | 13 | 3 |
| Animal matter | 1 | 2 | 1 |
| Vegetable matter | 2 | 6 | 2 |
| Textiles | 1 | 4 | 1 |
| Rubber and leather | 4 | 3 | 2 |
| Metals | 60 | 60 | 85 |
| Miscellaneous | 2 | 2 | 1 |

[a]All values reported are percent by weight.

consider the potential impact of heavy metals in ash residue in order to understand prospects for reutilization of ash residue, whether flyash and bottom ash streams should be managed separately and how ash residue that is not reutilized can be disposed.

## DIOXIN EMISSIONS FROM INCINERATION OF SOLID WASTE

Since about 1974,[5] it has been known that incineration of municipal solid waste can produce emissions of compounds such as chlorinated dioxins and furans (in the remainder of this report, when mention is made of dioxins, it should be understood that comments are applicable to furans also unless otherwise noted). While inventories of dioxins in the environment show that there must be other equally or more significant sources of dioxins in the environment besides incineration of municipal solid waste,[6-9] it is of interest to determine how to effectively minimize incinerator emissions of these compounds.

Consequently, various theoretical,[10,11] laboratory and pilot-scale, [12-24] and full-scale investigations of incineration practices[25-53] that offer prospects for minimizing emissions of these compounds have been carried out. Figure 1 reports the results of some findings from a recent study [29,31] carried out by Environment Canada at a full-scale operating incinerator facility in Quebec City in Canada. The emissions of dioxins from the tested facility were observed before and after extensive design modifications and before and after various adjustments of combustion conditions in the furnace of the incinerator. It was found that desirable design and operating condition options could be identified which can afford a dramatic reduction in dioxin emissions. These results are typical of present experience with pollution abatement control equipment. The author can provide readers with numerous other examples of such performance from tests that have been carried out worldwide.

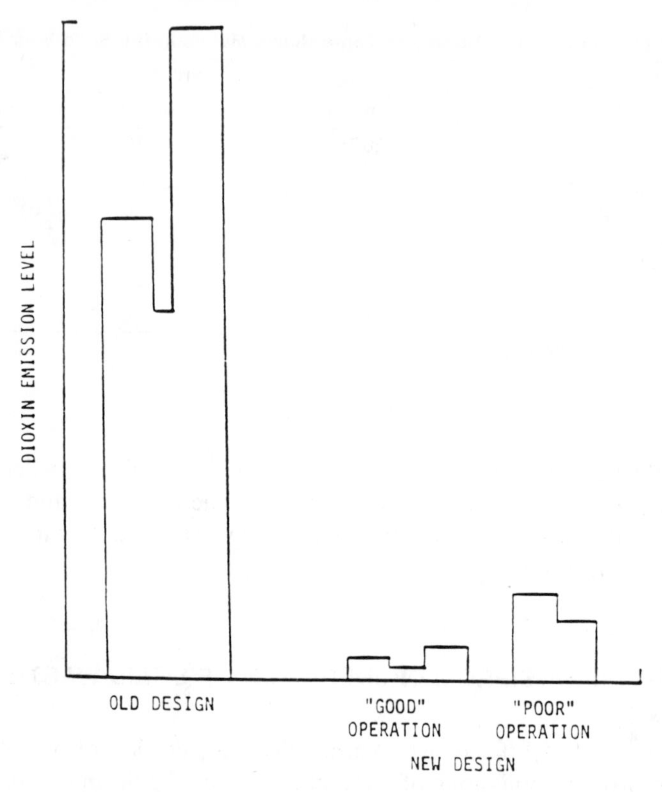


**Figure 1.** Recent results reported from the NITEP study of design and operating effects upon emissions of dioxins during combustion.

Taken as a whole, the results of recent studies appear to indicate that approaches to minimizing combustion-generated emissions of these compounds are available. A confounding aspect of incineration technology is that some significant amounts of dioxins can be produced by postcombustion reactions involving catalysis on flyash surfaces.[5, 10-12,14-16,18,19,21-23] The possibility that these processes are operative was first noted by this author in 1981.[54] Some refined understanding about how these processes take place has been recently emerging.[55] At the present time, strategies for mitigating postcombustion processes that lead to dioxin formation have concentrated upon reduction of precursor compounds during the combustion process.[10,11,31,32] It is thought that elimination of such compounds as chlorinated phenols can reduce prospects for catalytic formation of dioxins associated with such precursor compounds. Strategies for reducing emissions of those dioxins that are inadvertently formed despite attempts at combustion control of precursor compound formation have concentrated upon containment with pollution abatement control devices such as scrubber-baghouse or scrubber-ESP (electrostatic precipitation) combinations.[4,10,11,31,45]

## EMISSIONS CONTROL AND MONITORING

Data are available which indicate that very substantial reductions of many incinerator pollutant emissions will be realized when these facilties are equipped with scrubber-baghouse, scrubber-ESP, or equivalent control technologies.[11,31] It is useful to consider the present impacts of existing incinerators in order to determine future emissions reduction strategies:

- Lead: About 25 times more lead comes from all other known sources than is derived from incineration of municipal solid waste. About 20 times as much lead comes from automobiles as comes from incinerator emissions.[56]
- Sulfur dioxide: A recent source inventory survey conducted by HDR Techserve in Florida concluded that 1000 times more $SO_2$ comes from other sources than that which comes from incineration of municipal solid waste.[57]
- Nitrogen oxides: The California Air Resources Board recently determined that 1000 times more nitrogen oxides come from other sources in the Los Angeles area than come from emissions produced by incineration of municipal solid waste.[58]
- Polycyclic organic compounds: A recent study reported at a conference held in Oregon indicated that more than 10,000 times as much of these compounds are produced by wood burning in stoves and fireplaces than are produced by incineration of municipal solid waste.[59]
- Carbon monoxide: 93.6% of all carbon monoxide emissions come from sources other than the incineration of municipal solid waste.[56]
- Dioxins: Between 10 and 100 times as much dioxins in the environment in the United States come from sources other than those produced by incineration of municipal solid waste.[6-9]
- Hydrogen chloride: A recent survey in the United Kingdom and western Europe indicates that 93% of all HCl emissions are produced by the burning of coal in coal-fired power plants, despite the low chlorine content of coal compared to municipal solid waste.[37,60]

In order to ensure that emissions are adequately controlled or minimized, emissions are continuously monitored or samples of emissions are collected and analyzed. Analytical capabilities exist which allow for sampling and analysis of both air emissions and ash residue.[25-53] Presently, not all chemical species can be continuously analyzed. In addition, it is very expensive to analyze either inorganic compound emission levels or organic compound emission levels in either air or solid ash residue samples. Costs can easily be anywhere from $50,000 to $250,000 or even higher for a complete analysis of all compounds that are currently of interest to monitor.[61]

In consequence, there can be economies of scale effects upon different-sized incinerator facilities. Smaller-sized incinerator facilties, when subject to the same testing and analysis requirements as larger-sized facilities, are faced with higher pass-on costs to customers. Table 5 illustrates such effects. A factor of 20 times higher pass-on costs may be realized by smaller facilties, although the reader is cautioned that the method of estimation utilized to generate the

**Table 5. Testing and Analysis Costs per Family of Four per Month as Function of Size of Incinerator Facility[a]**

| Costs of monthly tests | Facility size (TPD): 100 | 500 | 1,000 | 2,000 |
|---|---|---|---|---|
| $ 50,000 | $ 5 | $ 1 | $0.50 | $0.25 |
| $100,000 | $10 | $ 2 | $1.00 | $0.50 |
| $250,000 | $25 | $ 5 | $2.50 | $1.25 |

[a]Assumes (1) number of persons whose waste is incinerated is proportional to size of incinerator facility and (2) the average person produces a metric ton of waste per year.

information presented in Table 5 makes several assumptions which may not be strictly valid, so that the estimates should only be viewed in a qualitative sense. In sum, high costs for testing and analysis may pose a significant disincentive against utilization of smaller-sized facilities if pass-on costs are deemed to be unacceptably high.

In addition to uncertainties about costs of monitoring emissions, there are variations in the types of measurements that are required for monitoring purposes. Of particular interest are indications of stable, long-term performance of the incinerator combustion system. Experience with combustion processes appears to indicate that one significant metric of stable combustion is an examination of the record of concentration levels of specific emissions that are characteristic of the combustion process. Figure 2 illustrates this concept through a presentation of recent test results that were produced during tests of a modular incinerator facility that is located at Pittsfield, MA.[62] Stability of the incineration process was examined in terms of the variability of carbon monoxide emissions. Unstable combustion was found to be described by marked excursions in the level of carbon monoxide compared to those levels that were observed during stable operation.

Another indirect approach to ensure effective minimization of pollutant emissions from incinerators is to establish "operating envelopes" (a menu of allowable ranges of specific operating parameters or surrogate species emissions) which define conditions which, if met, will result in minimal pollutant emissions. This is a lengthly exercise to carry out, but in principle is technologically feasible.[63,64] However, it should be noted that different incinerator designs may have characteristically different operating envelopes. For example, the operating envelope for a fluidized bed incinerator may be significantly different from that which could be established for a mass burn or modular-type incinerator which could, in turn,be different from that which could be established for a refuse-derived fuel incinerator. The concept that operating envelopes for different types of incinerators may not overlap is illustrated for some control parameters in Figure 3.

As noted, when emissions from the combustion or postcombustion chambers (the boiler, economizer, etc.) cannot be adequately minimized, it may be necesary to employ pollution abatement control equipment. Most pollution abatement control equipment presently in use, such as scrubbers, ESP, or

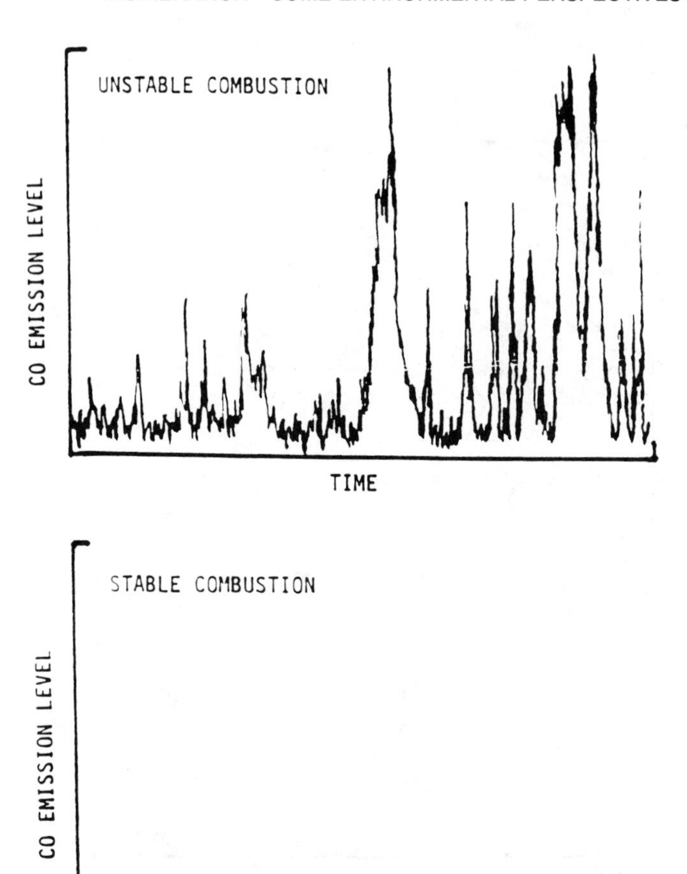


**Figure 2.** Illustrating that one indication of combustion conditions is the *fluctuation* in observed emission levels of chemical species.

baghouses, operate according to common basic principles, those being the adsorption or condensation of gaseous pollutants onto particulate matter (fly-ash or scrubber materials) and the subsequent containment of the particulate matter to prevent its emission out the stack of the incinerator. Figure 4, which is based upon both theoretical and experimental studies,[10,11,31,48] illustrates a basic aspect of the gas-to-solid transfer of pollutant emissions—that as the temperature of the control device is reduced, the extent of chemical adsorption or of condensation increases. Therefore, those species that are condensible can be readily deposited onto particulate matter which can subsequently be contained in the incinerator system as an ash residue.

At sufficiently low temperatures, it was found in recent studies carried out

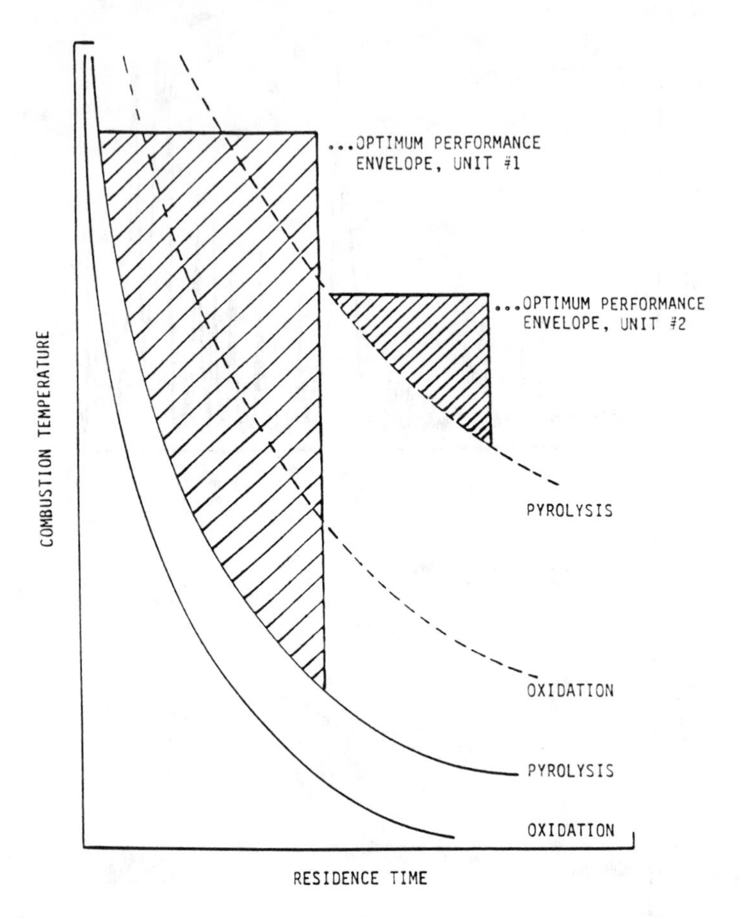


**Figure 3.** Illustrates map of operating envelopes established to ensure high combustion efficiency whether reaction is due to pyrolysis or is due to oxidation. The figure indicates that different facilities may have different operating envelopes.

by Environment Canada[11] that even mercury compounds could substantially be condensed onto particulate matter and subsequently contained. Tables 6 and 7 report some removal efficiencies and inlet-outlet control device pollutant concentrations that were observed in these studies.

## HEALTH RISKS ASSOCIATED WITH INCINERATOR AIR POLLUTANT EMISSIONS

Despite the high removal efficiency of chemical species by pollution abatement control devices, there remains public concern with emissions of those compounds that are not completely removed from stack emissions. Consequently, risk assessments of potential health and environmental impacts are carried out. These risk assessments invariably show that the potential risks

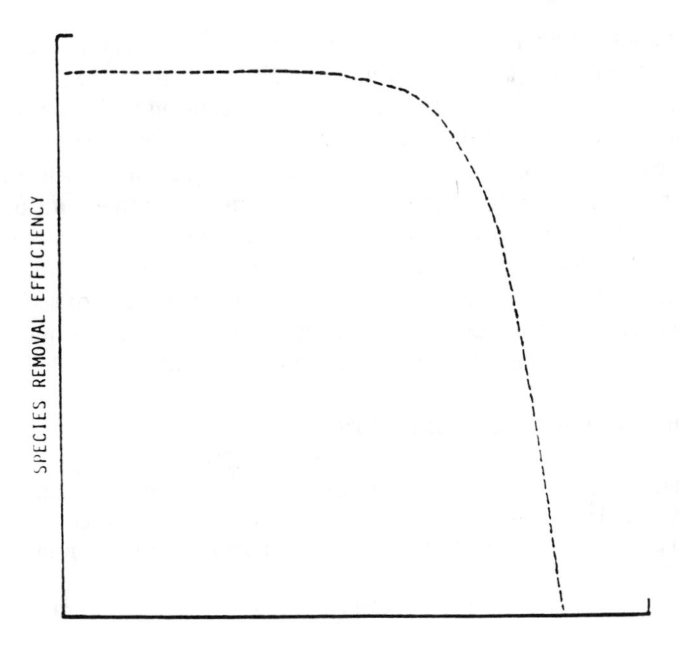


**Figure 4.** Illustrating that removal efficiency decreases when control device temperature is increased (for devices that operate based upon species condensation).

**Table 6. Illustrating Pronounced Tendency of Metals to Condense onto Particulates at Low Temperatures[11]**

| Metal | Inlet | Outlet |
|---|---|---|
| Zinc | 80,000—110,000 | 5—10 |
| Cadmium | 1,000—1,600 | ND—0.6 |
| Lead | 30,000—45,000 | 1—6 |
| Chromium | 1,400—3,000 | ND—1 |
| Nickel | 700—2,500 | 0.4—2 |
| Arsenic | 80—150 | 0.02—0.07 |
| Antimony | 800—2,200 | 0.2—0.6 |
| Mercury | 200—500 | 10—600 |

*Note:* All concentrations are in units of micrograms per normal cubic meter at 8% oxygen. The high outlet value for mercury (exceeds inlet value) may be measurement error.

**Table 7. Illustrating Pronounced Tendency of Species to Condense onto Particulates at Low Temperature[11]**

| | Reported removal efficiencies[a] | | | |
|---|---|---|---|---|
| **Temperature (°C)** | **110** | **125** | **140** | **>200** |
| Dioxins | >99.9 | >99.9 | >99.9 | 99.4 |
| Furans | 99.3 | >99.9 | >99.9 | 99.8 |
| HCl | 98 | 98 | 94 | 77 |
| $SO_2$ | 96 | 92 | 58 | 29 |

[a]Removal efficiencies expressed as percent removal.

associated with incinerator air pollution emissions are very small compared to the risks of everyday activities such as smoking, drinking, exposure to sunlight, driving a car, or the burden of naturally occuring toxic compounds that are found in food.[7,8,65–67] Where differences of opinion do occur is in the matter of just how small the risks posed by emissions associated with incinerators actually are. Different investigators who examine the same data often reach very different conclusions about risks, some perceiving the risks to be "serious" (though not serious compared to the everyday risks, smoking, etc. previously cited), while other scientists perceive the risks to be "not serious" either in an absolute or a comparative sense. The following remarks illustrate some recent opinions of some internationally known risk assessors:

### Regarding dioxins and incinerators

"If some politician says that you are insulting his intelligence by comparing the hazard of an incinerator to one cigarette a year per person, answer back without fear, that you are sorry that the people elected someone who is insulted by the truth."[68]

—Professor Richard Wilson (Harvard University)

"Results indicate that municipal waste incineration is not a major source of human exposure to dioxin, since the estimated *maximum* daily intake of incinerator emitted dioxins and furans is approximately 140 times lower than the estimated background daily intake of dioxin."[7]

—Dr. Curtis Travis (Editor-in-Chief, *Risk Analysis*)
—Dr. Holly Hattemer-Frey (Oak Ridge National Labs)

"The amount of TCDD (dioxin) emitted from modern incinerators is a completely trivial possible hazard to the public, in my scientific judgement."[65]

—Professor Bruce Ames (University of California, Berkeley)

### Regarding dioxins in general

". . . scientists who draw the diametrically different conclusion that dioxin has been shown to cause human cancers, birth defects and miscarriages . . . would argue that execution of still larger epidemiologic studies might reveal that dioxin has caused some severe health effects. A complete review of the literature does not support (such) conclusions. The highest exposures to dioxin have not caused detectable health effects; the small exposures now and in the future will not cause detectable effects either."[69]

—Dr. Michael Gough (Senior Fellow, Center for Risk Management and former Director of Special Projects, Congressional Office of Technology Assessment)

Considering potential impacts of pollutant emissions on human health and the environment, it is instructive to realize that regarding chemical species, besides dioxins and furans, that have been identified or as yet fall into the category of "unknown" identity . . .

**Table 8. Concentration of Dioxins (PCDDs) and Furans (PCDFs) Found in Leachate from Ash Monofills**

| Species | Concentration range (ppm) |
|---|---|
| T4CDD | .00000025—.000028 |
| PCDD | .00000015—.000093 |
| HCDD | .00000010—.000130 |
| 2,3,7,8-TCDD | .00000250—.000016 |
| Total PCDD | .00000060—.000543 |
| T4CDF | .00000025—.000065 |
| PCDF | .00000010—.000064 |
| HCDF | .00000005—.000076 |
| H7CDF | .00000015—.000060 |
| OCDF | .00000040—.000015 |
| 2,3,7,8-TCDF | .00000025—.000011 |
| Total PCDF | .00000004—.000280 |

Point 1. Most pollution abatement control technology is based on either adsorption or condensation of chemical species onto particulate matter. *These chemical-physical properties are not unique to dioxins and furans, but are instead generally characteristic and applicable, depending principally on the boiling point of the chemical species.*

Point 2. According to Professor Christoffer Rappe, Umea, Sweden,[70] it is illogical to consider either known or "unknown" chemicals that may be released to the environment from incineration if they are not found to bioconcentrate to toxicologically significant levels due to these emission sources.

Point 3. "Toxicological rule of thumb": In addition to the inherent toxicity of a substance, it is also necessary to consider the nature of an exposure, i.e., the concentration, the strength, the frequency, the duration, the atmospheric physics of transport of the substances, bioconcentration, compartmental transport dynamics, and the biological responses of exposed subjects.

## POLLUTANTS IN ASH RESIDUE—MEASUREMENTS, DATA, AND INDICATIONS

### Leaching

When dioxins are effectively removed from stack emissions, it is usually because they have been trapped onto particulate matter that ultimately becomes a part of the ash residue that is produced by the incinerator facility. This ash is either reused in part or as a whole for beneficial purposes such as in road-bed aggregate,[34,40,71,72] or it is disposed into a landfill. [43,73] When ash residue is disposed in the future, it appears that to a large extent this residue will be disposed into ash monofills.[30,40,73] Table 8 reports some recent findings about the extent of potential leaching of dioxins from ash monofills. It can be seen that the leaching of dioxins from such disposal sites is small, as is expected.[10,11,34,40]

**Table 9. Illustrating Interlaboratory Measurement Error. Results for Lead (mg/L) via EP Toxicity Test.**

| Sample No. | Laboratory #1 | Laboratory #2 | Laboratory #3 |
|---|---|---|---|
| 1 | 4.9 | 17.2 | 7.3 |
| 2 | 3.1 | 7.8 | 1.5 |
| 3 | 2.3 | 1.8 | 3.8 |
| 4 | 5.2 | 11.0 | 4.2 |
| 5 | 21.0 | 93.0 | 2.3 |
| 6 | 1.1 | 9.0 | 19.0 |
| 7 | 3.3 | 4.1 | 1.3 |
| 8 | 1.1 | 7.2 | 24.0 |
| 9 | 2.5 | 3.5 | 2.3 |

*Note:* EP toxicity test limit for lead is 5.0 mg/L.

Much attention has recently been focused upon heavy metals in ash residue fractions. In order to evaluate the presence of these compounds in ash residue and to assess the extent of leaching of these compounds that may occur when ash is disposed into landfills, tests have been devised. Nationally, no technologically sensible test has been developed for assessing the leaching behavior of heavy metals from ash residue at the present time. Some argue that at least for codisposal scenarios (the codisposal of incinerator ash residue with uncombusted municipal solid waste), one test, the U.S. EPA EP toxicity test, is appropriate as it was developed "exactly" for the purpose of determining the leaching behavior of heavy metals from codisposal sites.[75]

Unfortunately, while the EP Toxicity test was designed "exactly" for this purpose, it is not exactly reliable, not exactly precise, not exactly accurate, not exactly statistically representative, and in addition suffers from several procedural inconsistencies which render the test illogical.[50,52,63,64] Table 9 illustrates some of the problems associated with measurement errors which occur when this test is employed.[50] In the particular tests illustrated, it can be seen that in six out of nine cases, investigators would be unable to agree about the regulatory outcome of the test.

One other notable deficiency of the EP toxicity test and other similar tests is that they are batch tests, typically involving the utilization of liquid-to-solid ratios that are enormously higher than those found at actual disposal sites.[30,34,76] Consequently, these batch tests typically tend to predict much higher extents of leaching of heavy metals than in fact has been found to occur at disposal sites when actual samples of leachate have been analyzed for heavy metal content.

Another type of test is the column leach test, of which there are several variations reported.[30,34,76] There is no present consensus about which column leach test is most appropriate for any given disposal scenario, but it appears, based upon several recent studies of heavy metal leaching from column tests, that these tests can more realistically predict the likely nature and extent of leaching of heavy metals at disposal sites.[30,34,76]

Recently, EP toxicity test results, column leach tests results, and actual field monofill leachate data were collected for the same ash residue disposal site.[77,78] To illustrate how different investigators interpret the same data, compare how one environmental group[79] interpreted data, such as lead in leachate, that were obtained from tests and from the disposal site against the data as it is presented

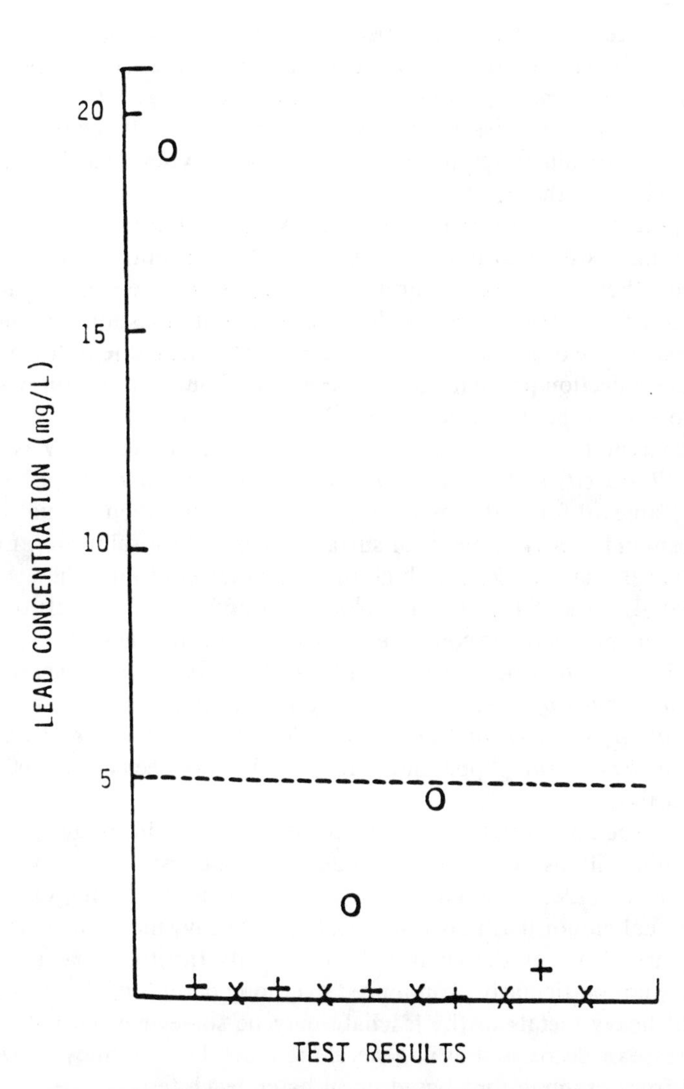


**Figure 5.** Results and comparison of EP Toxicity batch test (0), a column leach test (X), actual level of lead found in field leachate ( + ), and the U.S. EPA EP Toxicity test limit of 5.0 mg/L (———).

in Figure 5. What the environmental group reported about the data shown in Figure 5 is the following:

> "Recent monitoring of leachate from a New York ash-only landfill during its first year of operation found that many pollutants increased significantly over the monitoring period, and that average levels frequently exceeded drinking water standards; moreover, during this first year, pollutant levels in the leachate almost always exceeded often dramatically the highest levels predicted to occur at any time during the first 25 years of operation, based on laboratory simulations."

While some reported differences between column and actual field leachates were more striking for other heavy metal compounds such as zinc, similar plots of actual information show that *the fundamental point that batch and column leach tests can give strikingly different test results cannot be overlooked*. Table 10 illustrates more comparisons between field leachate and batch type (such as the EP Toxicity test) tests.

One apparently high value (though not exceeding the present EP toxicity test) that is inconsistent with all other reported leach results for ash monofills is reported—the value 2.92. A careful examination of the original literature (an EPA contractor study) reveals that the sample in question may not have been representative of ash monofill groundwater leachate due to the nature of the leachate collection procedures that were carried out. Future follow-ups are planned to extend the database by the EPA and others.

It is apparent from these results that, measurement reliability issues [63,64] aside, *the EP toxicity test is an inappropriate indicator of actual leaching*. Also shown in Table 10 for additional comparison are data about actual ambient levels of lead in the environment for surface water and for rainwater. Levels of lead in rainwater are sufficiently high that they may confound the determination of actual levels of lead leached from a monofill, as the levels of lead in rainwater can apparently be comparable to levels of lead that might leach from monofills if rainwater does not contain lead. *Therefore, it may be difficult to establish how much of the leachate that comes from an ash disposal site was derived from the site itself instead of from the rainwater that may have passed through the disposal site*. Apparently, this issue has not been clearly assessed at the present time.

Another issue concerned with levels of heavy metals in leachate produced from ash monofill disposal sites has to do with the question of how the levels of heavy metals in leachate may vary over long periods of time. One 17-year study of actual monofill ash disposal site levels of heavy metals in leachate that has been carried out by the Danish Water Quality Institute[76] has shown that changes are not particularly pronounced over long periods of time; that is, the amount of heavy metals in the leachate may be somewhat variable, may at times increase or decrease, but in general are much lower than what would be predicted from examinations based upon batch leach tests.

The Swedish government[40] has recently determined that the levels of heavy metals in leachate are so low that if capping of ashfills is carefully done,

**Table 10. EP Toxicity Test and Field Leachate Results for Lead (mg/L), (combined ash)**

| EP toxicity test results[78] | | | | | | Field leachate results |
|---|---|---|---|---|---|---|
| | | | | | | MSW site leachate |
| 8.4 | 4.3 | 0.5 | 9.1 | 11.1 | 9.0 | 0.22,[80] 0.37[80] |
| 5.31 | 7.2 | 0.71 | 6.7 | 19.1 | 3.0 | <0.1—0.50[80] |
| 24.3 | 5.1 | 19.1 | 5.2 | 6.9 | 4.8 | <0.1—0.90[80] |
| 1.53 | 4.1 | 4.46 | 11.3 | 12.8 | 2.0 | 0—14.20[81] |
| 1.8 | 2.1 | 1.1 | 5.8 | 6.4 | 10.3 | Co-disposal sites |
| 3.2 | 1.2 | 13.2 | 21.4 | 1.2 | 1.54 | 0.20[76] |
| 0.34 | 0.7 | 0.87 | 16.9 | | | <0.001—0.10[76] |
| | | | | | | <0.005—0.49[76] |
| | | | | | | <0.001—0.009[76] |
| | | | | | | <0.008—0.016[76] |
| | | | | | | 0.017[76] |
| | | | | | | 0.010—0.27[74] |
| Comparison data ambient levels in the environment[82,83] | | | | | | 0.10[76(a)] |
| | | | | | | 0.19[76(a)] |
| | | | | | | 0.06[76(a)] |
| | | | | | | 0.33[76(a)] |
| Surface water | | | | | | 0.12[76(a)] |
| Rural: 0.0021 mg/L | | | | | | |
| Urban: 0.0063—0.0695 mg/L | | | | | | Ash monofill sites |
| Rain water (32 U.S. stations) 0.0340 mg/L | | | | | | <0.01,[76] <0.003[76] |
| | | | | | | <0.002,[76] <0.005[76] |
| | | | | | | <0.01,[76] <0.012[76] |
| Maximum found = 0.300 mg/L | | | | | | <0.005,[76] <0.005[76] |
| | | | | | | <0.0029,[76] <0.013[76] |
| | | | | | | <0.001,[76] <0.0005[76] |
| | | | | | | <0.0007,[76] <0.0003[76] |
| | | | | | | 0.019,[76] 0.018[76] |
| | | | | | | 0.002—0.05[76(a)] |
| | | | | | | 0.1—0.6[76(a)] |
| | | | | | | 0.001—0.1[76(a)] |
| | | | | | | 0.012—2.92[74] |
| | | | | | | 0.12[78] |
| | | | | | | 0.16[78] |
| | | | | | | 0.16[78] |
| | | | | | | 0.16[78] |
| | | | | | | 0.005[78] |
| | | | | | | 0.65[78] |

unrestricted leaching of heavy metals out of sites having no bottom liners is acceptable if the leachate passes into outflow regions away from sites where drinking water is likely to be collected. Figure 6 illustrates recent findings of the Swedish government[34] that were based on a column-leaching simulation of very long-term (as much as 2000 years) leaching behavior at an ashfill.

## Dust blowing

When ash residue is managed, besides issues related to leaching, it is also of interest to examine environmental impacts of dust blowing in which flyash and other fine particulate matter may be entrained into air either onsite (of an

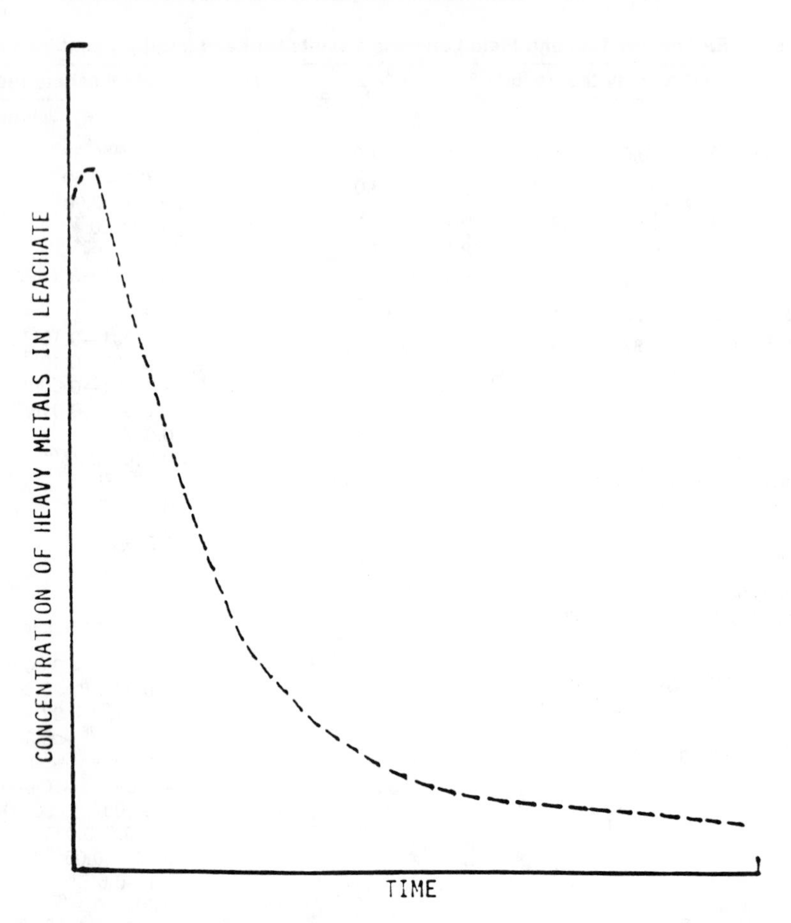


**Figure 6.** Result of Swedish laboratory simulation of long-term leaching behavior of heavy metals from an ash monofill.

incinerator, transfer station, ash disposal site, etc.) or offsite (such as from off-perimeter migration of blowing dust). Other issues involve releases such as from truck transport and surface water runoffs. However, leak-proof, covered trucks that maximize ash residue containment are apparently available, and means to mitigate surface water runoffs are technologically feasible.[84]

Recently, a detailed investigation of off-perimeter impacts of flyash dust blowing were reported.[35] According to the investigators who conducted the study:

> "Measurements of Cd and Pb indicate there is neither a marked nor an extensive contamination by these metals in the downwind area."

These observations were made even though the researchers noted evidence that fugitive releases of Cd and Pb had evidently occured onsite.

Input data and methodology can be used to make relatively simple, crude

**Table 11. Description of Parameters and Input Data Used in Equations 1 and 2 of this report[85]**

| | | |
|---|---|---|
| C | = | air-lead concentration ($\mu g/m^3$) |
| d | = | depth of penetration of leachate into soil (m) |
| p | = | soil density (1.6E + .06 $g/m^3$) |
| v | = | average particle settling velocity (0.01 m/sec) |
| t | = | duration of offsite migration [20 years (6.3E + .08 sec)] |
| D | = | lead deposition flux to soil ($g/m^2$) |
| S | = | lead in soil concentration ($\mu g/g$) |

*Note:* Numerical values either taken from Reference 85 or nominally assumed values assigned.

estimates of soil impacts due to fugitive releases of dust-blown flyash that contains lead.[64,85] Assume that an active incinerator, disposal, or ash manangement site will be in operation for at least 20 years. Then the potential impact of dust-blown emissions using assumed values for required input data can be estimated as follows.[85] The deposition flux of lead to soil from dust-blown ash particulate or other matter can be estimated as:

$$D = 10^{-6}Cvt \tag{1}$$

The lead in soil concentration can be estimated as:

$$S = D/(pd) \tag{2}$$

The significance of variables shown in Equations 1 and 2 and assumed values that have been assigned for making lead-to-soil impact estimates are reported in Table 11. Presumably, the source of fugitive emissions could come from the incinerator site, a transfer station, an ash disposal site, or some other ash management area. One characteristic to note in using the equations is that it is assumed that the source term is of constant strength. Modifications would be necessary if the source term could vary in strength.

Table 12 illustrates the results of the calculations outlined in Table 11. A comparison with actual data reported for soil-lead levels in rural areas and near urban freeways is instructive. The latter data are also reported in Table 12. From these comparisons, it can be concluded that offsite impacts to soil-lead levels due to dust blowing will be comparable to rural soil-lead levels if offsite dust blowing can be minimized so that lead concentrations in dust-blown particulate matter is at or below about 0.1 $\mu g$ of lead per cubic meter of air that moves offsite.

Some data are available which suggest that offsite concentrations of dust-blown flyash and consequently lead in air concentrations are significantly lower than dust levels onsite.[35,86] It is clear that precise estimates of off-perimeter dust migration can be confounded by high background levels, e.g., lead from automobile exhaust emissions, and other industrial sources.[35]

Besides settling onto soil and other surfaces, dust-blown particulate matter also can be inhaled. It is of interest to estimate potential impacts to humans from breathing this dust. Table 13 reports data that can be used in a simple

**Table 12. Lead in Soil Concentrations—Results from Use of Equations 1 and 2**

| C | d | D | S |
|---|---|---|---|
| 0.01 | 0.05 | 0.063 | 0.79 |
| 0.1 | 0.05 | 0.63 | 7.9 |
| 0.5 | 0.05 | 3.2 | 39.0 |
| 1.0 | 0.05 | 6.3 | 79.0 |
| 2.0 | 0.05 | 13.0 | 160 |
| 5.0 | 0.05 | 32.0 | 390 |
| 10.0 | 0.05 | 63.0 | 790 |
| 0.01 | 0.10 | 0.063 | 0.39 |
| 0.1 | 0.10 | 0.63 | 3.9 |
| 0.5 | 0.10 | 3.2 | 20.0 |
| 1.0 | 0.10 | 6.3 | 39.0 |
| 2.0 | 0.10 | 13.0 | 79.0 |
| 5.0 | 0.10 | 32.0 | 200 |
| 10.0 | 0.10 | 63.0 | 390 |
| 0.01 | 0.15 | 0.063 | 0.26 |
| 0.1 | 0.15 | 0.63 | 2.6 |
| 0.5 | 0.15 | 3.2 | 13.0 |
| 1.0 | 0.15 | 6.3 | 26.0 |
| 2.0 | 0.15 | 13.0 | 53.0 |
| 5.0 | 0.15 | 32.0 | 130 |
| 10.0 | 0.15 | 63.0 | 260 |

*Note:* Comparison data—ambient soil lead levels
Rural areas
0—5 ($\mu$g/g)
Near urban freeways
50—250 ($\mu$g/g)

compartmental model based upon inhalation of lead-containing particulate matter, partial retention of lead initially, and eventual loss of lead due to the reported half-life of lead (16 days) in humans.[87]

The compartmental model for human lead in blood levels due to inhalation of lead in airborne particulate matter that is produced by dustblowing can be expressed as:

$$C \xrightarrow{n} X \xrightarrow{m}$$

**Table 13. Description of Parameters Used in Equation 3[82,83,87]**

a = fraction of lead initially retained (0.2)
b = human blood volume:
adults—50 dL
children—12 dL
m = lead loss rate from blood (m = ln2/t; t = 16 days)
n = air inhalation rate:
adults—20 $m^3$/day
children—10 $m^3$/day
C = lead in air concentration ($\mu g/m^3$)
x = human blood lead level ($\mu g/m^3$)

*Notes:* X = 0.0 assumed initially.
From Equation 3, it can be seen that approximately X (children) / X (adults) = 2.

As an estimate, assume that the site (incinerator facility, transfer station, ash disposal site, etc.) acts as a constant source term, C, of entrained lead in flyash particulate matter. The strength of this term can be expressed in terms of lead in air concentrations. If the level of lead in human blood due to this constant source term is assumed to initially be zero prior to the beginning time of ash management operations, then the equation which describes lead in blood levels in humans is given by:

$$X = \frac{aCn}{bm}(1 - e^{-mt}) \quad (3)$$

All parameters and assigned values are reported in Table 13. Input data (assumptions) that can be used to make simple crude estimates of both short-term (8 hr) and long-term exposures to dust-blown lead-containing particulate matter are reported in Table 13.

Figure 7 illustrates the nominal time it takes to approach upper limit levels of lead in blood based on the simple model and input data that are found in Table 13. These calculations suggest that "long-term" exposure can be associ-

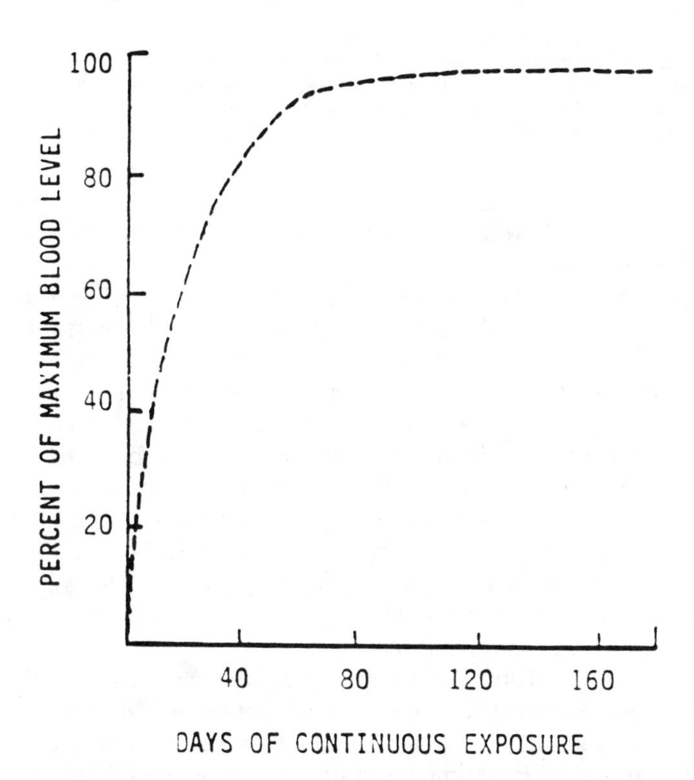


**Figure 7.** Illustrating the nominal time it takes to approach upper limit levels of lead in blood based on a simple compartmental model and assumed input data reported in this document.

**Table 14. Off-site Dust Blowing-Lead Inhalation Impacts Based on Equation 3**

| C | X (8 hr) | X[a] |
|---|---|---|
| 0.001 | 0.000026 | 0.0019[b] |
| 0.01 | 0.00026 | 0.019 |
| 0.1[c] | 0.0026 | 0.19[d] |
| 0.3[c] | 0.0078 | 0.56[d,e] |
| 1.0[c] | 0.026 | 0.93[e] |
| 1.5[f] NAAQS limit | 0.039 | 1.9[e] |
| 2.0[f] | 0.052 | 2.8[e] |
| 3.0[f] | 0.078 | 4.6[g] |
| 5.0[f] | 0.13 | 9.3[h] |
| 10.0[f] | 0.26 | 19.0 |
| 50.0 OSHA PEL limit | 1.3 | 93.0 |
| 100.0 | 2.6 | 190[i] |
| 250.0 | 6.5 | 460 |
| 500.0 | 13.0 | 930 |
| 1000 | 26.0 | 1900 |
| 2100 | 54.0 | 3900[j] |

[a]Long-term exposure (more than about 100 days).
[b]Maximum ground level impact from stack (estimated).
[c]Rural ambient air level, no incinerator present.
[d]Offsite dust blowing impact (estimated).
[e]Onsite occupational exposures reported at older sites (includes incinerators, transfer stations, and landfills).
[f]Urban ambient air levels, no incinerator present.
[g]Older facility, high-level onsite occupational exposure: boiler operator.
[h]Older facility, high-level onsite occupational exposure: boiler room cleaner.
[i]Older facility, high-level onsite occupational exposure: boiler grate inspector.
[j]Older facility, high-level onsite occupational exposure: precipitator cleaner.

ated with approximately 80 to 100 days of continuous exposure to a constant source of lead emissions.

Table 14 reports the results of calculations based on the use of the simple compartmental model and assumed levels of lead in air due to dust-blown particulate matter. For comparison, regulatory limits [NAAQS and Occupational Safety and Health Administration (OSHA)],[41,42] some estimates based on air monitoring done at an older facility that has since been extensively rebuilt,[42] and rural and urban ambient air concentrations that have been reported[83] are presented. It appears that at older facility sites, for some specific equipment maintenance tasks, high occupational exposures to lead may result in individuals who are exposed to high dust levels—e.g., boiler grate inspector and precipitator cleaner—if adequate respiratory protection is absent. Characteristically much lower occupational exposures may result from tasks involving other onsite management operations.

In one study of offsite dust blowing that was carried out in the United Kingdom, reported results appear to indicate that levels of dust-blown particulate matter are significantly lower offsite than onsite. Based on estimates obtained from use of Equation 3 and data presented in Table 13, it appears that if offsite dust blowing of lead-bearing particulate matter is below lead (in airborne particulate) levels of about 1 $\mu$g of lead per cubic meter of air, the

**Table 15. Airborne Lead Levels—Results Reported from Onsite, Seven Plant Study in 1982[89]**

| | |
|---|---|
| Tipping floor | 9.2 |
| Primary shredder | 20.2 |
| Secondary shredder | 2.0 |
| Reject load outs | 6.1 |
| Picking station | 3.7 |
| Aluminum separator area | 5.1 |
| Receiving pit area | 4.3 |
| Shredders | 1.1 |
| Metals separation | N.D. |
| Fuel processing | 1.0 |
| Crane deck | 6.0 |
| Process area | 1.1 |
| Boiler house | 2.8 |
| Charging floor | 7.2 |

*Note:* Airborne lead concentrations reported in units of micrograms per cubic meter.

*resulting exposures appear to be approximately comparable to those that would have resulted from exposure to rural ambient air.* A brief comparison of the results of the simple compartmental model calculation with a controlled lead exposure study[88] appears to predict similar results, suggesting that the compartmental model is reasonably predictive.

In 1982, an industry-wide survey of resource recovery facilities was reported. [89] This study was partially supported by the National Institute of Occupational Safety and Health (NIOSH), the Centers for Disease Control, and the U.S. Department of Health and Human Services to illustrate the situation as of 1982 regarding worker occupational exposure to lead in airborne particulate matter and for other purposes. The investigators who conducted the study reported:

> "The objective of this study was to characterize the resource recovery industry with regard to current or potential occupational health problems . . . the assessment described herein was based on in-depth industrial hygiene surveys at seven resource recovery facilities selected to represent the major large-scale technologies in the industry. Included were four facilities that produced refuse derived fuel (RDF) for off-site use, two mass burn incineration plants, and one RDF plant which had on-site combustion."

> "Trace metals such as lead, zinc, cadmium, chromium, and beryllium did not appear to present any air-borne health hazard in the facilities surveyed. These metals were either not detectable or were detected only at negligible concentrations. These particular metals were selected as being representative of the most potentially significant trace metal hazards to be found in resource recovery facilities. Therefore it is assumed that other trace metals can also be discounted as potential air-borne health hazards." (Shown in Table 15 are some of the lead in air occupational exposure levels that were reported by these investigators.)

It is instructive to consider, given attention to dust blowing of lead-bearing particulate matter such as flyash, how impacts from incinerator sites, transfer

**Table 16. Comparison Data—Lead Emissions from Incinerators, Recycling Operations, Smelters, Automobile Exhausts, and Rural and Ambient Air Levels**

| | |
|---|---|
| Annual reported emissions (kg) | |
| El Paso lead smelter[90] | 337,000 |
| Marion County incinerator[91] | 24 |
| Lead-air ground level impacts (ng/m$^3$) | |
| Bergsoe battery recycling plant[92] | 14,000.0 |
| Scrubber-baghouse-equipped incinerator[83] | 0.2 |
| Morning rush hour, L.A. freeway[83] | 29,000.0 |
| Rural ambient lead in air levels, U.S.[83] | 110.0 |
| Lead in soil impacts ($\mu$g/g) | |
| From smelter emissions[35] | 160.0 |
| From ESP-equipped incinerator[35] | 9.9 |
| Rural U.S. soil content[83] | 5.0—30.0 |
| 25 ft from major highway (cars)[83] | 47.0—266.0 |

stations, and landfills may compare to other sources of dust-blown lead-bearing particulate matter such as lead smelters, battery recycling facilities, automobiles, and rural soil. Table 16 illustrates some of these comparisons. This information indicates that if dust blowing of lead-bearing particulate matter is of concern, then attention should principally be focused to a large extent upon emission sources such as automobile exhausts, smelters, and lead battery recycling facilties.

If dust blowing is deemed to be in need of incrementally proscriptive management control practices, some attention beyond dust control by keeping it wet may be found advisable. It appears that dust blowing can be mitigated by such ash wetting practices, and in fact this is one element of proposed incinerator ash permit conditions that is advocated by one nationally recognized environmental group.[93]

The following description of ash management practices at a recently constructed and operating incinerator facility illustrates some approaches to mitigation of dust blowing problems[100]

> "The design of the (incinerator) facility encompasses the concept of closed systems to handle the process residue streams so that there will be no employee or public exposure to airborne dust from the ash. All hoppers from the boiler to the air pollution control device have sealed, air-locked valves that transfer the fly ashes and/or dry scrubber reagent and reaction products to sealed screw or drag conveyors. These, in turn, deliver all the ashes to the sealed ash discharger at the lower end of the furnace just below the level where the bottom ash drops off the stoker grate. As the bottom ash falls down the ash discharge chute into the water quench bath, the fly ashes are captured, wetted and made an integral part of the bottom ash before the combined ash is hydraulically ram discharged. The ram extrudes and dewaters the ash prior to dropping it on the conveyor system. The ash at this point has the consistency of wet concrete. As it moves outside the enclosed boiler building, an enclosed conveyor transfers it to the ash storage building or area to await transfer to covered, watertight trucks or containers for eventual utilization or burial in a landfill. Thus, at any point where the employ-

**Table 17. Estimated Costs for Managing Ash**

| | |
|---|---|
| Baseline | |
| | Landfilling ash, average cost:[94] $12.50/ton (any ash) |
| Incremental costs to add to baseline | |
| | 10% Portland or 15% Rutland cement:[95] $5—15/ton (combined ash) $10—25/ton (flyash) |
| | Excess lime addition to scrubber/flyash:[95] $50/ton |
| | Remedial disposal site action:[95] $1.0/ton and up |
| | Transportation cost per mile:[95] $0.50/ton |
| | Geopolymerization:[96,97] $20/ton (any ash) |
| | Brick manufacture, ceramics:[98] $40/ton (flyash) |
| | Vitrification (Takuma):[99] $150/ton (any ash) |

*Notes:* If flyash is 10% (weight) of all ash, all ash is 30% (weight) of municipal solid waste (MSW), and MSW disposed is 1 ton per person per year, then the costs per family of four per month for processing ash are 1% of cost of processing flyash and 10% of cost of processing combined ash.
Results are not absolute, but can be used comparatively to estimate differences in management approaches. Estimates assume that ash is managed as a non-hazardous waste stream.

ees or the public could be exposed, the ash system is either sealed or the ash wetted and enclosed in a building or conveyor."

Table 17 reports estimates of incremental costs for managing ash by various ash treatment approaches. It appears that except for vitrification processes, the costs are low for a family of four on a monthly basis of accounting for customer charges for services. Therefore, beneficial treated ash reutilization appears to be economically practical.

However, an important word of caution must be offered in regard to all treatment methods for ash management other than direct landfilling or beneficial reuse of untreated ash. In the United States, present legal requirements for managing wastes make a distinction between wastes that are deemed to be hazardous and wastes that are deemed to be not hazardous. The nature of the legal requirements is such that to treat waste that is deemed hazardous would require a hazardous waste treatment permit and would incur significant administrative, liability, and insurance cost burdens due to management of such material as a hazardous waste. The data reported in Table 17 have been developed assuming that to begin with, that is prior to treatment, ash is not deemed to be a hazardous waste, in which case a hazardous waste treatment permit would not be required and significant additional costs probably would not be realized. If ash is deemed to be hazardous waste, then the cost estimates that have been provided should not be referred to. Actual treatment costs would be different, presumably significantly higher. It is also conceivable that some variation in estimated costs, probably in an upward direction, would also result if treatment beyond just straight landfilling were mandatory as some offering treatment services would recognize a demand for services that could allow different profit margins to be established.

## CONCLUDING REMARKS

It is not feasible to address all areas of interest in a short report without risking the prospect of providing more information than can be absorbed in one brief reading. For a more extensive general overview, the reader is advised to consult References 44, 45, 53, and 64.

Taken as a whole, information that is presently available appears to indicate that some understanding and estimates of potential impacts of current and future incineration practices can be developed. Additional investigations designed to produce reliable data and involving utilization of comprehensive risk assessments, if done carefully, may enhance our knowledge base of *all* solid waste management practices. The author hopes that the information which has been presented in this report has been instructive.

## REFERENCES

1. NSWMA, "No Drop in Landfill Volume?" *Waste Age* (November 1986), pp. 27–28.
2. "The End-Use Market for 13 Non-Ferrous Metals," U.S. Department of Commerce (June 1986).
3. Linda Carrico (Bureau of Mines), Private communication to R. G. McInnes and N. H. Kohl (ERT). (June 1987).
4. National Swedish Environmental Protection Board, "Energy From Waste" (1987), p. 11.
5. Rappe, C., R. Andersson, P.-A. Bergqvist, C. Brohede, M. Hansson, L.-O. Kjeller, G. Lindstrom, S. Marklund, M. Nygren, S. E. Swanson, M. Tysklind, and K. Wiberg. "Sources and Relative Importance of PCDD and PCDF Emissions," *Waste Manage. Res.* 5:225–237 (1987).
6. Shaub, W. M. "Sources of Emissions in Resource Recovery Plants," from Appendix II of transcript of oral remarks made at the N.Y. Academy of Sciences Meeting: Resource Recovery in New York City: Science-Intensive Public Policy Issues, New York, December 18, 1984.
7. Travis, C. C., and H. A. Hattemer-Frey. "Human Exposure to Dioxin from Municipal Solid Waste Incineration," U.S. Government Report DE-AC05-840R21400 (1988).
8. Smith, A. H., "Infant Exposure Assessment for Breastmilk Dioxins and Furans Derived from Waste Incineration Emissions," Paper No. GP10, presented at DIOXIN 86—6th International Symposium on Chlorinated Dioxins and Related Compounds, Fukuoka, Japan, September 16 to 19, 1986.

9. California Air Resources Board, Engineering and Evaluation Branch Monitoring and Testing Laboratory Division, "Evaluation Test for PCDD and PCDF Emissions from Motor Vehicles," Preliminary Draft Report (October 1987).
10. Shaub, W. M. "Technical Issues Concerned With PCDD and PCDF Formation and Destruction in MSW Fired Incinerators," NBSIR 84-2975, National Bureau of Standards Report (November 1984).
11. Shaub, W. M. "Containment of Dioxin Emissions from Refuse Fired Thermal Processing Units—Prospects and Technical Issues," NBSIR 84-2872, National Bureau of Standards Report (February 1984).
12. Ballschmitter, K., R. Niemczyk, I. Braunmiller, J. Ehmann, U. Duwel, and A. Nottrodt. "Chemistry of Formation and Fate of PCDD/PCDF in the Municipal Waste Incineration Process," paper SE11 presented at Dioxin '87, University of Nevada, Las Vegas, October 4 to 9, 1987.
13. Golden, J. H. "Polyphenylene Oxides," in *High Temperature Resistance and Thermal Degradation of Polymers*, (London: Society of the Chemical Industry, 1961).
14. Hagenmaier, H., H. Brunner, R. Hagg, and M. Kraft. "Copper-Catalyzed Dechlorination/Hydrogenation of PCDDs, PCDFs, and Other Chlorinated Aromatic Compounds," *Environ. Sci. Technol.* 21(11):1085–1088 (1987).
15. Hagenmaier, H., H. Brunner, R. Haag, and M. Kraft. (Verfahren zur Zerstorung von Chloraromaten) Patent Pending DP 36 23 492.3.
16. Hagenmaier, H., and M. Kraft. "Studies Towards Validated Sampling of PCDDs and PCDFs in Stack Gas," presentation at the International Workshop on Municipal Waste Incineration, Montreal, Quebec, October 1 to 2, 1987.
17. Hay, A. G. "Aromatic Polyethers," *Adv. Polym.* Sci. 4:496-527 (1967).
18. Karasek, F. "Model Studies of PCDD Formation During Incineration," presentation at the International Workshop on Municipal Waste Incineration, Montreal, Quebec, October 1 to 2, 1987.
19. Stieglitz, L. "New Aspects of PCDD/PCDF Formation in Incineration," presentation at the International Workshop on Municipal Waste Incineration, Montreal, Quebec, October 1 to 2, 1987.
20. Stein, S. Private communication to W. M. Shaub, (January 19, 1988).
21. Stieglitz, L., and H. Vogg. "On Formation Conditions of PCDD/PCDF in Fly Ash From Municipal Waste Incinerators," paper presented at the 6th International Symposium on Chlorinated Dioxins and Related Compounds, September 16 to 19, 1986.
22. Vogg, H., H. Braun, M. Metzger, and J. Schneider. "The Specific Role of Cadmium and Mercury in MSW Incineration," *Waste Manage. Res.* 4:65–74 (1986).
23. Vogg, H., M. Metzger, and L. Stieglitz. Recent Findings on the Formation and Decomposition of PCDD/PCDF in Solid Municipal Waste Incineration," paper presented at a Specialized Seminar on Emission of Trace Organics from Municipal Solid Waste Incinerators, Copenhagen, Denmark, January 20 to 22, 1987.
24. Atkinson, R. "Estimation of OH Radical Rate Constants and Atmospheric Lifetimes for Polychlorobiphenyls, Dibenzo-p-dioxins, and Dibenzofurans," *Environ. Sci. Technol.* 21:305–307 (1987).
25. Brorstrom, E., P. Grennfelt, A. Lindskog, A. Sjodin, and T. Nielsen. "Transformation of Polycyclic Aromatic Hydrocarbons During Sampling in Ambient Air by Exposure to Different Nitrogen Compounds and Ozone," in *PAH: Formation,*

*Metabolism and Measurement*, M. Cooke and A. J. Dennis, Eds. (Columbus: Battelle Press, 1983).

26. Brorstrom-Lunden, E., and A. Lindskog. "Degradation of Polycyclic Aromatic Hydrocarbons During Simulated Stack Gas Sampling," *Environ. Sci. Technol.* 19:313–316 (1985).
27. Buekins, A., and P. K. Patrick. "Incineration," in *Solid Waste Management: Selected Topics*, M. J. Suess, Ed. (Copenhagen: World Health Organization, 1985), pp. 79–150.
28. Carlson, K. "Heavy Metals from Energy from Waste Plants—Comparison of Gas Cleaning Systems," *Waste Manage. Res.* 4:15–20 (1986).
29. Concord Scientific Corporation, "NITEP P.E.I. Testing Program, Volume II: A Report to Environment Canada," Downsview, Ontario, June 1985.
30. Eighmy, T. T., N. E. Kinner, and T. P. Ballestero. "Final Report —"Codisposal of Lamprey Regional Solid Waste Cooperative Incinerator Bottom Ash and Sommersworth Wastewater Sludges," prepared for Lamprey Regional Solid Waste Cooperative, Durham, NH, January 29, 1988.
31. Finkelstein, A., R. Klicius, and D. J. Hay. "The NITEP Program," presentation at the International Workshop on Municipal Waste Incineration, Montreal, Quebec, October 1 to 2, 1987.
32. Ghezzi, U., and M. Giugliano. "L'incenerimento ed i microinquinanti organoclorurati: effetti della post-combustione," *Ing. Ambientale* 14:175–179 (1985).
33. Greenberg, R. R., W. H. Zoller, and G. E. Gordon. "Composition and Size Distribution of Particles Released in Refuse Incineration," *Environ. Sci. Technol.* 12:566 to 573 (1978).
34. Hartlen, J., and P. Elander. "Residues from Waste Incineration," SGI Varia 172, Swedish Geotechnical Institute, Linkoping (1986).
35. Hutton, M., A. Wadge, and P. J. Milligan. "Environmental Levels of Cadmium and Lead in the Vicinity of a Major Refuse Incinerator," *Atmos. Environ.* 22(2):411–416 (1988).
36. Kox, W. M. A., and E. Van Der Vlist. "Thermal Treatment of Heavy-Metal-Containing Wastes," *Conserv. Recycling* 4:29–38 (1981).
37. Lightowlers, P. J., and J. N. Cape. "Sources and Fate of Atmospheric HCl in the U.K. and Western Europe," *Atmos. Environ.* 22(1):7–15 (1988).
38. Malte, P. C., and J. C. Kramlich. "Further Observations of the Effect of Sample Probes on Pollutant Gases Drawn from Flame Zones," *Combust. Sci. Technol.* 22:263–269 (1980).
39. Mozzon, D., D. A. Brown, and J. W. Smith. "Occupational Exposure to Airborne Dust, Respirable Quartz and Metals Arising from Refuse Handling, Burning and Landfilling," *Am. Ind. Hyg. Assoc.* 48(2):111–116 (1987).
40. Modig, S. "Swedish View of the Ash Issue," paper presented at a seminar jointly sponsored by the Swedish World Trade Association and the National Resource Recovery Association, Washington, DC, March 23, 1988.
41. National Institute of Occupational Safety and Health. Health Hazard Evaluation Report No. HETA 82–201–1365 (September 1983).
42. National Institute of Occupational Safety and Health. Health Hazard Evaluation Report No. HETA 85–041–1709 (July 1986).
43. NSWMA, "Incinerator Ash: Aggregate Tonnage and Cost," NSWMA Fact Sheet (prepared 1986).
44. Organization for Economic Cooperation and Development. "Fate of Small Quan-

tities of Hazardous Waste," OECD Environment Monograph No. 6, Paris (August 1986).
45. Organization for Economic Cooperation and Development. "Energy from Waste – A Brief Assessment of the Situation in OECD Member Countries," ENV/ WMP/86.6, Paris (drafted May 15, 1986).
46. Risby, T. H., and S. S. Lestz. "Is the Direct Mutagenic Activity of Diesel Particulate Matter a Sampling Artifact?" *Environ. Sci. Technol.* 17:621–624 (1983).
47. Roos, C. E. "Is Lead a Big Problem?" *Waste Age* (February 1988), pp. 54–56.
48. Schure, M. R. "The Effect of Temperature Upon the Transformation of Polycyclic Organic Matter," PhD Thesis, Colorado State University (1981).
49. Shaub, W. M. "Thermal Decontamination of Soils: A Theoretical Analysis of Soils that Contain PCDDs and PCDFs," Private document (prepared October 1984).
50. Sutherland, M. "Application of Statistical Analysis on Incinerator Test Data," oral presentation at the International Workshop on Municipal Waste Incineration, Montreal, Quebec, October 1 to 2, 1987.
51. Taylor, D. R., M. A. Tompkins, S. E. Kirton, T. Mauney, P. K. Hopke, and D. F. S. Natusch. "Analysis of Fly Ash Produced from Combustion of RDF and Coal Mixtures," *Environ. Sci. Technol.* 16:148–154 (1982).
52. University of Massachusetts. "Incinerator Ash Management Project," Draft Report for NYSERDA/Mass. DEQE (October 1987).
53. Yakowitz, H. "Waste Management Activities in Selected Industrialized Countries," paper presented at Wastech '87, San Francisco, October 26 to 27, 1987.
54. Shaub, W. M. "Physical and Chemical Properties of Dioxins in Relation to Their Disposal," oral presentation of report by W. M. Shaub and W. Tsang at the International Symposium on Chlorinated Dioxins and Related Compounds," Arlington, VA, October 25 to 29, 1981.
55. Shaub, W. M. "Incineration – Some Things I Would Like to Know About and Why," report prepared for the 3rd Chemical Congress of North America, Toronto, Canada, June 5 to 10, 1988.
56. U.S. Department of Commerce. *Statistical Abstract of the United States 1987*, 107th ed. (Washington, DC: Bureau of the Census, 1987.
57. H. D. R. Techserv, Inc. "Evaluation of Project Impacts to the City of Tampa, Pinellas County, and Hillsborough County, Florida," (April 21, 1987).
58. California Air Resources Board, "The Effects of Oxides of Nitrogen upon California Air Quality," Report No. TSD-85-01, (March 1986).
59. Cooper, J. A., and D. Malek, Ed. *Proceedings, 1981 International Conference on Residential Solid Fuels– Environmental Health Impacts and Solutions*, (Portland, OR: Oregon Graduate Center, 1981).
60. Robertson, C. A. M. *Solid Wastes Manage.* 64:139–154 (1974).
61. Velzy, C. O. "Statement of the Research Committee on Industrial and Municipal Wastes of the American Society of Mechanical Engineers on H.R. 2787", before the Subcommittee on Health and the Environment, Committee on Energy and Commerce, U.S. House of Representatives, Washington, DC, July 2, 1987.
62. Midwest Research Institute. "Results of the Combustion and Emissions Research Project at the Vicon Incinerator Facility in Pittsfield, Massachusetts," Draft Final Report, (February 18, 1987).
63. Shaub, W. M. "Implementation of Incinerator Technologies – A Snapshot of

Some Technical Issues," background document in support of oral presentation made at the National Science Foundation, April 19, 1988.
64. Shaub, W. M. "Air Emissions Control and Ash Problems," background document in support of an oral presentation made at the Solid Waste Management Options for Texas Conference, Austin, TX, May 19, 1988.
65. Ames, B. N. Personal communication with L. Carothers, Commissioner, Connecticut Department of Environmental Protection (October 22, 1987).
66. Ames, B. N., "Six Common Errors Relating to Environmental Pollution," *Water* 27:20–22 (1986).
67. Ames, B. N., R. MaGaw, and L. S. Gold. "Ranking Possible Carcinogenic Hazards," *Science* 236:271–280 (1987).
68. Wilson, R., Personal communication with W. M. Shaub (June 22, 1987).
69. Gough, M. "Health Risks from Dioxin," report prepared for the Connecticut Resources Recovery Authority, Hartford, CT (October 16, 1987).
70. Rappe, C., oral remarks made at the Incineration Seminar jointly sponsored by the National Resource Recovery Association and the Swedish Trade Council, Washington, DC, March 23, 1988.
71. Danish Ministry of the Environment. "Order Governing Utilization of Slag and Flyash," Order No. 568 (December 6, 1983).
72. Danish Ministry of the Environment. "Circular on Utilization of Slag and Flyash," Danish EPA Circular (December 8, 1983).
73. Institute of Resource Recovery. "Management of Municipal Waste Combustion Ash," NSWMA (July 10, 1987).
74. NUS Corporation. "Characterization of Municipal Waste Combustor Ashes and Leachates from Municipal Solid Waste Landfills, Monofills and Codisposal Sites," Final Report, Report No. R-33-6-7-1 (September 1987).
75. Denison, R. A., and E. K. Silbergeld. "Risks of Municipal Solid Waste Incineration: An Environmental Perspective," this journal.
76. Hjelmar, O. "Leachate from Incinerator Ash Disposal Sites," Danish Water Quality Institute, Horsholm, Denmark (1987).
77. Cundari, K. L., and J. M. Lauria. "The Laboratory Evaluation of Expected Leachate Quality from a Resource Recovery Ashfill," Internal Report, Malcolm Pirnie, Inc., White Plains, NY (1987).
78. Shaub, W. M., Ed. CORRE Information Bulletin 87-0002 (December 21, 1987).
79. Denison, R. A. "Fundamental Objectives of Municipal Solid Waste Incinerator Ash Management," Paper No. 88-26.6, presented at the 81st Annual Meeting and Exhibition of the Air Pollution Control Association, Dallas, TX, June 20 to 24, 1988.
80. Fichel, K., and W. Beck. "Auslaugverhalten von Ruckstanden aus Abfallverbrennungsanlagen (2)," *Mull Abfall*, (November 1984), pp. 331–339.
81. (a) Steiner, R. C., A. A. Fungaroli, R. J. Schoenberger, and P. W. Purdom. "Criteria for Sanitary Landfill Development," *Public Works* 102:77–79 (1971); (b) Environmental Protection Agency. "An Environmental Assessment of Potential Gas and Leachate Problems of Land Disposal Sites," EPA SW-110 (1973); (c) Environmental Protection Agency. "Gas and Leachate from Land Disposal of Municipal Solid Waste," Summary Report MERL, Cincinatti, OH (1975).
82. Smith, A. H., M. T. Smith, R. Wood, H. Goeden, P. Coyle, T. Chambers, and

E. T. Wei. "Health Risk Assessment of the Los Angeles City Energy Recovery (LANCER) Project," Vol. 1 to 3, Berkeley, CA (April 17, 1987).

83. (a) California Air Resources Board. "Source-Receptor Reconciliation of South Coast Air Basin Particulate Air Quality Data," Sacramento, CA (1981); (b) California Air Resources Board. "California Air Quality Data. Summary of 1982 Air Quality Data," Vol. 14, Sacramento, CA (1983); (c) Friberg, L., G. F. Nordberg, and V. B. Vouk. *Handbook on the Toxicology of Metals*, (New York: Elsevier New-Holland Inc. 1979); (d) Ratcliffe, J. M. *Lead in Man and the Environment*, (New York: John Wiley & Sons, Inc., 1981).
84. Forrester, K. oral remarks during panel presentation "Issues/Ash Characteristics and Disposal Criteria," New Jersey Department of Environmental Protection Solid Waste Conference, Spring Lake, NJ, April 21 to 22, 1988.
85. Kellermeyer, D. A. Private communication to D. Taam, Spokane Regional Solid Waste Disposal Project, (July 14, 1987).
86. Harper, M., K. R. Sullivan, and M. J. Quinn. "Wind Dispersal of Metals from Smelter Waste Tips and Their Contribution to Environmental Contamination," *Environ. Sci. Technol.* 21(5):481–484 (1987).
87. Chamberlain, A. C., W. S. Clough, M. J. Heard, D. Newton, A. N. B. Stott, and A. C. Wells. "Uptake of Lead by Inhalation of Motor Exhaust," *Proc. Roy. Soc. London Ser. B* 192:77–110 (1975).
88. Smith, A. H., M. T. Smith, R. Wood, H. Goeden, P. Coyle, T. Chambers, and E. T. Wei. "Health Risk Assessment of the Los Angeles City Energy Recovery (LANCER) Project," Vol. 3, Appendix 9, Berkeley, CA (April 17, 1987), p. 10.
89. Mansdorf, S. Z., M. A. Golembiewski, and M. W. Fletcher. "Industrial Hygiene Characterization and Aerobiology of Resource Recovery Systems," Midwest Research Institute Project No. 4856-L(9), Contract No. 210–79–0013, Final Report (June 1982).
90. Centers for Disease Control. *New England J. Med.* 292(1):1–6 (1975).
91. Zurlinden, R. A., H. P. VonDemFange, and J. L. Hahn. Environmental Test Report No. 105, Ogden Projects, Inc., regarding Marion County Solid Waste-to-Energy Facility— Boilers 1 and 2 (November 7, 1986).
92. (a) Oregon Department of Environmental Quality, report regarding the Bergsoe Battery Recycling Plant (1982) as cited in Draft Report of the Health Impact Review Panel, "Report on the Trash Incinerator Facility Proposed for Columbia County, Oregon," (January 1988); (b) Shaub, W. M. "More About Incinerators and Lead Smelters," *CORRE Newsl.* 2(2):3 (1988).
93. Herz, M. "Testimony of EDF at a hearing before the Subcommittee on Transportation, Tourism and Hazardous Materials of the House Committee on Energy and Commerce," Washington, D.C. (April 13, 1988; Attachment 3 dated January 20, 1988).
94. Lyman, J. in written record submitted at the time of testimony given before the Subcommittee on Transportation, Tourism and Hazardous Materials of the House Committee on Energy and Commerce, Washington, D.C. (April 13 1988).
95. Goodwin, R. Private communication with W. M. Shaub (April 25, 1988).
96. Comrie, D. C. Private communication with W. M. Shaub (April 26, 1988).
97. Comrie, D. C., and J. Davidovits, "Waste Containment Technology for Management of Uranium Mill Tailings," presented at the 117th Annual Meeting of the AIME/SME, Phoenix, AZ, January 25 to 28, 1988.

98. Mackenzie, J. D. Private communication with W. M. Shaub (April 25, 1988).
99. Hasselriis, F. Private communication with W. M. Shaub (April 25, 1988).
100. Sussman, D. B., and J. L. Hahn. "Municipal Waste Combustion Ash," paper presented at meeting entitled "New Developments in Incinerator Ash Disposal," New York, April 26, 1988.

## CHAPTER 15

# Concentrations of PCDD/PCDF in Ambient Air in the Vicinity of Municipal Refuse Incinerators

**D. J. Wagel, T. O. Tiernan, J. H. Garrett, J. G. Solch, L. A. Harden, and G. F. VanNess**

## INTRODUCTION

Determination of the concentrations of polychlorinated dibenzo-p-dioxins (PCDDs) and polychlorinated dibenzofurans (PCDFs) present in the ambient air is potentially important for identifying the sources from which these compounds are emitted and for understanding the transport and fate of these materials in the atmosphere. In addition, data on the ambient air concentrations of PCDD/PCDF measured at ground level are more readily utilized to estimate the exposure of the human population to such airborne compounds than are stack emission data obtained from sources such as incinerators. However, largely because of the uncertainties and difficulties associated with ambient air measurements of these and related compounds, comparatively few determinations of this type have been made. Recently, our laboratory has evaluated a standard polyurethane foam (PUF)-type ambient air sampler (designed to collect airborne organic compounds) for its applicability in collecting PCDDs/PCDFs present in ambient air at concentrations of 1 pg/m$^3$ or lower.[1] The present report describes further experiments aimed at determining the fate of particulate-bound PCDDs/PCDFs collected by such a sampler as well as the overall collection efficiency of this sampler for these compounds. The ultimate objective is to better understand the relative distribution of PCDD/PCDF isomers present in ambient air between the vapor phase and the particulate-bound components. The present study also involved the extension of an earlier preliminary survey conducted by our laboratory of PCDDs/PCDFs in the ambient air of the Dayton, OH, metropolitan area.[2] This previous assessment revealed the presence of concentrations of these compounds ranging from nondetectable to approximately 30 pg/m$^3$ of PCDDs and PCDFs at several sites within this area at which samples were collected. Not surprisingly, the prior study also indicated that the industrialized regions of the subject area are impacted by outfall from more than one source of PCDDs and

PCDFs. In the second phase of the ambient air monitoring study of the Dayton area, the results of which are described herein, additional samples have been collected to determine the day-to-day variability of PCDDs/PCDFs in the air at a site located near and within the outfall region of a municipal waste incinerator (MWI). The PCDD and PCDF total congener group concentrations have been measured and the patterns used to compare results for several different sampling sites within the metropolitan area of interest. These results, as well as patterns of specific PCDD/PCDF isomers observed in the ambient air samples, have also been used to distinguish and characterize different sources of these compounds within the area.

## SAMPLING APPARATUS AND COLLECTION PROCEDURE

The ambient air sampler employed for these studies was the General Metal Works Model PS-1 PUF Sampler. This sampler draws ambient air through a sampling module which contains a QM-A quartz filter* backed by a glass cartridge containing (in sequence) a 5-cm PUF plug, 25-$cm^3$ of XAD-2 resin,** and a second 2.5-cm PUF plug. The PUF plugs used here were manufactured by Olympic Products and were cleaned by Soxhlet extraction prior to use. The PS-1 sampler has been shown to retain 85 to 100% of each of the 2,3,7,8-substituted PCDD and PCDF isomers when the pure isomers are added to the filter prior to sampling 250 to 300 $m^3$ of air.[1] The sampler was also found to retain $^{13}C_{12}$-labeled PeCDD, PeCDF, HxCDD, and HxCDF isomers under field sampling conditions, when these isomers are added to the filter.

Ambient air samples were collected by drawing air through the sampler at the maximum flow rate (225 to 300 L/min) for 22 to 24 hr. Following sample collection, the filter and glass cartridge were removed from the sampling module, sealed in containers, and maintained at less than 20°C during transfer to the laboratory. Laboratory studies were conducted in which synthetic and MWI flyashes containing PCDD/PCDF isomers were distributed over the surface of the filter prior to air sample collection.[2] A second sampler with an unspiked filter was operated during each flyash experiment to monitor background levels of PCDDs and PCDFs. These background levels were subtracted from the spiked sample results before calculation of isomer distributions or total recovery. PCDDs and PCDFs were not detected in laboratory and field blanks prepared during the ambient air field sampling.

## AMBIENT AIR SAMPLE EXTRACTION, PREPARATION, AND ANALYSIS

The efficacy of the sample extraction, preparation, and analysis procedures was determined by adding one $^{13}C_{12}$–2,3,7,8-substituted isomer from each

*Whatman Laboratory Products, Inc.
**Supelco, Inc.

tetra-to octa-PCDD/PCDF congener group to the sample before Soxhlet extraction and measuring these compounds along with the native PCDDs/PCDFs. The Soxhlet extraction and liquid chromatographic separation procedures employed for analysis of field samples and for determining the distribution of PCDD and PCDF isomers between the filter and cartridge following the flyash experiments have been described previously.[1-3] The gas chromatography/mass spectrometry (GC/MS) analyses were performed on a 60-m DB-5 capillary column using a Carlo Erba gas chromatograph coupled to a Kratos MS-25 mass spectrometer.[3,4] Selected samples were also analyzed on a 60-m SP-2330 capillary column to provide more specific information on the 2,3,7,8-substituted PCDD and PCDF isomers.

## DISTRIBUTION OF MWI FLYASH-BOUND PCDD AND PCDF ISOMERS BETWEEN THE FILTER AND CARTRIDGE FOLLOWING SAMPLE COLLECTION

In earlier studies, it was demonstrated that pure 2,3,7,8-substituted PCDD and PCDF isomers added to the filter of the PS-1 sampler migrated from the filter to the PUF/XAD-2 cartridge during sample collection.[1] Losses of these compounds from the filter were nearly 100% for the lower chlorinated isomers and these losses decreased as the degree of chlorination of the PCDDs/PCDFs on the filter increased. It was also demonstrated that significant quantities of the 2,3,7,8-substituted PCDD and PCDF isomers, which had been added to a synthetic inorganic flyash, which was then distributed on the inlet filter prior to air sampling, were detected in the PUF/XAD-2 cartridge following sampling.[2] Also in this case, the extent of migration of PCDD/PCDF isomers from the filter to the sorbent cartridge depended on the degree of chlorination of the compounds. These experiments were extended in the present study to include naturally occurring MWI flyashes containing PCDDs/PCDFs in an effort to determine the fate of PCDDs and PCDFs incorporated in actual flyash following collection of such ash on the filter of an ambient air sampler. The two natural flyashes used in this study were obtained from the Ontario Ministry of the Environment, Toronto, Canada. Following analysis to determine the PCDD and PCDF content of each of these flyash samples, known quantities of approximately 100 mg of each flyash were distributed on the inlet filter of the PS-1 sampler in three separate experiments. The volume of air sampled, temperature, temperature range, and relative humidity during each experiment are presented in Table 1.

The percentages of PCDDs and PCDFs observed in the filter and cartridge, respectively, following spiking of the filter with Flyash 1 and sampling of ambient air at 23°C, are presented in Table 2. In contrast to the studies described earlier, in which significant quantities of pure and synthetic flyash-bound PCDD and PCDF isomers were observed to migrate from the filter to the cartridge, only 1% of the lower chlorinated PCDF congener groups were

**Table 1. Sampling Conditions for Investigation of PCDDs/PCDFs Sorbed on MWI Flyashes**

| MWI flyash matrix | Volume of air collected ($m^3$) | Flowrate (L/min) | Average temperature (°C) | Relative humidity (%) |
|---|---|---|---|---|
| Flyash No. 1 | 346 | 238 | 8 (4–12)[a] | 36 |
| Flyash No. 1 | 330 | 227 | 23 (11–28) | 44 |
| Flyash No. 2 | 334 | 238 | 17 (10–26) | 29 |

[a]Temperatures in parenthesis are the lowest and highest temperatures which occurred during the sampling period.

stripped from the natural MWI flyash and deposited in the cartridge. The experiment in which the filter of the sampler was spiked with Flyash 1 and air was then sampled at 8°C resulted in less than 1% of the lower chlorinated congeners being transferred to the cartridge. The observed distribution of the isomers between the filter and the cartridge when Flyash 2 was placed on the inlet filter followed by air sampling is presented in Table 3 and is similar to the distributions observed for Flyash 1 in that less than 1% of the isomers were found to migrate to the cartridge during sampling. In each of these flyash experiments, the combined recoveries of the PCDD and PCDF congeners in the filter and cartridge, which were observed following sampling, support the conclusions derived from the earlier sampler validation studies which were performed with pure isomers.[1] Obviously, the overall collection efficiency of this sampler for PCDDs/PCDFs is quite high. Table 4 presents a summary of the results obtained for the 2,3,7,8-TCDD isomer when this isomer was added to the filter as a pure compound, as compared to the case in which it was incorporated into a synthetic or MWI flyash.

These studies clearly demonstrate that the PCDD and PCDF isomers are not easily stripped from a MWI flyash matrix when sample volumes of 300 to 350 $m^3$ of ambient air are passed through the filter containing this ash. The results of these preliminary experiments therefore support the interpretation that

**Table 2. Distribution of PCDDs and PCDFs Following Air Sampling When the Filter was Spiked with Flyash No. 1**

| Congener | Percent recovered in filter | Percent recovered in cartridge | Total percent recovered |
|---|---|---|---|
| TCDD | 68 | 0 | 68 |
| PeCDD | 76 | 0 | 76 |
| HxCDD | 74 | 0 | 74 |
| HpCDD | 85 | 0 | 85 |
| OCDD | 94 | 0 | 94 |
| TCDF | 74 | 1 | 75 |
| PeCDF | 64 | 1 | 65 |
| HxCDF | 69 | 1 | 70 |
| HpCDF | 79 | 0 | 79 |
| OCDF | 75 | 0 | 75 |

**Table 3. Distribution of PCDDs and PCDFs Following Air Sampling When the Filter was Spiked with Flyash No. 2**

| Congener | Percent recovered in filter | Percent recovered in cartridge | Total percent recovered |
|---|---|---|---|
| TCDD | 77 | 0 | 77 |
| PeCDD | 85 | 0 | 85 |
| HxCDD | 116 | 0 | 116 |
| HpCDD | 126 | 0 | 126 |
| OCDD | a | a | a |
| TCDF | 75 | 0 | 75 |
| PeCDF | 80 | 0 | 80 |
| HxCDF | 84 | 0 | 84 |
| HpCDF | 126 | 0 | 126 |
| OCDF | a | a | a |

[a]Isomer not detected in flyash added to filter.

PCDDs and PCDFs recovered in the filter are representative of the portion of these isomers which are particulate-bound, and the PCDDs/PCDFs found in the cartridge are representative of the concentrations of these present in the vapor phase. However, PCDD and PCDF isomers bound to the smaller particles collected by the sampler from actual ambient air may not behave in the same manner as PCDD and PCDF isomers bound to the relatively large flyash particles used in the laboratory studies. The effects of filter pore size and retention efficiency on the results of experiments such as those described here also need to be investigated before definitely assigning the contents of the cartridge to the vapor phase component.

## DESCRIPTION OF DAYTON, OH, AMBIENT AIR MONITORING SITES

During an initial survey of the Dayton, OH, metropolitan area, eight ambient air samples were collected at four sampling sites.[2] Each of these sites is currently used by the Regional Air Pollution Control Agency in this area to monitor air quality (in terms of quantity of particulate only) in the metropolitan area. Samples from the downtown area were collected with the sampler located on the roof of a three-story building near the center of the downtown area. Background samples were collected at a rural site located approximately

**Table 4. Summary of 2, 3, 7, 8-TCDD Distribution Following Air Sample Collection**

| Matrix added to filter | Percent on filter | Percent in PUF | Temperature (°C) | Sample volume ($m^3$) |
|---|---|---|---|---|
| Pure Isomer | 0 | 100 | 26 | 271 |
| Synthetic Ash | 8 | 92 | 14 | 334 |
| MWI Flyash No. 1 | 100 | 0 | 8 | 346 |
| MWI Flyash No. 1 | 100 | 0 | 23 | 330 |
| MWI Flyash No. 2 | 100 | 0 | 17 | 334 |

15 km northeast of the downtown area. A roadside site in a Dayton suburb, which was located about 8 km southeast of the downtown area, was sampled in an effort to investigate the possible contribution of automobiles and other such vehicles to the ambient air levels of PCDDs and PCDFs. The latter sampler was located at ground level, about 3m from an intersection through which some 60,000 vehicles pass each day. The fourth site was located approximately 300 m south-southwest of a MWI which is positioned 6 km southwest of the downtown area. Another identical MWI is located 8 km north-northeast of the downtown area.

## INITIAL AMBIENT AIR RESULTS FOR DAYTON, OH, AREA

During the initial investigation of the Dayton area, the concentrations of PCDDs and PCDFs were near or below the limit of detection for the samples collected at the rural site.[2] During sampling, the wind originated from the southwest, assuring that the rural site was not impacted by sources located within the metropolitan Dayton area. The highest observed ambient air concentrations of PCDDs and PCDFs were found in the vicinity of the MWI where 29.4 pg/m$^3$ of PCDDs and 35.4 pg/m$^3$ of PCDFs were detected. Figure 1 shows that the PCDD and PCDF congener group patterns observed in samples collected at the downtown and MWI locations were quite similar. The congener profiles observed in air samples from the roadside area, however, were distinctly different because OCDD was present at high levels relative to the other PCDD and PCDF congeners. For each of the samples in which

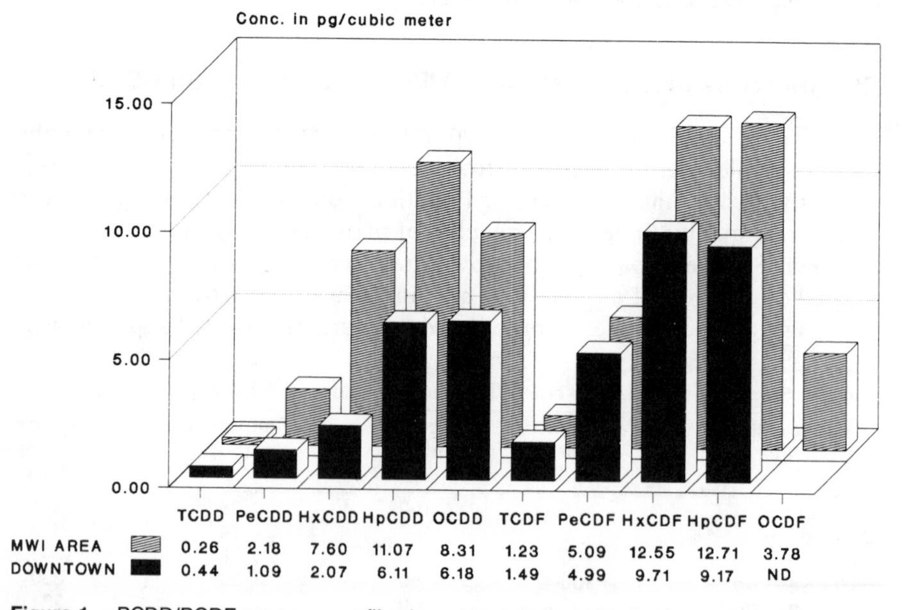


**Figure 1.** PCDD/PCDF congener profiles in ambient air from the downtown and MWI areas.

**Table 5. Sampling Conditions for Ambient Air Samples Collected May 16 to 20, 1988**

| Date | Sample volume ($m^3$) | Average wind direction | Average wind speed | Average temperature (°C) | Relative humidity (%) |
|---|---|---|---|---|---|
| | | MWI area (Site 8) | | | |
| May 16—17 | 324 | NNW (WNW–N) | 10 | 18 | 49 |
| May 17—18 | 349 | NNE (N–E) | 9 | 16 | 68 |
| May 18—19 | 365 | NE (N–ENE) | 12 | 16 | 64 |
| May 19—20 | 329 | NNE (N–NE) | 5 | 15 | 83 |
| | | Roadside area (Site 13) | | | |
| May 17—20 | 1025 | NNE (N–E) | 9 | 16 | 72 |

PCDDs and PCDFs were detected, the observed gas chromatographic patterns of specific isomers were nearly identical.[2]

## SAMPLING CONDITIONS FOR AMBIENT AIR SAMPLES

In order to investigate the daily variability of PCDD and PCDF concentrations in ambient air at the MWI area, air samples were collected on four consecutive days. During the first phase of the ambient air study, low concentrations of PCDDs and PCDFs were found in samples collected at the roadside area. Therefore, a larger volume of air was sampled at the roadside site in order to permit comparison of the congener and isomer patterns in this sample with those in samples from the other Dayton area sites. The collection dates, sample volumes, and weather conditions prevailing during sampling for each of the ambient air samples collected are presented in Table 5.

The south MWI located in the Dayton area, which is capable of incinerating 525 tons of refuse in a 24-hr period, was operating at 80% capacity during each of the four sequential sampling periods from May 16 to 20. This incinerator is a rotary kiln and employs an electrostatic precipitator to control emissions. An industrial facility which removes polyvinyl chloride (PVC) from metal surfaces by pyrolysis in ovens is also located in the vicinity of the MWI, less than 1 km north of the sampling site near the MWI. This facility was operating on May 17 to 18. Because our collection of air samples was started on the morning of the first date listed for the five samples in Table 5 and was completed on the morning of the second date, the samples which were collected on May 17 to 18 and May 18 to 19 could reflect contributions from the PVC processing plant as well as from the MWI.

## RESULTS OF AMBIENT AIR ANALYSES

The PCDD and PCDF total congener group patterns for the four sequential samples collected at the MWI area on May 16 to 20 are presented in Figure 2. The patterns observed in the samples collected on May 16 to 17 and May 19 to 20 are dominated by the OCDD congener and are clearly different from the

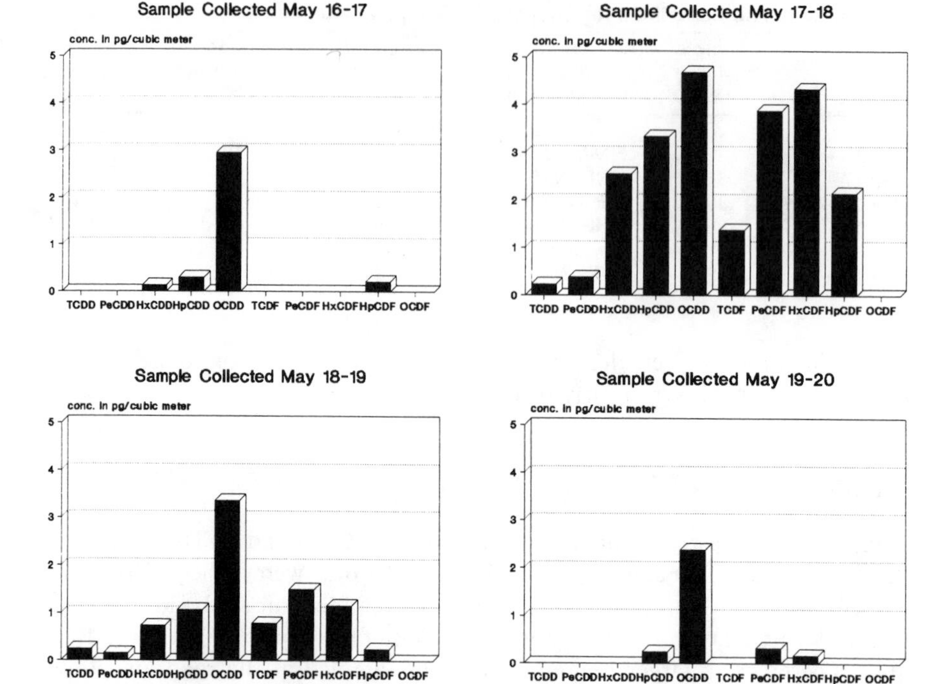


**Figure 2.** PCDD and PCDF total congener group patterns for four sequential samples collected at the MWI area.

profiles obtained on the other two days. The sample collected on May 18 to 19 also contains relatively high levels of OCDD, but significant quantities of the other congener groups are present and the pattern resembles the May 17 to18 sample. As mentioned earlier, the incinerator was operating at a constant level during the entire sampling period and the prevailing wind at the time of sampling should have carried the incinerator effluent over the sampling site during each sampling period. The event which most significantly correlates with the observed changes in the congener profiles is the operation of the PVC processing plant. Laboratory pyrolysis of PVC and other organochlorine polymers has previously been shown to yield PCDD and PCDF products.[5]

The PCDD/PCDF congener pattern in the sample from the roadside site obtained here is similar to the pattern observed in samples from this site collected during the first phase of our ambient air study.[2] The OCDD congener was again present in the highest concentration while much lower concentrations of the other congener groups were detected. The congener pattern in the roadside sample is compared in Figure 3 to the two distinct MWI area patterns found in samples collected on different days. The roadside sample congener pattern is observed to be similar to the congener pattern found in samples collected in the MWI area on May 19 to 20, the day when the PVC processing plant was not in operation. The volume of sample collected at the roadside on

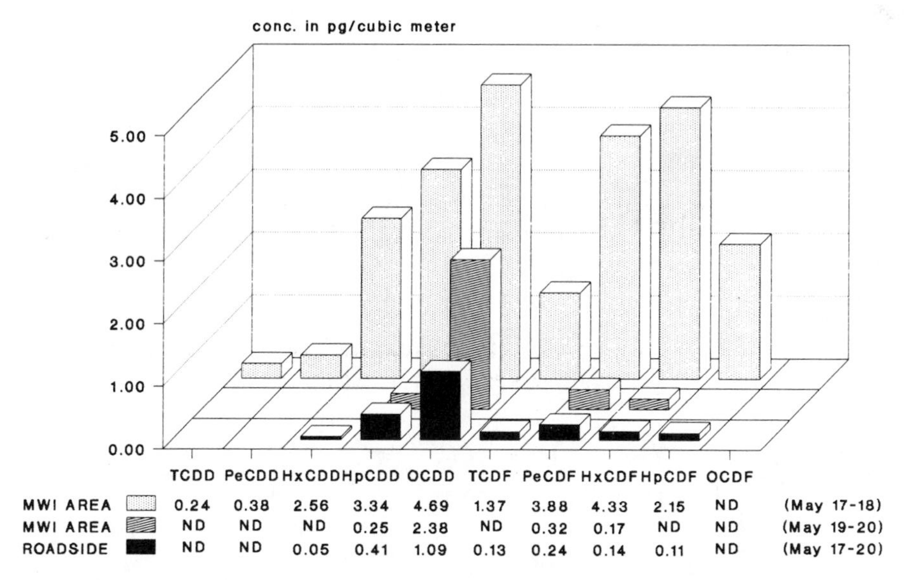


**Figure 3.** Comparison of PCDD/PCDF congener profiles in ambient air samples.

May 17 to 20 was three times greater than the volume of air collected at the MWI area and the increased sensitivity may account for more isomer groups being detected in the roadside sample relative to those found in the May 19 to 20 MWI area sample. The GC/MS mass chromatograms of the PeCDF region for the two MWI samples and for the roadside sample, the three samples for which total congener patters are illustrated in Figure 3, are presented in Figure 4. The PeCDF isomer patterns for the three samples are seen to be virtually identical in spite of the very different total congener patterns and concentrations observed for these samples.

## COMPARISON OF PRESENT RESULTS TO PUBLISHED STUDIES

The similarity of the total PCDD/PCDF congener patterns which were observed in samples collected during the first phase of this study to those reported in previous studies of incinerator stack emissions, as well as to the patterns in ambient air samples collected in Hamburg, downwind of a MWI has been discussed earlier.[2] Table 6 presents a comparison of the results of the isomer specific analysis of a MWI area sample collected in the present study with the previously reported results for an ambient air sample collected downwind of a MWI.[6] Except for the 1,2,3,7,8-PeCDD and OCDF isomers, the two sets of results are quite similar. In our more recent investigations, however, the PCDD/PCDF total congener pattern in samples from the MWI area which compares most favorably with MWI congener patterns reported by other investigators is that obtained when both the MWI and the PVC processing

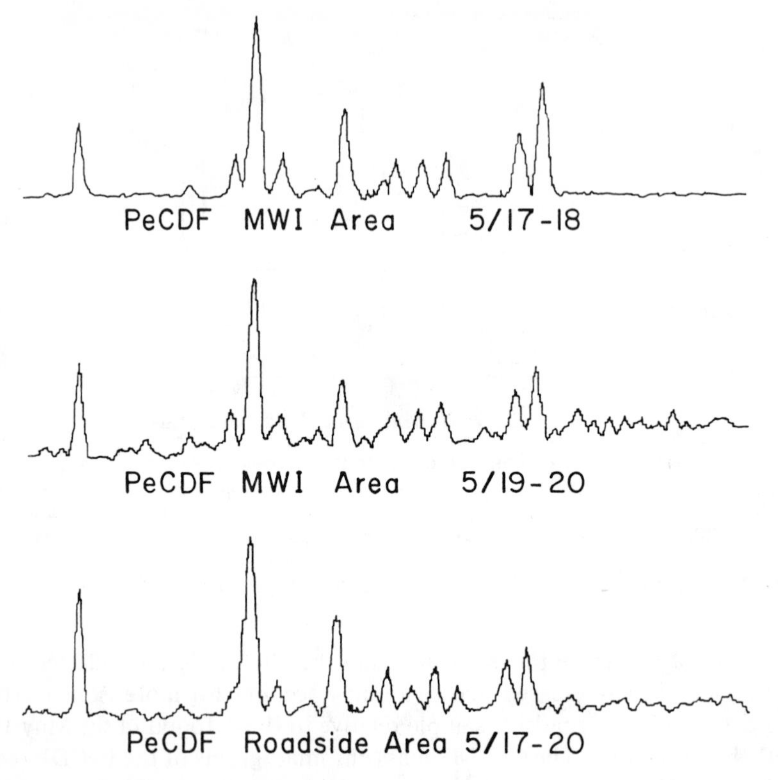


**Figure 4.** GC/MS mass chromatograms of the PeCDF region for ambient air samples collected at the MWI and roadside areas.

plant were in operation. Further sampling studies are planned at a more remote PVC processing plant and at the other MWI in the Dayton area in an effort to better distinguish the PCDD and PCDF emissions associated with each of these sources. Problems similar to those encountered here in obtaining reproducible results when sampling downwind of a MWI have also been reported by other researchers.[6]

## CONCLUSIONS

The laboratory experiments conducted in the present study demonstrated that PCDD and PCDF isomers bound to natural MWI flyash do not migrate to a significant degree from the inlet filter of an ambient air sampler to the backup sorbent cartridge under the sampling conditions which were evaluated here. Following further experiments to determine the effects of particle size and filtering media on the sampling results, additional air samples will be collected to determine the relative quantities of vapor phase and particulate-bound PCDDs and PCDFs present in ambient air. These studies will also

**Table 6. 2, 3, 7, 8-Substituted PCDD/PCDF Concentrations from the MWI Area Compared to Published Results**

| | Concentrations (pg/m³) | |
|---|---|---|
| **Isomer** | **MWI Area present study** | **Downwind of MWI[a]** |
| 2378-TCDD | ND (0.06) | 0.08 |
| 12378-PeCDD | 0.57 | 0.05 |
| 123478-HxCDD | 0.63 | 1.19 |
| 123678-HxCDD | 1.19 | 0.90 |
| 123789-HxCDD | 0.91 | 0.38 |
| 1234678-HpCDD | 6.02 | 4.40 |
| OCDD | 8.26 | 7.70 |
| 2378-TCDF | 0.11 | 0.50 |
| 12378-PeCDF | 0.46[b] | 0.75[b] |
| 23478-PeCDF | 0.53 | 0.47 |
| 123478-HxCDF | 1.18[b] | 0.50[b] |
| 123678-HxCDF | 2.27 | 0.50 |
| 123789-HxCDF | ND (0.06) | 0.08 |
| 234678-HxCDF | ND (0.41) | 0.36 |
| 1234678-HpCDF | 8.22 | 3.00 |
| 1234789-HpCDF | 0.56 | 0.32 |
| OCDF | 3.78 | <0.96 |

[a]Rappe et al. *Chemosphere* 17:3 (1988).
[b]Result is not specific for listed isomer.

examine the effects of temperature and different ambient air sources on this distribution. Although the specific isomer patterns detected in air samples collected in the present study were quite similar, the patterns of the PCDD/PCDF congener groups in these samples showed considerable variability. The interpretation of the results of the PCDD and PCDF measurements in air samples collected in the vicinity of a MWI have been complicated by the presence of other nearby industrial sources. Further studies, which will entail analyses of flyash samples from specific emission sources, as well as source modeling and examination of the additional ambient air samples noted earlier, are planned in an effort to differentiate the PCDDs and PCDFs in MWI emissions from those arising from other potential sources in the Dayton metropolitan area.

## REFERENCES

1. Wagel, D. J., T. O. Tiernan, M. L. Taylor, J. H. Garrett, G. F. VanNess, J. G. Solch, and L. A. Harden. "Assessments of Ambient Air Sampling Techniques for Collecting Airborne Polyhalogenated Dibenzo-p-dioxins (PCDD), Dibenzofurans (PCDF), and Biphenyls (PCB)," *Chemosphere*, 18:177–184 (1988).
2. Tiernan,T. O., D. J. Wagel, G. F. VanNess, J. H. Garrett, J. G. Solch, and L. A. Harden. "Evaluation of the Collection Efficiency of a High Volume Sampler Fitted With an Organic Sampling Module for Collection of Specific Polyhalogenated Dibenzodioxin and Dibenzofuran Isomers Present in Ambient Air," Proceedings of

the EPA/APCA Symposium: Measurement of Toxic and Related Air Pollutants, Raleigh, NC, May 1988.

3. Tiernan, T. O. "Analytical Chemistry of Polychlorinated Dibenzo-p-dioxins and Dibenzofurans: A Review of the Current Status," in *Chlorinated Dioxins and Dibenzofurans in the Total Environment*, G. Choudhary, L. H. Keith, and C. Rappe, Eds., (Boston: Butterworth Publishers, 1983), Chapter 13.
4. Solch, J. G., G. L. Ferguson, T. O. Tiernan, G. F. VanNess, J. H. Garrett, D. J. Wagel, and M. L. Taylor. "Analytical Methodology for Determination of 2,3,7,8-Tetrachlorodibenzo-p-dioxin in Soils," in *Chlorinated Dioxins and Dibenzofurans in the Total Environment II*, L. H. Keith, C. Rappe, and G. Choudhary, Eds., (Boston: Butterworth Publishers, 1985), Chapter 28.
5. Marklund, S., L.-O. Kjeller, M. Hansson, M. Tysklind, C. Rappe, C. Ryan, H. Collazo, and R. Dougherty. "Determination of PCDDs and PCDFs in Incineration Samples and Pyrolytic Products," in *Chlorinated Dioxins and Dibenzofurans in Perspective*, C. Rappe, G. Choudhary, and L. Keith, Eds., (Chelsea, MI: Lewis Publishers, Inc., 1986), Chapter 6.
6. Rappe, C., L.-O. Kjeller, P. Bruckmann, and K.-H. Hackhe. "Identification and Quantification of PCDDs and PCDFs in Urban Air," *Chemosphere*, 17:3–20 (1988).

## CHAPTER 16

# Polycyclic Aromatic Compounds in Steel Foundry Emissions

C. Tashiro, M. A. Quilliam, D. R. McCalla, C. Kaiser-Farrell, and E. Gibson

## INTRODUCTION

### PAHs in the Environment

Polycyclic aromatic hydrocarbons (PAHs) in the environment have been widely studied because they are ubiquitous and carcinogenic and/or mutagenic.[1-5] PAHs are the products of the incomplete combustion of organic materials and they have been detected in such sources as foundry emissions,[6,7] shale oils,[8,9] diesel particulates,[10-12] marine sediments,[13] air particulates,[14,15] and brown-coal-fired stoves.[16] PAH derivatives are also formed in combustion processes and by other atmospheric reactions of the parent PAHs.

Steel foundry emissions contain large numbers and quantities of alkyl–PAH derivatives, some of which are of interest due to their degree of mutagenicity.[17] Molten steel pouring conditions are favorable for the production of these alkyl-PAHs. Other mutagenic PAH derivatives of interest in environmental samples include polycyclic aromatic ketones (PAKs), polycyclic aromatic quinones (PAQs), polycyclic aromatic furans (PAFs), and the alkyl derivatives of these compounds.

Of specific interest in steel foundry samples, as will be shown here, are the nitro-PAH derivatives. One of these compounds, 1-nitropyrene (1NP), is known to be a very mutagenic compound. It is most commonly found in diesel air particulate samples and urban air particulates. 1NP is formed from pyrene in the presence of nitrogen oxides. 1NP and other nitro-PAHs have recently been detected by our laboratory in steel foundry emissions.[18]

Since such large quantities of PAHs and their derivatives are produced on a global scale, it is important that researchers are able to analyze environmental samples for these types of compounds. However, the analysis of these samples is difficult due to the complexity of the mixtures. The analytical approach generally taken is to separate the sample into a large number of simpler fractions using a combination of extraction and cleanup techniques, such as Sox-

**Table 1. Risk of Lung Cancer to Foundry Workers According to Job Location**

| Job | Risk |
|---|---|
| Crane operator | 7.14 |
| Finisher | 3.14 |
| Molder | 2.55 |
| Core maker | 2.08 |
| Electric furnace operator | 1.14 |

hlet extraction and open-column chromatography with various adsorbents. This can be followed by high-performance liquid chromatography (HPLC) separation. Mutagenic fractions can be tracked through the cleanup by use of the Ames mutagenicity assay. The use of this assay aids in the determination of which "hot" fractions (those exhibiting appreciable mutagenicity) require further cleanup and investigation. Final analysis of the target mutagenic fractions can be done using gas chromatography/mass spectrometry (GC/MS) or HPLC/MS.

## Lung Cancer in the Steel Industry

In North America, the steel industry at one time employed more than 400,000 people, and in Hamilton, Ontario, the industry is still one of the major employers.[19] Dominion Foundries and Steel Limited (DOFASCO) has been operational since 1912. DOFASCO's Department of Occupational and Environmental Health has undertaken epidemiology studies to determine health effects on its foundry workers. There was a specific concern that an excess risk of lung cancer existed for foundry workers. Excess lung cancer deaths had been determined in iron foundries, but not in steel foundries.[20]

Through routine pulmonary testing, it was determined that nonsmoking foundry workers had pulmonary function levels similar to nonfoundry office workers who were daily smokers. The number of people that had been exposed in the the foundry over the years that could be suffering from similar decreased pulmonary function or lung cancer was significant. An epidemiological study was undertaken and the results indicated a 2.5 times greater risk of lung cancer for the foundry workers as compared to a standard population.[21] Table 1 shows the excess risk in the foundry according to job location. The results showed a seven times greater risk of lung cancer for the overhead crane operators as compared to a normal population.

Gibson hypothesized that the excess lung cancer among the crane operators in particular was due to exposure to the pyrolysis products of the organic binders used in the steel casting molds.[21] As molten steel is poured into the sand/organic molds, the high temperatures and absence of oxygen causes the breakdown of the organic binders into a variety of organic compounds that rise as a plume from the pouring floor to the crane operator location. The exposure of other workers who have increased lung cancer incidences can be explained by the overall direction of airflow toward the center of the foundry.

The question that arose from these data was what carcinogenic agents are present that give rise to an increased risk of lung cancer, especially for the crane operators? In order to answer this question, air particulate samples were collected from various locations in the foundry and subjected to the Ames assay for mutagenicity testing.[22] Figure 1 shows the results obtained for these tests, indicating mutagenic activity in all areas of the foundry. The high mutagenic level associated with the crane operator area and the mold area and the corresponding high standardized mortality ratio (SMR) for these workers should be noted. The direct and indirect acting mutagens that were found are presumed to be produced by the interaction of the hot metal with the organic binding compounds used in the mold. An initial determination of marker mutagenic PAHs did not account for all of the mutagenicity that was observed. Binder materials have not changed over the years, therefore the mutagenic compounds giving rise to the excess risk were assumed to still be present.

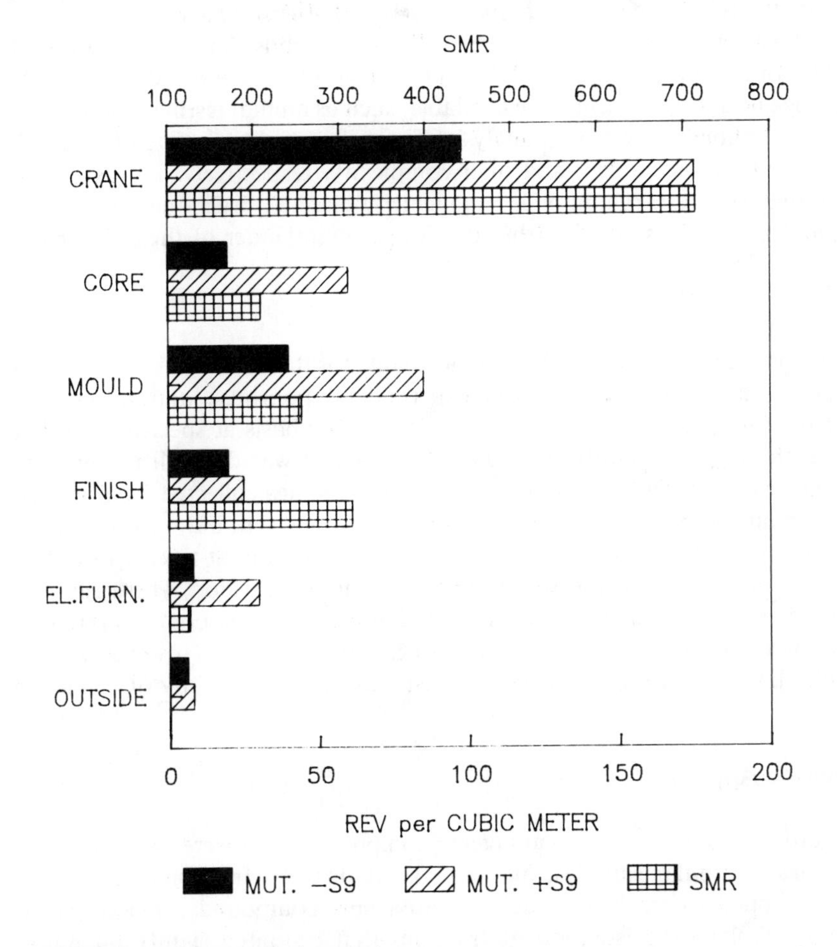


**Figure 1.** Air particulate mutagenicity—without (−S9) and with (+S9) activation.

Next, a link between the mutagenicity of the air particulate and the pyrolysis process had to be established. Experimental-size molds of various binder systems in sand were prepared at DOFASCO and 46 kg of molten steel was poured into each mold.[18] The composition of the various binder systems used is listed in Table 2. High-volume samplers were used to collect the particulate produced by the pyrolysis. Figure 2 shows the mutagenicity of the samples collected from the experimental pourings. The Shell Core and the Oil/Clay/Cereal molds both showed high levels of mutagenic activity and a corresponding high level of extractable organic material. Smaller particles had greater indirect acting mutagenicity as compared to the larger particles.

These experimental results lead to further questions, such as what are the compounds present in the foundry emission that are giving rise to the observed mutagenicity and is it possible to reduce the levels of mutagens present in the foundry to protect the workers? Alkyl-PAHs and other PAHs were determined in the emission samples.[18] A number of the compounds that were determined are known to be mutagenic.[2,10,23-25] A number of approaches to determine the carcinogenic agents present were available, such as animal testing, analysis for known carcinogens, or a total analysis. These approaches suffered from high cost, narrow target range, and high complexity, respectively. It was at this point that the research into a mutagenicity-directed fractionation scheme began. This is discussed in further detail in the remainder of the chapter.

## Available Foundry Samples

A number of foundry samples were available for the investigation, including air particulates, particulates from experimental pourings, and rafter dust samples. Air samples could be collected on a daily basis at specific sites, but mutagenicity varied greatly from day to day and it was difficult to correlate mutagenicity levels with foundry activity. The samples collected from experimental pourings provided a great deal of organic material, but were subject to high cost factors. It was therefore decided to do the initial research on dust samples collected from the rafters above the pouring floor. These samples represented an integrated sample that could indicate the compounds present in the foundry atmosphere and large quantities were available. However, a large sample size was required because the dust was made up mainly of inorganic materials.

## Research Approach

In order to determine the mutagenic compounds that were present in the emissions or similar samples, an Ames assay-directed fractionation scheme was developed to track and identify mutagenic compounds. This targeted analysis involved the isolation of fractions that exhibited significant mutagenicity from the bulk of the sample, followed by further fractionation and identification of mutagenic compounds as shown in Figure 3. For example, if

**Table 2. Typical Binder System Compositions**

| System | Organic constituents | Weight (kg) of organics in 100 kg of mix | Inorganic additives |
|---|---|---|---|
| Oil/Clay/ Cereal | Ceratex (pure grain product)<br>Linoil (vegetable oil) | 1.72 | Bentonite |
| Green Sand | Facetex (pure grain product)<br>Dextrine (wheat starch gum)<br>Linocel (cellulose) | 0.52 | Bentonite<br>Water |
| Sodium silicate | Ester activator (proprietary) | 0.20 | Silicate binder |
| Furan No Bake | Furruryl alcohol<br>Formaldehyde<br>Phenol<br>Xylene/toluene sulfonic acids | 1.30 | ——— |
| Kold Set | Kold Set oil (vegetable oil and petroleum distillates) | 1.45 | Activator (sodium perborate) |
| Shell Core | Phenol-formaldehyde resin<br>Hardener (hexamethylene-tetramine) | 0.90 | Kaolin clay |

an Ames assay is done on a set of three fractions and one is shown to be highly mutagenic, this fraction is further fractionated to isolate the mutagenic compounds. The iterative process yields simpler fractions at each stage, making the final confirmatory analysis much easier.

Complex foundry samples were fractionated using chromatographic methods, and at each stage, Ames assays were performed. The bioassay results were used to guide further fractionation until the fractions were reduced in complexity to allow GC/MS or HPLC/MS determination.

## EXPERIMENTAL

### Samples

Rafter dust samples from the DOFASCO foundry were collected from a variety of locations, including the pouring area where the overhead cranes are located, the shakeout area where the casting is removed from the mold, and the finishing area where sandblasting and grinding are carried out. Emissions from experimental pouring samples were also collected from castings into various custom-made small-sized molds. The two binder systems that produced the most mutagenic emissions were studied in this research.[26] These were the Oil/Clay/Cereal system and the Shell Core system. The molds were prepared from unused nonmutagenic silica, and between 18 and 43 kg of molten steel (1530 to 1540°C) was poured into the molds. The pouring was done in a freshly cleaned portion of the foundry. After the pouring was completed, a

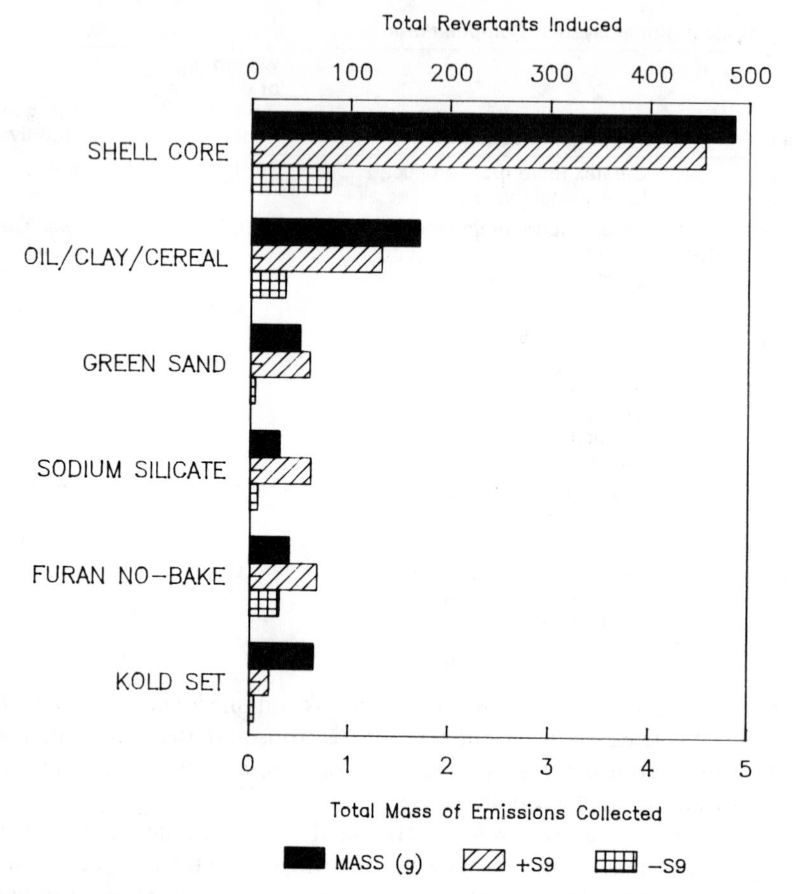


**Figure 2.** Experimental pouring results—total mass and mutagenicity (–S9 and +S9).

sampling hood with a HiVol motor and fan was placed 2.5 cm above the mold. The emissions were collected on tared type AE glass fiber filters. A number of filters were required for each sample due to the large amount of particulates material generated. A background HiVol sampler was also set up a short distance from the experimental pouring. Both emission samples and background samples were collected for approximately 3 hr.

## Extraction and Cleanup

All samples were Soxhlet extracted for 24 hr using either methylene chloride or methylene chloride followed by methanol for another 24 hr. Extra long all-glass extraction thimbles were required for the extraction due to the slow drainage of the thimbles when filled with dust. The samples were protected from light during the extraction.

The extracts were evaporated to approximately 5 mL and then filtered using

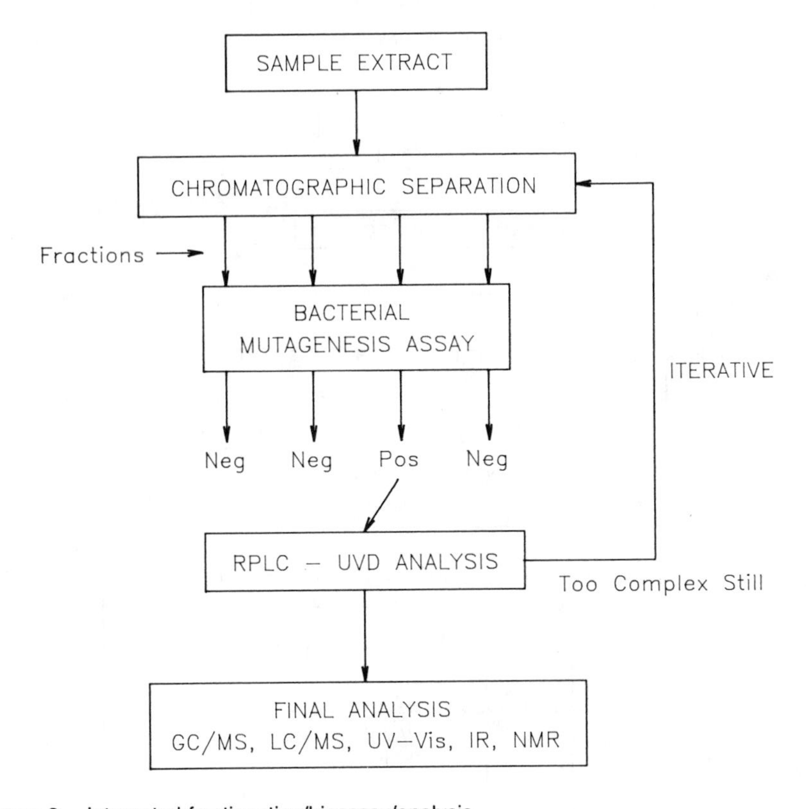


**Figure 3.** Integrated fractionation/bioassay/analysis.

a 0.5-$\mu$m Teflon®* filter to remove any particulate material that had carried over in the extraction. The filtered extracts were dried and weighed to determine percentage of extractable material. A portion of the extract was submitted for bioassay analysis.

The next step in the clean-up procedure, as shown in Figure 4, was an alumina column cleanup with sequential elutions with hexane to remove aliphatic material, benzene and chloroform to isolate the mutagenic polycyclic aromatic material, and methanol/water to remove the more polar compounds. Once again, a portion of each fraction was submitted for Ames assay analysis to determine where the mutagenic material was located.

The mutagenic aromatic material was further fractionated using preparative normal-phase HPLC using a Whatman Magnum-9 PAC column and gradient elution with methylene chloride/isopropanol. The resulting fractions are indicated in Figure 4. Mutagrams were developed for each chromatographic separation. A mutagram is a chromatogram of Ames assay results for the individual fractions collected (revertants versus fraction number) and can be matched to the high-performance liquid chromatography/ultraviolet (HPLC/UV) trace

*Registered trademark of E. I. du Pont de Nemours and Company, Inc., Wilmington, Delaware.

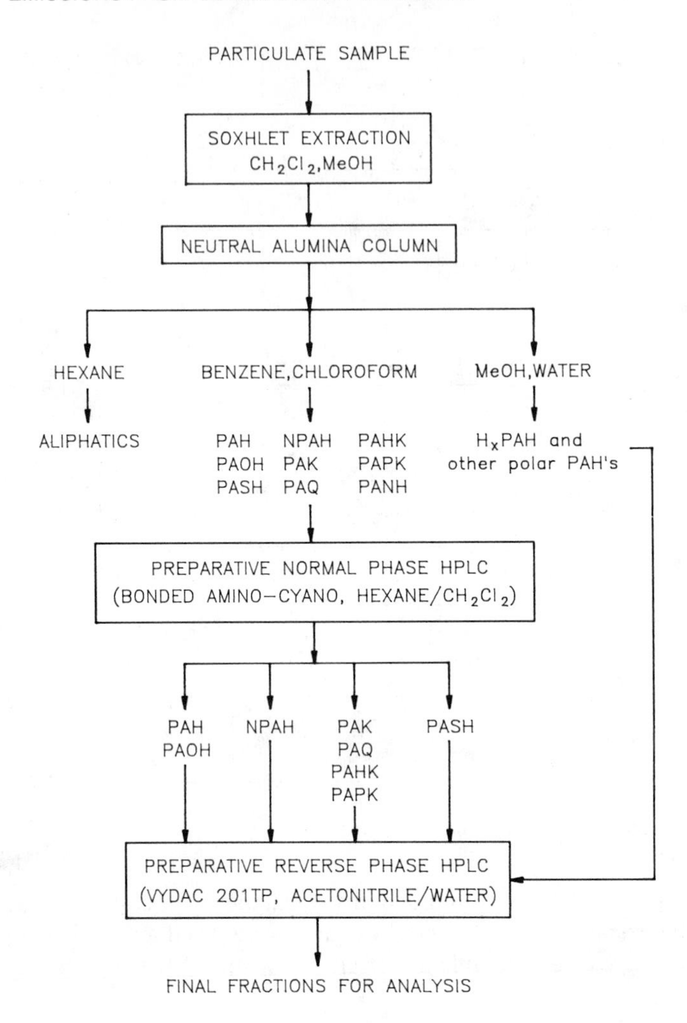


**Figure 4.** Sample fractionation scheme.

to determine where the fractions of interest are eluting. Further fractionation was done using reverse-phase HPLC with a Vydac C18 bonded phase column, and corresponding reverse-phase HPLC/UV chromatograms and mutagrams were generated.

## Analysis

GC analysis was one of the identification techniques used as indicated in Figure 3. A Varian 3700 GC with a cold on-column injector and a 30-m DB-5 column was used. The GC was equipped with a flame ionization detector (FID) and a nitrogen phosphorus detector (NPD). An Apple II+ computer was used for data acquisition.

Reverse-phase HPLC analysis was performed on either a Spectra Physics 8000 system with a UV and/or fluorescence detector or on a Hewlett-Packard 1090 HPLC with a diode array detector. Vydac RP-18 columns were used with a variety of acetonitrile-water gradients.

Low resolution GC/MS and HPLC/MS were done using a VG 7070F double-focusing mass spectrometer. The HPLC/MS interface was a moving belt type. High-resolution mass spectral data was acquired using a VG ZAB-E double-focusing mass spectrometer.

### Ames Assay

The Ames assay was used to test the mutagenicity of the samples at all stages in the fractionation scheme. The Ames assays were performed using the procedure described by Ames et al.[22] Salmonella typhimurium TA98 bacterium was used both with and without S-9 rat liver homogenate. A known portion of the samples obtained at each stage throughout the fractionation was submitted for Ames assay. The samples were dissolved in a small amount of dimethylsulphoxide and assayed at three or more levels to determine dose response. In the case of HPLC fractionation, all fractions were dissolved in equal volumes of dimethylsulphoxide and were deposited on the plates at a single concentration. The resulting data were used to generate mutagrams.

## RESULTS

Figure 5 shows the distribution of mass and mutagenic activity of the fractions obtained from the alumina column separation. Extract refers to the extract obtained from Soxhlet extraction. The alumina fractionation gave rise to fractions A1 through A6. A1 refers to the aliphatic material, A2 and A3 refer to the benzene and chloroform soluble aromatic material, and A4 to A6 are polar fractions. The bar entitled "total" refers to the total mass or total mutagenicity from the alumina column. This allows for a comparison with the mass and mutagenicity that was originally put on the column. About 50% of the mass was recovered and the remainder was assumed to be left on the alumina column. Recovery of the mutagenic material was more successful. Almost 100% recovery of mutagenic activity obtained with S9 activation was observed and 60 to 70% recovery of activity without S9 activation was seen. Due to the high mutagenic activity of fractions A2 and A3 and due to the fact that the two fractions contained a number of the same compounds, it was decided that these fractions should be combined and subjected to further analysis even though the remaining fractions contained a significant fraction of the total mass remaining.

The capillary GC/FID analysis of the aromatic fraction (A2 + A3) showed a complex pattern of compounds. The fraction was further cleaned up using semipreparative normal-phase HPLC and a corresponding mutagram was gen-

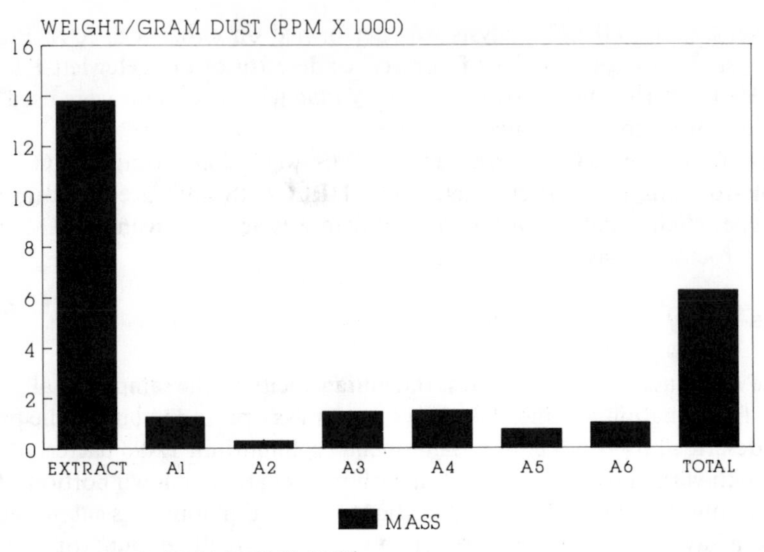


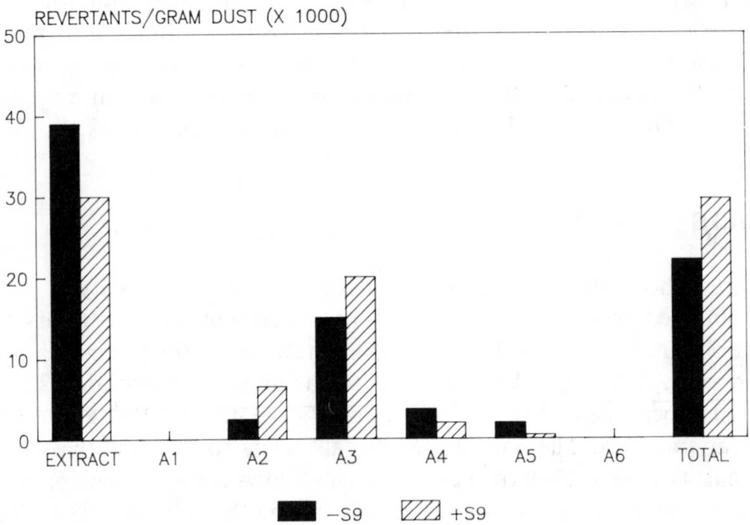


**Figure 5.** Neutral alumina separation.

erated. The aliphatic and PAH fractions that were generated contained relatively low levels of direct and indirect acting mutagenicity. The majority of the mutagenicity appeared in the nitro-PAH/oxy-PAH fraction and the dinitro-PAH/polyaromatic quinone fraction. Lower levels of mutagenicity were determined in fractions containing polyaromatic ketones and other quinones and polyaromatic nitrogen compounds.

The highly mutagenic fraction containing the nitro-PAHs and oxy-PAHs was subjected to an intense investigation. This fraction was separated on a reverse-phase HPLC column. A mutagram was generated as seen in Figure 6.

One sharp peak was observed in the mutagram in the absence of S9 activation. This peak was diminished in the presence of S9. This mutagram peak corresponded to one of the smaller peaks present in the HPLC/UV trace. The peak was identified using RPLC/MS and RPLC/UV data as shown in Figure 7. The upper trace is the diode array UV spectrum obtained for the shaded peak in Figure 6. The corresponding mass spectrum for this peak is also shown with the high-resolution mass spectral data indicated. The characteristic masses, the unique UV trace, and the retention time confirmed that the peak was 1NP. Nitro-PAHs have not previously been reported in steel foundry samples. Several other oxidized derivatives of PAHs were also detected in this fraction using LC/MS analysis.

The fraction containing dinitro-PAHs and polyaromatic quinones was subjected to a similar RPLC/UV and RPLC/MS analysis and mutagrams were generated. Once again, a large single peak in the mutagram was observed without S9 activation. This peak disappeared in the presence of S9 activation. The UV trace revealed a small, almost non-UV absorbing peak that had a retention time similar to 1,6-and 1,8-dinitropyrene. Mutant strains of Salmonella, other than the TA98 strain used in the majority of the research, were used to further confirm the presence of these compounds along with mass spectral data. The levels of dinitropyrene were estimated to be approximately 0.5 ppb in the original dust, compared to 3 to 5 ppb of 1NP in the dust.

Later-eluting fractions contained oxidized PAHs (quinones and ketones) and nitrogen heterocyclic compounds. A partial listing of some of the compounds determined in one of the dust samples is given in Table 3.

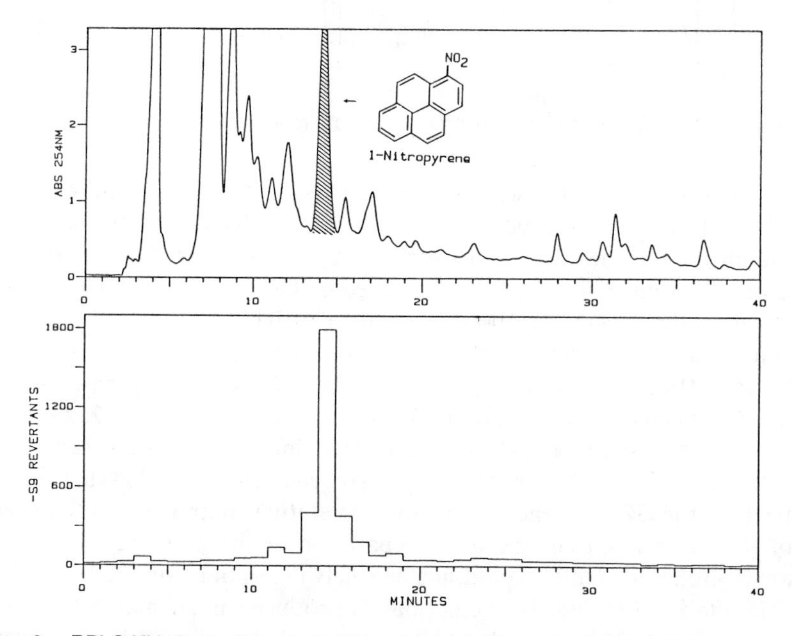


**Figure 6.** RPLC-UV chromatogram and mutagram of mutagenic NPLC fraction.

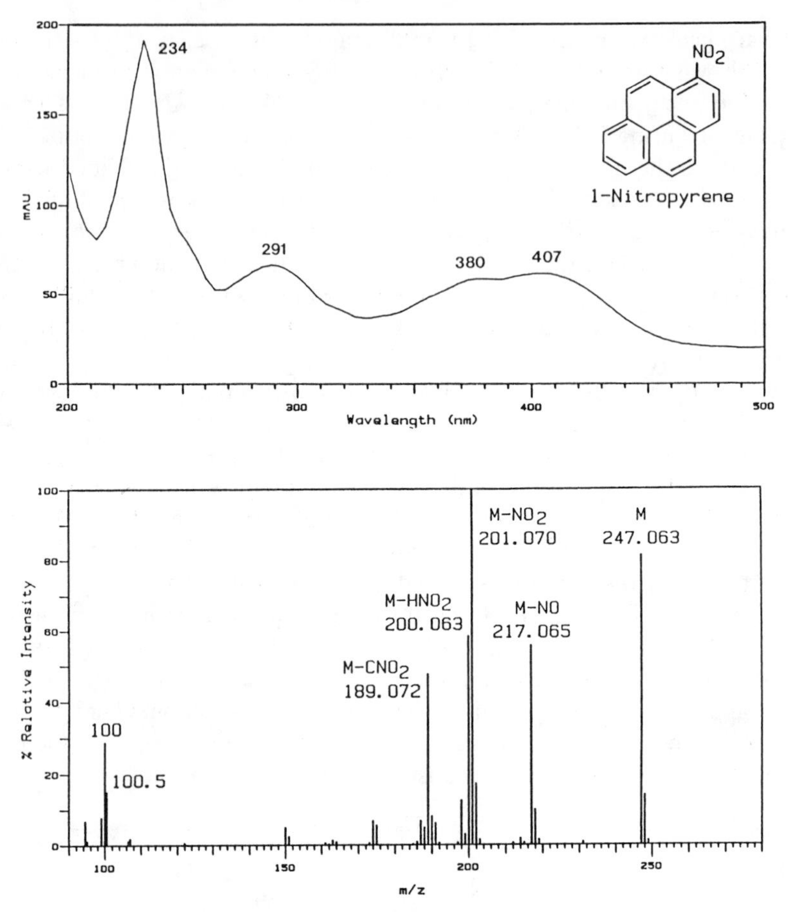


**Figure 7.** 1-Nitropyrene—high-resolution mass spectrum and UV spectrum.

An important factor that was noted in this research was the requirement for both GC/MS and LC/MS determination in the analysis of PAHs because of the number of isomers possible for a single molecular weight. In some instances, the isomer sets are fully separated by GC and not by HPLC and, in some cases, the reverse is observed. The selectivity of the HPLC column sometimes provides a better separation than a GC column that has a higher separation efficiency. An example of the use of HPLC to separate m/z 252 compounds in an environmental sample is described.[27] In the m/z 252 GC/MS mass chromatogram, five peaks were observed, but when the same sample was analyzed using HPLC/MS, seven peaks were seen. This indicated that some of the peaks in the GC/MS trace represented co-eluting compounds. The selectivity of the LC column enhanced the separation of the m/z 252 compounds. However, isomeric identity could not be solely determined on the basis of the LC/MS analysis because these compounds produce similar mass spectra. All 25, five-ring, m/z 252 compounds were not available for retention time deter-

**Table 3. Partial List of Dust Sample Components**

| Molecular weight | Compound Name |
|---|---|
| 128 | Naphthalene |
| 129 | Quinoline |
| 129 | Isoquinoline |
| 130 | 1,1′-Biphenyl |
| 142 | 2-Methylnaphthalene |
| 142 | 1-Methylnaphthalene |
| 144 | 3-Methyl-biphenyl |
| 154 | Biphenyl |
| 154 | Acenaphthene |
| 156 | C2-Alkyl-naphthalene |
| 158 | C2-Alkyl-biphenyl |
| 166 | Fluorene |
| 168 | Dibenzofuran |
| 170 | C3-Alkyl-naphthalene |
| 178 | Phenanthrene |
| 180 | 9-Fluorenone |
| 180 | Methyl-fluorene |
| 184 | C4-Alkyl-naphthalene |
| 192 | Methyl-178PAH |
| 192 | Methyl-178PAH |
| 194 | Methyl-9-fluorenone |
| 194 | Methyl-9-fluorenone |
| 202 | Fluoranthene |
| 202 | Pyrene |
| 206 | C2-Alkyl-178PAH |
| 208 | 9,10-Anthraquinone |
| 208 | C2-Alkyl-9-fluorenone |
| 220 | C3-Alkyl-178PAH |
| 222 | Methyl-9,10-anthraquinone |
| 228 | Chrysene/triphenylene |
| 230 | Benzofluorenone |
| 234 | Unknown ketone |
| 234 | C4-Alkyl-178PAH |
| 247 | 1-Nitropyrene |
| 252 | Benzofluoranthenes (3) |
| 252 | Benzo(e)pyrene |
| 252 | Benzo(a)pyrene |
| 258 | 228PAH quinone |
| 276 | Indeno(123-cd)pyrene |
| 276 | Benzo(ghi)perylene |
| 292 | 1,6-and/or 1,8-Dinitropyrene |
| 292 | 1,3-Dinitropyrene |

mination; therefore, diode array UV spectra were used to confirm the identity of each component with a high degree of certainty. The UV spectra were collected simultaneously with the mass spectral data by use of a HPLC column effluent splitter. The UV spectra that were collected were compared with library spectra and six of the seven peaks were identified, three of which had co-eluted in the GC/MS trace. There are other isomer sets where GC provides a better separation; therefore, GC/MS and LC/MS can provide complementary information for PAH identification.

## DISCUSSION

The results shown here indicate that the integrated chemical fractionation/bioassay/analysis approach was successful in tracking the mutagens in the foundry dust samples; however, a large amount of material was required for this developmental work. Foundry air particulates are now being analyzed using the fractionation scheme developed.

The complementary data provided by GC/MS, HPLC/MS, and HPLC/UV enabled us to successfully identify a number of compounds present in the foundry emissions, some of which had not been previously determined in these samples. In particular, nitro-PAHs and dinitro-PAHs were determined. These compounds gave rise to the majority of the mutagenicity, but represented only a small fraction of the total organic material. The presence of these compounds was puzzling since the mold conditions are reducing rather than oxidizing and diesel engines are not used in the foundry. The nitroarenes may have been produced in the pyrolysis process itself or by the nitration of PAHs by nitrogen oxides in the foundry air, but this has not been confirmed yet.

This research has shown that high levels of mutagenic activity from airborne particulates still exists in the foundry. The mutagenic activity is due to a number of potent mutagenic compounds detected. Therefore, it can not be assumed that current foundry levels are safe and that exposure has decreased over the years. DOFASCO has provided protection for its workers by enclosing the cranes and providing fresh air supplies to the crane cabs.

## REFERENCES

1. Hermann, M. "Synergistic Effects of Individual Polycyclic Aromatic Hydrocarbons on the Mutagenicity of Their Mixtures," *Mutation Res.* 90:399–409 (1981).
2. LaVoie, E. J., L. Tulley-Freiler, V. Bedenko, and D. Hoffmann. "Mutagenicity of Substituted Phenanthrenes in Salmonella Typhimurium," *Mutation Res.* 116:91–102 (1983).
3. LaVoie, E. J., L. Tulley-Freiler, V. Bedenko, and D. Hoffmann. "Mutagenicity, Tumor-Initiating Activity, and Metabolism of Methylphenanthrenes," *Cancer Res.* 41:3441–3447 (1981).
4. McFall, T., G. M. Booth, M. L. Lee, Y. Tominaga, R. Pratap, M. Tedjamulia, and R. N. Castle. "Mutagenic Activity of Methyl-Substituted Tri- and Tetracyclic Aromatic Sulfur Heterocycles," *Mutation Res.* 135:97–103 (1984).
5. Moller, M., I. Hagen, and T. Ramdahl. "Mutagenicity of Polycyclic Aromatic Compounds (PAC) Identified in Source Emissions and Ambient Air," *Mutation Res.* 157:149–156 (1985).
6. Verma, D. K., D. C. F. Muir, S. Suncliffe, J. A. Julian, J. H. Vogt, J. Rosenfeld, and A. Chovil. "Polycyclic Aromatic Hydrocarbons in Ontario Foundry Emissions," *Ann. Occup. Hyg.* 29:17–25 (1982).
7. Schimberg, R. W. "Industrial Hygenic Measurements of Polycyclic Aromatic Hydrocarbons in Foundries," in *Polynuclear Aromatic Hydrocarbons: Chemical*

*Analysis and Biological Fate*, M. Cooke and A. J. Dennis, Eds. (New York: Academic Press, 1981), p. 755.

8. Wiley, C., M. Iwao, R. N. Castle, and M. L. Lee. "Determination of Sulfur Heterocycles in Coal Liquids and Shale Oils," *Anal. Chem.* 53:400–407 (1981).
9. Hertz, H. S., J. M. Brown, S. N. Chesler, F. R. Guenther, L. R. Hilpert, W. E. May, R. M. Parris, and S. A. Wise. "Determination of Individual Organic Compounds in Shale Oil," *Anal. Chem.* 52:1650-1657 (1980).
10. Lee, F. S-C., and D. Schuetzle. "Sampling, Extaction, and Analysis of Polycyclic Aromatic Hydrocarbons from Internal Combustion Engines," in *Handbook of Polycyclic Aromatic Hydrocarbons*, A. Bjørseth, Ed. (New York: Marcel Dekker, Inc., 1983), p. 27.
11. Campbell, R. M., and M. L. Lee. "Capillary Column Gas Chromatographic Determination of Nitro Polycyclic Aromatic Compounds in Particulate Extracts," *Anal. Chem.* 56:1026–1030 (1984).
12. Choudhury, D. R. "Characterization of Polycylic Ketones and Quinones in Diesel Emission Particulates by Gas Chromatography," *Environ. Sci. Technol.* 16:102–106 (1982).
13. Ramos, L. S., and P. G. Prohaska. "Sephadex LH-20 Chromatography of Extracts of Marine Sediment and Biological Samples for the Isolation of Polynuclear Aromatic Hydrocarbons," *J. Chromatog.* 211:284–289 (1981).
14. May, W. E., and S. A. Wise. "Liquid Chromatographic Determination of Polycyclic Aromatic Hydrocarbons in Air Particulate Extracts," *Anal. Chem.* 56:225–232 (1984).
15. Konig, J., E. Balfanz, W. Funcke, and T. Romanowski. "Determination of Oxygenated Polycyclic Aromatic Hydrocarbons in Airborne Particulate Matter by Capillary Gas Chromatography and Gas Chromatography/Mass Spectrometry," *Anal. Chem.* 55:599–603 (1983).
16. Grimmer, G., J. Jacob, K-W. Naujack, and G. Dettbarn. "Determination of Polycyclic Aromatic Compounds Emitted from Brown-Coal-Fired Residential Stoves by Gas Chromatography/Mass Spectrometry," *Anal. Chem.* 55:892–900 (1983).
17. Quilliam, M. A., M. S. Lant, C. Kaiser-Farrell, D. R. McCalla, C. P. Sheldrake, A. A. Kerr, J. N. Lockington, and E. S. Gibson. "Identification of Polycyclic Aromatic Compounds in Mutagenic Emissions from Steel Casting," *Biomed. Mass Spectrom.* 12:143–150 (1985).
18. McCalla, D. R., M. A. Quilliam, C. Kaiser-Farrell, C. Tashiro, K. Hoo, E. S. Gibson, N. J. Lockington, A. A. Kerr, and C. Sheldrake. "Integrated Approach to the Detection and Identification of Steel Foundry Mutagens," in *Carcinogenic and Mutagenic Responses to Aromatic Amines and Nitroarenes*, C. M. King, L. J. Romano, and D. Schuetzle, Eds. (New York: Elsevier Science Publishing Co., Inc., 1988), p. 47.
19. Gibson, E. S., D. R. McCalla, C. Kaiser-Farrell, A. A. Kerr, J. N. Lockington, C. Hertzman, and J. M. Rosenfeld. "Lung Cancer in a Steel Foundry: A Search for Causation," *J. Occup. Med.* 25:573-578 (1983).
20. Palmer, W. G., and W. D. Scott. "Lung Cancer in Ferrous Foundry Workers: A Review," *Am. Ind. Hyg. Assoc. J.* 42:329–340 (1981).
21. Gibson, E. S., R. H. Martin, and J. N. Lockington. "Lung Cancer Mortality in a Steel Foundry," *J. Occup. Med.* 19:807–812 (1977).
22. Ames, B. N., J. McCann, and E. Yamasaki. "Methods for Detecting Carcinogens

and Mutagens with the Salmonella/Mammalian-Microsome Mutagenicity Test," *Mutation Res.* 31:347–364 (1975).

23. Santella, R., T. Kinoshita, and A. M. Jeffrey. "Mutagenicity of Some Methylated Benzo[a]pyrene Derivatives," *Mutation Res.* 104:209–213 (1982).
24. Bjørseth, A., and G. Eklund. "Analysis for Polynuclear Aromatic Hydrocarbons in Working Atmospheres by Computerized Gas Chromatography-Mass Spectrometry," *Anal. Chim. Acta* 150:119–128 (1979).
25. Daisey, J. M., and M. A. Leyko. "Thin-Layer Gas Chromatographic Method for the Determination of Polycyclic Aromatic and Aliphatic Hydrocarbons in Airborne Particulate Matter," *Anal. Chem.* 51:24–30 (1979).
26. McCalla, D. R., M. A. Quilliam, C. Kaiser-Farrell, M. Lant, E. S. Gibson, J. N. Lockington, A. A. Kerr, and C. Sheldrake. "Formation of Bacterial Mutagens from Various Mould Binder Systems Used in Steel Foundries," in *Polynuclear Aromatic Hydrocarbons: Chemistry and Biology*, M. Cooke and A. J. Dennis, Eds. (Columbus, OH: Batelle Press, 1984).
27. Quilliam, M. A., P. G. Sim, C. Tashiro, J. C. Marr, and K. H. Hoo. "Determination of Polycyclic Aromatic Compounds by Liquid Chromatography, Diode Array Detection and Mass Spectrometry," presented at the 11th International PAH Symposium, Gaithersburg, MD, 1987.

CHAPTER 17

# The National Incinerator Testing and Evaluation Program (NITEP) Mass-Burning Technology Assessment

A. Finkelstein, R. Klicius, and D. Hay

## INTRODUCTION

This chapter discusses the important design modifications and subsequent extensive combustion test program completed on a state-of-the-art mass-burning incinerator located in Quebec City, burning municipal solid waste.

Today's modern industrialized society is a generator of substantial quantities of municipal solid waste (MSW). Conservative estimates put per capita generation at about 1.8 kg per day or approximately 16 million t per year for all of Canada, most of which ends up in landfills. The increasing cost and complexity of landfilling MSW, combined with the difficulties of locating new sites, have forced municipalities to seriously consider alternatives such as incineration in energy-from-waste (EFW) facilities. Incineration reduces the volume of waste to be landfilled by up to 90%, extending the life of existing sites and reducing the need for new ones. These facilities also offer the option of generating revenue through energy recovery.

One question that is raised repeatedly with both existing and proposed EFW facilities is how safe are the emissions produced? Many synthetic and toxic chemicals, ingredients which support contemporary lifestyles, end up in the waste stream as used and unwanted consumer waste. Uncontrolled burning of these products and their packaging can release emissions of trace organics, heavy metals, and acid gases.

Addressing these issues and many others is the objective of the multifaceted National Incinerator Testing and Evaluation Program (NITEP). The five-year program is mandated to identify EFW technologies in Canada; to assess relationships among state-of-the-art designs, operations, energy benefits, and emissions; to examine effectiveness of emission control systems; and to develop national guidelines for emissions. To date, reports have been published on the extensive test programs successfully completed on a modular two-stage design in Prince Edward Island (PEI) and on two different types of air pollution control systems in Quebec City. Test results of mass-burning technology is the subject of this technical paper.

## PLANT BACKGROUND

The Quebec Urban Community Municipal Solid Waste Incinerator Plant is a mass-burning design developed in the early 1970s to burn as-received refuse (i.e., without any preparation) in a water-wall furnace. The plant produces steam using flue gas heat recovery boilers. The incinerator plant is owned by the Quebec Urban Community (QUC) and is located in an industrial area of Quebec City, adjacent to residential and commercial zones. It receives municipal and commercial solid waste collected by the QUC, as well as from several other municipalities and private contractors. All of the steam generated by the plant's four incinerator units is sold to a local paper company.

The incinerator plant employs technology developed in Europe and represented a contemporary design when built in 1974. Throughout the years, a number of design changes were made to improve the operational problems in the plant, such as furnace slagging and emissions of large unburned material. However, some of the design changes compounded existing emissions problems.

## PLANT OPERATION

The principle elements of the QUC municipal waste incineration plant are shown in Figures 1 and 2, and include a refuse storage pit and crane system; four incinerators/boilers, each rated at 227 t per day; electrostatic precipitators; ash quenching systems with storage pit and crane; and a common stack. Each incinerator consists of a vibrating feeder-hopper, feed chute, drying/burning/burn-out grates, refractory-lined lower burning zone, water-walled partially lined upper burning zone chamber, a vertical tube mechanically wrapped waste heat recovery boiler with superheater and economizer, a two-stage electrostatic precipitator, an induced draft fan, and a wet ash quench/removal system.

## DESIGN MODERNIZATION

In May 1985, a comprehensive study was completed under NITEP on the modernization of the Quebec incinerators. The study detailed the many changes that would be required to transform the incinerators into a "state-of-the-art" mass-burning design and to reduce the existing emission problems. Based on the study findings, the QUC decided to experiment by upgrading one of the incinerator units to assess the impact of the proposed changes before modifications were made to all the units.

To ensure that the most appropriate furnace configuration was developed, a

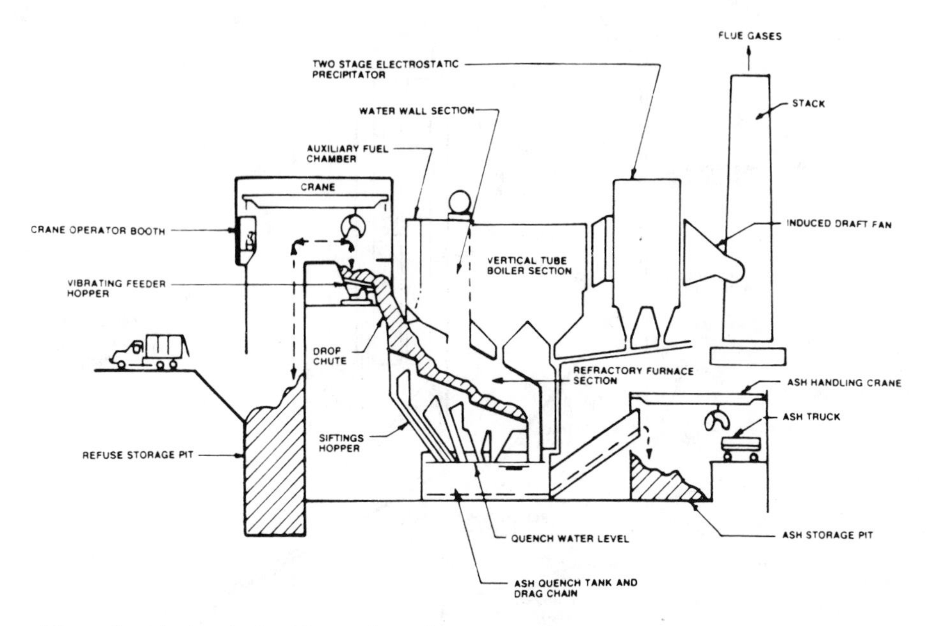


**Figure 1.** Quebec incinerator—schematic cross-section.

one-sixth scale, three-dimensional flow model of the furnace was fabricated of wood and plexiglass. Through the plexiglass panels, trained observers were able to visually assess the mixing pattern of the various furnace configurations being investigated. Sawdust and air were used to simulate a number of combustion conditions. In addition, video tapes were made of the model studies to allow for a more detailed assessment at a later date. In all, 52 different conditions were run which looked at (1) the original furnace design (see Figure 2), (2) a partially modified design, (3) and a completely new furnace design.

Based on a thorough assessment of the various conditions investigated, a final furnace design configuration, as shown in Figure 3, was selected which provided the highest degree of turbulence and mixing, both of which are necessary for good combustion. The following important design modifications were provided: (1) an improved primary air distribution system was installed; (2) the nozzle design and location of the secondary air system were selected, (3) the configuration and positioning of the front and rear furnace "bull noses" were determined to optimize mixing, (4) combustion gas residence time was increased 30% by enlarging the upper furnace chamber; and (5) automatic computer controls were installed.

The primary air system for the modified unit was redesigned to provide independently controlled airflow to nine zones located beneath the grates, each fitted with individual motorized and flow-controlled dampers. The first primary air zone is located at the downstream end of the drying grate and provides combustion air required for early ignition of the refuse. Air to the burning grate is now distributed into six hopper zones of approximately the

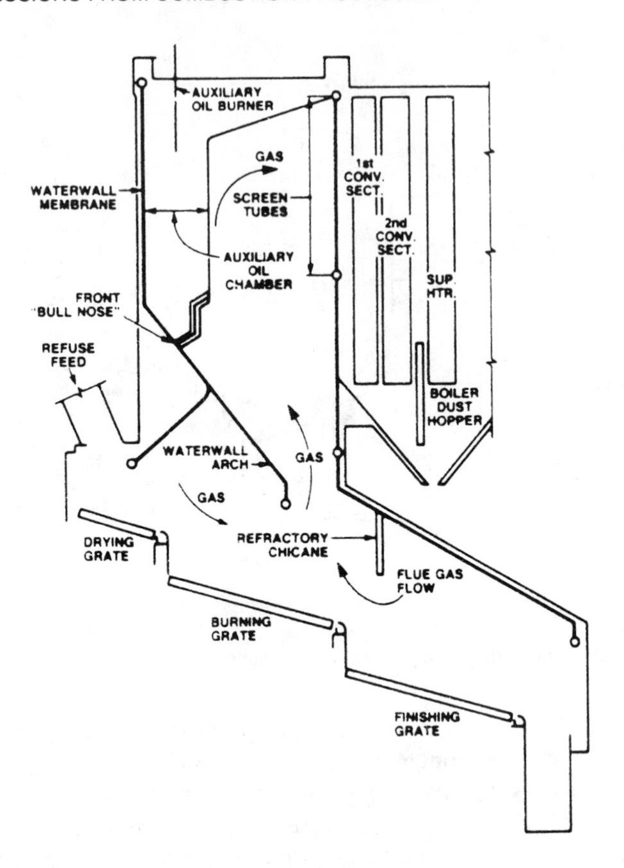


**Figure 2.** Quebec incinerator—original design.

same size. Each of the hopper zones is equipped to automatically control airflows and pressures through separate ducts from the primary air header. This feature prevents excessive amounts of combustion air from bypassing large refuse piles on the grate and carryover of fine material from sections of the grate where the piles are smaller. The finishing grate air is supplied through two hopper zones with individual dampers and ensures that all the refuse is burned prior to discharging into the ash quench tank.

Secondary air for the modified design is introduced into the furnace radiation chamber by a series of nozzles at the front "bull nose" and another series at the rear "bull nose." The nozzles have individual manual dampers for control purposes and to ensure that all provide equal flows. To achieve the required air distribution between the front and rear nozzles, a control damper is employed to adjust the distribution. Good control of airflows through these nozzles is important to ensure suitable flue gas temperatures and the desired excess air level while achieving sufficient mixing for completing combustion.

The concept of the "bull nose" configuration shown in Figure 3 is essential to ensure that upper flue gases velocities are maintained relatively low, that

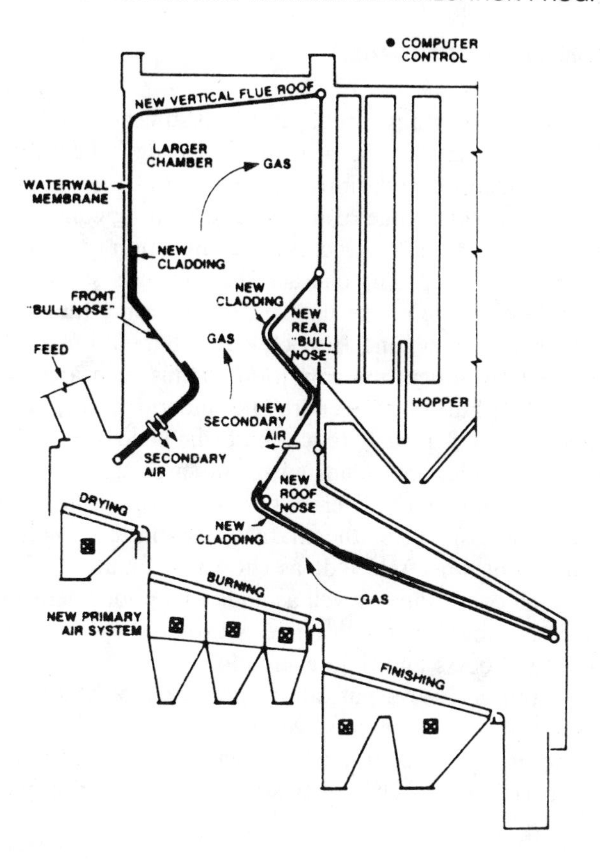


**Figure 3.** Quebec incinerator—modified design.

good mixing of combustion gases occurs with overfire air in the lower portion of the chamber, and that improved gas distribution into the boiler is provided. In addition, the shape of the "bull noses" addressed the directional flow of the flue gas in the furnace and provided an excellent area where overfire (i.e., secondary) air could be strategically located. The final "bull nose" configurations and overfire air jet locations were primarily based on the airflow modeling previously discussed.

In order to bring the Quebec City test unit up to the "state of the art" in automatic controls, the existing control panel was replaced by a computer control system which displays and controls all process elements. The new control system included (1) a grate-speed control system, (2) steam flow control, (3) automatic airflow control of the primary air distribution system, and (4) secondary air supply control system. All the upgrading was completed by March 1986 and commissioning trials began in early April 1986.

## COMBUSTION TEST PROGRAM

An extensive combustion test program was established to evaluate the performance and effectiveness of the new furnace design. The objective of this program was to develop correlations between various operating parameters and emissions over a wide range of different operating conditions.

Two levels of testing were involved in the combustion test program: characterization and performance. The characterization tests served as a basis for developing a understanding of the incinerator operating range, debugging of all systems, facility logistics, and field crew familiarization.

To thoroughly assess operating conditions, items such as refuse feed rate, excess air levels, distribution of overfire and undergrate air, and temperature profiles were recorded for each test. These included both poor and good operating practices under low-, normal-, and high-steam loads. In all, 22 characterization tests under 18 different operating conditions were conducted. Upon a thorough assessment of the characterization test results, five performance test conditions were selected, as shown in Figure 4. These represent three different load conditions as well as good and poor operations (i.e., high CO levels) at design load. These latter tests involved extensive sampling and analysis as well as process and data evaluation.

To compare input versus output, all incoming refuse and all the ashes collected were weighed and representative samples analyzed. Each sample was analyzed for two groups of compounds: organics, which consisted of dioxins (PCDDs), furans (PCDFs), chlorobenzene (CB), polychlorinated biphenyls

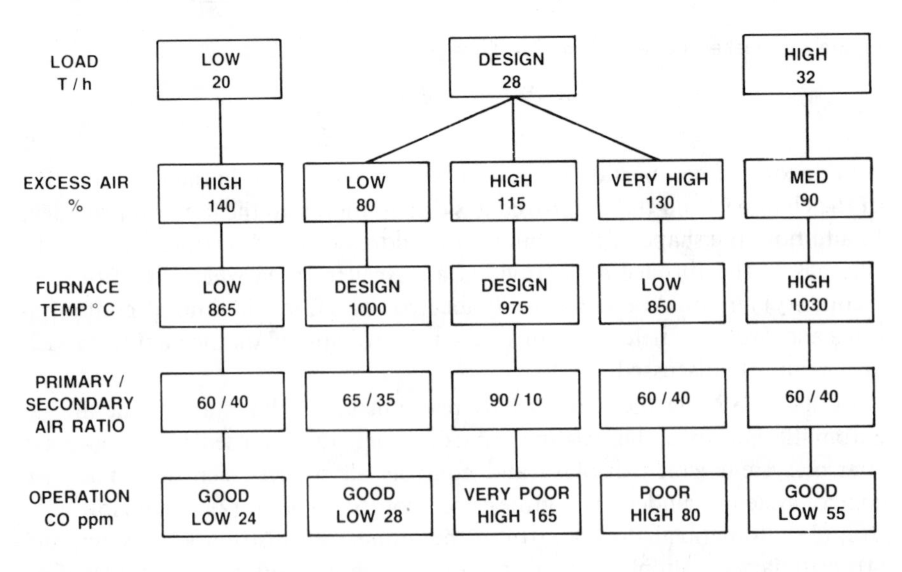


**Figure 4.** Performance test conditions.

(PCBs), polycyclic aromatic hydrocarbons (PAHs), and chlorophenols (CPs); and metals, consisting of cadmium (CD), lead (Pb), mercury (Hg), and 27 others. In addition, the combustion gases leaving the stack were sampled for the compounds listed, as well as for acid gases, such as hydrogen chloride (HCl), sulphur dioxide ($SO_2$), and nitrogen oxides ($NO_x$); and conventional products of combustion, such as carbon monoxide (CO), carbon dioxide ($CO_2$), oxygen ($O_2$), and total hydrocarbons (THC). All sampling and analyses were completed using recognized North American protocols.

Three separate microcomputer-based systems, linked into a network, were chosen to log, store, and analyze all the data collected during each test day. The first system was used to collect continuous gas data from the combustion exhaust gases. The second system collected and analyzed incinerator process data, while the third system collected readings from two temperature thermocouple grids installed in the radiation chamber zones of the furnace. Each of the computer systems provided a "real-time" graphic and statistical analysis of selected logged data. This provided on-the-spot monitoring of the progress of the tests and enabled project managers to make decisions quickly. In addition, plots of concentrations, temperatures, and process data versus time were displayed on the computer screen on an hourly interval to observe trends.

Upon completion of each test day, the data generated were reviewed by a second crew overnight to verify and correct all data and to produce detailed tables as well as summary tables and graphics. This allowed the field project managers the opportunity to review the previous day's results to assess the adequacy of each test.

An extensive amount of test data is available on the old furnace design. The QUC incinerator has been the subject of acceptance testing, annual testing for particulates, particle size determination, testing for HCl and metals emissions, as well as dioxin and furan emissions, all as part of a Provincial assessment program. All this information has provided an excellent reference for the comparison and assessment of the improvements made to the furnace design.

## PCDD/PCDF EMISSIONS

In 1984, Environment Quebec undertook an assessment of the dioxin and furan emissions from the QUC incinerator plant. Three tests were completed which indicated relatively high concentrations of dioxins (i.e., 800 to 4000 ng/$Sm^3$) and furans, which is not uncommon for facilities of this vintage. However, after implementing the modifications as previously discussed, a significant reduction in dioxin emissions was observed (as shown in Figure 5). By employing good operating practices as defined in modern incinerator facilities, dioxin concentrations were reduced between 40 and 100 times below the 1984 test results (i.e., from 800 to 4000 to 20 to 50 ng/$Sm^3$). Furthermore, under poor operating practices, the reduction achieved was an order of magnitude (i.e., below 300 ng/$Sm^3$). This suggests that the new design had a significant

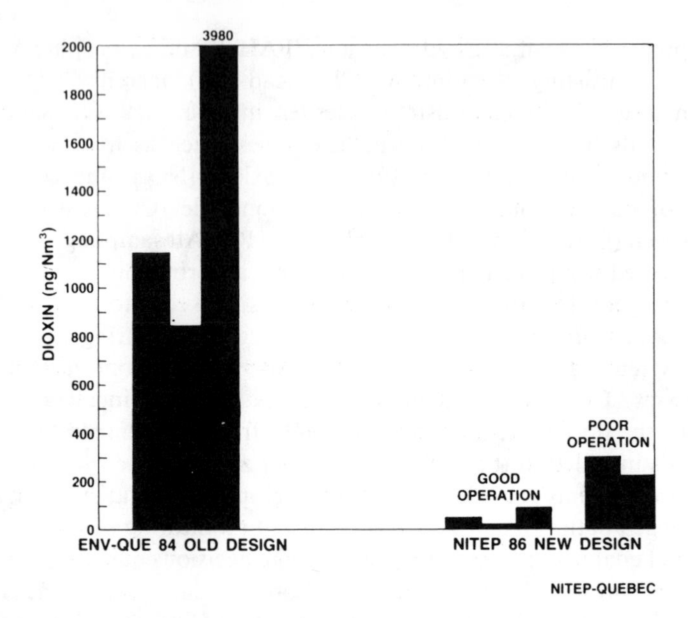


**Figure 5.** Dioxin emission comparison.

influence on dioxin emissions; however, good operating practice was also essential to minimize dioxin formation and emission.

## OTHER TRACE ORGANIC EMISSIONS

Table 1 presents a summary of the emissions of trace organic compounds for each of the five performance test conditions. The results indicate that trace organic emissions of each major compound group were lowest under good operation at the design rate. Conversely, trace organic emissions were highest at one or both of the poor operating conditions.

**Table 1. Trace Organic Emissions per Performance Test Mode**

| | Good conditions burning rate | | | Poor conditions design burn rate | |
|---|---|---|---|---|---|
| | Low | Design | High | Low temperature | Poor distribution |
| PCDD (ng/Sm$^3$) | 52.6 | 18.8 | 55.4 | 298.5 | 218.9 |
| PCDF (ng/Sm$^3$) | 114.5 | 44.5 | 100.7 | 298.3 | 306.4 |
| CP ($\mu$g/Sm$^3$) | 9.5 | 5.1 | 8.0 | 22.5 | 23.7 |
| PAH ($\mu$g/Sm$^3$) | 7.1 | 4.0 | 5.4 | 21.9 | 3.2 |
| CB ($\mu$g/Sm$^3$) | 3.5 | 3.3 | 4.4 | 9.9 | 9.5 |
| PCB ($\mu$g/Sm$^3$) | 4.3 | 3.0 | 4.9 | 7.0 | 1.7 |

*Note:* Values corrected to 12% $CO_2$.

**Table 2. Particulate/Metal Emissions per Performance Test Mode**

| | | Good conditions burning rate | | | Poor conditions design burn rate | |
|---|---|---|---|---|---|---|
| | | Low | Design | High | Low temperature | Poor distribution |
| Particulate | (mg/Sm$^3$) | 26 | 22 | 36 | 55 | 62 |
| Particle size | (% 2.5 μm) | 33 | 29 | 24 | 26 | 24 |
| Cadmium | (μg/Sm$^3$) | 26 | 24 | 41 | 90 | 76 |
| Lead | (μg/Sm$^3$) | 978 | 673 | 1599 | 2039 | 2495 |
| Chromium | (μg/Sm$^3$) | 11 | 7 | 15 | 21 | 14 |
| Nickel | (μg/Sm$^3$) | 9 | 5 | 8 | 8 | 7 |
| Mercury | (μg/Sm$^3$) | 783 | 704 | 872 | 810 | 622 |
| Antimony | (μg/Sm$^3$) | 35 | 36 | 44 | 112 | 87 |
| Arsenic | (μg/Sm$^3$) | 2 | 3 | 5 | 7 | 6 |
| Copper | (μg/Sm$^3$) | 39 | 33 | 54 | 91 | 89 |
| Zinc | (μg/Sm$^3$) | 1619 | 1130 | 2061 | 5122 | 3429 |

*Note*: Values corrected to 12% $CO_2$.

## PARTICULATE/METAL EMISSIONS

Emission concentrations of particulates and nine priority metals are summarized in Table 2. Particulate emissions were unexpectedly low (below 62 mg/Sm$^3$), considering that the incinerator was equipped with a two-field electrostatic precipitator. Under good operating conditions, particulates were 22 to 36 mg/Sm$^3$. These favorable results are attributed to the low particulate load to the precipitator that resulted from modernization of the furnace design.

Metal emissions (except mercury and nickel) were generally highest under the poor operating conditions and lowest for good operation at the design rate, which corresponds to particulate emissions. The new furnace design resulted in a large reduction in particulate emissions compared to the old furnace design. In Figure 6, the particulate concentrations at the stack are compared between the old and new design at 28 and 32 t/hr steam flow. On the average, reductions by one to two orders of magnitude were achieved with the new design concept at both of these steam rates. The degree of reduction in both dioxins and particulates was approximately the same, which suggests a possible interrelationship. Based on current theory that dioxins are mostly bound to particulates in the combustion gases, it is postulated that good combustion conditions that reduce the carryover of particulates from the furnace will also minimize the formation and emission of dioxins into the environment.

## TRACE ORGANICS IN ASHES

A comparison of the concentrations of trace organics in each type of ash is provided in Table 3 for the design rate and good operation. Trace organic concentrations (except PAH) were highest in the precipitator ash, followed by boiler, and lowest in the incinerator (bottom) ash. Similar trends have been

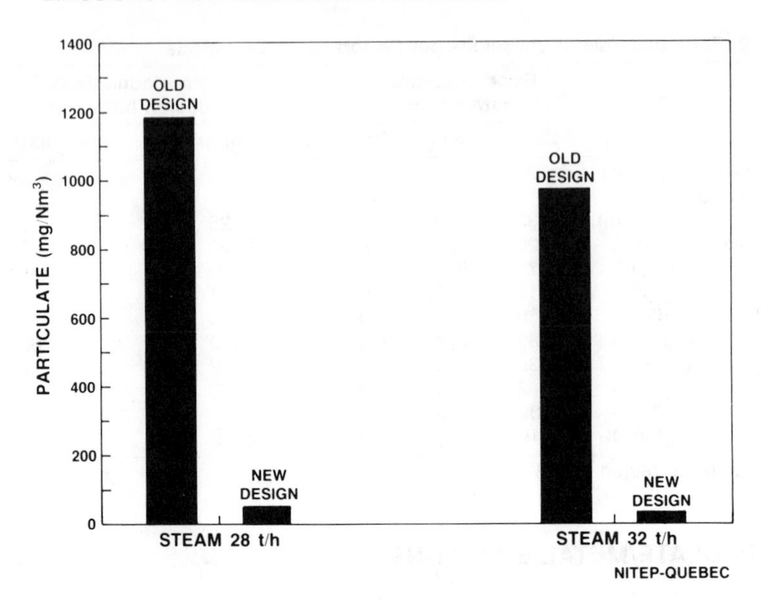


**Figure 6.** Particulate emission comparison.

observed in other NITEP studies (i.e., highest concentrations in the finer ash).

No significant differences in trace organic concentrations in the ashes were observed at different operating modes. However, it is relevant to note that the mass rates of boiler and precipitator ash were significantly less under good operation versus poor operation. Accordingly, the output rate of organics in ashes were highly dependent on operating mode.

## METALS IN ASHES

Concentrations of metals in the ashes were highest in the precipitator ash for the more volatile metals, such as mercury, lead, antimony, arsenic, cadmium, and zinc. Nickel concentrations were similar in all three ashes. Copper concentrations were highest in incinerator ash. As for trace organics, no significant differences in metal concentrations in ashes were observed at different operat-

**Table 3. Trace Organic Concentrations (ng/g) in Ashes for Good Operation at Design Rate**

| | Bottom ash | Boiler ash | Precipitator ash |
|---|---|---|---|
| PCDD | 0.2 | 37 | 584 |
| PCDF | 1.0 | 31 | 186 |
| CB | 45 | 356 | 892 |
| PCB | ND | ND | ND |
| CP | 16 | 80 | 1820 |
| PAH | 538 | 25 | 111 |

**Table 4. Metals Input/Output (g/ton of refuse)**

| Metal | Input | Output | Out/in Ratio |
|---|---|---|---|
| Cr | 102 | 70 | 0.7 |
| Pb | 314 | 580 | 1.9 |
| Ni | 23 | 44 | 1.9 |
| Sb | 3 | 10 | 3.4 |
| As | 1 | 3 | 3.2 |
| Cd | 5 | 8 | 1.8 |
| Zn | 218 | 981 | 4.5 |

ing modes; however, the higher ash rates under poor operating conditions would result in higher metal output rates as well.

## INPUT/OUTPUT OF METALS AND ORGANICS

In Table 4, the input to the incineration system of several priority metals in the refuse feed is compared to the output of these metals in all output streams (i.e., all ashes and stack emissions). In general, the calculated metal output was higher than the calculated metal input by approximately a factor of two. In view of the difficulty in obtaining representative samples of bottom ash and refuse, this mass balance for metals can be considered to be acceptable.

An input/output comparison for PCDDs and PCDFs is illustrated in Figure 7 and indicates a substantial decrease by a factor of 0.3 for dioxins (expressed per ton of refuse). For furans, the output/input ratio was 2, showing a small net increase. Similar results were obtained for the NITEP PEI tests. A significant net decrease was also found for PCBs and CPs, and a very slight increase for CB.

## STATISTICAL ANALYSES

A key objective of the tests was to identify relevant relationships between the emissions of selected trace organics and inorganics versus process operating variables and versus continuous emission data. Simple and multiple linear regression analyses were completed to identify these relationships by calculating correlation coefficients between selected dependent and independent variables. These relationships were used to generate "prediction models," which predict trace organic emissions based on readily measurable variables (surrogates), and "control models," which determine how to control the incinerator to minimize emissions.

Based on simple linear correlations, the following relationships are important to note:

1. Dioxin and furan exhaust emissions showed strong correlations with the following parameters:
   a. CO emissions

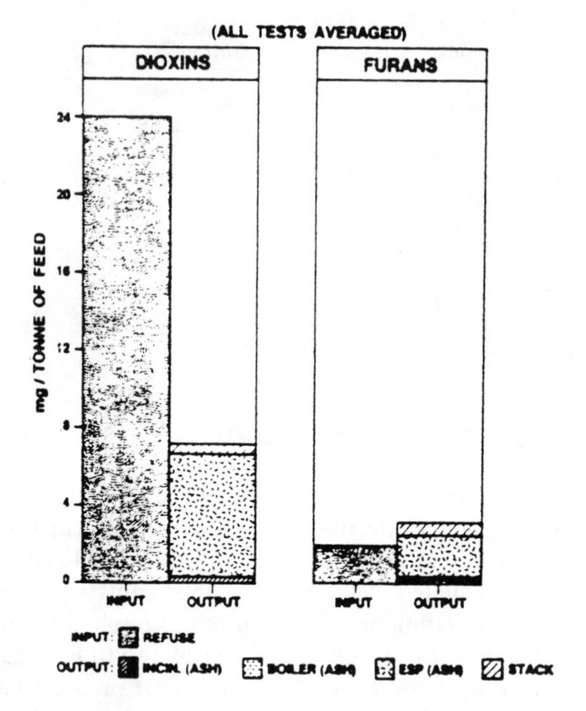


**Figure 7.** Organic input/output.

   b. particulate emissions
   c. exhaust gas flow and primary airflow
   d. CB and CP emissions
   e. copper emissions
2. No correlation was found between dioxin or furan emissions and:
   a. dioxin and furan concentrations in the refuse
   b. PAH and PCB emissions.
3. Particulate emissions correlated with the following:
   a. CO emissions
   b. exhaust gas flow and primary airflow
   c. some of the trace organics emissions, namely dioxin, furan, and CB
   d. most of the priority metal emissions, such as cadmium, copper, and lead
4. Emissions of selected metals correlated well with CB, CP, dioxin, and furan emissions. It is believed that the reason for the correlations between the priority metals and trace organics is that these contaminants are generally associated with the particulate material and will considerably vary up or down depending on particulate loads. No correlations were found between trace organics and mercury emissions.
5. Lower radiation chamber temperature correlated well with combustion parameters such as $O_2$ and $CO_2$ concentrations. $O_2$ concentrations correlated inversely with process temperatures, while $CO_2$ correlated directly with process temperatures.

Although simple regression analyses provided a good first-order screening, as indicated by the above results, multiple linear regression analyses resulted in much stronger prediction models and control models:

1. Multiple linear regression analysis produced prediction and control models with coefficients of determination ($r^2$), which are indicators of the model's strength, ranging from 0.74 to 0.90 for the best fit models.
2. The prediction models which best characterized trace organics emissions of dioxins (Figure 8), furans, CB, and CPs used two or three of the following monitoring parameters:
   a. CO
   b. $NO_x$
   c. $O_2$
   d. $H_2O$ in flue gas
   e. lower radiation chamber temperature
3. The best control models to minimize trace organic emissions (Figure 9 for PCDDs) used three of the following operational settings:
   a. total airflow
   b. primary/secondary air ratio
   c. steam rate or refuse rate
   d. secondary air front/rear ratio
4. Carbon monoxide was determined to be the best single surrogate in the one variable (simple) prediction model for most of the trace organics, with the exception of PAH and PCB.
5. Although $NO_x$ was the best second variable to improve the prediction capabilities of the models, its significance is not clearly understood at this time.
6. Primary airflow was the most influential operational setting for the one-variable control models for dioxins, furans, CB, and CPs.

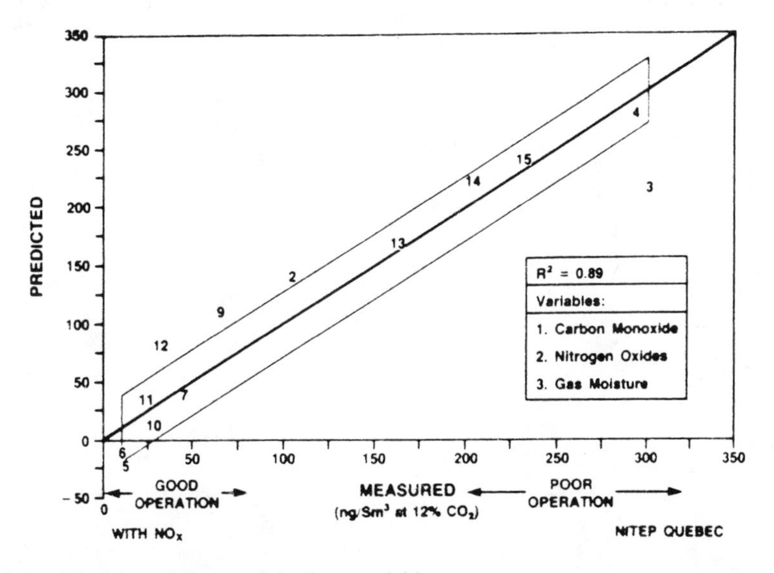


**Figure 8.** Dioxin prediction model—three variables.

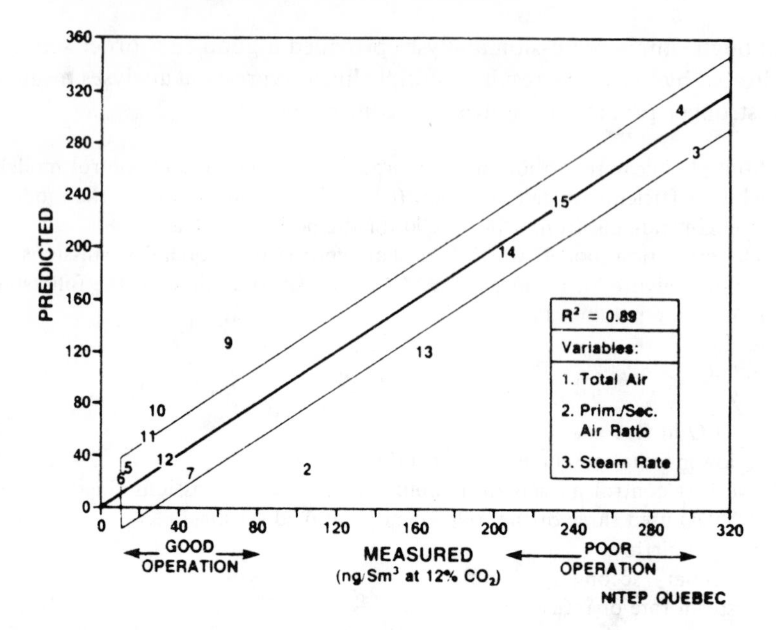


**Figure 9.** Dioxin control model—three variables.

7. The PCB and PAH models examined contained either a relatively low $r^2$ value or the data scatter implied in a poor predictive model. Thus, no useful models were found for these two groups of trace organics.
8. The models developed for dioxins, furans, CB, and CPs, are consistent with the fact that poor incinerator operating conditions resulted in higher emission concentrations.

## CHAPTER 18

# The National Incinerator Testing and Evaluation Program (NITEP) Air Pollution Control Technology Assessment Results

**R. Klicius, A. Finkelstein, and D. Hay**

## SUMMARY

Environment Canada, in co-operation with Flakt Canada Ltd. conducted a series of comprehensive tests on two pilot-scale air pollution control systems. The tests were designed to determine the capability of these systems to control emissions of dioxin, other trace organics, metals, acid gases, etc. Field testing was completed in August/September 1985 for two different designs: one using a wet-dry scrubber (spray-dryer) plus fabric filter and the other using a dry scrubber plus fabric filter. The systems were constructed by Flakt Canada and were erected adjacent to the mass-burning incinerator in Quebec City.

Many different operating conditions were tested to identify optimum removal efficiency for all pollutants of concern. Final results were very significant. Flue gas temperature was found to be an important operating variable for achieving high removal efficiency for many of the pollutants measured. Appropriate operating conditions were identified to obtain extremely high removal efficiencies for dioxins and furans (>99%); other trace organics (>98%); heavy metals, including arsenic and lead (>99%); and up to 97% for mercury. Excellent removal (>95%) was also obtained for acid gases, i.e., HCl and $SO_2$.

## INTRODUCTION

The literature has identified substantial variations in the quality of air emissions from municipal energy-from-waste facilities. It is evident that modern well-operated units tend to have lower emissions than older and less effectively operated facilities. Modern air emissions control technologies play an important role in reducing emissions; however, there is very little data, particularly with respect to dioxins and other organic compounds, which quantify control capabilities.

Under the National Incinerator Testing & Evaluation Program (NITEP), Environment Canada, in co-operation with Flakt Canada, established an extensive test program to conclusively evaluate the capability of two control systems to remove particulate, acid gases, heavy metals, dioxins, furans, and other organic compounds. In addition, the optimum operating conditions to minimize these contaminants were also of great interest.

In order to undertake this work, Flakt Canada constructed a large-scale pilot-plant facility at the energy-from-waste plant owned by the Quebec Urban Community (CUQ) and operated by Montenay Inc. in Quebec City.

## INCINERATOR DESCRIPTION

The CUQ incinerator is a mass-burning design, developed in the early 1970s to burn as-received refuse in a water-wall furnace. There are four incinerators, each rated at 227 t/day with a common refuse storage pit and stack. As illustrated in Figure 1, each incinerator consists of a vibrating feeder-hopper, feed chute, drying/burning/burn-out grates (Von-Roll design), refractory-lined burning zone, water-walled, partially lined upper burning zone, a waste heat recovery boiler with superheater and economizer (Dominion Bridge), a two-field electrostatic precipitator, an induced draft fan, and a wet ash quench/removal system.

The incinerator receives municipal, commercial, and suitable industrial solid

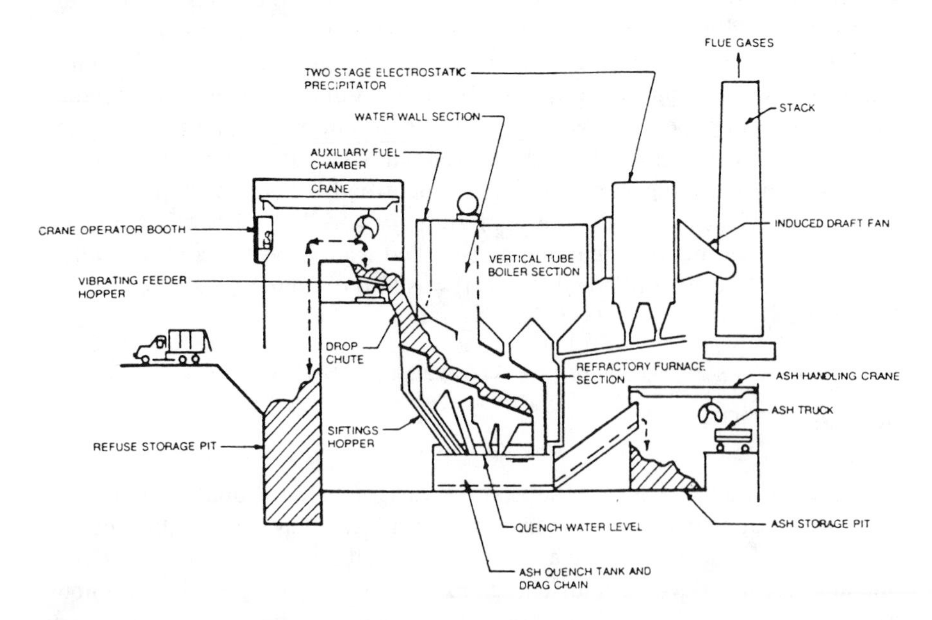


**Figure 1.** Quebec incinerator schematic cross-section.

waste. Each of the four units is capable of independent operation and is rated to produce 37,000 kg/hr of steam when burning 227 t/day of refuse with a heating value of 13,950 kJ/kg. All the steam generated is sold to Reed paper Ltee at a guaranteed steady (±7%) flow and specified pressure range.

## PILOT-PLANT DESCRIPTION

The principal components of the pilot plant, as shown in Figure 2, are as follows:

- A flue gas stream take-off duct (with eight nozzles) from the electrostatic precipitator inlet of incinerator unit No. 3. The arrangement was employed to obtain a representative sample of flue gas to the pilot plant
- A wet-dry scrubber—(also used as a gas cooler) with slurry spray nozzle and bottom screw conveyor
- A dry scrubber—with a single dry lime injection nozzle and internal cyclone integral with the scrubber at the entrance
- A pulse-jet fabric filter—using high-temperature Teflon bags as the filtering media with an air-to-cloth ratio of 4.4. Instrumentation included a pulse pressure controller, as well as duration and timer controls
- An induced draft fan with flow venturi
- A stack

Ancillary equipment included bypass ducts across the dry scrubber and fabric filter, removable ash hopper drums on each of the three vessels, flue gas duct isolating dampers, an i.d. fan flow control damper, two lime slurry variable speed pumps with associated mix tanks, agitators, controls, and a venturi/eductor dry lime feed system with variable control and lime silo. It

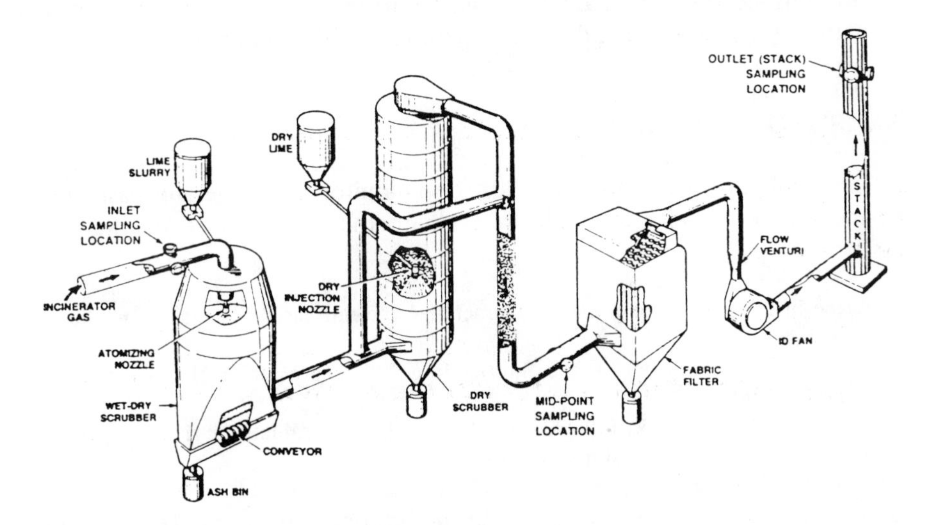


**Figure 2.** Flakt's pilot plant main equipment components.

must be noted that there are certain proprietary aspects to the design of the scrubbers and fabric filter which would distinguish this equipment and its performance from similar equipment of other manufacturers.

The pilot plant was operated in one of two modes, as follows:

1. Dry system—the hot flue gas from the incinerator entered at about 255 to 270°C into the wet-dry scrubber (now in the role of a gas cooler) where it was cooled to the desired temperature with a water spray. It then entered the bottom of the dry scrubber tangentially into an internal cyclone. Dry hydrated lime was then air injected through a single nozzle countercurrent into the gas stream. The flue gas entrained with particulates and lime was then directed into the fabric filter dust collector.
2. Wet-dry system—Hot flue gas from the incinerator entered the top of the wet-dry reactor at 255 to 270°C, where it was intimately mixed and cooled with a finely atomized lime slurry spray. The flue gas was then directed into the fabric filter dust collector.

For both systems, final removal of particulates, including lime, was accomplished with a fabric filter dust collector. Seventy-two bags arranged in six rows of twelve bags each were contained in a one-compartment design. The bag material was Teflon with Gortex scrim, which is chemically inert to acid gases and has a maximum operating temperature of 250°C. The bags were supported on cylindrical wire cages with particulate collection occurring on the outside of the bag.

The pilot-plant process control system was relatively uncomplicated. Temperature control was accomplished by either adjusting the rate of slurry into the wet-dry system or by water spray into the dry system. Flue gas flow was controlled by adjusting the damper at the inlet of the induced draft fan. To ensure constant flow was maintained during the tests, the pressure drop across a venturi upstream of the i.d. fan was continuously monitored.

## SAMPLING LOCATIONS

Three sampling locations were selected to monitor the incinerator gases and to measure pollutants in order to evaluate the performance of the two systems:

1. the inlet to the pilot plant to characterize the untreated raw gas from the incinerator
2. the midpoint prior to the fabric filter to characterize the scrubber performance
3. the outlet (i.e., stack) following the fabric filter to evaluate fabric filter performance and overall system performance

In addition, all material settling out in the bottom hoppers of the scrubbers and the fabric filter were collected for analysis.

**Table 1. Summary of Characterization Test Conditions**

| Operating mode | Temperature after cooling Target (°C) | Actual (°C) | Lime feed rate | Test no. |
|---|---|---|---|---|
| Dry system | | | | |
| Normal temperature operation | 140 | 145 | Normal | 11 |
| | 140 | 151 | High | 4 |
| | 140 | 152 | High (+ recycle) | 5 |
| | 140 | 151 | Very high | 3 |
| High temperature operation | >200 | 222 | Very high | 2 |
| Low temperature operation | 125 | 133 | Normal (+ recycle) | 12 |
| Wet-dry system | | | | |
| Normal temperature operation | 150 | 156 | Normal | 6, 10 |
| | 150 | 158 | High | 8 |
| | 150 | 158 | High (+ recycle) | 9 |
| | 150 | 156 | Very high | 7 |
| Low temperature operation | 135 | 141 | Normal | 13 |
| | 135 | 144 | Normal (+ recycle) | 14 |

## QUALITY ASSURANCE/QUALITY CONTROL

To ensure that the data was collected in a manner which would minimize any negative effects on data quality, an extensive third-party quality assurance/quality control program (QA/QC) was set up. This program component represented about 10% of the overall effort and was in addition to the internal QA/QC that is a routine part of all sampling programs.

## TEST PROGRAM DESCRIPTION

As in previous NITEP tests, the project was divided into two parts. The first was a characterization phase to familiarize the project staff with the facility and to assess, using conventional parameters, the effect of the various operating variables on system performance. Table 1 summarizes the characterization test conditions. The characterization-phase test results identified that flue gas temperature and the ratio of lime to acid gas were the key parameters influencing removal efficiency of the pollutants measured.

Based on the first-phase results, six performance test conditions were selected for a more detailed assessment of all parameters of interest (organics, metals, acid gases, and conventional gas parameters). Table 2 summarizes these conditions with a rationale for selection. Each condition was tested twice to assess repeatability of results.

For each of the performance test conditions selected, sampling and data collection were conducted for the parameters shown in Figure 3. Ten manual sampling trains and thirteen continuous gas monitors were operated simultaneously during these tests, requiring over 30 engineers and technicians.

**Table 2. Selection of Performance Test Conditions**

| Operating Mode | Target temperature before fabric filter | Rationale |
|---|---|---|
| Dry system | >200 | To collect data with no gas cooling for contrast to the three other temperatures |
| | 140 | Benchmark for comparison with other tests; considered normal for full-scale |
| | 125 | To observe potential improvements in organic removal at lower than normal temperatures |
| | 110 | To push the limits of low-temperature operation, particularly for impact on organics removal |
| Wet-dry system | 140 | Benchmark for comparison with other tests |
| (+ Recycle) | 140 | To investigate whether lime recycle enhances the removal efficiency of organics and acid gases |

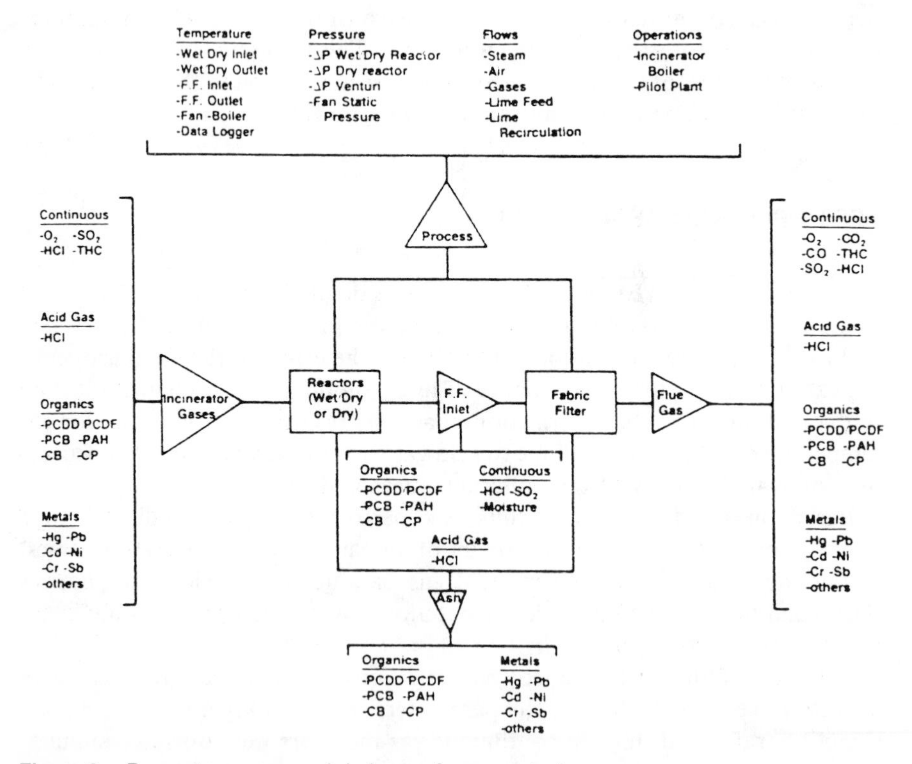


**Figure 3.** Parameters measured during performance tests.

**Table 3. Key Operating Conditions**

| | |
|---|---|
| | Incinerator |
| Steam | 32,000 kg/hr |
| Gas at boiler outlet | 280—300°C |
| Gas at pilot-plant inlet | 250—270°C |
| | Pilot Plant |
| Lime $Ca(OH)_2$ | 3.4—3.8 kg/hr |
| $\Delta$ P fabric filter | 14—16 cm $H_2O$ |
| Outlet gas flow | 3,600 $Sm^3$/hr for >200°C |
| | 4100—4300 $Sm^3$/hr for 110°C, 125°C, and 140°C |
| CO | 120—200 ppm (dry) |
| Oxygen | 12—13% (dry) |
| Inlet particulate | 6000—8000 mg/$Sm^3$ @ 8% $O_2$ |

## SUMMARY OF RESULTS

The most significant operating parameters measured during the performance tests are summarized in Table 3. The incinerator was operated as it normally operates at a steam production rate of 31,000 to 34,000 kg/hr. The average flow rate of the flue gas slipstream for the pilot plant was measured as 3500 $Nm^3$/hr (dry). The operating parameters for the pilot plant such as fabric filter pressure drop, lime flow rate, and flue gas temperature after cooling were carefully controlled at the selected conditions for each test.

## DIOXIN/FURAN RESULTS

For the parameter of greatest concern, dioxins, Table 4 summarizes the inlet, midpoint, and outlet concentrations and overall removal efficiencies. As is clearly evident, the greatest proportion of dioxin removal occurs across the fabric filter for both systems and under all temperature conditions tested. Although the outlet concentration of dioxins was low for all operating conditions, there appeared to be some temperature effect at operating conditions below 140°C. Slightly higher removal efficiencies were observed at the lower temperatures.

The dioxin homologue distribution is similarly important and a typical example is shown in Figure 4 for the dry system operating at 140°C. The bell

**Table 4. PCDD Concentration (ng/$Nm^3$ @ 8% $O_2$) in Flue Gas and Efficiency of Removal**

| | Dry system | | | | Wet-dry system | |
|---|---|---|---|---|---|---|
| Operating condition | 110°C | 125°C | 140°C | >200°C | 140°C | 140°C + Recycle |
| Inlet (ng/$Nm^3$) | 580 | 1400 | 1300 | 1030 | 1100 | 1300 |
| Midpoint (ng/$Nm^3$) | 310 | 570 | 540 | 1140 | 840 | 1270 |
| Outlet (ng/$Nm^3$) | 0.2 | ND | ND | 6.1 | ND | 0.4 |
| Efficiency (%) | | | | | | |
| Inlet/Midpoint | 47 | 60 | 57 | (11) | 24 | 2 |
| Overall | >99.9 | >99.9 | >99.9 | >99.9 | >99.9 | >99.9 |

shape of the distribution curve is similar for all test conditions. The tetra's are the least prevalent with the 2,3,7,8 isomer being generally less than 0.5% of the total dioxins.

Furan removal efficiencies were similarly high (over 99.3%) as for dioxins. Table 5 provides the average data for each performance test condition. Figure 5 shows a typical furan homologue distribution. It is interesting to note that, in the furan homologue distribution, there is a greater prevalence of the tetra-isomers.

Tables 6 and 7 provide the concentrations of dioxins and furans in the hopper ashes collected from the three vessels. The concentrations are by far the greatest in the fabric filter ash. This was expected since the fabric filter had the greatest impact on dioxin and furan removal. It is significant to note that

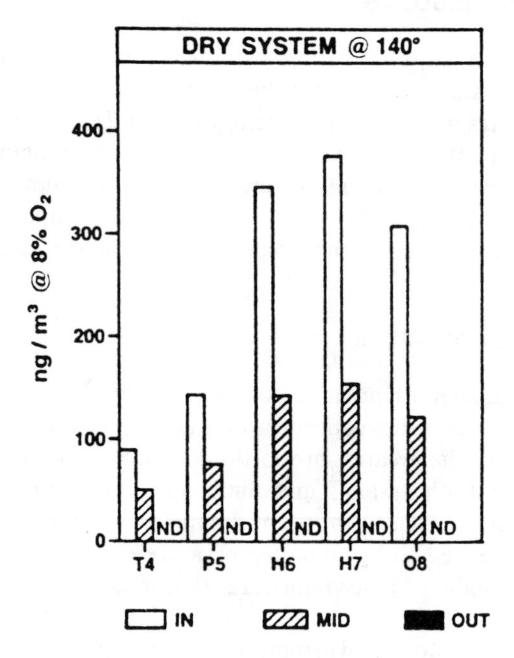


**Figure 4.** Typical PCDD homologue in gas.

**Table 5. PCDF Concentrations (ng/Nm$^3$ @ 8% $O_2$) in Flue Gas and Efficiency of Removal**

| Operating condition | Dry system 110°C | Dry system 125°C | Dry system 140°C | Dry system >200°C | Wet-dry system 140°C | Wet-dry system 140°C + Recycle |
|---|---|---|---|---|---|---|
| Inlet (ng/Nm$^3$ | 300 | 940 | 1000 | 560 | 660 | 850 |
| Midpoint (ng/Nm$^3$) | 270 | 440 | 630 | 490 | 690 | 1030 |
| Outlet (ng/Nm$^3$) | 2.3 | ND | 1.0 | 1.2 | ND | |
| Efficiency (%) | | | | | | 0.9 |
| Inlet/Midpoint | 11 | 54 | 37 | 13 | −4 | −21 |
| Overall | 99.3 | >99.9 | 99.9 | 99.8 | >99.9 | 99.9 |

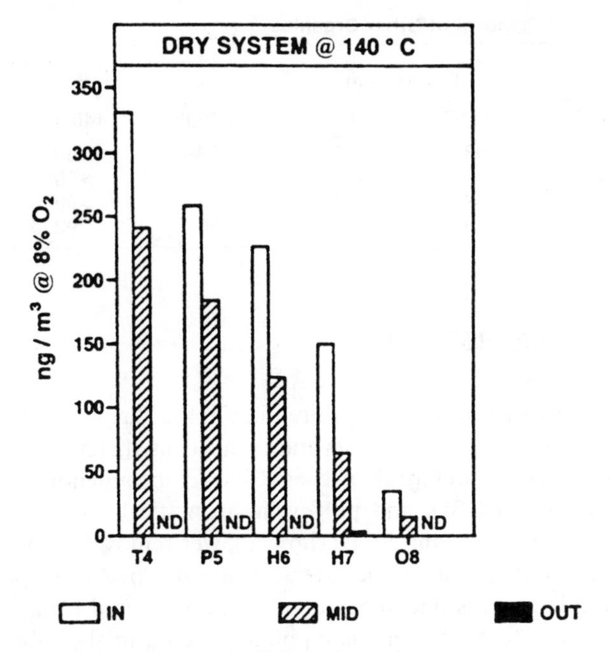


**Figure 5.** Typical PCDF homologue in gas.

the fabric filter ash was much finer in appearance than the scrubber ashes. The homologue distribution of dioxins and furans in the ash phase is essentially identical to that noted for the gas phase.

**Table 6. PCDD (ng/g) Concentration in Ash**

| | Dry system | | | | Wet-dry system | |
|---|---|---|---|---|---|---|
| Operating condition | 110°C PCDD | 125°C PCDD | 140°C PCDD | >200°C PCDD | 140°C PCDD | 140°C + Recycle PCDD |
| Wet-dry scrubber ash | 13 | 14 | 6 | 6 | 6 | 12 |
| Dry scrubber ash | 160 | 64 | 94 | 31 | N/A[a] | N/A |
| Fabric filter ash | 280 | 570 | 570 | 740 | 160 | 230 |

[a]N/A = Not applicable.

**Table 7. PDF (ng/g) Concentration in Ash**

| | Dry System | | | | Wet-dry system | |
|---|---|---|---|---|---|---|
| Operating condition | 110°C PCDF | 125°C PCDF | 140°C PCDF | >200°C PCDF | 140°C PCDF | 140°C + Recycle PCDF |
| Wet-dry scrubber ash | 10 | 12 | 4 | 5 | 5 | 8 |
| Dry scrubber ash | 87 | 36 | 68 | 22 | N/A[a] | N/A |
| Fabric filter ash | 160 | 320 | 380 | 320 | 130 | 170 |

[a]N/A = Not applicable.

**Table 8. Percent of Removal of Other Organics**

| | Dry System | | | | Wet-dry system | |
|---|---|---|---|---|---|---|
| | 110°C | 125°C | 140°C | >200°C | 140°C | 140°C + Recycle |
| CB | 95 | 98 | 98 | 62 | >99 | 99 |
| PCB | 72 | >99 | >99 | 54 | >99 | >99 |
| PAH | 84 | 82 | 84 | 98 | >99 | 79 |
| CP | 97 | 99 | 99 | 56 | 99 | 96 |

## OTHER ORGANICS RESULTS

The removal effectiveness of other organics in the flue gas by the pilot plant is outlined in Table 8. There are several significant differences between the removal effectiveness for chlorobenzenes (CBs), chlorophenols (CPs), polychlorinated biphenyls (PCBs), and polycyclic aromatic hydrocarbons (PAHs). Essentially, the CBs, CPs, and PCBs show a lower removal at higher temperatures, whereas the PAHs show the reverse relationship. At this point, there is no firm rationale for this phenomena. As for dioxins and furans, the highest concentration for all the organic compounds occurs in the fabric filter ash. Figure 6 illustrates the concentrations of CBs in the different ashes. For the other organic compounds, the distribution is similar, with highest concentrations occurring in the fabric filter ash.

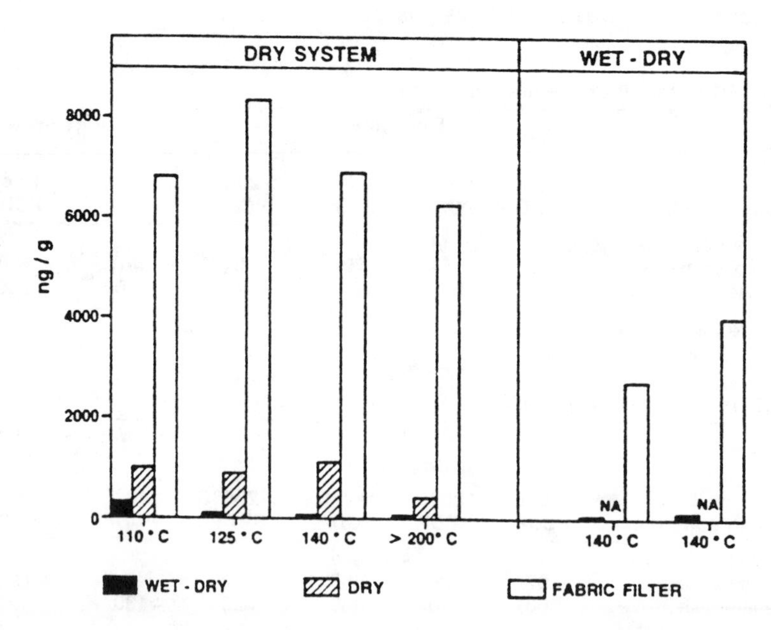


**Figure 6.** CP concentration in ashes.

**Table 9. Inlet/Outlet Metal Concentrations ($\mu g/Nm^3$ @ 8% $O_2$)**

| Metal | Inlet | Outlet |
|---|---|---|
| Zinc | 80,000—110,000 | 5—10 |
| Cadmium | 1,000—1,600 | ND—0.6 |
| Lead | 30,000—45,000 | 1—6 |
| Chromium | 1,400—3,000 | ND—1 |
| Nickel | 700—2,500 | 0.4—2 |
| Arsenic | 80—150 | 0.02—0.07 |
| Antimony | 800—2,200 | 0.2—0.6 |
| Mercury | 200—500 | 10—600 |

**Table 10. Inlet/Outlet Mercury Concentrations ($\mu g/Nm^3$ @ 8% $O_2$)**

| Operation dry system | Inlet | Outlet |
|---|---|---|
| 100°C | 440 | 40 |
| 125°C | 480 | 13 |
| 140°C | 320 | 20 |
| >200°C | 450 | 610 |
| Wet-dry system | | |
| 140°C | 190 | 10 |
| 140°C + Recycle | 360 | 19 |

## METAL RESULTS

Metal emissions are presented in Table 9, indicating the overall removal effectiveness of the control system for 8 of the 27 metals tested. As shown in more detail in Table 10, the only notable exception occurs with mercury at the higher (200°C) temperature tested, at which there is no removal of mercury. Mercury removal is significantly enhanced by operating the system at a flue gas temperature below 140°C.

In order to ensure that trace quantities of metals and organics in the lime did not contribute significantly to the observed levels in the ashes, an analysis was carried out and is reported in Table 11. In general, it may be concluded that metals present in the lime did not contribute significantly to the metal concentration in the ash. Organic analysis showed the lime to be a noncontributor to the presence of organic substances in the ash.

**Table 11. Metal Concentration in Lime Versus Metals in Hopper Ash ($\mu g/g$)**

| | Dry system | | Wet-dry system | |
|---|---|---|---|---|
| Metal | Lime | Fabric filter | Wet-dry scrubber | Fabric filter |
| Zinc | 40 | 17,000 | 4,100 | 13,000 |
| Cadmium | 10 | 320 | 35 | 180 |
| Lead | 250 | 7,800 | 1,500 | 6,000 |
| Chromium | 40 | 170 | 190 | 200 |
| Nickel | 30 | 47 | 100 | 95 |
| Antimony | ND | 330 | 40 | 200 |
| Arsenic | 4 | 42 | 5 | 20 |
| Mercury | ND | 220 | 4 | 95 |

**Table 12. HCl and $SO_2$ Concentrations (@ 8% $O_2$) and Collection Efficiency**

| Operating condition | Dry system | | | Wet-dry system | | |
|---|---|---|---|---|---|---|
| Flue gas temperature at fabric filter | 110°C | 125°C | 140°C | >200°C | 140°C | 140°C + Recycle |
| Stoichiometric ratio | 1.16 | 1.03 | 1.04 | 1.49 | 1.19 | 1.10 |
| **Hydrogen chloride** | | | | | | |
| Inlet (ppm) | 423 | 464 | 425 | 392 | 366 | 470 |
| Midpoint (ppm) | 15 | 69 | 129 | 196 | 149 | 152 |
| Outlet (ppm) | 7 | 9 | 29 | 91 | 29 | 42 |
| Efficiency to midpoint (%) | 96 | 85 | 73 | 50 | 59 | 69 |
| Efficiency overall (%) | 98 | 98 | 94 | 77 | 92 | 91 |
| **Sulfur dioxide** | | | | | | |
| Inlet (ppm) | 119 | 118 | 99 | 117 | 106 | 106 |
| Midpoint (ppm) | 24 | 65 | 64 | 103 | 67 | 70 |
| Outlet (ppm) | 4 | 10 | 41 | 83 | 35 | 43 |
| Efficiency to midpoint (%) | 80 | 45 | 35 | 11 | 37 | 35 |
| Efficiency overall (%) | 96 | 92 | 58 | 29 | 67 | 60 |

## ACID GAS RESULTS

The ability of the two systems to reduce the emission of acid gases was evaluated for several different operating modes. Tables 12 demonstrates the impact of operating conditions on acid gas removal. It is clearly evident that temperature is a major factor for acid gas removal, which is consistent with observations for organics and metals. At lower operating temperatures, the removal efficiencies are significantly higher than at higher temperatures. There appears to be no significant difference between the dry and wet-dry systems for acid gases removal; however, the recycle condition (wet-dry mode) did show that less lime was required to achieve the same removal efficiency as without recycle. This is a significant finding as it has the potential to reduce operating costs by reducing lime consumption.

## CONCLUSIONS

The following key conclusions have been drawn from the extensive test data on the Flakt pilot-plant system:

1. Both the wet-dry system using lime slurry and the dry system using powdered lime, followed by a fabric filter, are capable of high removal efficiencies for all pollutants of concern, with no significant difference in removal efficiency by either system.
2. Cooling of the flue gas temperature was a key operating parameter for effective removal of HCl, $SO_2$, and mercury for both systems.
3. The removal efficiencies for dioxins and furans were very high, exceeding 99%. For most test runs, the concentrations of dioxins and furans after the control system approached the detection limits of the sampling and analytical methods employed.

4. The highest dioxin and furan concentrations in the ashes occurred in the fabric filter ash and the lowest concentrations were found in the scrubber ashes. These results were anticipated based on the high removal efficiency of dioxins and furans across the fabric filter collector.
5. Other trace organics, such as chlorobenzenes (CBs), polyclorinated biphenyls (PCBs), chlorophenols (CPs), and polycyclic aromatic hydrocarbons (PAHs) were also efficiently removed (80 to 99%) by both systems when operated under cooled flue gas conditions.
6. Concentrations of trace organics in the various hopper ashes followed a similar pattern as for dioxins and furans, whereby the highest concentrations occurred in the fabric filter ash.
7. Metal collection efficiencies generally exceeded 99.9% with both systems except for mercury, for which flue gas cooling was essential to maintain a high removal efficiency.
8. Low emissions of acid gas can be accomplished by either increasing the ratio of lime to acid or by cooling the flue gases. However, flue gas cooling is a more economical approach since increasing lime utilization is more costly.
9. Ash recycle (i.e., fabric filter ash containing some residual unreacted lime) added to the fresh makeup lime was beneficial in providing the same $SO_2$ removal efficiency as the no-recycle condition. By recycling ash, less fresh lime was used, which reduces operating cost.
10. Two key independent variables tested were flue gas temperature control and the ratio of lime to acid gas concentration. Temperature was found to be the most significant variable in affecting removal efficiency.

CHAPTER 19

# Dioxin Recovery by Chemisorption—Control of Incineration Process Emissions

A. J. Teller

## ABSTRACT

Chlorinated dibenzo dioxins are emitted at significant levels from incinerator boiler combinations and have not been reduced to regulatory levels by control of combustion conditions only. Utilizing the chemisorptive process, the dioxins were reduced to toxic equivalent levels of 0.009 ng/Nm$^3$, inclusive of the nondetectable limits, in commercial operation. The pseudo-vapor pressure of the 4CDD group in the vapor-chemisorptive equilibrium with flyash and a silica-alumina matrix reflected an enthalpy of chemisorption of the order of 22 KCal/mol. The kinetic requirements for effective recovery reflect a diffusion controlling mechanism.

## INTRODUCTION

The presence of dioxins in the flue gas emitted from the incineration of municipal waste has been well documented.[1,3,12,14,16] It has been further established that the ambient levels of dioxins to which the proximate community is exposed are directly related to the rate of emissions of dioxins from the local municipal waste incinerator.[19]

The requirement for the reduction of dioxin emissions because of the implied carcinogenic behavior of these compounds has resulted in studies concerning the source, formation, and/or modes of destruction or containment of the dioxins.[4,9,22] It has been established that the dioxins form during the combustion and postcombustion processes and that although the presence of chlorinated plastics in municipal waste does contribute to the formation of dioxins, the removal of these materials does not result in the elimination of the dioxins.[6,32]

Theoretical and laboratory studies[21,23,30,31] concerning the formation and thermal destruction of dioxins led to a recommendation by the American Society of Mechanical Engineers (ASME) solid waste committee and many regulatory agencies that establishing combustion conditions of 900 to 1000°C for a period of 1 to 2 sec would result in the elimination of the dioxins.

However, as a result of the heterogeneity in the fuel, both in composition and water content, there exists a very wide variation in the combustion conditions within the furnace, precluding the implied result of the "average" combustion conditions.[5,16,30,33]

It has been further established that thermal destruction of dioxin cogeners can create fragments containing chlorinated aromatic precursors to dioxins or that precursors, such as chlorophenols, form in the combustion process. These compounds are more refractory to thermal attack than dioxins and combination of these compounds to form dioxins occurs on catalytic surfaces, such as flyash, at lower temperatures in the range of 200 to 500°C. Continued chlorination of dioxins does occur on catalytic surfaces, such as flyash, at temperatures as low as 100°C[4] and dimerization of chlorophenols on silica surfaces to form dibenzo-p-dioxins (PCDD) has been observed.[34]

The data on dioxin emissions obtained by Environment Canada at Prince Edward Island serve as an example of the phenomenon of recombination.[7] In this case, an afterburner was employed between the combustion zone and the boiler. The PCDD emissions postboiler significantly exceeded the post afterburner emissions, corroborating studies[23] that the formation of dioxins occurs at postcombustion temperatures with flyash as the catalyst.

The phenomenon of "puffing" or nonsteady-state combustion, inherent in the burning of heterogeneous municipal waste, resulted in observation of PCDD emissions from 700 ng/Nm$^3$ from a furnace operating at a statistical average temperature of 1065°C.[5] Nittrod and Ballschmitter[16] measured emissions of TCDD equal to 0.9–1.6 ng/Nm$^3$ (in solid phase) and 5.1 ng/Nm$^3$ (in gas phase) from an incinerator operating at 880 to 1088°C. This type of observed behavior has resulted in expressions of concern by Clarke,[5] Tsang,[30] and Volskow[33] that the statistical temperature-time conditions in the combustion zone are not sufficient parameters to serve as the only criteria for reduction of dioxin emissions.

A further combustion condition for dioxin control proposed by Hasselriss,[12] the maintenance of low CO levels, has also proved to be inadequate. Tests conducted by the California Air Resources Board reported by Clarke[5] found incinerator outlet PCDD and polychlorinated dibenzofurans (PCDF) emissions exceeding 1000 ng/Nm$^3$ where the CO content was less than 50 ppm.

A recent study by Getter[11] included the dioxin emissions from two Martin incinerator-boiler installations operated by the same owner-operator. The only difference in design and operation is that one system uses an electrostatic precipitator (ESP) for particulate removal only and the other uses the quench reactor-dry venturi-baghouse (QR-DV-BH) system (Table 1).

The significantly lower emissions achieved by the QR-DV-BH system compared with the ESP system implies that the QR-DV-BH system removed a minimum of 94% of the PCDD emitted by the furnace operating in the prescribed range of 900 to 1000°C for 1 to 2 sec. This reduction is a minimum

**Table 1. Dioxin Emissions—Waste-to-Energy Facilities—Martin Furnace**

| System | Emission control system | Dioxin Emissions ($ng/nm^3$) | |
|---|---|---|---|
| | | TCDD | PCDD |
| Tulsa | Electrostatic precipitator | 1.613 | 18.93 |
| Marion | QR-DV-BH | 0.189 | 1.131 |

inasmuch as the scrubber-ESP does remove a significant portion of the dioxins adsorbed on the flyash, equivalent to approximately 60% of the total dioxins.[15] This was confirmed in the Marion tests where the incinerator outlet contained in excess of 25 $ng/Nm^3$ of dioxins.

The data obtained from full-scale systems thus confirms that time-temperature and CO criteria are inadequate for consistent achievement of the effective reduction of dioxin emissions from municipal waste incineration.

## PHYSICAL FORM OF DIOXIN EMISSIONS

The dioxin emissions in the flue gas from the incineration of municipal solid wastes are present in both the gaseous and solid states. Although the early assumptions were that the dioxins were primarily present in the solid state because of their low volatility, the more recent studies indicate that the lower chlorinated cogeners, particularly the more toxic TCCD, are present primarily in the gaseous state[1,2] (Table 2).

Although both the condensate and absorber catches are normally considered to represent the vapor-phase content, indicating that 82% of the 2,3,7,8 TCDD is in the vapor form, Ballschmitter[1] has suggested that the condensate may contain 0.3 μm particulate as a result of nucleation capture. Therefore, the vapor-phase 2,3,7,8 TCDD may represent from 62 to 82% of the TCDD emitted by the incineration process.

Removal of the dioxins, therefore, requires treatment of both the gaseous and solid phases. With respect to the treatment of the solid phase, particular attention must be given to the removal of the 0.3 μm particulates, if the hypothesis by Ballschmitter is correct.

**Table 2. Physical State of Dioxin Emissions[2]**

| | Concentration in flue gas ($ng/nm^3$) | | | |
|---|---|---|---|---|
| | Particulate | Condensate | Absorber | Total |
| 2,3,7,8 TCDD | 0.1 | 0.17 | 0.28 | 0.55 |
| TCDD | 6.7 | 5.1 | 6.9 | 18.7 |
| PCDD (3–7) | 34 | 68 | 14.4 | 116.4 |
| OCDD | 11 | 3.1 | 0.1 | 14.2 |

**Table 3. Vapor Pressure and Concentration Equivalent of TCDD[24]**

| Temp °C | Vapor pressure (Pa) (average) | Gas-phase concentration (ng/nm³) |
|---|---|---|
| 30.2 | $4.68 \times 10^{-7}$ | 66 |
| 54.6 | $1.82 \times 10^{-5}$ | 2,566 |
| 62 | $4.97 \times 10^{-5}$ | 7,008 |
| 71 | $1.59 \times 10^{-4}$ | 22,419 |
| 108 (extrap) | $10^{-2}$ | $1.4 \times 10^{6}$ |

## VAPOR-SOLID PHASE BEHAVIOR OF DIOXINS

The dioxins, because of their low volatility and their apparent formation in the incineration-combustion process and in the condensed-adsorbed phase, exist in incinerator emissions in both the vapor and adsorbed on solid surfaces. The vapor-condensed-phase equilibria can be achieved in three conditions of the condensed phase: (1) pure solid-vapor, (2) physically adsorbed-vapor, and (3) chemisorbed-vapor.

Schroy et. al. [24] measured the vapor pressure of 2,3,7,8 TCDD in the apparent pure solid compound-vapor condition. The reported vapor pressures and the equivalent concentration in ng/Nm³ are indicated in Table 3.

It is evident that these vapor concentration values exceed by six to nine orders of magnitude the measured vapor-phase concentrations in the presence of an adsorbed phase on flyash.[1-3,11] There exist, therefore, pseudo-vapor pressures in the presence of adsorbents, either via physical or activated chemical adsorption.

For example, Politski et al.[20] reported on the comparison of vapor pressure of pure organic substances with pseudo-vapor pressure when adsorbed on silica gel. A reduction in "vapor pressure" of the order of $10^{-4}$ was achieved at 23°C for 2,4,6 trichlorophenol and 1,2,4 trichlorobenzene, potential precursors of dioxins. The silica gel also bears a similarity to the structure indicated for flyash.[23]

If chemisorption occurs, there occurs a significantly greater reduction compared with physical adsorption in the pseudo-vapor pressure of the adsorbate relative to the pure compound vapor pressure.

A unique behavior was noted in the data published by Cavallaro[3] reflecting six tests on incinerator emissions at 220°C (Table 4). The concentrations of TCDD in both the vapor and solid phases were reported as an equivalent gas-phase concentration.

The vapor-phase concentration, within the range of accuracy and reproducibility of sampling and analysis, was relatively constant while the concentration in the particulate phase varied by a factor as great as 5000. The arithmetic mean for the vapor concentration of the TCDD is 19 ng/Nm³. If the apparently anomalous value of 60 ng/Nm³ is withdrawn, the mean remains 19 ng/Nm³ with an average of 17 ng/Nm³.

The relative constancy of the vapor-phase TCDD concentration at a given temperature coupled with extreme variation of the concentration in the solid

**Table 4. TCDD Emissions from Incinerators (220°C)[3]**

| Incinerator | TCDD concentration ($ng/nm^3$) Gas phase | Solid phase | Ratio gas-phase concentration: solid-phase concentration |
|---|---|---|---|
| 1 | 19.6 | 1.6 | 12.2 |
| 2 | 17 | 172.2 | 0.1 |
| 3 | 19 | 0.037 | 513 |
| 4 | 60 | 10.9 | 5.5 |
| 5 | 9.6 | 0.34 | 18.2 |
| 6 | 19 | ND | >500 |

phase establishes a temperature-controlled vapor pressure mechanism with a pseudo-vapor pressure equivalent to 19 $ng/Nm^3$ at 220°C. This is many orders of magnitude less than the pure substance vapor pressure and further implies a chemisorptive mechanism because of the high temperature associated with the adsorption.

The chemisorptive hypothesis is further supported by Eichman and Rghei,[10] who established that 1,2,3,4 TCDD, adsorbed on flyash, in the presence of 0.1% HCl (typical of incinerator emissions) formed higher chloride cogeners of PCDD. This confirmed the anticipated catalytic-adsorbed-phase behavior of flyash, proposed[23] as a mechanism for the formation of dioxins in the incineration process.[10] A plot of the Eichman-Rghei data (Figure 1) provides the form of curve of the combined process of physical and chemisorptive mechanisms as proposed by Smith.[25]

The observation by Dickson et al.[8] that the rate of equilibrium desorption of PCDD depended mainly on the amount of the homolog originally present on the flyash further supports the physico-chemisorptive process.

Schaub[23] anticipated a chemisorptive mechanism for the reduction in vapor pressure of dioxins and projected that the mechansim would be (Figure 2), a C••O bonding subsequent to dechlorination.

This process of attachment and activation is not consistent with the data obtained by Eichman and Rghei, who found increased chlorination activity after adsorption rather than dechlorination. The more probable attachment providing activation of the 4 CDD molecule is that proposed by Kislev (in the text by Oscik)[17] for the adsorption of oxygenated organics. Kislev proposed the attachment of the ether-oxygen bond on silica-activated surfaces exhibiting an O-H active surface (Figure 3A).

In an analogous manner, the 2,3,7,8 TCDD would be chemisorbed in accordance with the Kislev hypothesis (Figure 3B). The formation of the silicic acid surface structure is a result of hydration of the silica by the in situ-developed water due to HCl neutralization by the calcium component in the matrix.

This chemi-physical association permits the activation of the remaining C-H sites to achieve the further chlorination observed by Eichman and Rghei.[10]

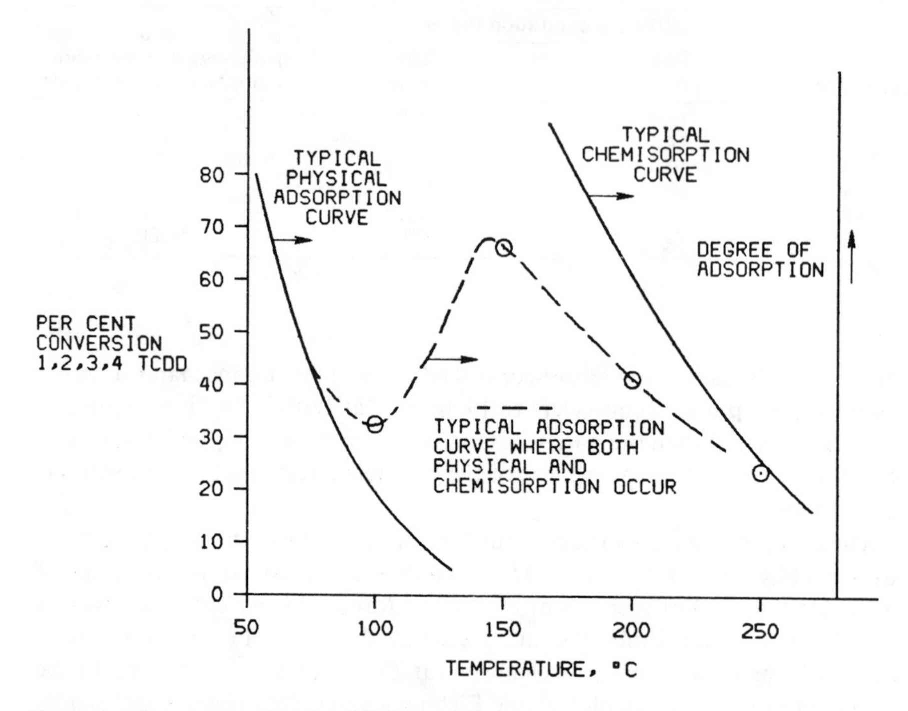


**Figure 1.** Conversion of 1,2,3,4, TCDD to higher chlorinated CDD cogeners superimposed on typical adsorption characteristics.

## ADSORBED-PHASE VAPOR PRESSURE OF DIOXINS

Utilizing the data obtained by Cavallaro,[3] Teller[27] and Teller and Lauber,[26] in 1983, the Cox-Othmer procedure was used to estimate the pseudo-vapor pressure-temperature relationship of the 4 CDD homologous group adsorbed

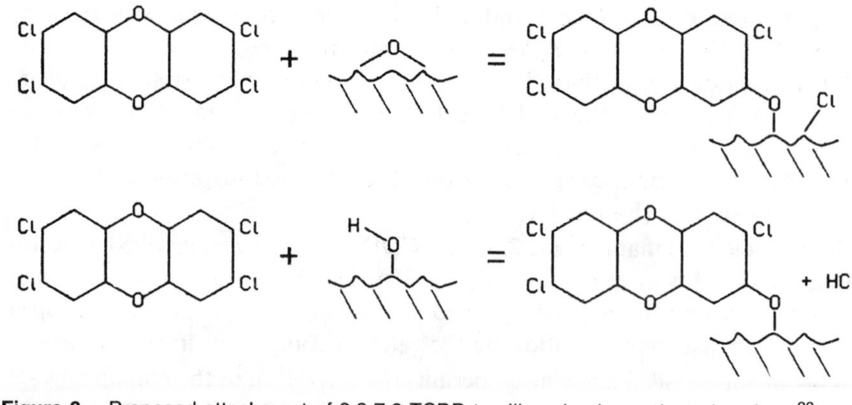

**Figure 2.** Proposed attachment of 2,3,7.8 TCDD to silica-alumina-activated surface.[23]

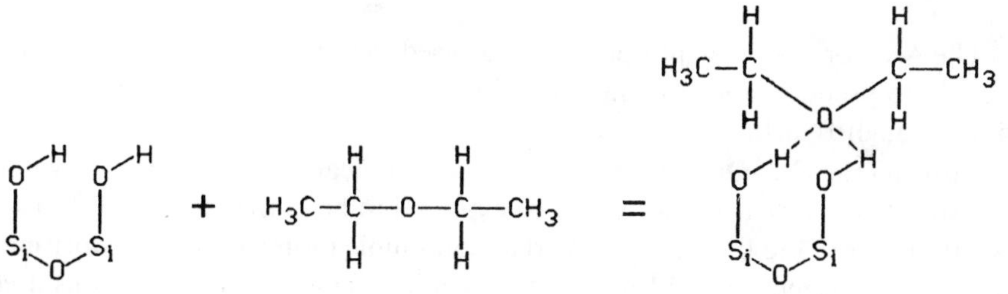

**Figure 3a.** Attachment of C-O-C bond to silica surfaces.[17]

on flyash. The basis for that estimate was the hypothesis that the heat of sublimation was of the same order of magnitude as the heat of vaporization of a group of substances having similar structural characteristics. The group included PCB—Aroclor 1,2,4,8, PCB—Aroclor 1,2,5,4, 2 Chlor-ethoxy-methyl benzene ether, and Pentachlorophenol.

The Cox-Othmer relationship is based on the Van't Hoff or Clausius-Clapeyron equation relating the vapor pressure of a substance with the heat of vaporization or sublimation and temperature, with the assumption that over a narrow range in temperature, less than 200°C, that the heat of vaporization is essentially constant:

$$\ln p = \frac{-\Delta H}{RT} + C$$

For comparison of the volatility of two substances, the Cox-Othmer relationship therefore established for a specific temperature:

$$\ln P_1 \; \frac{\Delta H_1}{\Delta H_2} \ln P_2 + a$$

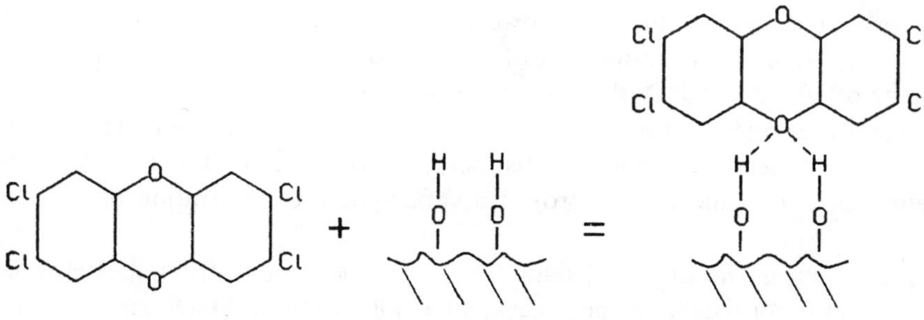

**Figure 3b.** Proposed attachment of a 4 CDD to silica-alumina surfaces.

The Aroclor 1,2,5,4 vapor pressure was used as the comparison base and the Cavallaro pseudo-vapor pressure of $1.35 \times 10^{-7}$ Pa at 220°C based on filtration through flyash.

Utilizing the Cox-Othmer procedure, it was indicated that the pseudo-vapor pressure of 4 CDD at 120°C would be expected to be in the range of 1.2 to 4.3 $\times 10^{-9}$ Pa or 0.3 to 0.6 ng/Nm$^3$. Verbal communication with Olie[18] confirmed that in tests of incinerator flue gas, where a less than adequate ESP was used for emission control, in-stack filtration at 220°C resulted in gaseous emissions of 15 to 25 ng/Nm$^3$ for TCDD. The use of the EPA-5 test method providing for filtration at 120°C resulted in nondetectable levels of 4 CDD. The limit in that test method was of the order of 1 ng/Nm$^3$. Subsequent analysis was based on the chemisorption hypotheses with the mechanism of linkage established in Figure 3, where the coordinate covalent O---H association with the catalyst surface occurs.

The estimation of the heat of activated adsorption utilized the heat of unassociated sublimation in the range 13 to 17 KCal/mol based on the data for substances having similar molecular structural components, the polychlorinated biphenyl (PCB) compounds, 2-chlor-ethoxy methyl benzyl ether, pentachloro phenol, and 2,6 dichlorobenzo quinone, and the estimate by the Trouton rule.

An additional thermal release occurs in the formation of a coordinate covalency with the hydrogen on the surface of the flyash. Two methods of estimation were used—the bond dissociation energy of the O-H group and the bond energy exchange in the quinone-hydroquinone equilibrium. The range was 9 to 16 KCal/g atom. Thus, the activated heat of sublimation was estimated at 22 to 33 KCal/mol.

The Cox-Othmer plot (Figure 4) was established using the PCB compound Aroclor 1,2,5,4 as the reference compound, indicating the established relationship with a lower molecular PCB, Aroclor 1,2,4,8 and with pentochlorophenol. The estimated vapor pressure of 4 CDD was based on the data of Cavallaro for three heats of sublimation—15 KCal/mol for physical adsorption and both 22 KCal/mol and 30 KCal/mol for chemisorption.

The estimated vapor pressures of pure 4 CDD by Schroy are compared with the proposed method (Figure 5) for enthalpies of activated adsorption for 22 and 30 KCal/mol. The estimates of this method and that of Schroy differ by a factor of $10^{10}$. At 120°C, the Schroy[24] estimate would predict an equilibrium vapor-phase concentration of 2,3,7,8 TCDD of $4.3 \times 10^6$ ng/Nm$^3$. The procedure utilizing observed constant temperature data for 4 CDD and the estimated heats of vaporization project a vapor-phase concentration of 0.01 to 0.045 ng/Nm$^3$.

Inasmuch as the observed data for 4 CDD emissions from incineration establish the coexistence of both gas- and adsorbed-phase 4 CDD, it is evident that the vapor pressures established by Schroy[24] reflect only the sublimation of

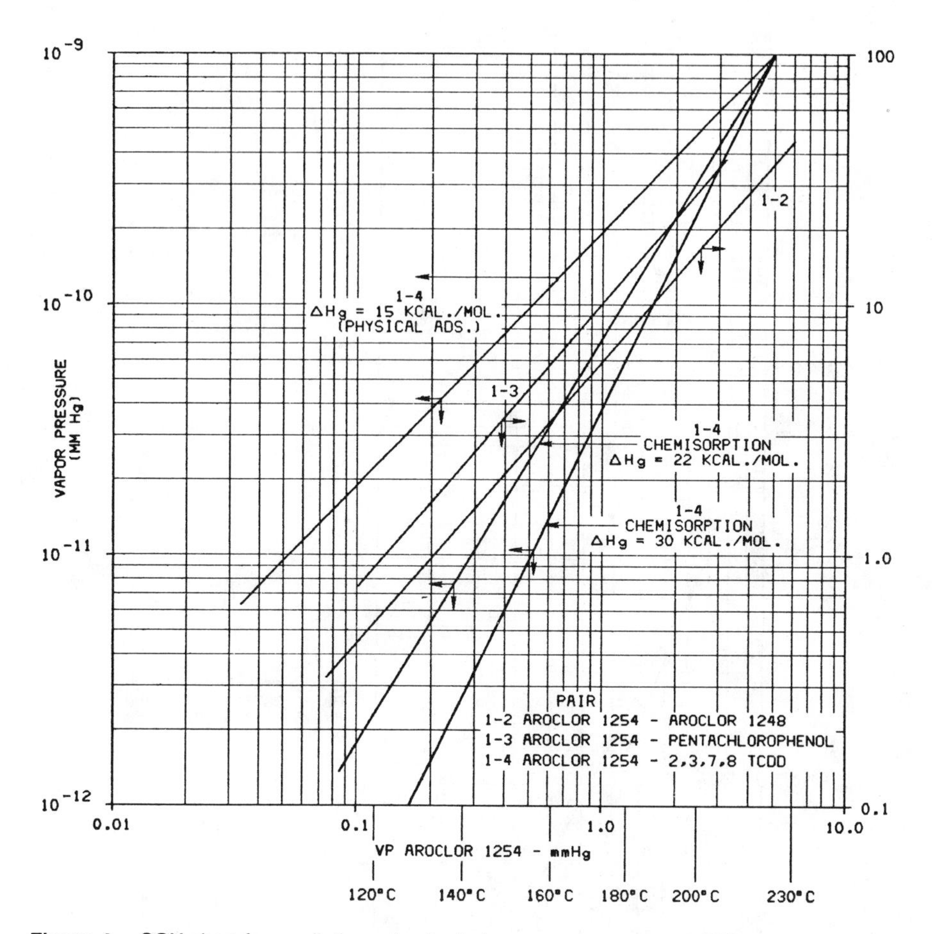


**Figure 4.** COX chart for predicting adsorbed-phase vapor pressure 4 CDD.

the pure compound and not the effect of activated adsorption on the vapor pressure of the 4 CDD homologous group.

## KINETICS OF ACTIVATED ADSORPTION

The model by Schaub[23] for the time required for achievement of solid-phase equilibrium of the vapor-adsorbed phase 4 CDD:

$$t_{(99\%)} = \frac{4N_p}{Z_p da_o} \frac{(\sigma_p)^2}{(\sigma_a)} \ln\left[1 - \frac{0.99ds(eg)}{4N_p} \frac{(ao)}{(do)} \frac{(\sigma a)^2}{(\sigma p)}\right] e$$

implies an equilibrium time of the order of $10^{-3}$ sec exclusive of the time necessary for diffusion of the dioxin and competitive molecules both in the

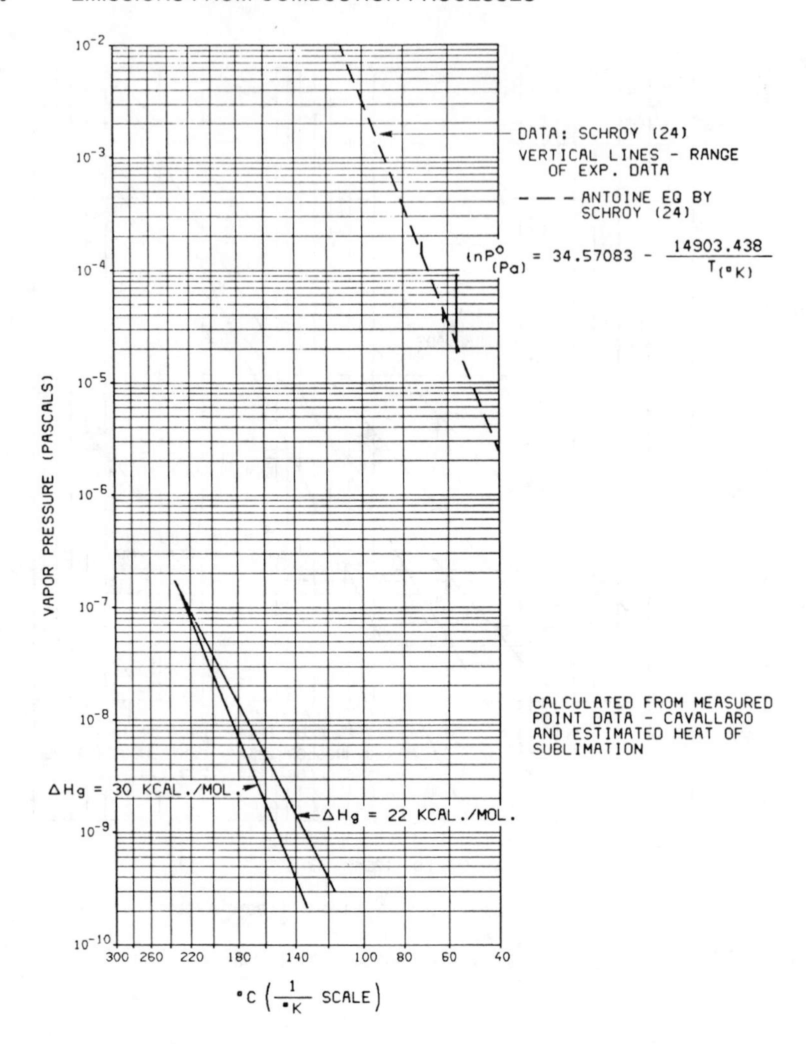


**Figure 5.** Pseudo-vapor pressure of 4 CDD.

bulk and pore areas. This extremely short time interval for surface concentration equilibration implies that the diffusion is rate controlling.

Previous studies[27] related to the recovery of acid gases emitted in the incineration process on solid surfaces established that for rapid surface reaction equilibration, effective adsorption would occur primarily in concentrated-phase solid adsorbents rather than on solids dispersed in a gas stream. This minimizes the gas-phase diffusion resistance by establishing a relative velocity between the gas stream and particulates and by providing a high concentration of the solid surface.

The wave front behavior for fixed-bed adsorption exhibits a pattern as indicated in Figure 6.

The change in the gas-phase composition with depth of bed is generally represented by the function:

$$X = f\left(\frac{Fa}{d_p^{\,b}}, \frac{v}{F}, \frac{V}{v}\right)$$

Essentially, the variables are represented by the wave front number, WF, where

$$WF = f\,(v, D_v, d_p, d, q)$$

In the procedure developed, using baghouse cake as the fixed adsorption bed, the thickness of the bed is increased at a rate equivalent to the rate of addition of both the reagent and the adsorbate. Thus, a moving wave front is created (Figure 7). This behavior may be represented as efficiency of adsorption in the relationship:

$$\frac{\chi_i - \chi_c}{\chi_i} = f\,(MWF)$$

The performance behavior of this relationship for recovery of HCl by adsorption–rapid reaction on a baghouse cake in an incinerator emission control system is indicated in Figure 8.

Inasmuch as the velocities, temperatures, and physical characteristics of the cake are the same for both the recovery of HCl, HF, $SO_2$, and the dioxins, and

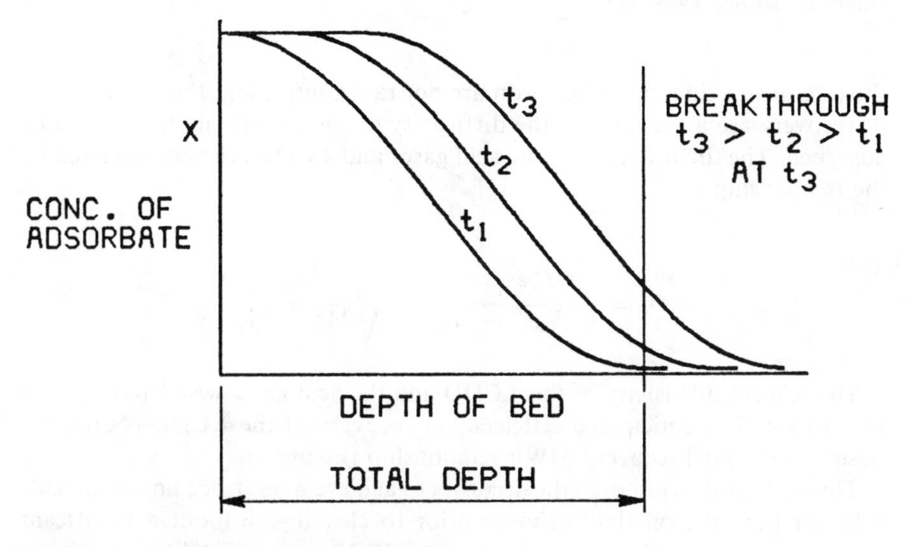


**Figure 6.** Adsorption breakthrough behavior.

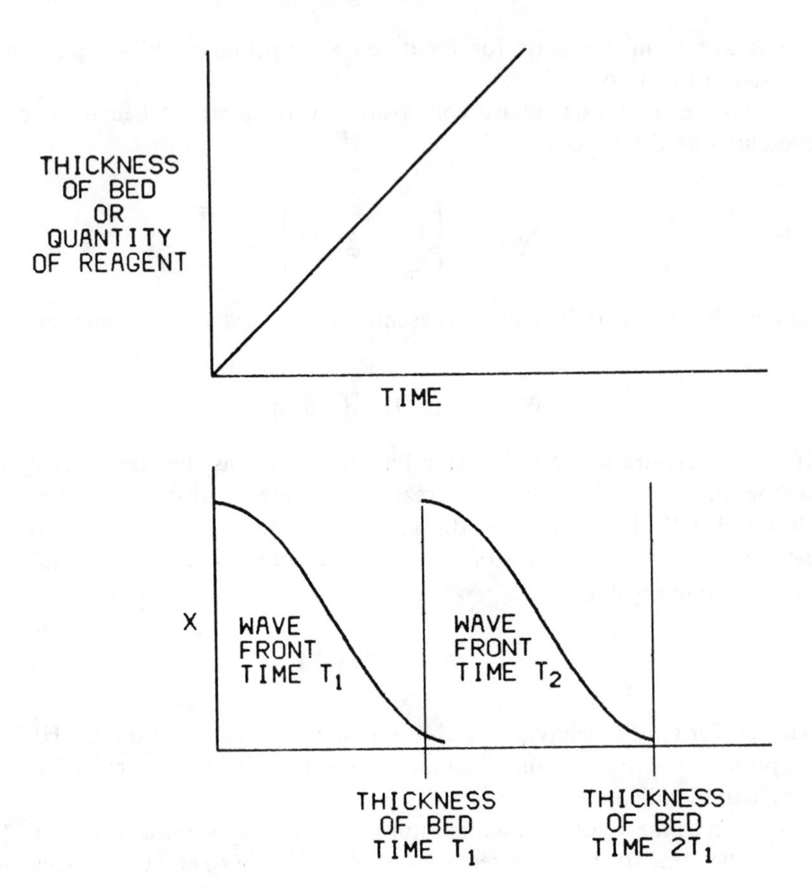


**Figure 7.** Moving wave front.

the rate of reaction and adsorption are not rate controlling, the relative rates of recovery are a function of the diffusivity of the molecular structure to be adsorbed. The diffusivities for the acid gases and 4 CDD are approximated by the relationship:

$$D_v = \frac{CT^{3/2}}{P\,(V_A^{1/3} + V_B^{1/3})^2} \sqrt{\frac{1}{M_A} + \frac{1}{M_B}}$$

The relative diffusivity of the 4 CDD and the acid gases was determined to be 1 to 1.4. The anticipated efficiency of recovery of the 4 CDD is superimposed on the HCl recovery-MWF relationship (Figure 8).

This relationship imposed the necessity to achieve a residence time of at least 5 hr for the cake on the baghouse prior to cleaning, without a significant increase in pressure drop, in order to provide the proper fixed-bed depth.

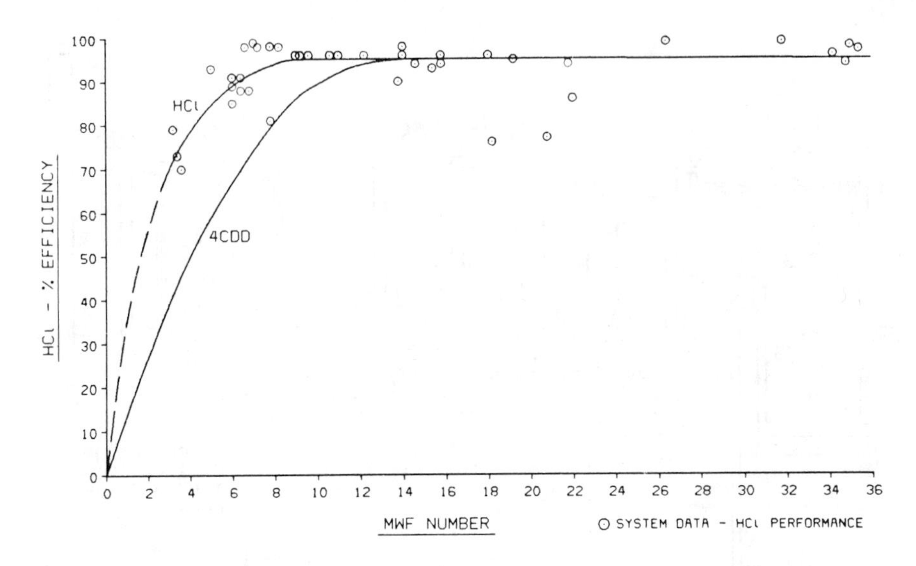


**Figure 8.** Moving wave front—efficiency of recovery—baghouse cake adsorption.

## OPERATING SYSTEM

The system, incorporating a dry venturi,[28] achieves this requirement. Thus, the parameters of design for the recovery of vapor phase dioxins in the order 95% were established as follows:

| | |
|---|---|
| Operating temperature of the baghouse | 120°C |
| Residence time for the filter cake prior to cleaning | 6 hr |

The system temperature conditions are indicated in Figure 9.
The system consists of:

1. A quench reactor where a slurry of lime reagent suspended in water is mixed with the incoming gas stream for initial neutralization and gas cooling.[29]
2. A dry venturi for addition of crystalline material to change the characteristics of the particulate and the filter cake so that increased depth of cake is achievable at a low-pressure drop.[28]
3. A fabric filter where depth of filter cake in the range of 1 to 5 mm is achieved to provide for improved filtration of fine particulate and to act as a fixed bed for adsorption-reaction processes.

Two commercial systems installed in the United States to achieve simultaneous reduction in dioxins, and in gases, fine particulates, and heavy metals were at the City of Commerce, California, CA, and Marion County, OR. The Commerce facility consists of a 350 TPD incinerator-boiler train, followed by the Teller semi-dry scrubbing train. The pollution control system inlet temperature ranges from 200 to 290°C. The temperature is reduced in the quench reactor to 135 to 140°C and further to 135°C as the gas passes through the dry

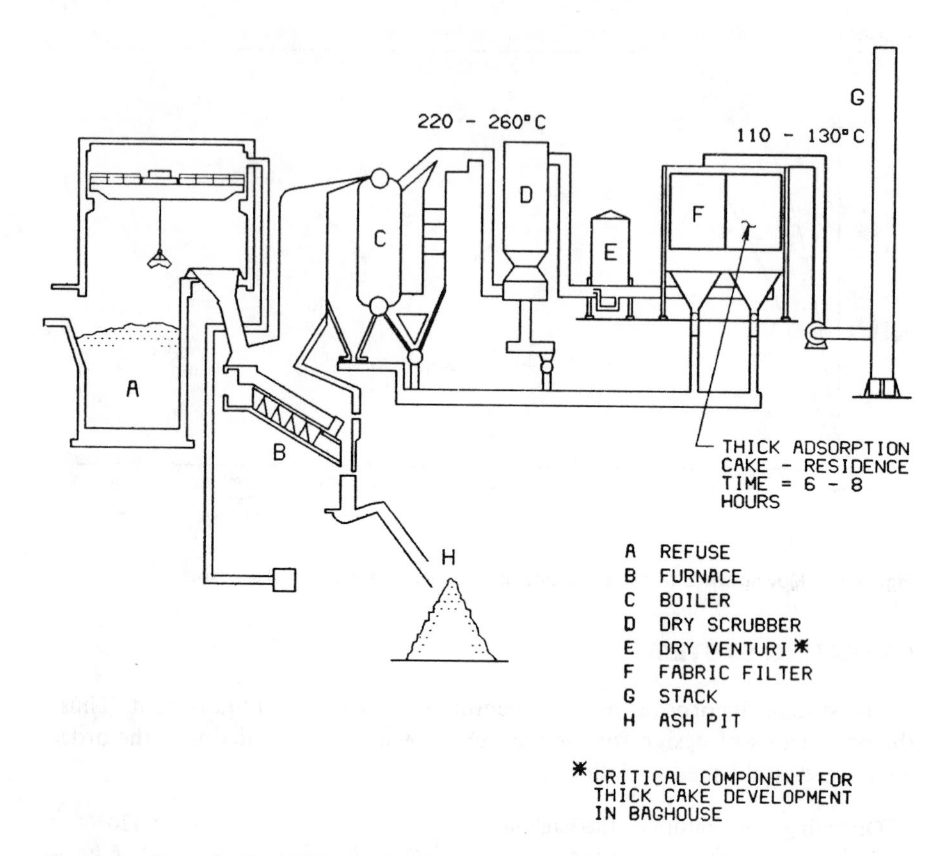


**Figure 9.** Example of a MSW resource recovery incineration bact dry scrubbing installation.

venturi and leaves the system in the range of 120 to 130°C. The Tesisorb feed material is a $SiO_2$-$Al_2O_3$ complex so that it serves not only as a particulate modifier to permit increased residence time in the baghouse at a low pressure drop, but also adds to the chemisorptive capability of the flyash.

## SYSTEM PERFORMANCE

The dioxin emissions from the boiler (Table 5) confirms the anticipation that combustion conditions of 900 to 1000°C for 1 to 2 sec does not destroy the dioxins in the furnace. The total PCDD emissions from the boiler were 8.20 ng/Nm$^3$ (corr. 12% $CO_2$). The exit concentration of TCDD ranged from 0.032 ng/Nm$^3$ (corr. 12% $CO_2$) to 0.27 ng/Nm$^3$ (corr. 12% $CO_2$) with an arithmetic average of 0.112 ng/Nm$^3$ and a mean of 0.033 ng/Nm$^3$ (corr. 12% $CO_2$). Assuming equilibrium occurred inasmuch as HCl and $SO_2$ reductions in the fabric filter were 95%, the outlet conditions imply that the heat of sublimation is 23 KCal/mol for the 4 CDD-flyashTesisorb chemisorption.

**Table 5. PCDD/PCDF Emissions from Commerce Refuse-to-Energy Facility**

| | Concentration (ng/Nm$^3$ @ 12% $CO_2$) | | | | |
|---|---|---|---|---|---|
| Location | Boiler exhaust | Stack | | | |
| test | 17 | 15 | 16 | 17 | Average |
| 2378-TCDD | ND < 0.097[a,b] | ND < 0.003[a,b] | ND < 0.003[a,b] | ND < 0.003[a,b] | ND < 0.003 |
| Total TCDD | 0.865 | 0.033 | 0.032 | 0.27 | 0.112 |
| 12378-PCDD | 0.097 | ND < 0.002 | ND < 0.001 | ND < 0.005[a] | ND < 0.003 |
| Total PCDD | 0.448 | 0.011 | 0.011 | 0.130 | 0.051 |
| 123478-HxCDD | 0.078 | ND < 0.003[a] | ND < 0.001 | 0.005 | <0.003 |
| 123678-HxCDD | 0.124 | ND < 0.008[a] | ND < 0.006[a] | ND < 0.017[a] | ND < 0.011 |
| 123789-HxCDD | 0.124 | ND < 0.008[a] | ND < 0.010[a] | 0.015 | <0.011 |
| Total HxCDD | 1.261 | 0.055 | 0.056 | 0.195 | 0.102 |
| 1234678-HpCDD | 1.067 | ND < 0.058[a] | 0.059 | 0.127 | <0.081 |
| Total HpCDD | 2.16 | 0.062 | 0.059 | 0.254 | 0.125 |
| OCDD | 3.47 | 0.153 | 0.213 | 0.31 | 0.225 |
| Total PCDD[c] | 8.20 | 0.314 | 0.370 | 1.15 | 0.611 |
| 2378-TCDF | 0.59[b] | ND < 0.020[a,b] | ND < 0.024[a,b] | ND < 0.041[a,b] | ND < 0.028 |
| Total TCDF | 11.5 | 0.227 | 0.51 | 1.56 | 0.77 |
| 12378-PCDF | 0.78 | ND < 0.003[a] | 0.003 | 0.012 | <0.006 |
| 23478-PCDF | 0.51 | ND < 0.015[a] | ND < 0.016[a] | 0.051 | <0.027 |
| Total PCDF | 2.83 | 0.059[d] | 0.057 | 0.27 | 0.129 |
| 123478-HxCDF | 0.64 | 0.03 | 0.016 | 0.056 | 0.034 |
| 123678-HxCDF | 0.37 | ND < 0.012[a] | 0.011 | 0.029 | <0.052 |
| 234678-HxCDF | 0.03 | 0.015 | 0.011 | 0.032 | 0.019 |
| 123789-HxCDF | ND < 0.006 | ND < 0.001 | ND < 0.001 | ND < 0.0005 | ND < 0.001 |
| Total HxCDF | 2.91 | 0.077 | 0.089 | 0.24 | 0.135 |
| 1234678-HpCDF | ND < 0.0006 | ND < 0.001 | ND < 0.001 | 0.159 | <0.054 |
| 1234789-HpCDF | 0.156 | ND < 0.001 | ND < 0.0011 | ND < 0.008[a] | ND < 0.003 |
| Total HpCDF | 2.18 | 0.086 | 0.075 | 0.22 | 0.127 |
| OCDF | 0.88 | ND < 0.032[a] | ND < 0.041[a] | 0.071 | <0.048 |
| Total PCDF[c] | 20.3 | 0.480 | 0.771 | 2.36 | 1.20 |
| Total PCDD/PCDF | 28.5 | 0.794 | 1.14 | 3.51 | 1.82 |
| Surrogate recovery | | | | | |
| 13C12-TCDF | 98.3% | 97.7 | 104.0 | 92.8 | 98.1 |
| 37C1-TCDD | 94.0% | 101.1 | 103.2 | 95.3 | 99.9 |
| 13C12-HxCDF | 90.4% | 86.9 | 91.2 | 85.4 | 87.8 |

[a]EMPC (estimated maximum possible concentration.
[b]Confirmation result.
[c]Sum of total values for $Cl_4$ thru $Cl_8$ subtotals; ND values are included in totals.
[d]Ether interference.

The reduction in total 4 CDD emissions of both vapor and particulate phases was 87% average and 96% mean. Total PCDD reduction was 92%. The outlet toxic equivalent (Table 6) was reduced to 0.009 ng/Nm$^3$ (corr. 12% $CO_2$) for the PCDD of which 0.006 ng/Nm$^3$ represents the lowest detectable limit, and 0.075 ng/Nm$^3$ (corr. 12% $CO_2$) total toxic equivalent (PCDD + PCDF) of which 0.043 ng/Nm$^3$ represents the lowest detectable limit.

**Table 6. Average Toxic Equivalent Emissions at Stack by CA Dohs Method IV (ng/Nm$^3$ @ 12% $CO_2$)—Commerce Refuse-to-Energy Facility**

| Compound | Multiplying factor | Commerce stack average | Toxic equivalent |
|---|---|---|---|
| 2378 TCDD | 1.00 | 0.003[a] | 0.003 |
| Other TCDD | 0.00 | 0.109 | 0.000 |
| 12378 PCDD | 1.00 | 0.003[a] | 0.003 |
| Other PCDD | 0.00 | 0.048 | 0.000 |
| 123478 HxCDD | 0.03 | 0.003[b] | 0.0001 |
| 123678 HxCDD | 0.03 | 0.011[a] | 0.0003 |
| 123789 HxCDD | 0.03 | 0.011[b] | 0.0003 |
| Other HxCDD | 0.00 | 0.077 | 0.000 |
| 1234678 HpCDD | 0.03 | 0.081 | 0.0024 |
| Other HpCDD | 0.00 | 0.044[c] | 0.000 |
| OCDD | 0.00 | 0.225 | 0.000 |
| Total PCDD | | | 0.009 |
| 2378 TCDF | 1.00 | 0.028[a] | 0.028 |
| Other TCDF | 0.00 | 0.74 | 0.000 |
| 12378 PCDF | 1.00 | 0.006[c] | 0.006 |
| 23478 PCDF | 1.00 | 0.027[b] | 0.027 |
| Other PCDF | 0.00 | 0.096 | 0.000 |
| 123478 HxCDF | 0.03 | 0.034 | 0.0010 |
| 123678 HxCDF | 0.03 | 0.052[c] | 0.0016 |
| 234678 HxCDF | 0.03 | 0.019 | 0.0006 |
| 123789 HxCDF | 0.03 | 0.001[a] | 0.0000 |
| Other HxCDF | 0.00 | 0.029 | 0.000 |
| 1234678 HpCDF | 0.03 | 0.054[b] | 0.0016 |
| 1234789 HpCDF | 0.03 | 0.003[a] | 0.0001 |
| Other HpCDF | 0.00 | 0.070 | 0.000 |
| OCDF | 0.00 | 0.048 | 0.000 |
| Total PCDF | | | 0.066 |
| Total toxic equivalent (2,3,7,8 TCDD equivalents) | | | 0.075 |

[a]Not detected—the detection limit is used for these calculations.
[b]Not detected on two of three samples.
[c]Not detected on one of three samples.

These concentrations represent a reduction by the emission control system of the toxic equivalents from the boiler (Table 7) of 96% in the PCDD and the PCDD + PCDF toxic equivalents. The PAH emissions (Table 8) were reduced essentially to the nondetectable limit except for naphthalene. The chlorophenols were present in the boiler exhaust, representing the precursors to the catalytic formation of dioxins (Table 9). The simultaneous reductions of acid gases, heavy metals, and particulates are indicated in Table 10.

Significant in the reduction of the particulate emissions is the low fraction of pulmonary respirable particulates in the solid (first-half) particulate emissions, less than 18% below 2 $\mu$m. Thus, the emission of the particulates containing

**Table 7. Average Toxic Equivalent Measurements at Boiler Exit by CA DOHS Method IV ($ng/Nm^3$ @ 12% $CO_2$)**

| Compound | Multiplying factor | Commerce stack average | Toxic equivalent |
|---|---|---|---|
| 2378 TCDD | 1.00 | 0.097[a] | 0.097 |
| Other TCDD | 0.00 | 0.768 | 0 |
| 12378 PCDD | 1.00 | 0.097 | 0.097 |
| Other PCDD | 0.00 | 0.351 | 0 |
| 123478 HxCDD | 0.03 | 0.078 | 0.002 |
| 123678 HxCDD | 0.03 | 0.124 | 0.004 |
| 123789 HxCDD | 0.03 | 0.124 | 0.004 |
| Other HxCDD | 0.00 | 0.935 | 0 |
| 1234678 HpCDD | 0.03 | 1.067 | 0.032 |
| Other HpCDD | 0.00 | 1.09 | 0 |
| OCDD | 0.00 | 3.47 | 0 |
| Total PCDD | | 8.201 | 0.236 |
| 2378 TCDF | 1.00 | 0.59[a] | 0.59 |
| Other TCDF | 0.00 | 10.9 | 0 |
| 12378 PCDF | 1.00 | 0.78 | 0.78 |
| 23478 PCDF | 1.00 | 0.51 | 0.51 |
| Other PCDF | 0.00 | 1.54 | 0 |
| 123478 HxCDF | 0.03 | 0.64 | 0.019 |
| 123678 HxCDF | 0.03 | 0.37 | 0.011 |
| 234678 HxCDF | 0.03 | 0.03 | 0.001 |
| 123789 HxCDF | 0.03 | 0.0006 | 0 |
| Other HxCDF | 0.00 | 1.87 | 0 |
| 1234678 HpCDF | 0.03 | 0.0006[a] | 0 |
| 1234789 HpCDF | 0.03 | 0.156 | 0.005 |
| Other HpCDF | 0.00 | 2.02 | 0 |
| OCDF | 0.00 | 0.88 | 0 |
| Total PCDF | | | 1.916 |
| Total toxic equivalent (2,3,7,8 TCDD equivalents) | | | 2.152 |

[a]Not detected—the detection limit is used for these calculations.

the preponderance of the heavy metals and dioxins was reduced to 0.55 mg/ $Nm^3$. This behavior was also evident in the same system at Tsushima, Japan, where the pulmonary respirable particulates was reduced to zero.[33] It is implied, therefore, that the dry venturi effect on filter cake thickness achieves essentially total removal of particulate during filtration and discharges to the atmosphere only during the bag cleaning operation. The heavy metals, as a result of the removal of the fine particulates, are generally reduced to below the detectable limit. The HCl, $SO_2$, HF, $SO_3$ removal efficiency was of the order of 98 to 99%, with the concentration of emissions below fog-forming potential.

**Table 8. PAH Emissions**

| Location test species | Concentration ($ng/Nm^3$ @ 12% $CO_2$) | | | | |
|---|---|---|---|---|---|
| | Boiler Exhaust | Stack | | | |
| | 17 | 15 | 16 | 17 | Average |
| Naphthalene | 3.0[a] | 0.86[a] | 2.15[a] | 0.76[a] | 1.26[a] |
| 2-Methylnaphthalene | ND < 0.64 | 0.29 | ND < 0.29 | ND < 0.29 | 0.29 |
| 2-Chloronaphthalene | ND < 0.64 | ND < 0.24 | ND < 0.29 | ND < 0.29 | ND < 0.27 |
| Acenaphthalene | ND < 0.64 | ND < 0.24 | ND < 0.29 | ND < 0.29 | ND < 0.27 |
| Acenaphthene | ND < 0.64 | ND < 0.24 | ND < 0.29 | ND < 0.29 | ND < 0.27 |
| Fluorene | ND < 0.64 | ND < 0.24 | ND < 0.29 | ND < 0.29 | ND < 0.27 |
| Phenanthrene | ND < 0.64 | ND < 0.24 | ND < 0.29 | ND < 0.29 | ND < 0.27 |
| Anthracene | ND < 0.64 | ND < 0.24 | ND < 0.29 | ND < 0.29 | ND < 0.27 |
| Fluoranthese | ND < 0.64 | ND < 0.24 | ND < 0.29 | ND < 0.29 | ND < 0.27 |
| Pyrene | ND < 0.64 | ND < 0.24 | ND < 0.29 | ND < 0.29 | ND < 0.27 |
| Benzo-A-anthracene | ND < 0.64 | ND < 0.24 | ND < 0.29 | ND < 0.29 | ND < 0.27 |
| Chrysene | ND < 0.64 | ND < 0.24 | ND < 0.29 | ND < 0.29 | ND < 0.27 |
| Benzo-B-fluoranthese | ND < 0.64 | ND < 0.24 | ND < 0.29 | ND < 0.29 | ND < 0.27 |
| Benzo-K-anthracene | ND < 0.64 | ND < 0.24 | ND < 0.29 | ND < 0.29 | ND < 0.27 |
| Benzo-A-pyrene | ND < 0.64 | ND < 0.24 | ND < 0.29 | ND < 0.29 | ND < 0.27 |
| Indeno (1,2,3-cd) pyrene | ND < 0.64 | ND < 0.24 | ND < 0.29 | ND < 0.29 | ND < 0.27 |
| Dibenz (a,h) anthracene | ND < 0.64 | ND < 0.24 | ND < 0.29 | ND < 0.29 | ND < 0.27 |
| Benzo (g,h,i) perylene | ND < 0.64 | ND < 0.24 | ND < 0.29 | ND < 0.29 | ND < 0.27 |

[a]Blank corrected (1.2 $\mu$g napthalene in field blank).

**Table 9. Chlorobenzene, Chlorophenols, and PCBs from Commerce Refuse-to-Energy Facility**

| Location test | Concentration ($ng/Nm^3$ @ 12% $CO_2$) | | | | |
|---|---|---|---|---|---|
| | Boiler Exhaust | Stack | | | |
| | 17 | 15 | 16 | 17 | Average |
| DiCl Benzene | 0.65 | 0.44 | 0.18 | 0.15 | 0.26 |
| TriCl Benzene | 1.18 | 0.47 | 0.65 | 0.17 | 0.43 |
| TetraCl Benzene | 0.5 | 0.12 | 0.18 | 0.09 | 0.13 |
| PentaCl Benzene | 0.3 | ND < 0.3 | ND < 0.3 | ND < 0.3 | ND < 0.3 |
| HexaCl Benzene | ND < 0.6 | ND < 0.3 | ND < 0.3 | ND < 0.3 | ND < 0.3 |
| CL Phenol | ND < 0.6 | ND < 0.3 | ND < 0.3 | ND < 0.3 | ND < 0.3 |
| DiCl Phenol | ND < 0.6 | ND < 0.3 | ND < 0.3 | ND < 0.3 | ND < 0.3 |
| TriCl Phenolne | 0.5 | ND < 0.3 | 0.11 | 0.83 | <0.41 |
| TetraCl Phenol | 0.13 | ND < 0.3 | ND < 0.3 | <0.05 | <0.22 |
| PentaCl Phenol | ND < 0.6 | ND < 0.3 | ND < 0.3 | ND < 0.3 | ND < 0.3 |
| MonoCl PCB | ND < 0.6 | ND < 0.3 | ND < 0.3 | ND < 0.3 | ND < 0.3 |
| DiCl PCB | ND < 0.6 | ND < 0.3 | ND < 0.3 | ND < 0.3 | ND < 0.3 |
| TriCl PCB | ND < 0.6 | ND < 0.3 | ND < 0.3 | ND < 0.3 | ND < 0.3 |
| TetraCl PCB | ND < 0.6 | ND < 0.3 | ND < 0.3 | ND < 0.3 | ND < 0.3 |
| PentaCl PCB | ND < 0.6 | ND < 0.3 | ND < 0.3 | ND < 0.3 | ND < 0.3 |
| HexaCl PCB | ND < 0.6 | ND < 0.3 | ND < 0.3 | ND < 0.3 | ND < 0.3 |
| HeptaCl PCB | ND < 0.6 | ND < 0.3 | ND < 0.3 | ND < 0.3 | ND < 0.3 |
| OctaCl PCB | ND < 0.6 | ND < 0.3 | ND < 0.3 | ND < 0.3 | ND < 0.3 |
| NonaCl PCB | ND < 0.6 | ND < 0.3 | ND < 0.3 | ND < 0.3 | ND < 0.3 |
| DecaCl PCB | ND < 0.6 | ND < 0.3 | ND < 0.3 | ND < 0.3 | ND < 0.3 |

**Table 10. Acid Gas and Particulate Emissions**

| Total | | Inlet | Outlet |
|---|---|---|---|
| Particulate | $mg/Nm^3$ | 3801 | 7.3 |
| | $mg/Nm^3$ (12% $CO_2$) | 4602 | 9.8 |
| | kg/hr | 284 | 0.65 |
| Percent distribution in solid particulate | >10 μm | | 71.4 |
| | <10 μm | | 28.6 |
| | <2 μm | | 17.4 |
| Metals | kg/hr | | |
| | Antimony | 0.08—0.095 | ND < 0.002 |
| | Arsenic | 0.01—0.012 | ND < $1.2 \times 10^{-4}$ |
| | Beryllium | $3.4—4.0 \times 10^{-4}$ | ND < $3.9 \times 10^{-5}$ |
| | Cadnium | 0.12—0.14 | ND < $1.8 \times 10^{-4}$ |
| | Chromium | 0.033—0.039 | ND < $3.9 \times 10^{-4}$ |
| | Copper | 0.25—0.30 | ND < $2.4 \times 10^{-4}$ |
| | Lead | 2.21—2.62 | $1.1—2.4 \times 10^{-4}$ |
| | Mercury | 0.02—0.024 | 0.015—0.033 |
| | Nickel | 0.031—0.035 | ND < $1.6 \times 10^{-3}$ |
| | Selenium | $7.7—8.6 \times 10^{-4}$ | ND < $8.2 \times 10^{-5}$ |
| | Silver | $6.4—7.3 \times 10^{-3}$ | ND < $2 \times 10^{-4}$ |
| | Thallium | ND < $5.5 \times 10^{-4}$ | ND < $1.6 \times 10^{-4}$ |
| | Zinc | 3.36—3.95 | < $4.5 \times 10^{-3}$ |
| $SO_2$ | ppm | 213 | 0.98 |
| | ppm at 3% $O_2$ | 300 | 1.56 |
| $SO_3$ | ppm | 37 | 0.09 |
| | ppm at 3% $O_2$ | 52 | 0.15 |
| $SO_x$ ($SO_2$ + $SO_3$) | kg/hr | 50.4 | 0.26 |
| HCl | | | |
| | ppm | 731 | 7.15 |
| | ppm at 3% $O_2$ | 1152 | 11.35 |
| | kg/hr | 84.4 | 1.0 |
| HF | ppm | 13 | 0.031 |
| | ppm @ 3% $O_2$ | 20 | 0.049 |
| | kg/hr | 0.86 | $2.3 \times 10^{-3}$ |

## CONCLUSIONS

Dioxins are emitted at significant levels from incinerator-boiler combinations in spite of combustion operation at temperature-time exposure recommended for thermal reduction of dioxins. Dioxins can be removed from the flue gas stream to toxic equivalent levels of 0.009 $ng/Nm^3$ inclusive of the nondetectable limit, of the order of 96% reduction from the boiler emissions levels. The mechanism of removal is chemisorption on the flyash and silica-alumina matrix particulate introduced into the gas stream. The reduction of dioxins occurs preferentially in the region of 120°C. The estimated heat of chemisorption is 22,000 cal/mol. The chemisorptive process occurs by passage of the gas through a fixed-bed filter cake inherent in the total pollution control process, thus not requiring additional equipment.

## NOMENCLATURE

$a_o$ – initial molecular number density for competitively adsorbed species for Np particles/$cm^3$
$a_d$ – for dioxins
$d_p$ – distance between particles
Dv – diffusivity
E – energy of activation for adsorption
M – molecular weight
MWF – moving wave front number
Np – number density of flyash or solid particulate
P – vapor pressure or pseudo
t – time
T – temperature °C or K
$V_A$, $V_B$ – molecular diameter
Zp – collision frequency of gas-phase molecules per Np particles
$\triangle H$ – heat of vaporization, physical adsorption, or chemisorption
$r_p$ – radius of flyash particle
$r_g$ – radius of polychlorinated phenol
$r_d$ – radius of dioxins

## REFERENCES


1. Ballschmitter, K. et al. *Chemosphere* 14(6–7):851 (1985).
2. Berlincioni, M., and A. Domenico. *ES&T* 21(11):1063 (1987).
3. Cavallaro, A. et al. *Chemosphere* 11(8):859 (1982).
4. Choudry, G. G., K. Olie, and O. Hutzinger. *Mechanisms in The Formation of Chlorinated Compounds* (Elmsford, NY: Pergamon Press, Inc. 1982).
5. Clarke, M. *Waste Age* (November 1987), pp. 102–117.
6. Cooper Engineers. Draft Report – Inv. of Tsushima Inc. (1983).
7. Concord Scientific Corporation. "National Inc. Testing and Evaluation Program," Environment Canada Report EPS 3/WP/1 (1985).
8. Dickson, L. C. et al. *Int. J. Environ. Anal. Chem.* 24(1):55 (1986).
9. Duvall, D. S. and W. A. Ruby. "Laboratory Evaluation of High Temperature Destruction of Polychlorinated Biphenyls and Related Comp.," U.S. Department of Commerce, NTIS, PB-279139 (1977).
10. Eiceman, G. A., and H. O. Rghei. *Chemosphere* 11(9):833 (1982).
11. Getter, R. "Dioxin Emissions and Regulations for Modern Resource Recovery Facilities," APCA Atlantic City Conference 1987.
12. Gizzi, F. et al. *Chemosphere* 11(6):577 (1982).
13. Husselriss, F. "Relationship Between Municipal Refuse Combustion Conditions and Trace Organic Emissions," *Proc. APCA* (June 1985).
14. Hutzinger, O. "Recycling International," Umwelteknik, pp. 624–628.
15. Nielsen, K. et al. *Chemosphere* 15(9–12):1247 (1986).
16. Nittrod, A., and K. Ballschmitter. *Chemosphere* 15(9–12):1225 (1986).
17. Oscik, J. *Adsorption*, (New York: Halsted Press, 1982), p. 192.
18. Olie, K. Personal communication (1983).

19. Pastoretti, G. et al. *Acqua Aria* 3:271 (1986).
20. Politski, G. R. et al. *Chemosphere* 11(12):1217 (1982).
21. Rordorf, B. F. *Chemosphere* 14(6–7):885 (1985).
22. Sawyer, R. F. "Formation and Destruction of Pollutants in Combustion Proceedings," in *18th Symposium of Combustion* (Pittsburgh: The Combustion Inst., 1981).
23. Schaub, W. M. "Containments of Dioxin Emissions from Refuse Fired Proc Units," NBSIR 84-2872, NTIS PB 84-217090 (1984).
24. Schroy, J. et al. "Aquatic Toxicology Hazard Assessment," ASTM Special Technical Publication 891 (1985), pp. 409–421.
25. Smith, J. M. *Chemical Engineering Kinetics* (New York: McGraw-Hill Book Company 1956).
26. Teller, A. J., and J. M. Lauber. *Proceedings 76th Annual Meeting of the APCA* (1984).
27. Teller, A. J. "Baghouse Chemisorptive Mechanisms," Teller Environment Systems (1984).
28. Teller, A. J. U.S. Patent 4,319,890, 1982; and 3,957,464, 1976.
29. Teller, A. J. U.S. Patent 4,293,524, 1981; and 4,375,455, 1983.
30. Tsang, W. Environmental Quality Assurance With Respect to Organic Emissions," DOE/CE/30790-TI (1986).
31. U.S. Environmental Protection Agency. "Incineration of PCB," U.S. EPA Deer Park, TX (1981).
32. Visall, J. *J. Air Poll. Control Agency* 37(12):1451 (1987).
33. Volskowk, *Muellerei Abfull* 17(4):122 (1985).
34. Zoller, W. and K. Ballschmitter. *Chemosphere* 15(9–12):2129 (1986).

## CHAPTER 20

# Control of Air Emissions of Total Reduced Sulfur Compounds from Combustion Processes in Pulp and Paper Plants

**J. Marson, V. M. Ozvacic, G. Wong, and A. Melanson**

## INTRODUCTION

The air emissions of reduced sulfur compounds are responsible for the characteristic odor at kraft pulp mills. The odorous compounds most commonly encountered in these emissions are hydrogen sulfide, methyl mercaptan, dimethyl sulfide, and dimethyl disulfide, collectively called TRS.

The U.S. Environmental Protection Agency published TRS emission data from 80 kraft mills in 1973.[1] Two combustion processes – recovery boilers and lime kilns – were identified as the major sources of TRS air emissions in kraft mills. The TRS emission factors for these sources spanned over a wide range. Also, there were indications that TRS emissions at the same source may change considerably in time.

The Ontario Ministry of the Environment initiated a program to measure TRS air emissions from kraft processes in the 1980s. Field studies at six out of the total of nine kraft pulp mills in Ontario were conducted by the personnel of the Air Resources Branch in the summers of 1984 and 1985. The objective of the program was to test measurement methods for TRS in air emissions, to audit the existing TRS stack emission monitoring systems in the industry, and to assess the sources of TRS emissions in kraft mills. This program produced about 500 hr of simultaneous stack emission monitoring data for TRS, oxygen ($O_2$), carbon dioxide ($CO_2$), and carbon monoxide (CO) from combustion sources at kraft mills. In the case of recovery boilers, monitoring was performed at the furnace exhausts downstream of scrubbers and direct contact evaporators. At lime kilns, the monitoring data were obtained from downstream scrubber locations, except at mill "E," where the feed end of the kiln was also monitored.

The purpose of this chapter is to review TRS air emission data from combustion sources in kraft processes, both published by others and measured in this program; to present the measured data; to interpret these data in terms of

**Table 1. Average TRS Concentrations at Recovery Boilers (ppm, dry basis)[a]**

| Mill | Furnace | DCE/scrubber[b] | Observations |
|---|---|---|---|
| A | — | 34 | Overloaded, TRS monitor |
| C | — | 400 | Scrubber problems |
| D | — | 139 | Overloaded, TRS monitor |
| E | 4 | 23 | Conventional DCE |
| F | 59 | — | "Low-odor" type |
| | 488 | — | Low $O_2$ level |

[a]ppm = parts per million by volume.
[b]DCE = direct contact evaporator.

mechanism of TRS release; and to assess the emission reduction achievable by improved combustion control.

## RECOVERY BOILER AIR EMISSIONS

The recovery boiler is a vital part of the kraft process and potentially its largest TRS source. The boiler recovers chemicals used in the digestion of wood chips and burns organic compounds in the spent digestion liquor. After digestion, the fibers and the spent liquor are separated. The spent liquor is concentrated to about 65% solids in a series of multiple-effect evaporators. The concentrated spent liquor is sprayed into the furnace of the recovery boiler where organic compounds are burned, forming a smelt of sodium sulfide and sodium carbonate. Combustion air is injected into the furnace at two or three levels. Combustion in the lower level is maintained at substoichiometric conditions and is completed at higher levels by adding sufficient air. The smelt flows from the furnace into a tank, where it is quenched and dissolved in water.

Most recovery boilers built before 1970 are equipped with direct contact evaporators. A direct contact evaporator enables a contact between the concentrated spent liquor and the boiler's exhaust gases, thus allowing additional heat recovery and fewer steam evaporators. This contact may result in TRS release by thermal stripping and a reaction between acidic flue gases and sodium sulfide. TRS release and the resulting TRS air emissions can be reduced by oxidizing the liquor with air, which converts sodium sulfide into a more stable sodium thiosulfate.

Average TRS levels measured at the recovery boilers in this program are listed in Table 1. Taking as a reference the typical emission level of less than 20 parts per million (ppm) by volume of TRS reported for newly designed boiler units, the values recorded in the program were high and varied from site to site. However, the magnitude of the measured TRS levels seemed to depend more on the operating conditions than on the type, make, or age of the boilers.

The boiler at mill "A" was an old overloaded unit with a direct contact evaporator, generating only 34 ppm of TRS in the air emissions. Such low concentrations were partially a result of adjustments made to the boiler operation on the basis of readings from the TRS emission monitor installed in the stack. At the time of the survey, the boiler at mill "C" had operational problems which were reflected in high TRS levels in the air emissions—400 ppm of TRS on the average. The boiler had been retrofitted with an alkaline scrubber for the control of emissions from the furnace and the direct contact evaporator. However, the packing of the scrubber deteriorated, adversely affecting the TRS scrubber removal efficiency during the measurements.

The boiler at mill "D" was operated approximately 34% above the rated capacity. The TRS levels in the air emissions were almost constant at 139 ppm on the average. The readings of a continuous TRS monitor were used to avoid TRS surges in emissions from this boiler. TRS air emissions from the boiler at mill "E" were typical of a conventional boiler unit equipped with a direct contact evaporator and black liquor oxidation. The furnace exhaust gas contained 4 ppm of TRS on the average, which increased to 23 ppm after passing through the evaporator. The exhaust gas from the smelt dissolving tank also contributed to this increase in TRS concentrations.

Mill "F" had two modern "low-odor-type" of boilers. However, one of the boilers was operated at full capacity without excess combustion air, producing high TRS levels of 488 ppm. It was interesting to note that the boiler was not equipped with a TRS monitor and that the oxygen levels in the furnace were kept low to prevent corrosion and pluggage of the air preheater. TRS emissions from the other boiler averaged 59 ppm.

Emissions from a given boiler varied considerably and rapidly with time. Figure 1 illustrates typical TRS levels monitored at the mill "A" boiler: each data point represents an 1-min average reading. The boiler had a direct contact evaporator; hence, the shown TRS levels were the result of combined emissions from the furnace and the evaporator. Since evaporator emissions depend on black liquor oxidation and are relatively constant for a given batch of black liquor, the TRS surges shown on the graph must have been caused by fluctuations in the furnace operation. The TRS levels varied by up to 10-fold in 1 min, the time corresponding to the TRS monitor's response time. The actual rate of change may have been more rapid, possibly indicating that TRS emissions were produced by fast gas-phase reactions. Carbon monoxide fluctuated similarly and simultaneously with TRS.

Figure 2 shows the CO and TRS concentrations in the air emissions measured at the mill "E" boiler when the overfire combustion air was deliberately throttled to observe the effect of this change on TRS levels. Monitoring was performed at the exit of the economizer where the $O_2$ levels varied from 0.2 to 2.4%. A correlation between the TRS and CO levels is obvious from Figure 2 and not surprising since both of these compounds are the products of incomplete combustion in the lower part of the furnace. Their relative concentrations could be expected to behave similarly as a result of variation in the combustion efficiency of the furnace.

A correlation between the CO and TRS levels may be used to control TRS emissions from recovery boilers indirectly by monitoring and controlling the

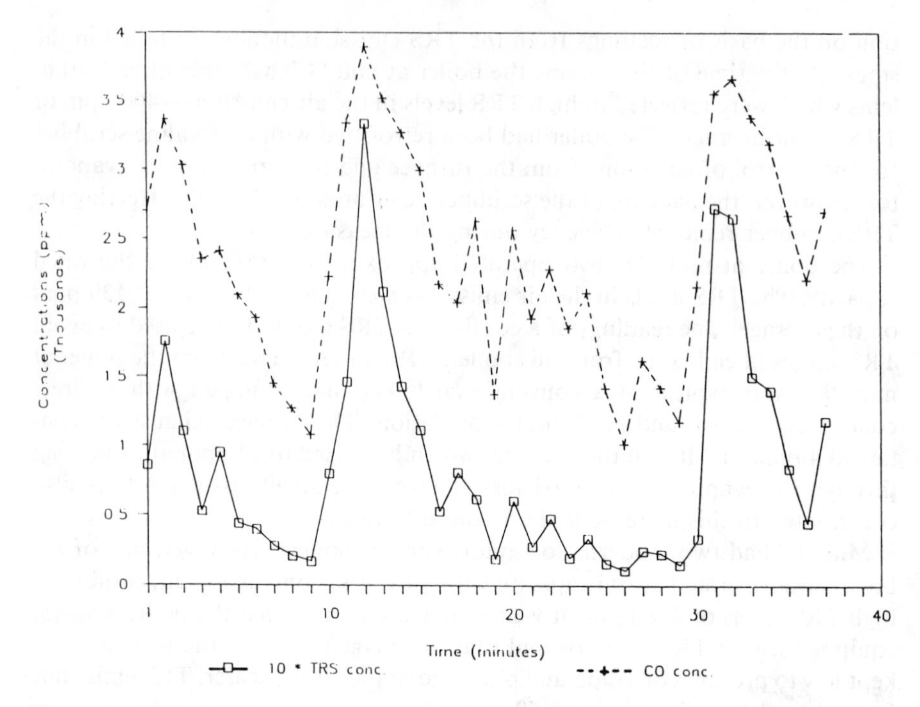


**Figure 1.** TRS and CO concentrations after a direct contact evaporator at cite "A".

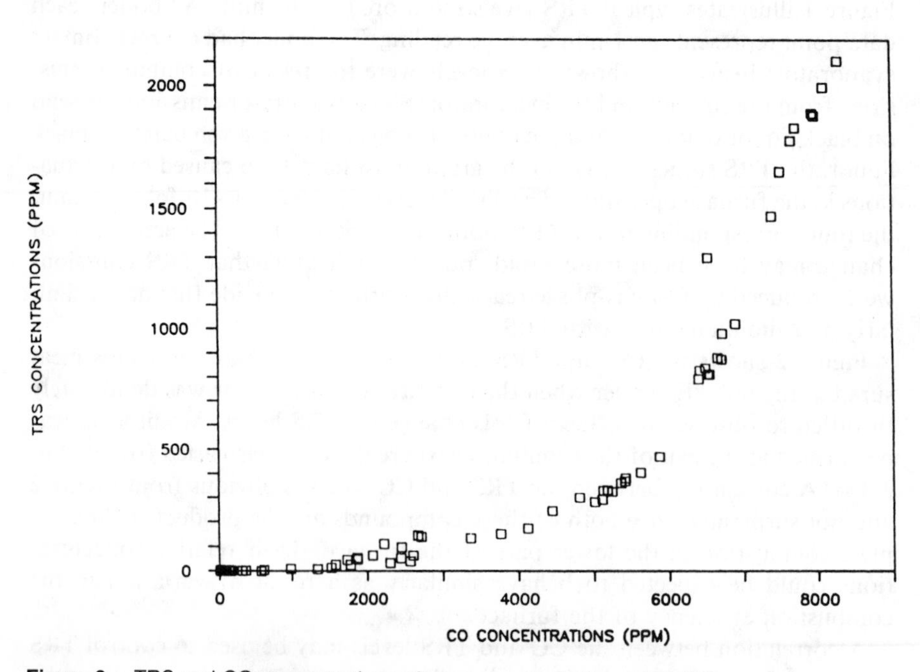


**Figure 2.** TRS and CO concentrations after a recovery boiler at Mill "E".

**Table 2. Average TRS Concentrations at Lime Kilns (ppm, dry basis)[a]**

| Mill | Kiln | Scrubber | Observations |
|---|---|---|---|
| B | — | 8[b] | Intermittent low $O_2$ |
| C | — | 100 | Contaminated scrubber |
| D | — | 18 | |
| E | 24 | 30 | |
| F | — | 25[b] | Intermittent low $O_2$ |

[a]ppm = parts per million by volume.
[b]Upsets excluded.

CO levels in the furnace exhaust gas. CO monitoring systems are simpler to operate and easier to maintain than TRS monitors. The use of CO monitors for TRS control has also been proposed by others.[2,3] This study confirms the basic findings of the two referenced papers and provides additional experimental evidence of TRS/CO correlation at kraft combustion sources obtained with modern instruments of fast response.

The analysis of data from all sites (A to F) in this study indicates that it is necessary to establish the actual relationship between CO and TRS levels in air emissions at each site in order to effectively use the measurements of CO for TRS control. In addition, the CO/TRS correlation should be re-established periodically at each site.

## LIME KILN AIR EMISSIONS

The solution formed in the smelt dissolving tank is treated with calcium oxide, resulting in a precipitate of calcium carbonate and a solution of sodium sulfide and sodium hydroxide. The solution is recycled to the digester while the calcium carbonate mud is rinsed, dewatered, and converted to calcium oxide in a rotary kiln. The kiln is also used to incinerate odorous gases from various sources in the mill. Venturi scrubbers are typically used to control the air emissions of particulate matter from these kilns.

The average TRS levels measured at lime kilns are listed in Table 2. They were generally lower than at the recovery boilers and did not exceed 30 ppm except for mill "C," where the increased TRS levels were due to the presence of contaminated condensate in the scrubber—the condensate contained TRS compounds which were stripped off upon contact with the kiln exhaust gas.

All lime kilns were operated with low excess combustion air to increase their thermal efficiencies. Some kilns were equipped with oxygen monitors, but none with CO or TRS monitors. The intermittent periods of low oxygen recorded at sites B, C, and F coincided with high TRS emissions.

The effect of small variations in oxygen levels on the TRS levels in the lime kiln emissions at mill "C" is shown in Figure 3. The TRS levels at the site exceeded 80 ppm as a result of contaminated scrubber water, as mentioned earlier. The measured $O_2$ levels averaged 4%, which represents a reasonable level of excess combustion air. However, this level of $O_2$ included oxygen from

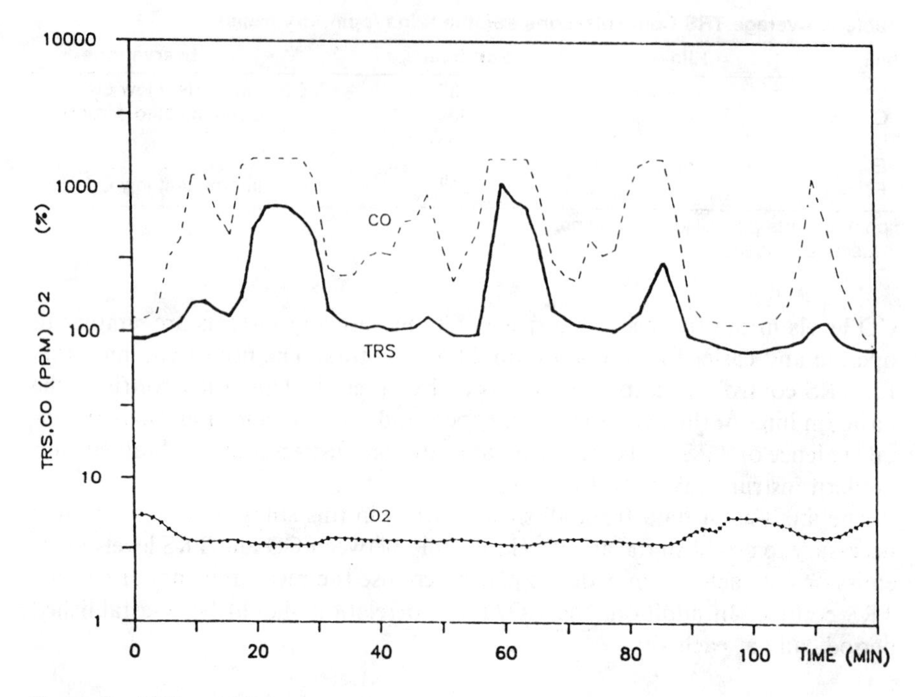


**Figure 3.** TRS and CO concentrations during an upset condition in a lime kiln at Mill "C".

the combustion gas as well as that from infiltrated ambient air. This infiltration was responsible for the appearance of very small decreases in $O_2$ concentrations measured at the time when large CO and TRS surges were recorded, as shown in Figure 3. The measured $O_2$ concentrations did not represent the actual level of excess combustion air, illustrating a common shortfall of using $O_2$ monitoring for combustion control. On the other hand, CO monitoring is more sensitive and less affected by ambient air infiltration.

Figure 4 shows a correlation between the CO and TRS concentrations in the air emissions from the kiln at mill "D," obtained by deliberately curtailing combustion air in progressive steps over a period of approximately 2 hr. As in the case of recovery boilers, higher TRS and CO levels were measured when the excess air in combustion was low.

In a typical gas-fueled kiln, such as at mill "D", the main sulfur input is the residual sodium sulfide in the calcium carbonate mud fed into the kiln. The effect of mud drying on TRS emissions has been the subject of various experimental studies carried out by the National Council of the Paper Industry for Air and Stream Improvement. However, those studies were rather inconclusive. Our data suggest that combustion conditions, and specifically the availability of combustion air, may be more important in the control of TRS emissions from lime kilns than mud drying.

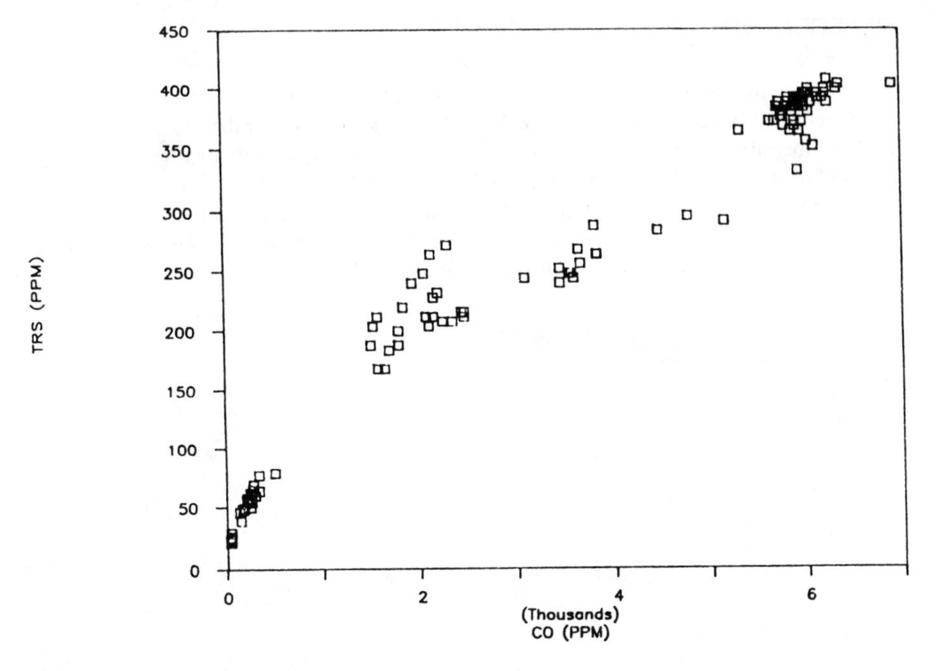


**Figure 4.** TRS—CO correlation in a lime kiln at Mill "D".

## CONCLUSIONS

In summary, this study confirms previous findings regarding the air emissions of TRS from combustion processes in kraft pulp mills. Recovery boiler emissions originate in the furnaces as well as in the direct contact evaporators. Lime kilns and their scrubbers are other potentially significant TRS sources in those mills. Incomplete combustion is the predominant cause of air emission of TRS from combustion sources. Proper operation of recovery furnaces and lime kilns is an essential part of an effective abatement strategy for these sources.

CO and TRS levels in the air emissions from combustion sources showed a definite correlation. Monitoring of TRS and/or CO concentrations in the stack gases of recovery boilers and lime kilns can be an effective tool in the control of TRS air emissions from kraft pulp mills.

## ACKNOWLEDGMENTS

The authors express their thanks to the staff of the mills included in this study as well as the staff of the regional offices of the Ministry of the Environment for their cooperation. Thanks are also extended to Dr. R. S. Rowbottom for his helpful comments in the preparation of this paper.

## REFERENCES

1. U.S. Environmental Protection Agency. "Atmospheric Emissions from the Pulp and Paper Manufacturing Industry," EPA–450/1-73-002 (September 1973).
2. Chamberlain, R. et al. "Eliminate Recovery Furnace $H_2S$ Emissions by Controlling CO," *Pulp Paper Mag.* (Canada), 79(2):44–48 (1978).
3. Bostrom, G. C., P. Grennfelt, and I. Malmstrom. "Odour Control at Recovery Furnaces by Monitoring Carbon Monoxide," in *Proceedings of the International Conference on Recovery of Pulping Chemicals* (1981).

# CHAPTER 21

# Control of Acid Rain Precursors

**H. Whaley, E. J. Anthony, and G. K. Lee**

## ABSTRACT

Research sponsored under an International Energy Agency Agreement between Canada, Denmark, and Sweden has shown that $NO_x$ and $SO_x$ emissions from pulverized coal flames can be reduced significantly by staged combustion concepts. An evaluation of 45 coals determined that much of the fuel nitrogen in the volatile matter can be transformed to $N_2$ instead of NO. The conversion of fuel nitrogen to NO was from 30 to 45% in conventional flames, but only 7 to 14% in staged flames. Nitrogen retained in the char showed about 20% conversion to NO and appeared to be relatively independent of local oxygen concentration. Reduction in sulfur emissions of 50% by sorbent injection into tertiary combustion air was achievable with a Ca/S molar ratio of 2. These research results have been incorporated into two field demonstrations involving retrofits of staged combustion systems in a front-wall-fired and a tangentially-fired boiler.

An alternative to conventional boilers is provided by fluidized bed combustion systems. These units operate at temperatures below that at which the nitrogen in the air is fixed and are inherently low producers of $NO_x$. Considerable work has been carried out in Canada under the auspices of CANMET/EMR on a number of pilot- and full-scale fluidized bed combustors in Canada burning a wide range of Canadian fuels. The results suggest that typically only 10 to 25% of the fuel nitrogen is actually converted to $NO_x$, which results in low $NO_x$ emissions (typically less than 258 ng/J), while removing 90% of the $SO_2$ produced when burning high sulphur fuels. By adroit use of secondary air, $NO_x$ emissions can be kept to even lower levels in staged bubbling bed combustors, while circulating fluidized bed combustors achieve staging by the nature of their design, and are also associated with extremely low $NO_x$ emissions.

## INTRODUCTION

One of the major concerns associated with the expanded use of coal for heat and electricity is the emission of acid rain precursors, $NO_x$ and $SO_x$, to the atmosphere. In 1980, North American utility boilers most of which are coal fired, emitted over 5.9 million t of $NO_x$ and 16.6 million t of $SO_x$ with Canadian utility sources accounting for about 5% of both the $NO_x$ and $SO_x$ emissions.

Current developments on abatement technology indicate that these pollutants can be controlled by flue gas scrubbing, by combustion system modifications, or by the use of novel combustion technologies, such as fluidized bed combustion (FBC). Flue gas scrubbing generally has proven energy intensive, expensive, and complex. For this reason, increased emphasis has been placed on combustion modification or the development of novel combustion technologies. Combustion modification now appears to be the most cost-effective method for controlling $NO_x$ and/or $SO_x$ from pulverized coal-fired boilers.

This paper reviews some of the fundamental aspects of $NO_x$ and $SO_x$ control in flames and describes two Canadian initiatives to suppress simultaneously the generation of $NO_x$ by staged-burner aerodynamics and sorbent injection, respectively. The results of both bubbling and circulating fluidized bed combustion pilot- and full-scale work with a wide range of Canadian fuels are also described.

## THEORETICAL CONSIDERATIONS

Flame-generated $NO_x$ is produced from two sources: (1) thermal $NO_x$ formed by high-temperature reaction of atmospheric oxygen and nitrogen, and (2) fuel $NO_x$ formed by the reaction of fuel nitrogen with available oxygen during combustion.

Thermal $NO_x$, although a major source of the total $NO_x$ generated when low-nitrogen fuels are burned, is a minor source of $NO_x$ when high-nitrogen fuels are burned in suspension-firing systems or when FBC technology is employed. Typical high-nitrogen fuels include residual oil, coke, and many coals. Thermal $NO_x$ forms fairly slowly and is favored by flame temperatures above 1400°C, low rates of heat extraction, high excess air levels (with intense turbulent air/fuel mixing), and high volumetric heat release rates.

Fuel $NO_x$, on the other hand, appears to be only weakly dependent on temperature and forms very rapidly. Formation is promoted by fuel-rich conditions during pyrolysis, a large evolution of volatile nitrogen species, and low rates of heat extraction.

Burners, or physical staging of combustion air, are being widely studied as a cost-effective means of minimizing the oxidation of fuel nitrogen. One popular technique has been to design burners, combustion chambers, or combinations of both with a fuel-rich primary stage followed by controlled additions of secondary air and injections of tertiary air to complete combustion slowly as shown in Figure 1.

During coal combustion, upwards of 70% of the total $NO_x$ emissions is derived from fuel nitrogen which forms $N_2$, HCN, NO, and $NH_3$ during devo-

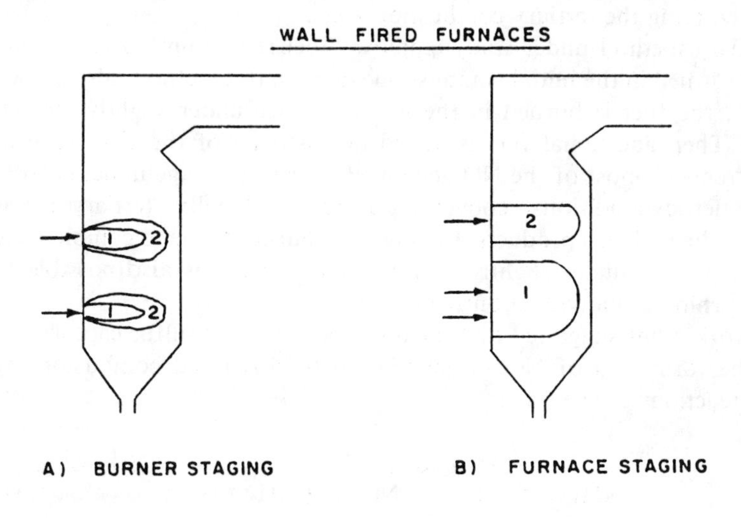


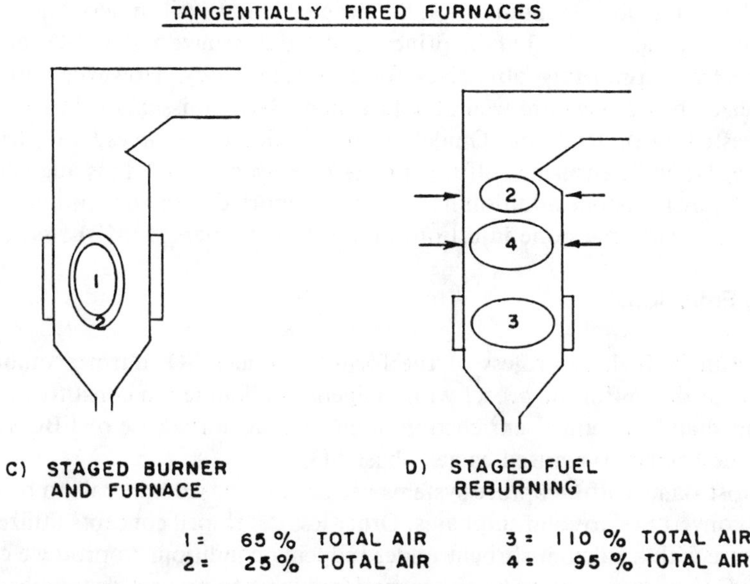


**Figure 1.** Staged combustion: burner, furnace, and fuel reburning.

latilization and substoichiometric combustion. About 20 to 35% of the fuel nitrogen can be retained in the char, but only about 20% of this char nitrogen is converted to $NO_x$ regardless of stoichiometry, i.e., whether conditions are fuel rich or fuel lean, or mixing intensity. Good char burnout can therefore be achieved under high excess air conditions without increasing significantly the $NO_x$ levels entering the tertiary combustion zone.

A secondary method under study employs "fuel reburning" to chemically reduce NO formed in the initial combustion stage. In this method, also shown in Figure 1, the fuel is burned in the lower furnace under slightly air-rich conditions. Then additional fuel is added downstream of the first zone to reduce or "reburn" most of the NO generated in the first stage under slightly substoichiometric conditions, converting it to $N_2$. Finally, tertiary air is injected into the fuel-rich products, leaving the reburning zone to complete the combustion process under slightly air-rich conditions. It is also possible to employ reburning in individual burners.[1]

Recent work[2–4] has suggested that the limestone used for sulfur capture can augment the formation of $NO_x$ in circulating fluidized bed combustors by catalyzing reactions of the type:

$$NH_3 + 5/4O_2 \overset{CaO}{=} NO + 3/2H_2O \quad (1)$$

This effect has not been reported in normal bubbling bed units, although it had been observed with a two-staged unit.[5] For circulating fluidized beds, however, it appears that changing the Ca/S molar ratio from 0 to 3 can cause the $NO_x$ emissions to double.[3,4] This process is not well understood and will require further study, but in principle it would suggest that $SO_x$ and $NO_x$ control are competing objectives for this technology. However, circulating fluidized bed boilers are well able to achieve $NO_x$ emissions below the limits suggested by the National Guidelines for Stationary Sources,[6] i.e., less than 258 ng/J, while ensuring sulfur captures of 90% or more. This augmentation of $NO_x$ by limestone addition has not been reported with conventional boilers equipped with limestone injection, multistaged burners (LIMB).

## $SO_x$ Emissions

Sulfur in fuel, regardless of the form, generates $SO_x$ during combustion. Some of this sulfur may react with indigenous alkaline ash constituents, or be trapped and sometimes enriched in unburnt char in the case of FBC systems, but the balance is emitted as gas-phase $SO_x$.

Most flame sulfur-capture systems use calcium-based minerals to react with and convert $SO_x$ to solid sulphates. Other less developed concepts utilize sulfur capture with a calcium sorbent under reducing conditions to produce calcium sulfide, which must then be either rendered inert to prevent decomposition of the sulfide, or oxidized to sulfate during the final combustion stage under excess air conditions. Therefore, the effectiveness of sulfur neutralization, with simultaneous $NO_x$ reduction, will depend strongly on sorbent injection into a furnace region with the desired stoichiometry, temperature, and residence time.

In FBC systems, sulfur is trapped by means of either a stationary or entrained bed of calcined limestone or dolomite, depending on whether a bubbling or circulating bed system is employed. The primary limitation with this technology is the relatively low sorbent utilization which is normally in the order of 30%. Recent results suggest that more advanced fluidized bed designs may be capable of achieving utilizations approaching 50%. Since this technology is likely to be used with high sulfur fuels ($S \geq 3\%$), optimizing the efficiency of the sorbent capture process to reduce both sorbent use and disposal costs for the waste residues is a major consideration to the economics of the process. Small amounts of sulfides have also been found in the solid residues produced by these processes (typically $S \leq 0.5\%$). These amounts do not appear to limit significantly fluidized bed combustion technology.

When injected into an oxidizing flame zone or a fluidized bed, the sulfur/sorbent reactions are as follows:

$$CaCO_3 = CaO + CO_2 \quad \Delta H = 42 \text{ kcal/gmole} \tag{2}$$

$$CaO + 1/2O_2 + SO_2 = CaSO_4 \quad \Delta H = -120 \text{ kcal/gmole} \tag{3}$$

$$CaSO_4 = CaO + 1/2O_2 + SO_2 \quad \Delta H = 120 \text{ kcal/gmole} \tag{4}$$

For pulverized-coal combustion, calcination must occur via Reaction 2 above 1000°C and below 1350°C. If the calcination does not occur above 1000°C, it is too slow and the unreacted $CaCO_3$ is swept out of the combustion zone. However, above 1350°C, the CaO is rapidly deactivated by dead burning via bulk diffusion of ions within the crystal lattice of the CaO. This reduces significantly the specific surface for reaction with sulfur. Overheating of the $CaSO_4$ end product is also undesirable because above 1350°C, CaO and $SO_2$ are regenerated, as indicated by Reaction 4. The best capture will occur when the sorbent is injected into an oxidizing region which allows calcination without deactivation. Sulfation (Reaction 3) can then occur at reaction temperatures from 750 to 1200°C (Figure 2).

In FBC systems, the residence times are inherently long for solids (typically minutes or hours) compared with pulverized coal-fired systems (typically seconds). Thus, they can be operated between 800 and 900°C, which appears to be the optimum temperature for the sulfation reaction for these systems. The calcination reaction can be regarded as going to completion and as effectively instantaneous for FBC systems unless they are operated at elevated pressure, i.e., pressurized fluidized bed combustion. In that case, dolomite is often employed as a sorbent since the calcination temperature of the $MgCO_3$ compo-

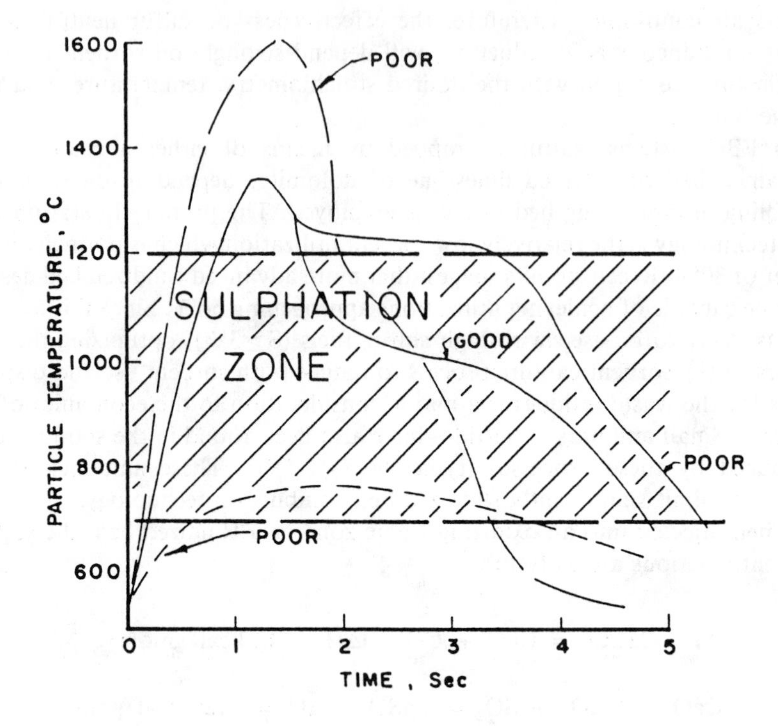


**Figure 2.** Time-temperature effect on limestone calcination.

nent is several hundred degrees lower than that of the $CaCO_3$ component. The $MgCO_3$ component will thus calcine, producing a porous and reactive sorbent, regardless of whether the $CaCO_3$ component calcines. This allows the sulfation reaction to proceed directly with the $CaCO_3$, in a reaction analogous to Reaction 3. At this time, pressurized fluidized bed combustion is still an experimental technology.

At the temperatures employed for fluidized bed combustion, Reaction 2 dominates and the primary reasons for sorbent utilization inefficiency are elutriation of unreacted sorbent from the FBC or pore plugging of the sorbent by the $CaSO_4$, which has a significantly larger molar volume than that of the original $CaCO_3$ or the CaO produced by calcination. These low temperatures also ensure that no thermal $NO_x$ is formed and that slagging and fouling problems are eliminated.

The mechanism for the sulfur reaction in reducing atmospheres is less clear, but may occur by means of these two endothermic reactions:

$$CaCO_3 + H_2S = CaS + CO_2 + H_2O \tag{5}$$

$$CaCO_3 + COS = CaS + 2CO_2 \tag{6}$$

The CaS formed may then, if required, be sulfated or oxidized as follows:

$$2CaS + 3O_2 = 2CaO + 2SO_2 \quad (7)$$

$$CaS + 2O_2 = CaSO_4 \quad (8)$$

In FBC systems, the sulfide formation does not seem to be via Reactions 5 or 6, or the analogous reactions with CaO. Detailed measurements of sulfide concentrations in the lower region of a pilot-scale circulating FBC unit indicated quite low concentrations of sulfide in the reducing region (beneath the region where secondary air enters the combustor). Instead, the highest levels of sulfides are found in regions where the char components are concentrated, i.e., secondary cyclone and baghouse. A possible mechanism is reduction of the sulphate via reactions of the type:

$$CaSO_4 + 4CO = CaS + 4CO_2 \quad (9)$$

though other possible mechanisms exist. In any case it appears that sulfide contamination of residues from FBC systems is not a serious problem.

During sulfation any sulfur in char that escapes the reducing zone, in a pulverized-coal combustion unit, must be captured in the subsequent oxidation stage. Application of physical-staged combustion systems, which are in the embryonic stage of development, will depend on further elucidation of critical process parameters, such as mixing, stoichiometric ratios of the primary and secondary combustion zones, sorbent injection points, volumetric heat release rates and char carryover from each stage.

Sulfur capture is also enhanced by increased thermal loading of the reaction zone, increased sorbent fineness down to about 50 $\mu$m. Ca/S molar ratios above 2, and the presence of halogens.

Three important factors that are strongly system dependent but which have not yet been fully investigated are:

1. the slagging and fouling propensity of the fuel ash and sorbent in the various combustion and downstream heat transfer zones
2. the amount of char emitted into the post flame gases and dust collectors
3. the flame shape with respect to furnace temperature gradients and geometry

Research data indicate that 50% sulfur capture is possible with a Ca/S molar ratio of 2 for sorbent injection.

## CURRENT CANADIAN PROJECTS FOR SIMULTANEOUS REDUCTION OF $NO_x$ AND $SO_x$ IN PULVERIZED COAL FIRED SYSTEMS

Two major Canadian initiatives in the simultaneous reduction of $SO_x$ and $NO_x$ from pulverized coal flames have been sponsored by the Canada Centre for Mineral and Energy Technology (CANMET). The first was an International Energy Agency (IEA) research project, co-funded by Canada, Denmark and Sweden with United States guidance and participation on the validation and optimization of advanced burner concepts. The second is a demonstration project in cooperation with the Canadian Department of National Defence (DND) at Canadian Forces Base Gagetown, New Brunswick to accelerate the application of state-of-the-art LIMB technology to operating boilers. In addition to the above two projects, Energy, Mines and Resources (EMR) Canada collaborates with the Flue Gas Desulphurization Panel of the Canadian Electrical Association on utility R&D programs for reducing $NO_x$ and $SO_x$ emissions from flames.

## INTERNATIONAL ENERGY AGENCY PROJECT

The International Energy Agency (IEA) project, which has been conducted by the Environment Energy Research Corporation in the United States was planned in three stages.[7] Stage 1 consisted of small-scale furnace trials on 45 coals, including nine from Canada, to elucidate the mechanism of $NO_x$ formation from fuel nitrogen under premixed and staged-combustion conditions. Illustrations of the small-scale furnace and burners are given in Figure 3.

Highlights from this research indicate that for unstaged flames with 5% $O_2$ in the exhaust gases:

1. More fuel nitrogen was converted to NO from premixed flames than from axial diffusion flames, with the difference in NO production between the two systems decreasing from high-volatile bituminous to lignite to medium-volatile bituminous coals (Figure 4).
2. Conversion of fuel nitrogen with premixed flames ranged from 30 to 45% regardless of the total fuel nitrogen content (Figure 5).
3. With premixed flames, high-volatile coals produced lower NO emissions from coarse particles than with fine. The converse was true for axial flames with fine particles producing the lowest NO levels. Evidently, fuel nitrogen, which is evolved early in the flames from fine particles, is available for conversion to NO in premixed flames if oxygen is present (Figure 6).

For staged flames, i.e., a substoichiometric first stage and an air-rich second stage with 5% oxygen in the exit gases:

1. All high-volatile coals, bituminous and lignite, produced minimum NO emissions at a stoichiometric ratio of about 0.6 in the first stage; the medium-volatile coals yielded progressive increases in NO as the first-stage stoichiometry increased.
2. Fuel nitrogen conversion ranged from 7 to 14%, with the lowest NO emissions coming from high-volatile coals (Figure 7).
3. Fine particles produced less NO than coarse particles for a high-volatile coal.
4. The lower-rank coals, lignite and subbituminous, produced large amounts of

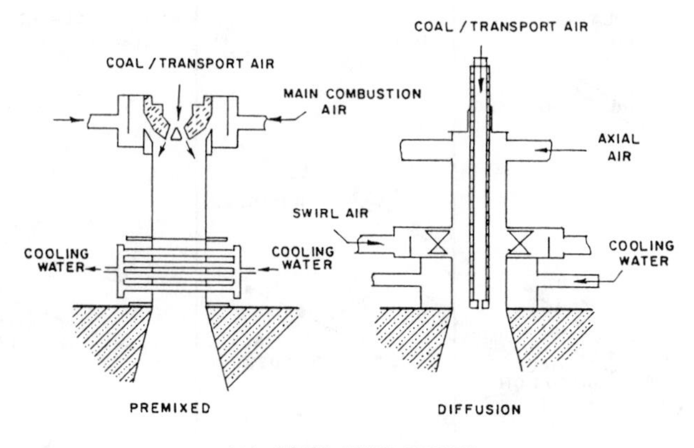


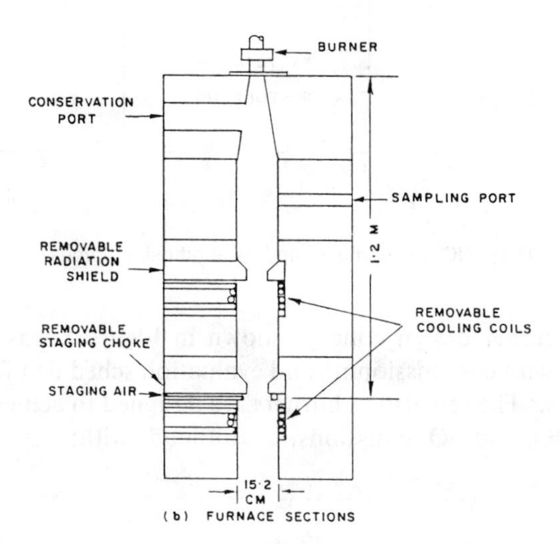


**Figure 3.** Modified furnace and burner sections for staged-combustion studies.

$NH_3$ and HCN at stoichiometric ratios below 0.6. These species appeared to favor transformation to $N_2$, rather than NO, during second-stage burnout.

5. Heat extraction lowered NO emission from low-rank coals, particularly for first-stage stoichiometric ratios above 0.7.

## GAGETOWN PROJECT

This project involved retrofitting LIMB burners on a 17 $MW_{th}$ front-wall-fired heating boiler at Gagetown, New Brunswick, to reduce simultaneously emissions of $SO_x$ and $NO_x$. The boiler had been previously fired with two conventional pulverized coal burners, each having its own pulverizer. Figure 8 illustrates the new installation.

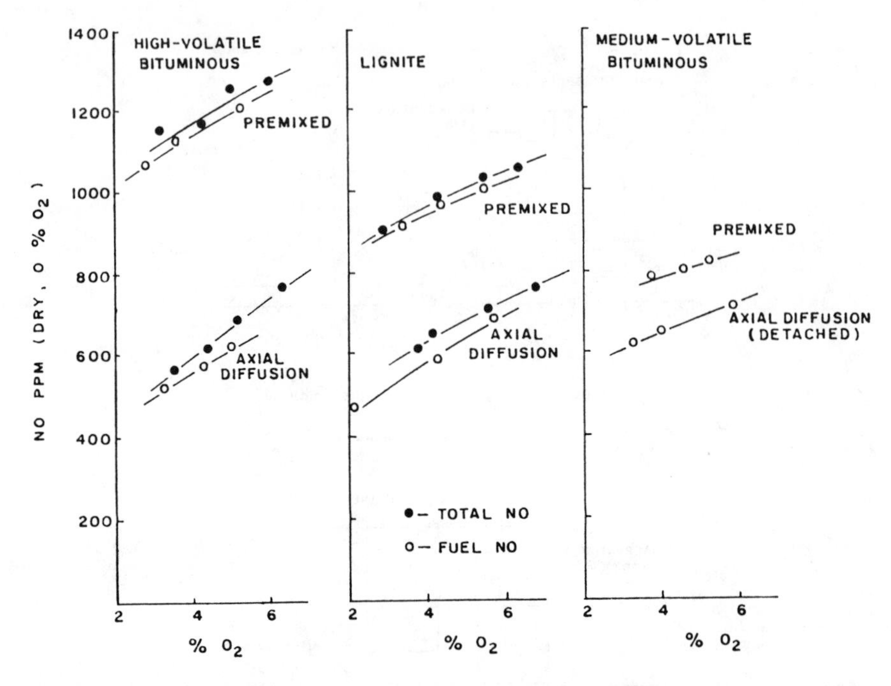


**Figure 4.** Fuel NO/total NO for premixed and axial diffusion flames.

The staged-burner design concept shown in Figure 8 was installed in the spring of 1984 with commissioning and evaluation scheduled for the following heating seasons.[8] The retrofitted burners are designed to achieve a 50% reduction in both $NO_x$ and $SO_x$ emissions, at full load, with:

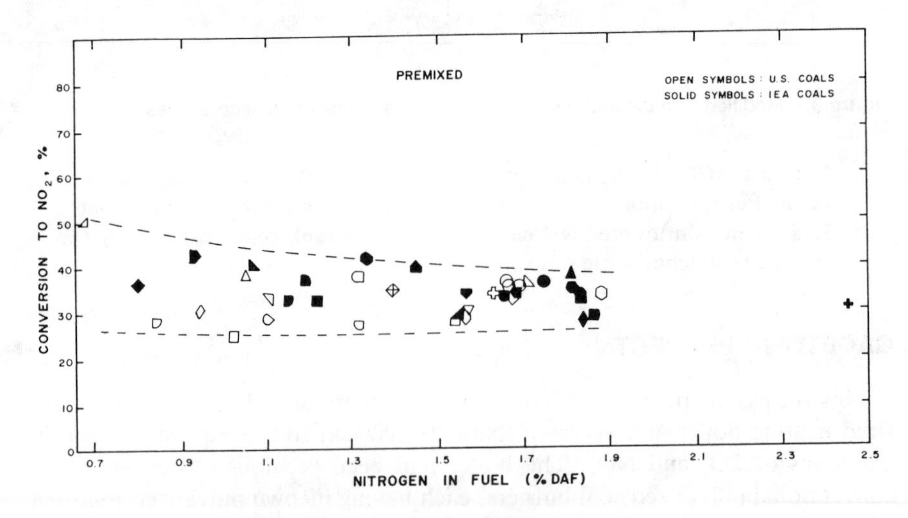


**Figure 5.** Fuel nitrogen conversion in premixed flames.

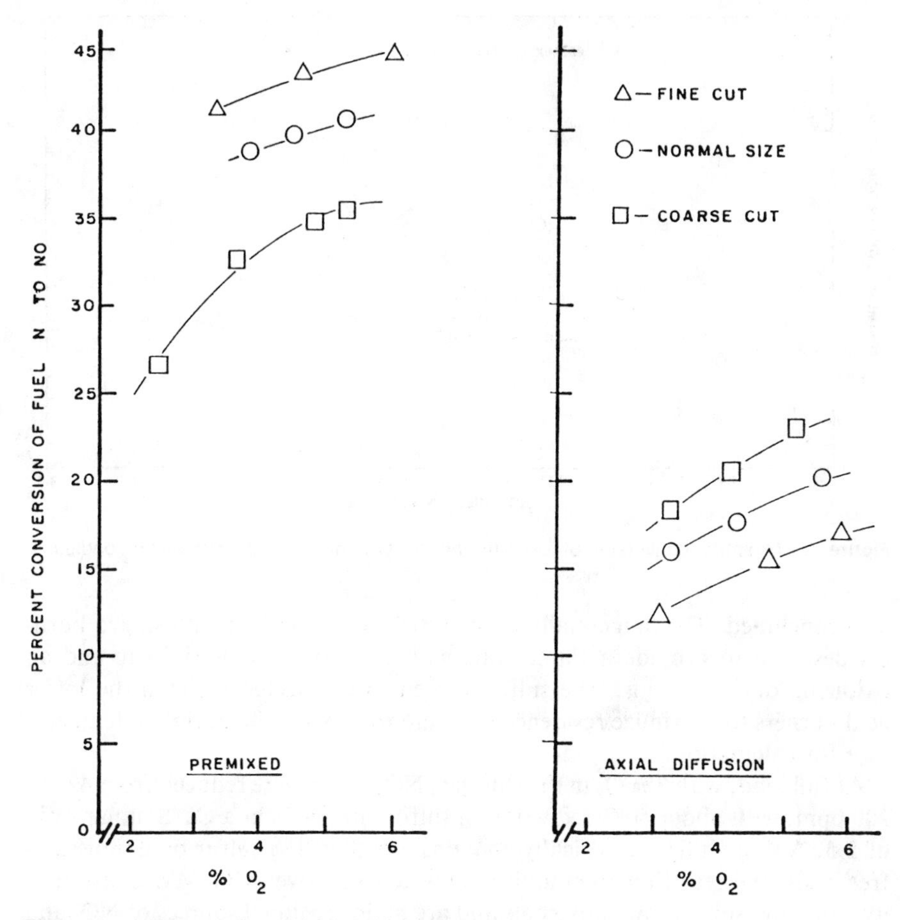


**Figure 6.** Influence of particle-size, high-volatile bituminous coal.

1. 3% high-volatile bituminous coal
2. Ca/S molar ratios of 2.5 or less
3. combustible in ash of 7% or less
4. no flame impingement on walls
5. boiler efficiency of 82% or more
6. no furnace slagging with routine soot blowing

The unit had completed commissioning trials at the time of writing and is now operated routinely by DND.

## CANADIAN ELECTRICAL ASSOCIATION PROJECT

A research project to study the feasibility of retrofitting a 350 $MW_e$, lignite-fired utility boiler in Western Canada with a low $NO_x/SO_x$ burner system was

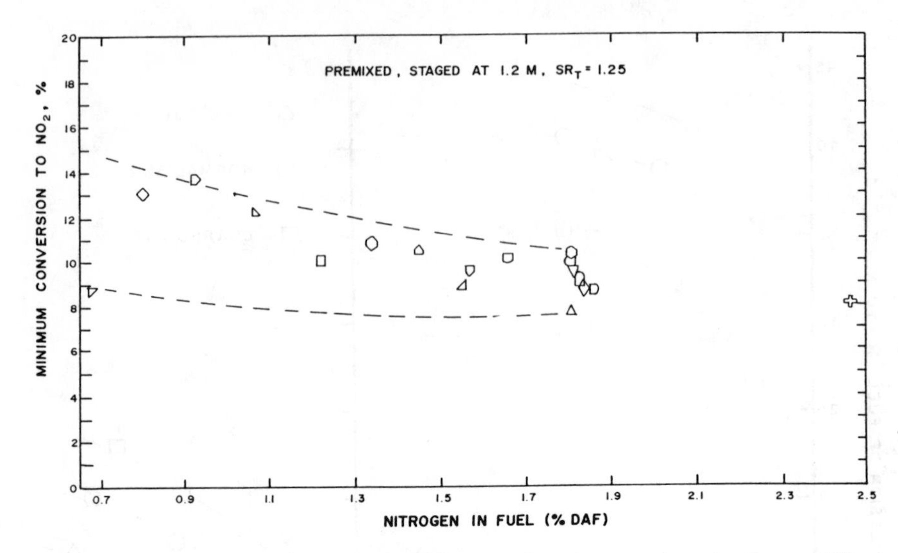


**Figure 7.** Minimum conversion of fuel nitrogen to NO under staged-combustion conditions.

also concluded. The tangentially fired system was modified with staged burners designed to provide a flame zone with a substoichiometric core and an oxidizing outer annulus. The sulfur sorbent was injected through the lower coal burners to maximize residence time and to provide the optimum temperature for calcination.[9]

At full load, with 3% $O_2$ in the flue gas, $NO_x$ levels were reduced from 425 to 200 ppm, with about 40% reduction in sulfur emissions at a Ca/S molar ratio of 2.5. Although lignite typically contains less than 1% sulfur on a moisture-free basis, Western Canadian utilities now generate over 5500 MW of electricity from low-sulfur, low-rank coals and are major regional sources of $NO_x$ and $SO_x$ emissions.

The IEA research program indicates that staged-combustion concepts, together with sorbent injection, can be used to reduce simultaneously $SO_x$ and $NO_x$ emissions in boiler furnaces. Staged combustion may however be only marginally effective in reducing $NO_x$ emissions from coals in which most of the fuel nitrogen is present in the fixed carbon.

The demonstration projects involving burner retrofits are directed at accelerating the transfer of research results to commerical application and in validating staged-burner concepts, together with sorbent injection, for reducing $NO_x$ and $SO_x$ emissions under operational conditions.

## FLUIDIZED BED COMBUSTION

In 1976, CANMET started a program of research on FBC technology which simultaneously achieves low $SO_2$ and $NO_x$ emissions. It is the most mature of

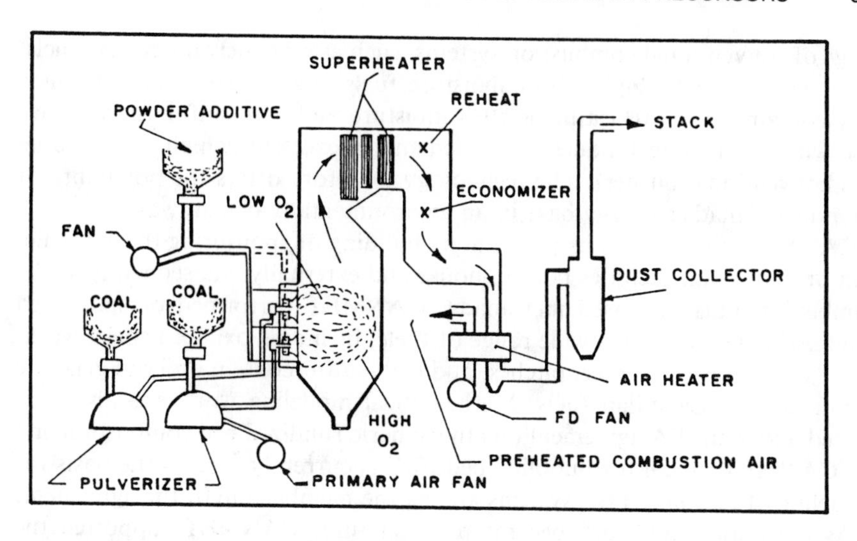


a) Schematic of Gagetown boiler system

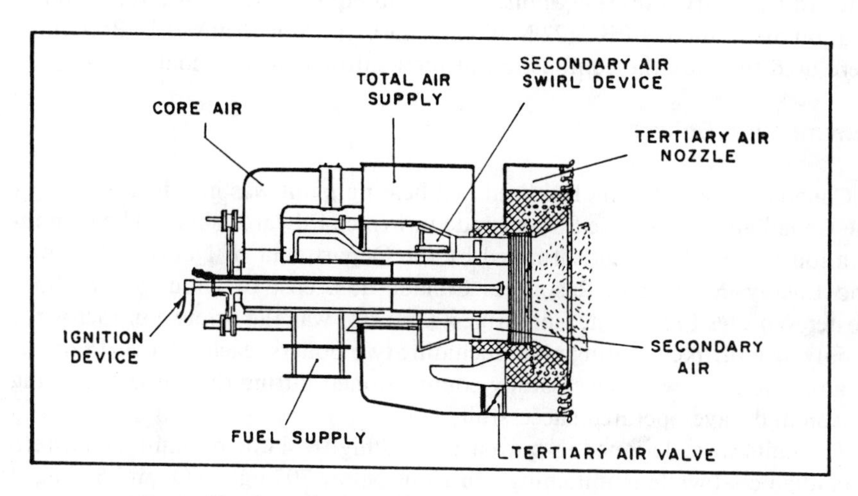


b) Staged mixing burner (SM burner)

**Figure 8.** Illustration of boiler and burner systems for Gagetown, New Brunswick.

all the novel combustion technologies and is being increasingly used across North America. The combustion of efficiency associated with FBC systems is somewhat lower than that of pulverized combustion systems (95 to 99% compared with 99% plus). Typically the technology is also more expensive per megawatt of power, although costs are strongly site and fuel specific, and if the pulverized combustion system must be used in conjunction with flue gas desulfurization technologies such as scrubbers, the costs per megawatt become similar. However, FBC systems can operate with fuels which are outside the range of conventional combustion systems, such as extremely unreactive fuels, e.g., oxidized coals, high ash or moisture fuels, e.g., washery rejects which may contain up to 70% ash, or 50% moisture and high-sulfur fuels (above 3%, where sorbent techniques would demand an excessively high solid loading in a conventional burner). The technology therefore offers the possibility of extending Canada's energy base in an environmentally benign way.

CANMET commenced its program by building or sponsoring the construction of pilot-plant facilities both in-house and externally. Subsequently as the number of units increased in Canada, CANMET performed or sponsored combustion research on a wide range of fuels, including oxidized coals, washery rejects, petroleum coke, pitches, and high-sulfur coal, in order to generate a database on Canadian fuels. Mathematical modeling work was also supported under an IEA agreement on atmospheric fluidized bed combustion and the CANMET model for bubbling bed FBC is currently used as the basis for modeling of bubbling FBC systems among the membership of the agreement.

As the pilot-plant work became more mature, CANMET supported the development of the first industrial bubbling bed boilers at Summerside, Prince Edward Island (two 15-$MW_{th}$ boilers) and subsequently the first utility demonstration of circulating FBC (22 $MW_e$) at Chatham, New Brunswick. Both units were used to study the combustion of high-sulfur maritime coal.

## Summerside Boilers

Canada's first bubbling fluidized bed heating plant was installed in 1982 at the Canadian Forces Base Summerside in Prince Edward Island. The demonstration project was jointly funded by the Department of National Defense, and Energy, Mines and Resources Canada at a cost of about $18 million. Foster Wheeler Ltd. of St. Catharines, Ontario, was the prime contractor for supplying a turnkey heating plant including two boilers, each rated at 18 tph of steam. The units were successfully commissioned during the 1984/85 heating season and have operated successfully since.

The units were designed to burn a 5.5% S high-volatile bituminous Eastern Canadian coal while maintaining emissions below 705 ng/J $SO_2$ and 516 ng/J $NO_x$. These limits were easily achieved and $NO_x$ emissions below 258 ng/J were also achieved, although the unit had to be operated with about 10% $O_2$ in the flue gas rather than the design value of 3.6% in order to prevent bed overheating and clinkering.

In a series of tests on five high-sulfur (4 to 11% S) maritime coals, sulfur captures of 76 to 84% were obtained with Ca/S molar ratios from 1.7 to 3.3, while $NO_x$ emissions were consistently below 250 ng/J for all loads.[10] In 1986, Energy, Mines and Resources Canada collaborated with the U.S. Environmental Protection Agency to study the behavior of these units in a series of tests to maximize sulfur capture at boiler loads of 70% or more.[11] These tests included a 30-day trial to establish the Ca/S molar for 90% capture. Operating data averaged over the 30 days yielded a boiler load of 70.6% while burning a 5.9%

sulfur coal. Average flue gas analyses indicated that 93% capture could be achieved with a Ca/S molar ratio of 4. This corresponded to emissions of 258 ng/J for $SO_2$ and 267 ng/J for $NO_x$. In other shorter tests, carried out during this series of trials, $SO_2$ captures up to 99.4% were achieved. The units have thus demonstrated their ability to cope with high-sulfur fuels while meeting the most stringent $SO_2$ capture requirements. They can also meet $NO_x$ emission guidelines of 258 ng/J providing combustion conditions are optimized, despite design problems which forced the units to operate with higher excess air than originally planned.

### Chatham Boiler

Canada's first industrial-scale circulating fluidized bed boiler was constructed at New Brunswick Power's Chatham Generating Station. The boiler system was supplied by Combustion Engineering Superheater Ltd., Canada, using technology developed by Lurgi of West Germany. The new boiler supplies steam to the existing 22-$MW_e$ turbine generator. The $38 million plant expansion was funded largely by the government of Canada.

Construction of the plant started in 1985, and despite some delays, the unit was ready for commissioning trials in 1987. Initially, test work was carried out on a high-sulfur (7% S) bituminous coal from New Brunswick (Minto) using a blend of Albert County oil shale and limestone for sulfur capture. Subsequently the unit has been used to burn a high-sulfur (5% S) Nova Scotia coal and a petroleum coke. The unit has been shown to be able to achieve 90% sulfur capture with Ca/S molar ratios of about 2.5. The $NO_x$ emissions have proven to be very low indeed with values of as low as 60 ppm for 3.5% $O_2$ in the flue gas, which corresponds to emissions below 100 ng/J. Work is continuing with this unit in order to optimize the sulfur capture performance. It is planned to burn a wide variety of fuels including coal wash-plant tailings, high-ash coals, and other high-sulfur coals. The unit has nevertheless been shown to be able to achieve emission guidelines for $SO_2$ capture and to easily meet the guidelines for $NO_x$ emissions.

## NITROUS OXIDE EMISSIONS

This gas is not normally thought of as an atmospheric pollutant. However, it appears to be a major source of NO in the stratosphere by means of the following reaction:

$$N_2O + O(^1D) = 2NO \qquad (10)$$

The NO so formed can then destroy ozone by means of the cyclic reactions:

$$NO + O_3 = NO_2 + O_2 \qquad (11)$$

$$NO_2 + O = NO + O_2 \tag{12}$$

This gas appears to be increasing on a global basis at about 0.2% /a.[12] The causes of this increase are almost certainly anthropogenic, and manmade nitrogenous fertilizers and combustion systems have been identified as likely contributors.

There are very few studies on $N_2O$ emissions from combustion systems and estimates of their overall contribution to $N_2O$ production are somewhat uncertain. This situation has recently been worsened by the realization that $N_2O$ can be produced from $NO_x$ during attempts to measure $N_2O$ emissions. Specifically, it has been shown that zeolite materials, which are often used in molecular sieves in gas chromatography, can convert $NO_x$ to $N_2O$[13] via the reaction:

$$4NO = N_2O + N_2O_3 \tag{13}$$

Gas chromatography is the standard technique for measuring $N_2O$ formation.

It has also been shown that in the presence of moisture and $SO_2$, NO can react to produce $N_2O$ in significant concentrations (up to several hundred parts per million by volume) over periods as short as 2 hr where none had originally existed. Even complete drying of the gas does not completely eliminate this effect.[14] It seems likely that many of the previous studies of $N_2O$, which relied on taking grab samples of combustion gases for subsequent analysis from combustion sources, have been affected by these phenomena. It is probable that the contribution of combustion sources to $N_2O$ formation has been overestimated in the literature.

As a result of these uncertainties, Environment Canada undertook an extensive sampling program of $N_2O$ emissions from boilers across Canada, the results of which are not yet available. In addition, CANMET is also supporting research to study the formation of $N_2O$ from circulating fluidized bed combustion.

## CONCLUSIONS

Studies of 45 coals undertaken in laboratory furnaces using premixed and diffusion flames in staged and unstaged modes indicate:

1. For unstaged flames, more fuel nitrogen was converted to $NO_x$ in premixed compared with diffusion flames. The conversion of fuel nitrogen was between 30 and 45% for premixed flames regardless of the coal nitrogen content. High-volatile coals produced lower $NO_x$ emissions from coarse grinds with premixed flames than with diffusion or axial flames.
2. For staged flames, all high-volatile coals produced minimum $NO_x$ levels at a stoichiometric ratio of 0.6 and medium-volatile coals produced progressively more $NO_x$ as the stoichiometric ratio was increased. The fuel-nitrogen conversion was between 7 and 14% with the lowest $NO_x$ produced from high-volatile coal. Finer pulverized coal grinds produced less $NO_x$ than coarser grinds.

3. High heat fluxes from the flames lowered $NO_x$ levels for low-rank coals, particularly at higher stoichiometric ratios.

Investigations conducted on a wide range of high-sulfur coals using bubbling and circulating fluidized bed combustion technology have indicated that it is possible to achieve $NO_x$ emissions below 258 ng/J without difficulty while achieving over 90% sulfur capture. Conversions of fuel nitrogen to $NO_x$ range from 5 to 25%, with circulating fluidized beds producing lower $NO_x$ emissions than bubbling beds.

The issue of $N_2O$ emissions from combustion systems is now being addressed. It seems likely that many combustion studies in the past overestimated the emissions of this species from both conventional and fluidized bed combustion systems. This is an area, however, that merits further study.

## REFERENCES

1. Knill, K. J. "A Review of Fuel Staging in Pulverized Coal Combustion Systems," International Flame Research Foundation, IFRF Document G13/a/3 (September 1987).
2. Johnsson, J. E. "A Kinetic Model for $NO_x$ Formation in Fluidized Bed Combustion," Proceedings of the Tenth International Conference on Fluidized Bed Combustion, San Francisco, CA, April 30 to May 3, 1989.
3. Leckner, B. "Tasks for Modelling of the Emissions of Nitrogen Oxides," presented at the IEA AFBC Mathematical Modelling Meeting, Siegen, October 26 to 27, 1987.
4. Zhao, J., J. R. Grace, C. J. Lim, C. Brereton, R. Legros, and E. J. Anthony. "$NO_x$ Emissions in a Pilot Scale Circulating Fluidized Bed Combustor," EPRI/U.S. EPA Symposium on Stationary Combustion, $NO_x$ Control, San Francisco, March 6 to 9, 1989.
5. Hirama, T., M. Tomita, M. Horio, T. Chiba, and H. Kobayashi. "A Two-Staged Fluidized Bed Coal Combustor for Effective Reduction of $NO_x$ Emission," in *Fluidization (Proceedings of the Fourth International Conference on Fluidization)*, D. Kunni, and R. Toei, Eds. (1983).
6. "Thermal Power Generation Emissions—National Guidelines for New Stationary Sources," *Canada Gazette* (April 25, 1981).
7. Chen, S. L., D. W. Pershing, and M. P. Heap. "International Energy Agency Project for Control of Nitrogen Oxide Emissions during Coal Combustion: Stage 1," prepared for the Executive Committee of the Contracting Parties for Canada, Denmark, and Sweden (December 1981).
8. L & C Steinmuller GmbH. "Design Study for Pulverized Coal Burner for the Combined Reduction of Nitrogen and Sulphur Oxides," Gummersbach, West Germany (March 1983).
9. Winship, R. D., S. A. Morrison, B. R. Clements, and B. Y. Lee. "$SO_2$ and $NO_x$ Reduction by Combustion Modification and Furnace Dry Sorbent Injection," Canadian Electrical Association 1984 Spring Meeting, Toronto, March 1984.
10. Lee, G. K., V. V. Razbin, and F. D. Friedrich. "Control of Acid Rain Emissions

from Canada's First Fluidized Bed Heating Plant," Energy Research Laboratories, Division Report ERP/ERL 87-31(OPJ) (April 1987).
11. "Statistical Analysis of Emission Test Data from Fluidized Bed Combustion Boiler at Prince Edward Island, Canada," U.S. Environmental Protection Agency Report EPA-450/3-86-015 (December 1986).
12. Weiss, R. F. "The Temporal and Spatial Distribution of Tropospheric Nitrous Oxide," *J. Geophys. Res.* 86:7185–7195 (1981).
13. Weiss, R. F., and H. Craig. "Production of Atmospheric Nitrous Oxide by Combustion," *Geophys. Res. Lett.* 3(12):751–753 (1976).
14. Muzio, L. J., and J. C. Kramlich. "An Artifact in the Measurement of $N_2O$ from Combustion Sources," *Geophys. Res. Lett.* 15(12):1369–1372 (1988).

## CHAPTER 22

# In Situ Sampling from an Industrial-Scale Rotary Kiln Incinerator

**Arthur M. Sterling, Vic A. Cundy, Thomas W. Lester, Alfred N. Montestruc, John S. Morse, Christopher Leger, and Sumanta Acharya**

## INTRODUCTION

Rotary kilns were initially developed as process furnaces for treating cement, lime, dolemite, and similar products. Their adaptability to a wide variety of continuous process operations, however, has led to widespread use in many other applications. In particular, rotary kiln incineration has been used for over 20 years in the United States and Europe for destroying waste organics. This technology is now generally regarded as the preferred method of treatment for many hazardous organic wastes. Solids, sludges, and liquids can all be treated simultaneously with continuous mixing at high temperatures. A comprehensive review of the role of rotary kiln incineration in the current and future management of hazardous wastes was recently given by Oppelt.[1]

The major improvements in the design of rotary kiln incinerators have focused on air pollution control equipment. Secondary combustion chambers, quenches, and scrubbers or absorbers have been added to clean and deacidify combustion gases.[2] Hazardous waste incinerators are now required to meet all applicable Federal regulations as well as state and local regulations. These regulations set the destruction and removal efficiences (DRE) for the principal organic hazardous constitutents (POHC), limit the emission rates of particulates and hydrogen chloride, and establish rigid permitting procedures.

Despite the proven ability of rotary kiln incinerators to meet these regulations,[3] there remains a high degree of public skepticism about the efficiency of incineration as a hazardous waste treatment modality. To a large measure, this skepticism is grounded in a concern about incinerator stack emissions. A better understanding of incinerator performance is a necessary prerequisite for more informed public comment on operations and site location. At a minimum, this will require that a believable risk assessment be provided. Furthermore, a better understanding of the chemical and physical phenomena that underlie existing units would accelerate the development of a new generation of incinerators.

Because of the extreme complexity of the incineration process, an improved understanding will not emerge solely from a "black-box" approach to incinerator performance analysis. The initial measurements inside rotary kiln incinerators, to date, indicate that $O_2$, CO, and $CO_2$ levels at the kiln exit are nonuniform.[4] This implies that hydrocarbon levels are nonuniform as well. We simply do not know how the pattern factors at the entrance to the afterburner (e.g., velocity, temperature, spatial and temporal species concentration) are affected by kiln operating parameters (firing rate, excess air, temperature, rotation rate, bed depth, etc.). A better understanding of and a predictive capability for rotary kiln and afterburner performance, as influenced by basic design and operational parameters, is needed if we are to optimize rotary kiln operation and improve rotary kiln design.

## OVERVIEW OF IN-DEPTH STUDY

A schematic depiction of the principal processes in a rotary kiln is shown in Figure 1. Auxillary fuel and/or combustable liquid wastes provide the primary heat source for the process and create a highly turbulent, anisotropic, strongly three-dimensional reacting flow. Radiant, convective, and conductive heat transfer to a solids bed induces volitilization of the absorbed wastes, which can

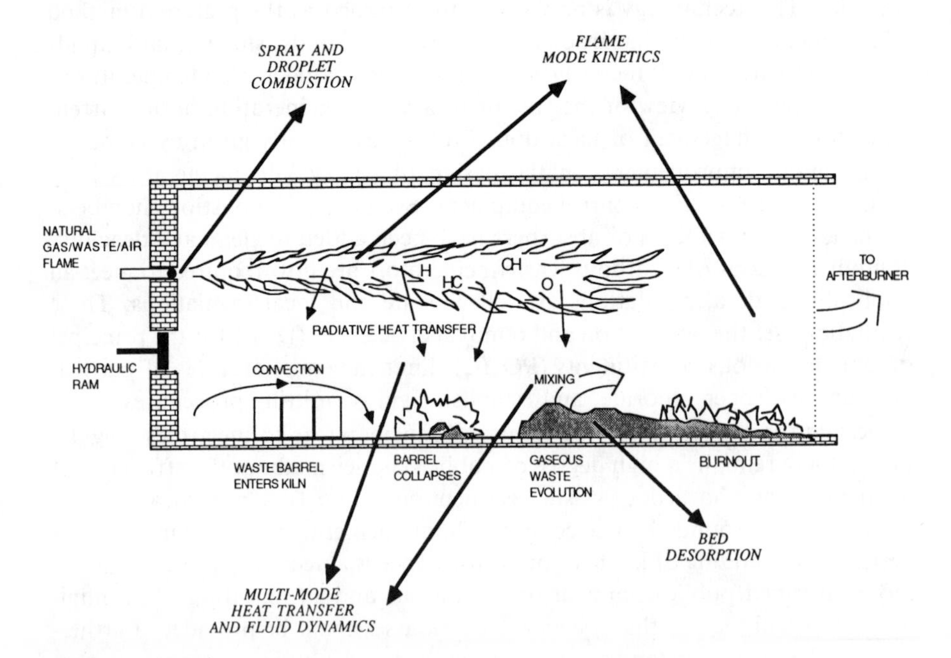


**Figure 1.** Depiction of the fundamental processes of importance in rotary kiln incineration.

either ignite and burn at the bed surface or be convected into the high-temperature primary flow where they react. The resulting spatial and temporal distributions of the variables describing the primary flow at the kiln exit constitute the input pattern factor for the secondary combustion chamber.

A program centered at Louisiana State University is being directed toward obtaining a fundamental understanding of these complex phenomena. It represents a combined effort between universities, governmental agencies, and private industry. The details of this program have been described previously,[4] but will be reviewed briefly here.

At the core is the development of a mathematical model for a three-dimensional, turbulent reacting flow, which integrates various experimental studies and will eventually provide the desired predictive capability. Providing input to the model are investigations of flame-mode kinetics and droplet combustion, which have been underway at Louisiana State University for several years.[5-8] More recently, fundamental bed desorption and pilot-scale rotary kiln studies underway at the University of Utah have been added to the program.[9-11] It is anticipated that in future work, pilot- and intermediate-scale rotary kilns at Research Triangle Park, NC, and Pine Bluff, AK, will be available for additional experimental work. Finally, a full-scale rotary kiln incinerator has been made available for measurements. It is the last of these components—the full-scale measurements—that we describe here. In particular, we will review some pertinent results obtained from sampling of gases near the kiln exit during quasi-steady incineration of carbon tetrachloride and compare this behavior with our initial results obtained during periodic loading of toluene-laden absorbant in plastic packs.

## FACILITIES AND SAMPLING EQUIPMENT

Details of the incinerator, located on the Louisiana Division of Dow Chemical USA site in Plaquemine, LA, have been given elsewhere.[4] A schematic diagram of the incinerator is shown in Figure 2. The design capacity of the 3.2-m i.d. and 10.7-m long kiln is 17 MW with a 800°C outlet temperature. The design capacity of the secondary combustion chamber is 7 MW with a 1000°C outlet temperature and 2-sec minimum residence time. Access ports for collection of gas samples from the kiln and afterburner are also indicated on Figure 2.

Gas samples are drawn through two, water-cooled, stainless-steel sampling probes of identical design, but with lengths of 7.8 m and 3.8 m, respectively. Axially centered, 4-mm o.d. sheaths that run through the length of the probe and extend approximately 5 cm beyond their ends contain connection wires for type K thermocouples at the tip. The thermocouples are enclosed by ceramic radiation shields with a 5-mm diameter side hole, as shown in Figure 3.

A jet-pump ejector, driven by 0.55 MPa air, provides sufficient vacuum to draw approximately 1 SCMM combustion air through the side hole in the

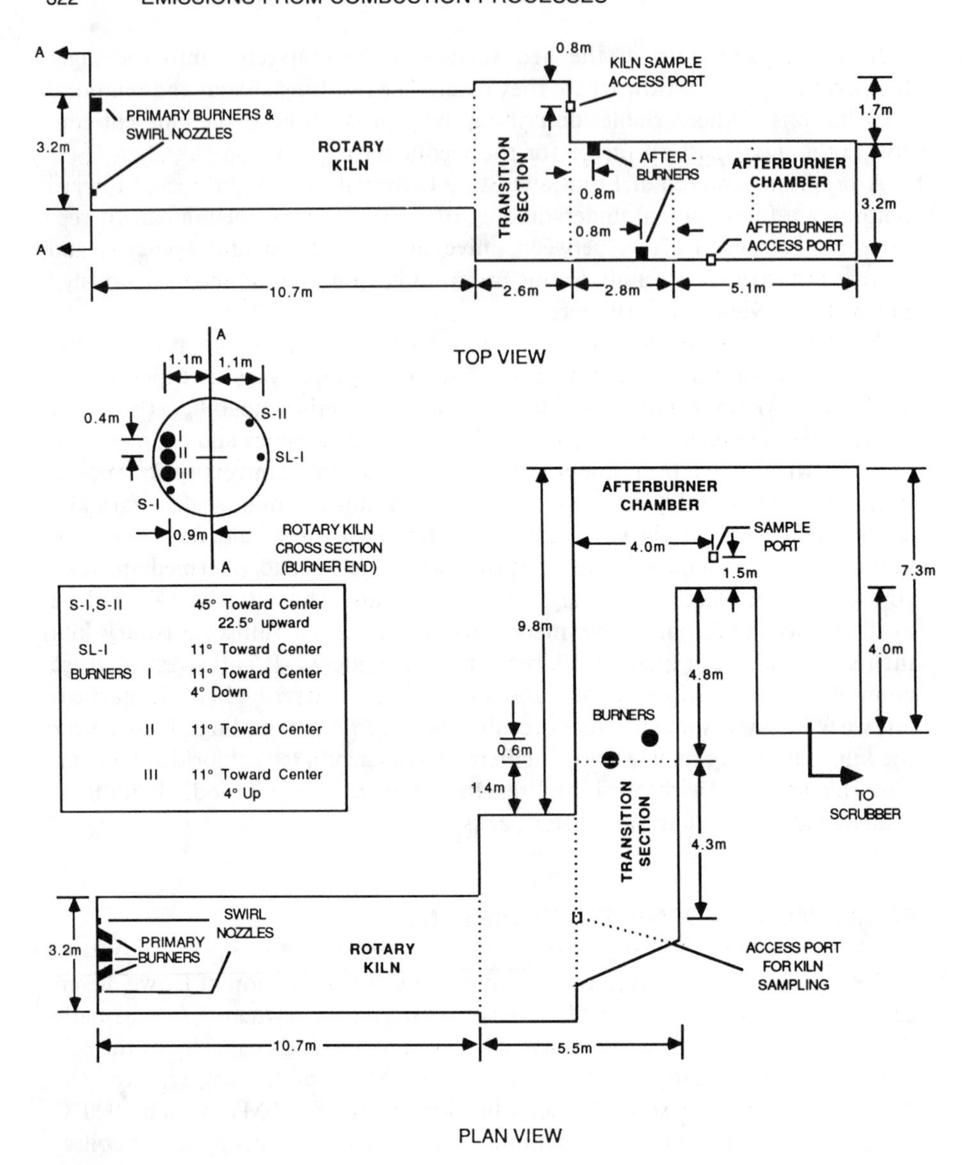


**Figure 2.** Layout of rotary kiln at the Louisiana Division of Dow Chemical USA.

radiation shield, across the thermocouple, and down the inner annulus of the probe, where it is rapidly quenched (~1000K/sec).

As shown in Figure 4, a portion of the quenched gas is drawn through a parallel flow path where, after condensate is removed, it passes through a 1-L sample bomb and/or a continuous total hydrocarbon (THC) analyzer. During sample collection, the flow rate through a bomb is nominally 1.7 SCMH (about 3% of the flow through the probe). Response tests on the sampling

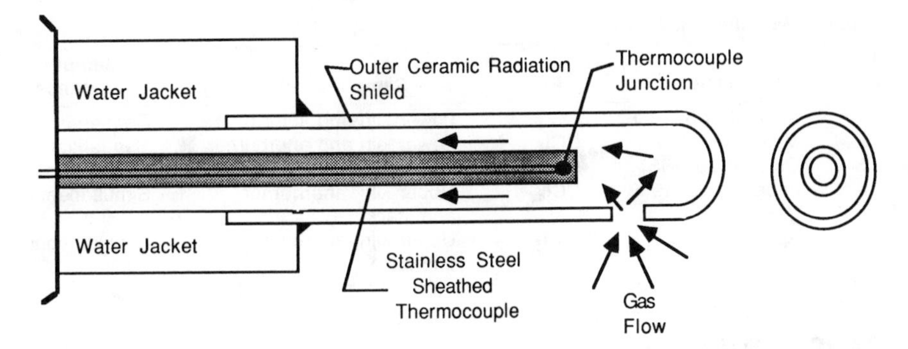


**Figure 3.** Detail of probe tip showing configuration of radiation shield for thermocouple and gas inlet port.

bomb indicated a time constant of approximately 4 to 5 sec for flow rates typically used during sampling. This value is about 30% greater than the time constant expected for perfect mixing and is small compared to the 2 to 5 min times typically used for sample collection. The THC has its own vacuum pump; the residence time in the sample line is estimated to be approximately 3 sec. The output of the THC is fed continuously to a strip-chart recorder.

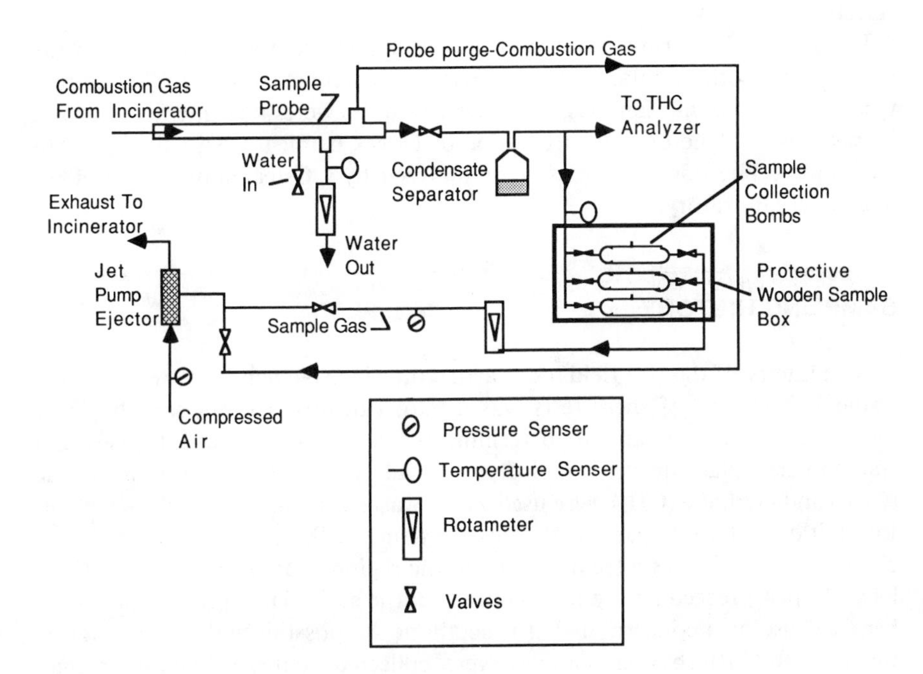


**Figure 4.** Schematic diagram of sampling train.

**Table 1. Sampling History**

| Test no. | Date | Feed | Sample locations | Sample sequence |
|---|---|---|---|---|
| 1 | November 5, 1986 | $CCl_4$ | Lower kiln and afterburner | Sequential |
| 2 | March 19, 1987 | $CCl_4$ | Lower kiln and afterburner | Sequential |
| 3 | October 29, 1987 | $CCl_4$ | Transition and afterburner | Simultaneous |
| 4 | February 25, 1988 | $CCl_4$ | Upper kiln, afterburner, and stack | Simultaneous |
| 5 | April 12, 1988 | $C_6H_5CH_3$ | Upper kiln, afterburner, and stack | Simultaneous |

## SAMPLE ANALYSIS

The bomb samples are transported to the analytical laboratories in the Department of Mechanical Engineering at LSU for analysis. The gas analysis system consists of two Varian Vista 6000 gas chromatographs (GCs) and a Varian Vista data aquisition system. One of the GCs, equipped with a thermal conductivity detector (TDC), is used for the analysis of permanent gases ($O_2$, $N_2$, $CO_2$, and CO) and high-level methane ($CH_4$). The second GC is equipped with a flame ionization detector (FID) for analysis of selected stable organic intermediates. The GC procedures were developed in the course of a companion project and are described in detail elsewhere.[12] Approximately every 10th sample is also subjected to mass spectrometric analysis by the analytical team at the Louisiana Division of Dow Chemical USA to verify the LSU GC results.

During our later runs, described later, samples were withdrawn from the stack using both standard volatile organic sampling train (VOST) capture techniques and bomb samples. The VOST samples were subsequently analyzed by personnel of the Louisiana Division of Dow Chemical USA using GC/MS techniques. The bomb samples were analyzed by GC techniques in the LSU analytical laboratory.

## SAMPLING HISTORY

A summary of the five field tests carried out since November 1986 is given in Table 1. The first of these tests was a pilot run to work out sampling and analytical procedures and to determine the levels of permanent gases and stable intermediates in the sampled gas. Steady flows of carbon tetrachloride ($CCl_4$) and methane ($CH_4$) were used as the waste and auxillary fuel in the kiln. In addition, $CH_4$ was used as the auxillary fuel in the secondary combustion chamber. Gas samples were drawn from the region near the kiln exit and at a location just preceeding the lower burner in the secondary combustion chamber. Because of a concern at that time about the possibility of sample aging, the kiln and afterburner samples were collected sequentially, rather than simultaneously.

One limitation to our sampling capability was immediately apparent. Because of substantial droop of the 10.7-m long probe, we were able to collect samples only from a horizontal plane located approximately in the lower third of the kiln exit.

The second field test replicated the first test conditions and provided our first complete set of data. As was the case in the first test, sampling from the kiln and afterburner was sequential. The afterburner sampling location, however, was moved to the top of the arch, approximately midway through the secondary combustion chamber (see Figure 2). Analysis of these data gave convincing evidence that sample aging was not a problem and allowed us to proceed with simultaneous sampling in subsequent runs.[4] As discussed later, the results also suggested the presence of highly nonuniform conditions in the vertical direction at the kiln exit and indicated the need to sample in the upper regions of the kiln as well.

Prior to the third test, an additional kiln port was created approximately 2 m above the existing one to provide access to the upper regions of the kiln exit. A second sampling system and additional sample bombs were also prepared. On the day of the test, however, we discovered that we were unable to insert the probe beyond the transition region and into the kiln because of limitations of the new port size and orientation. We were, however, able to collect samples from the transition region between the kiln exit and the afterburner. This run also provided an opportunity to develop techniques for simultaneous sampling with the two probes. The firing conditions of the third test replicated the conditions of the first two as nearly as possible.

The fourth test provided the desired data from the upper regions of the kiln. Conditions were maintained as close as possible to the conditions of the second test so that the data would be comparable. The feed rate of $CCl_4$, however, was only about 60% that of the second test. In addition to simultaneous sampling from the kiln and afterburner, simultaneous stack samples were also collected. This test completed our initial series on steady firing of $CCl_4$.

The latest test, in March 1988, initiated our test series on transient loading of wastes. Plastic packs of absorbant laden with toluene ($C_6H_5CH_3$) were fed to the kiln in 5-min intervals. Simultaneous samples were taken from the upper regions of the kiln exit, from the arch of the afterburner, and from the stack. In addition, a video recorder was used to obtain images of the inside of the kiln over several periods of pack loading. These images provide rough visual correlation with the temperature and concentration measurements. Because of the transient conditions, the collection time for the bomb samples was reduced to 30 sec to provide better temporal resolution. Sampling times were correlated to pack drop times, as described later. Such temporal resolution was not possible with the VOST samples from the stack. The required VOST sampling time of approximately 25 min resulted in stack conditions being integrated over approximately five pack loadings.

## PERTINENT RESULTS FROM CARBON TETRACHLORIDE TESTS

The set of four tests on the steady firing of $CCl_4$ (tests 1 to 4) included a matrix of sampling positions in the kiln, the transition region between the kiln and the afterburner, and in the afterburner. This set of measurements, which are described in detail elsewhere,[4,13,14] elucidate the spatial variations of temperature and species concentration. Two of the sampling locations in this matrix (as shown in Figure 5) are of particular interest here since they provide a basis for interpreting the results from the transient test discussed later. The $CCl_4$ results for these two points were obtained in tests 2 and 4, respectively.

Typical operating conditions for Tests 2 and 4 are given in Table 2. With reference to Figure 3, a $CH_4$/$CCl_4$/air mixture was fired from the upper burner (I), while a $CH_4$/air mixture was fired from the lower burner (III). The middle burner (II) was inactive. Note that in each test, measurements were made both with and without turbulence air activated. Turbulence air can be activated intermittently at the face of this kiln to enhance bulk mixing.

The measured concentrations of major stable species and of temperatures in

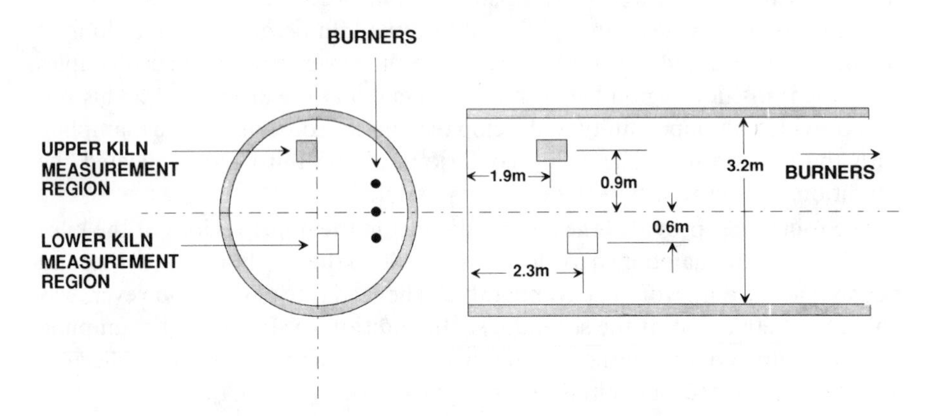


**Figure 5.** Sampling locations near the kiln exit used for the $CCl_4$ and $C_6H_5CH_3$ test results described in the text.

**Table 2. Kiln Operating Conditions During Tests**

| Test no. | Feed | Turbulence air activated | Feed rate (L/min) | Cl/H ratios | | Firing rate (kW) | |
|---|---|---|---|---|---|---|---|
| | | | | Kiln | AB | Kiln | AB |
| 2 | $CCl_4$ | No | 6.8 | 0.25 | 0.18 | 3220 | 4050 |
| | | Yes | 7.9 | 0.21 | 0.17 | 4100 | 4000 |
| 4 | $CCl_4$ | No | 4.2 | 0.11 | 0.09 | 5050 | 3540 |
| | | Yes | 4.2 | 0.09 | 0.06 | 5050 | 3550 |
| 5 | $C_6H_5CH_3$ | No | 3.8[a] | – | – | 3060 | 3840 |
| | | Yes | 3.8 | – | – | 3520 | 3830 |

[a]Average of 19 L every 5 min.

the kiln, at two vertical positions and as influenced by the activitation of turbulence air, are shown in Figures 6 and 7.

Observe first the measured concentrations with the turbulence air turned off (rear towers in Figure 6) and note the differences between the lower (left towers) and upper (right towers) regions of the kiln. In the lower kiln, the concentration of $O_2$ is high (near the ambient air level), the concentration of $CO_2$ is small, the concentration of CO is below detection limits, and $CH_4$ is present at ppm levels. In contrast, the upper region of the kiln contains reduced levels of $O_2$ and elevated levels of $CO_2$, CO, and $CH_4$. Carbon tetrachloride concentrations were below our detection limits in both the lower and upper regions. These results indicate that, in the absence of turbulence air, the lower kiln is an unreactive region dominated by combustion air. The upper kiln, however, appears to be a reactive region as indicated by the reduced level of $O_2$ and increased levels of $CO_2$ and CO. Recall that $CH_4$ and $CCl_4$ were fired from the upper burner, while only $CH_4$ was fired from the lower burner. The elevated level of $CH_4$ in the upper kiln suggests a quenching effect of chlorine on methane combustion, as has been observed in other studies.[7,15]

The effect of turbulence air is to reduce the concentration differences

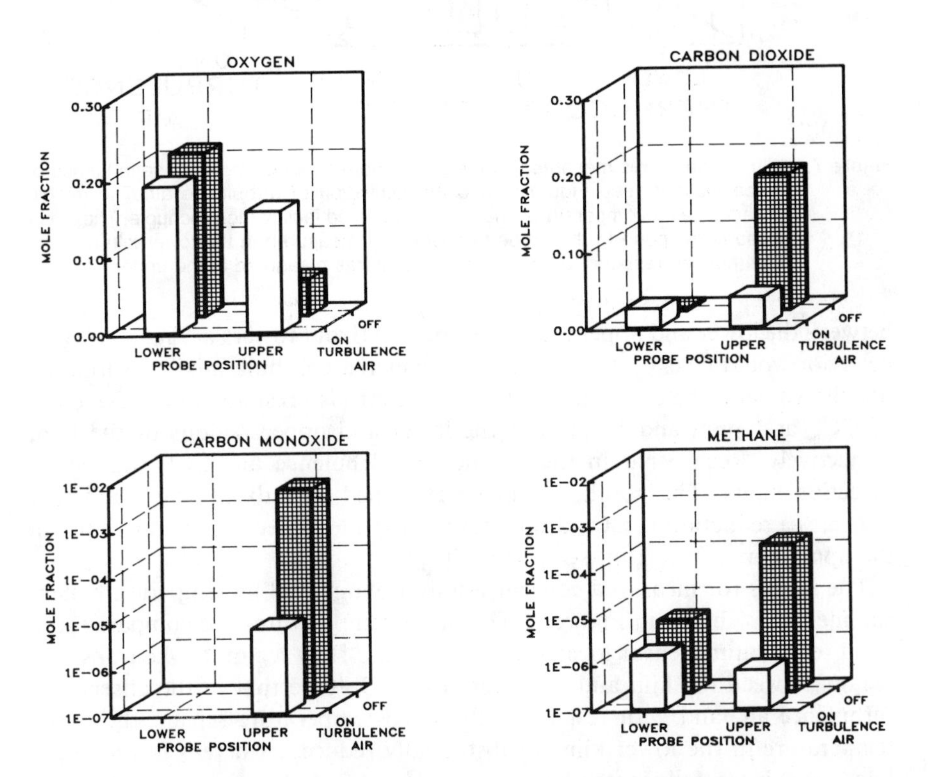


**Figure 6.** Concentrations of major stable species near the kiln exit showing the influence of probe position (see Figure 5) and the activation of turbulence air.

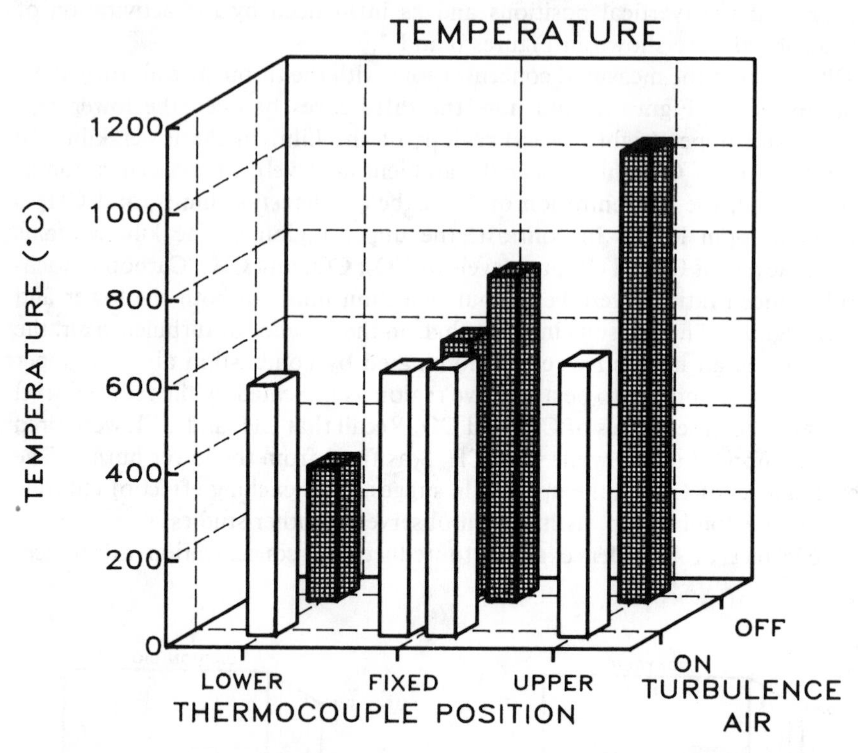


**Figure 7.** Kiln temperatures measured by the probe thermocouple showing the influence of probe location (see Figure 5) and the activation of turbulence air. The middle towers give the temperatures measured by a fixed facility thermocouple. The left of these correspond to the probe temperatures measured in the lower kiln and the rightmost correspond to the probe temperatures measured in the upper kiln.

between the lower and upper regions of the kiln (front towers in Figure 6). This behavior would be expected because of the increase in bulk mixing. Although not shown here, the activation of turbulence air also resulted in the detection of $CCl_4$ at 1 ppm and 12 ppm in the lower and upper regions of the kiln, respectively. Recall that in the absence of turbulence air, $CCl_4$ was below detection limits. This result cannot be explained by bulk mixing. Rather, it indicates a reduction in $CCl_4$ destruction rates, due to decrease temperature in the upper part of the kiln (see Figure 7).

The results for measured concentrations are mirrored by temperature measurements, as shown in Figure 7. The probe temperatures are compared to a fixed temperature sensor located in the bulk gas flow region of the transition region between the kiln and the afterburner. Observe that in the absence of turbulence air, and with respect to the fixed temperature sensor, the probe temperature in the lower kiln is substantially reduced, whereas in the upper kiln, it is substantially increased. The addition of turbulence air has little effect on the fixed temperature, but reduces the difference between the fixed and probe temperatures, demonstrating stratification at the kiln exit in the absence of turbulence air.

The picture that emerges from these kiln-exit measurements is that, in the absence of turbulence air, the flow is primarily axial with minimal radial mixing. The $CCl_4$ fired from the top burner appears to remain and react primarily in the upper regions of the kiln. The addition of turbulence air promotes bulk mixing between the upper and lower regions of the kiln. An interesting consequence is the reduction of temperature in the upper kiln, where the combustion processing appears to be significant. This, in turn, reduces the processing rates and implies that the demand on the afterburner, to achieve the same overall level of waste destruction, will be greater.

## PRELIMINARY RESULTS FOR TOLUENE TESTS

The conditions for test 5 are also given in Table 2. The toluene was fed to the incinerator at 5-min intervals, in 53-L plastic packs filled with oil sorb. Nineteen liters of $C_6H_5CH_3$ were added to each pack. Gas samples were taken from the upper level of the kiln near the exit, from the afterburner arch, and from the stack, as described earlier for the $CCl_4$ run 4.

Since we were particularly interested in temporal variations, we coordinated the initiation of the sampling in the kiln and afterburner with the start of each cycle, i.e., the time each pack was inserted into the kiln. The first sample was taken 30 sec after the insertion of the first pack, the second sample was taken 60 sec after the insertion of the second pack, and so on up to the fourth sample being taken 120 sec after the insertion of the fourth pack. A continuous record of total hydrocarbon concentration in gases drawn from the kiln was available from the facility THC analyzer. In addition, for several 5-min periods, we observed the temperature response in the kiln and afterburner by manually recording the reading of the probe temperature displays at 30-sec intervals.

The correlations between the temperature in the upper kiln and the total hydrocarbons are shown in Figures 8 and 9 with and without turbulence air, respectively. Notice that in both cases, there is an initial decrease in temperature as the pack is loaded and the influx of air increases, a rise to a temperature peak at about 3 min, then a second temperature peak at just over 4 min. As we observed with steady $CCl_4$ feed, turbulence air reduced the temperature and increased the total hydrocarbon concentration in the upper part of the kiln. Note that the range for THC in Figure 9 is one-half that in Figure 8. This overall inverse relationship between temperature and total hydrocarbon concentration, observed upon activation of the turbulence air, is also reflected in the temporal patterns—the total hydrocarbon concentration tends to increase as the temperature decreases, and vice versa.

The peaks in the total hydrocarbon concentration correspond roughly with visual observations of luminous intensity in the kiln. When a pack is first inserted, the pack is easily visible since the residual from the previous pack had

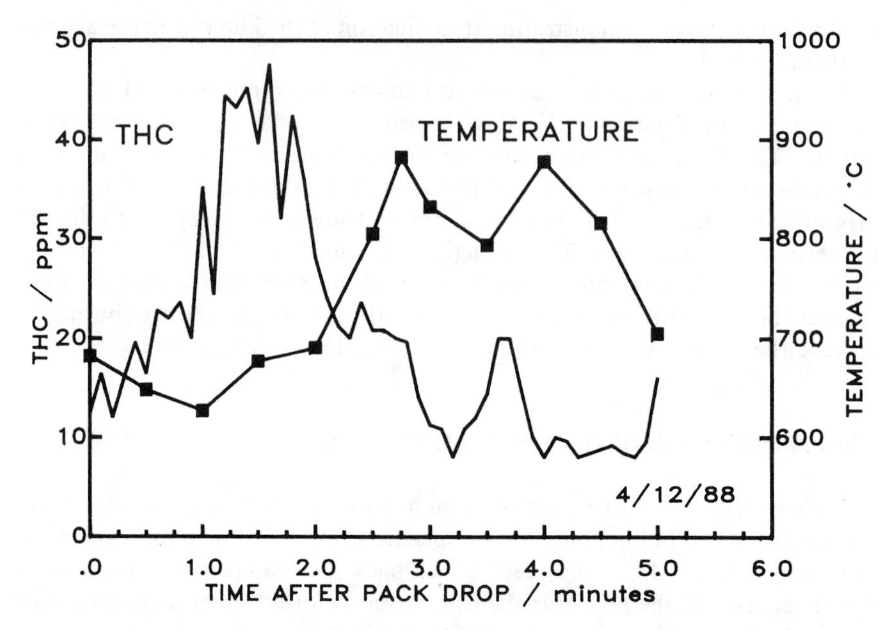


**Figure 8.** Temporal variations of simultaneous temperature and total hydrocarbon concentration (THC) in the upper region near the kiln exit with turbulence air activated. The THC measurements are continuous, whereas the temperature measurements are taken at 30-sec intervals. The solid line connecting the temperature data points is intended only to show trends and aid visualization.

been largely consumed. Just over 1 min after pack drop, a luminous, particulate-laden flame rapidly develops. Eventually, the entire kiln is totally obscured. This activity continues for about 1 min and then gradually subsides. At about 3 min, the bed of solids can usually be seen again, having moved with the rotation of the kiln to a position of about seven o'clock. The bed then rolls over, exposing fresh surface, and the highly luminous flame again fills the kiln. This second burst of activity occurs between 3½ to 4 min after the pack is inserted and is much shorter in duration than the first burst. These bursts of flame appear to be analogous to the intermittent excusions observed by Linak et al.[16,17]

Notice also in Figures 8 and 9 that a maximum lag of 2 min between pack insertion and sample collection is insufficient to characterize the temporal concentration variation. This is further evidenced by the variation of $O_2$ and $CO_2$ as shown in Figure 10 (turbulence air on) and Figure 11 (turbulence air off) up to a maximum 2-min lag between insertion and sample collection. The corresponding temperature measurements are also shown. Recall also that, in contrast to Figures 8 and 9 (which are continuous measurements), Figures 10 and 11 are a composite of measurements taken over four pack insertion intervals.

Observe first the concentration response with the turbulence air off (Figure

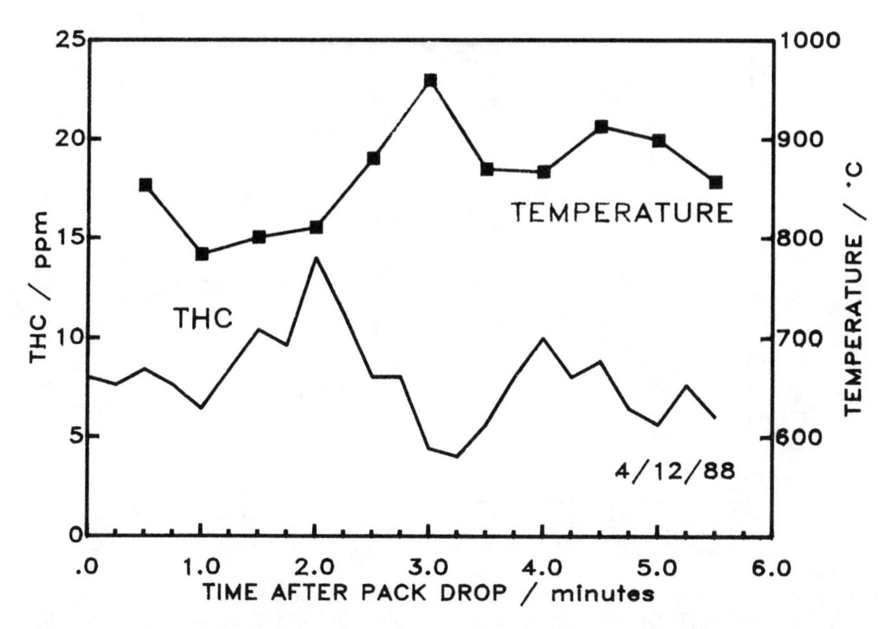


**Figure 9.** Temporal variations of simultaneous temperature and total hydrocarbons concentration (THC) in the upper region near the kiln exit with turbulence air off. The THC measurements are continuous, whereas the temperature measurements are taken at 30-sec intervals. The solid line connecting the temperature data points is intended only to show trends and aid visualization.

10) and recall that the first peak in the THC concentration occurred at about 2 min. The decrease in $O_2$ and increase in $CO_2$ with time indicate that substantial combustion is occurring in the top part of the kiln while the temperature remains relatively constant. Now compare this behavior to the behavior when the turbulence air is on (Figure 11). Recall in this case that the first peak in the THC concentration occurred at about 1½ min. The turbulence air reduces the temperature in the upper part of the kiln, and the decrease in $O_2$ and the increase in $CO_2$ with time is also less than that observed with the turbulence air off. Evidently, the reduction in temperature brought about by the introduction of turbulence air also slows the combustion reactions. One might argue that the smaller temporal decrease in $O_2$ (compared to the case when the turbulence air is off) owes soley to the transport of the oxygen-rich gas from the bottom of the kiln. If this were the case, however, the concentration of $CO_2$ should decrease also, as it did for the $CCl_4$ run. This implies that substantial waste processing is also occurring in the lower part of the kiln, although this speculation must await measurements from the lower part of the kiln for confirmation. Such behavior should be expected for the $C_6H_5CH_3$ tests since the waste is fed into the lower kiln rather than the upper kiln, as was the case in the $CCl_4$ runs.

At this point, our picture of the behavior in the kiln during transient loading

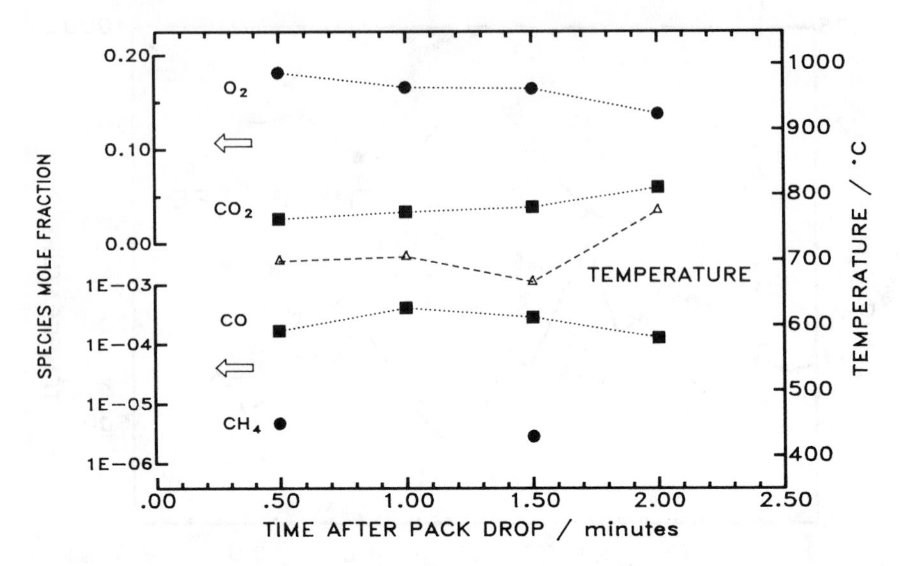


**Figure 10.** Temporal variations of simultaneous temperature and total hydrocarbons concentration (THC) in the upper region near the kiln exit with turbulence air activated. The dotted line connecting the temperature data points is intended only to show trends and aid visualization. No line connects the data points for $CH_4$, since the samples at 1 and 2 min did not contain any observable amounts of $CH_4$.

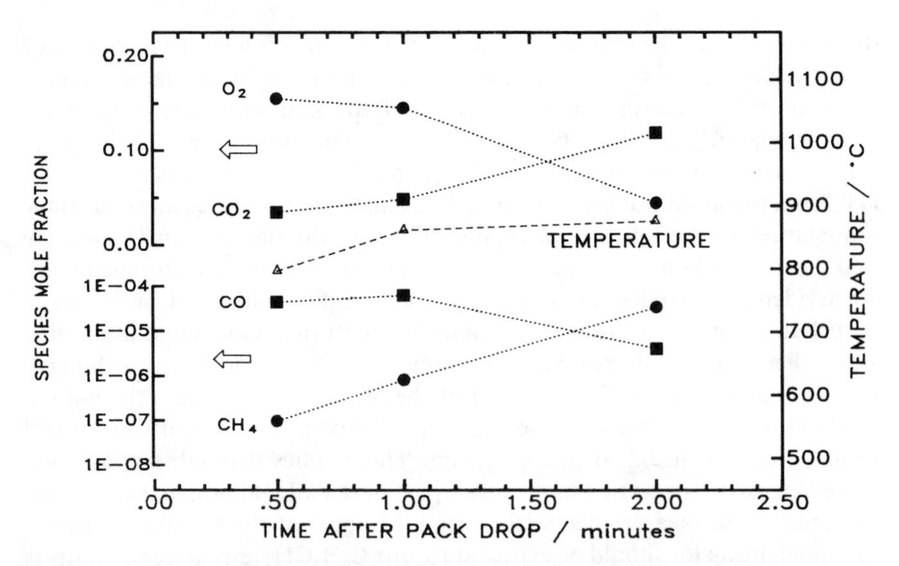


**Figure 11.** Temporal variations of simultaneous temperature and total hydrocarbons concentration (THC) in the upper region near the kiln exit with turbulence air off. The dotted lines connecting the data points are intended to show trends and aid visualization. No sample was taken at 1.5 min.

is beginning to develop, but additional measurements (in the lower part of the kiln and over longer time intervals, for example) are needed. It appears, however, that when the toluene-laden absorbant is exposed to the high-temperature environment of the kiln, there is a rapid evolution of vapor that ignites as it is convected, in part, to the upper regions of the kiln by buoyancy forces. The result is in a rapid increase in total hydrocarbon concentration and a slight reduction in temperature in the upper kiln. As the toluene volatilization decreases, the temperature in the upper kiln returns to near the natural gas/air flame temperature and the total hydrocarbon concentration decreases. As the solids bed rolls over and exposes new surface, the behavior is repeated. The addition of turbulence air accentuates this behavior.

This picture, however, is still incomplete. We have not yet been able to identify the components which comprise the THC concentrations. Curiously, we did not observe $C_6H_5CH_3$ in any of the gas samples collected from the kiln and afterburner at levels above our minimum delection. The high level of the temperature peaks in the presence of turbulence air is as yet unexplained. Future measurements should help clarify the picture.

## SUMMARY

In situ measurements from a rotary kiln incinerator during steady state and transient loading are helping to clarify the local behavior inside an operating kiln. This information is needed to develop reasonable mathematical models of the process for use in incinerator evaluation and incinerator design. Ultimately, the model could be developed to the point that knowledge of the flame-mode chemistry could be combined with knowledge of transport rates thoughout the kiln to tailor kiln operation and design to achieve optimum destruction performance.

An example of the utility of this approach can be drawn from the study of $NO_x$ formation during coal combustion. A recognition that there are numerous chemical pathways for $NO_x$ formation and that at moderate temperatures $NO_x$ is derived primarily from fuel-bound nitrogen has led to the concept of reburning for $NO_x$ control. A similar approach to hazardous waste incineration should also lead to new concepts for maximizing waste destruction and minimizing the formation of hazardous products of incomplete combustion.

Our results to date show clearly the existence of spatial and temporal variations in species concentration and temperature in the kiln. The addition of turbulence (to promote mixing) reduces, but does not eliminate, these variations. A consequence of the enhanced mixing, however, appears to be a reduction in processing rates in the upper regions of the kiln and an increase in the demand on the afterburner. In the particular incinerator under study, the afterburner is more than capable of handling this increased demand. Overall destruction rates of $CCl_4$ and $C_6H_5CH_3$ in all cases exceeded 99.99%.

## ACKNOWLEDGMENTS

The research described in this article has been funded in part by the United States Environmental Protection Agency (EPA) through Cooperative Agreement No. CR809714010 granted to the Hazardous Waste Research Center of Louisiana State University. Although this research has been funded by the EPA, it has not been subjected to agency review and therefore does not necessarily reflect the views of the agency and no official endorsement should be inferred.

The authors would like to acknowledge the assistance and cooperation of the Louisiana Division of Dow Chemical USA, in Plaquemine, LA. In particular, the assistance of Mr. Tony Grant, Mr. Darryl Sanderson, and Mr. Terry Wilks is deeply appreciated. Although this research has been undertaken in a cooperative nature with Dow Chemical USA, it does not necessarily reflect the views of the company and therefore no endorsement should be inferred.

The support offered by Professor L. J. Thibodeaux, Director of Louisiana State University Hazardous Waste Research Center, is appreciated. Assistance provided by Mr. Robert Breaux, Mr. J. S. Tsai, Mr. Greg Arbour, Mr. Nathan Adams, Mr. Earl Crocket, and Mr. David Tate is acknowledged.

## REFERENCES

1. Oppelt, E. T. "Incineration of Hazardous Waste: A Critical Overview," *J Air Poll. Control Assoc.* 37(5): 558–586 (1987).
2. Theodore, L., and J. Reynolds. *Introduction to Hazardous Waste Incineration* (New York: John Wiley & Sons, Inc. 1987).
3. Trenholm, A., T. Lapp, G. Scheil, G. Cootes, S. Klamm, and C. Cassady. "Total Mass Emissions from a Hazardous Waste Incinerator," National Technical Information Service, Order No. PB 87–228 508/AS (1987).
4. Cundy, V. A., T. W. Lester, J. S. Morse, A. N. Montestruc, C. Leger, S. Acharya, A. M. Sterling, and D. W. Pershing. "Rotary Kiln Incineration. I. An Indepth Study—Liquid Injection," *J Air Poll. Control Assoc.* 39(2):63–75 (1989).
5. Cundy, V. A., J. S. Morse, T. W. Lester, and D. W. Senser. "An Investigation of a Near Stoichiometric $CH_4/CCl_4$/Air Premixed Flat Flame," Chemosphere 16(5):989–1001 (1987).
6. Cundy, V. A., D. W. Senser, and J. S. Morse. "Practical Incinerator Implications from a Fundamental Flat Flame Study of Hazardous Waste Combustion," *J Air Poll. Control Assoc.* 36(7):824–828 (1986).
7. Senser, D. W., V. A. Cundy, and J. S. Morse. "Chemical Species and Temperature Profiles of Laminar Dichloromethane-Methane-Air Flames. I. Variation of Chlorine/Hydrogen Loading," *Combust. Sci. Technol.* 51(4–6):209–233 (1987).
8. Tsai, J. S., and A. M. Sterling. "Burning Rates for Linear Droplet Arrays," in *First Annual Symposium on Hazardous Waste Research* (Baton Rouge: Hazardous Waste Research Center, 1987).
9. Lighty, J. S., G. D. Silcox, D. W. Pershing, and V. A. Cundy. "On the Fundamentals of Thermal Treatment for the Cleanup of Contaminated Soils," 81st Air Pollu-

tion Control Association Annual Meeting and Exibition, Dallas, TX, June 19 to 24, 1988.

10. Lighty, J. S., W. D. Owens, G. D. Silcox, D. W. Pershing, and V. A. Cundy. "Thermal Effects and Heat Transfer in the Rotary Kiln Incineration of Contaminated Soils and Sorbents," in *Emissions from Combustion Processes* (Ann Arbor: Lewis Publishers, 1990),Chap. 23.
11. Lighty, J. S., R. Britt, D. W. Pershing, W. D. Owens, and V. A. Cundy. "Rotary Kiln Incineration. II. Laboratory-Scale Desorption and Kiln-Simulator Studies—Solids," *J Air Poll. Control Assoc.* 39(2):187–193 (1989).
12. Senser, D. W., and V. A. Cundy. "Gas Chromatographic Determination of $C_1$ and $C_2$ Chlorinated Hydrocarbon Species in Combustion Products," *Hazardous Waste J.* 4(1):99–110 (1987).
13. Cundy, V. A., T. W. Lester, A. N. Montestruc, J. S. Morse, C. Leger, S. Acharya, and A. M. Sterling. "Rotary Kiln Incineration. III. An Indepth Study—$CCl_4$ Destruction in a Full-Scale Rotary Kiln Incinerator," *J Air Poll. Control Assoc.* 39(7):944–952 (1989).
14. Cundy, V. A., T. W. Lester, A. N. Montestruc, J. S. Morse, C. Leger, S. Acharya, and A. M. Sterling. "Rotary Kiln Incineration. IV. An Indepth Study—Kiln Exit and Transition Section Sampling During $CCl_4$ Processing," *J Air Poll. Control Assoc.* 39(8):1073–1084 (1989).
15. Senkan, S. M. "On the Combustion of Chlorinated Hydrocarbons. II. Detailed Chemical Kinetic Modeling of the Intermediate Zone of the Two-Stage Trichloroethylene-$O_2$-$N_2$ Flames," *Combust. Sci. Technol.* 38:197 (1984).
16. Linak, W. P., J. D. Kilsore, J. A. McSorley, J. O. Wendt, and J. C. Dunn. "On the Occurance of Transient Puffs in a Rotary Kiln Incinerator Simulator," *J Air Poll. Control Assoc.* 37(1):54–65 (1987).
17. Wendt, J. O. L., and W. P. Linak. "Mechanisms Governing Transients from the Batch Incineration of Liquid Wastes in Rotary Kilns," *Combust. Sci. Technol.* 61:169–185 (1988).

CHAPTER 23

# Thermal Effects and Heat Transfer in the Rotary Kiln Incineration of Contaminated Soils and Sorbents

G. D. Silcox, J. S. Lighty, W. D. Owens, D. W. Pershing, and V. A. Cundy

## ABSTRACT

Bench-scale isothermal studies on the rate of desorption of toluene from calcined clay show a strong dependence of rate on temperature. Additional experiments conducted in a bench-scale rotary kiln show a similar dependence of flue toluene concentrations on average kiln gas temperature. The relationship between temperature and desorption and evaporation rates is exponential. Similarly, combustion rates, whether homogeneous or heterogeneous, depend exponentially on temperature. In view of the importance of thermal environment on these key processes, an understanding of heat transfer in rotary kilns is important. This paper briefly examines two experimental studies of thermal effects and presents a mathematical model of heat transfer in rotary kilns. The model is used to explore the effects of the following variables on bed solids temperatures: kiln length, feed moisture content, and feed rate.

## INTRODUCTION

Two series of experiments were performed in order to determine the waste evolution rates from calcined clay particles as a function of thermal environment. The first series involved the use of an isothermal, particle characterization reactor in which the controlling resistances are largely intraparticle. Hot, preheated nitrogen gas was passed through a bed of 2-mm particles that were contaminated with toluene. The gases leaving the bed were analyzed by gas chromotography (Hewlett Packard 5840A).

A second series of experiments, designed to simulate the actual combustion environment of a rotary kiln, were performed in a 73-kW natural gas-fired rotary kiln simulator. In the simulator, a batch of solid is loaded into the kiln and combusted for a given length of time. The inside dimensions of the kiln are 0.61 m by 0.61 m. Waste charges (toluene on calcined clay particles) remained in the kiln for about 30 min. The kiln rotated at 1 rpm.

In addition to these experiments, a mathematical model of the heat transfer in rotary kilns was developed. The heat transfer model is part of a larger modeling effort shown schematically in Figure 1. Three separate models are currently used to describe the kiln: a heat transfer model, a bed solids model, and a planar flow model. Direct coupling of the fluid and heat transfer models is currently impossible due to their different formulations, that is, differential versus zonal. The bed solids and heat transfer model can be directly coupled.

## EXPERIMENTAL RESULTS

The effects of temperature on toluene evolution rates from clay particles in the particle characterization reactor are shown in Figure 2. The particles were initially contaminated with 0.44%-by-weight toluene and desorbed in the reactor at temperatures ranging from 293 to 573 K. The concentrations of toluene in the exhaust sweep gases are shown in Figure 2 as a function of time. At the end of 5 hr, the concentration of toluene in the sweep gas at 293 K approaches 1 ppm and the rate of evolution levels off. At 443 K, the gas concentration is roughly 1 ppm after 1 hr, while at 573 K, the concentration falls to less than 0.2 ppm in less than 1 hr.

The sensitivity of desorption rates to temperature is consistent with the Boltzmann law which gives the fraction of absorbed molecules possessing the

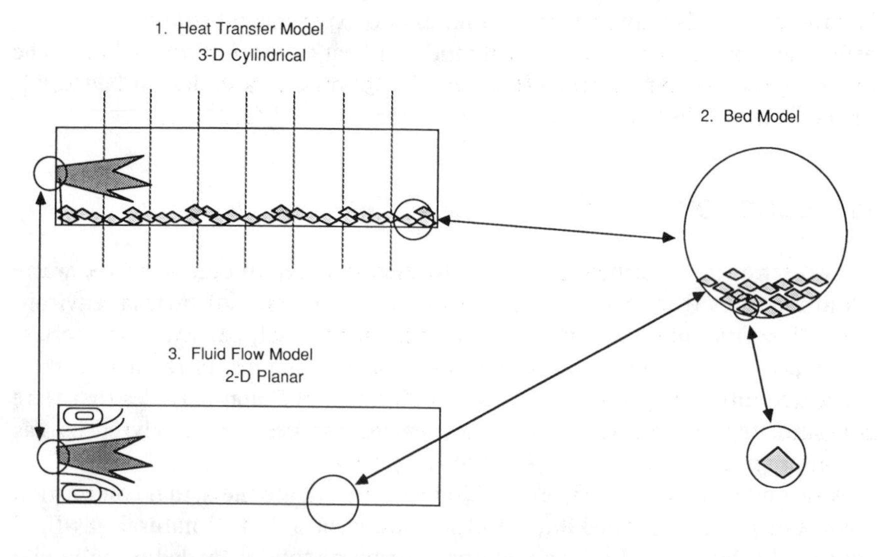


**Figure 1.** The three aspects of an integrated approach to kiln modeling including heat transfer, fluid flow, and bed modeling.

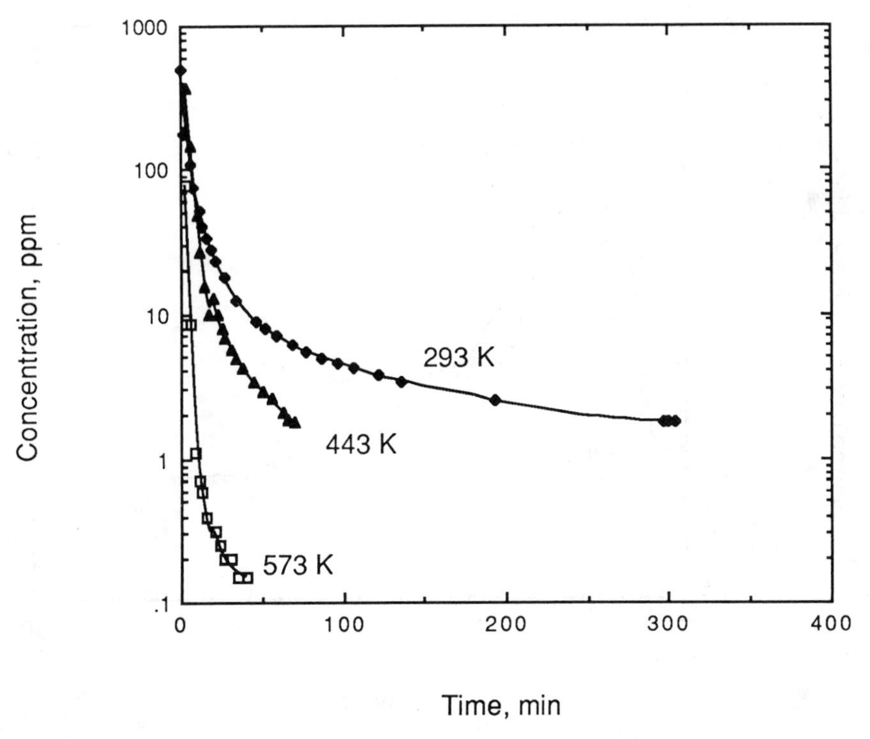


**Figure 2.** The effect of temperature on toluene desorption rates as measured in the particle characterization reactor.

required activation energy E for desorption, $e^{-E/R_gT}$, where $R_g$ is the gas constant. The leveling off in desorption rates is consistent with the observation that the activation energy E increases with decreasing coverage of absorption sites.

The effects of temperature on the flue gas flow rates of toluene from a rotary kiln simulator are shown in Figure 3 as a function of time. The sorbent clay particles were contaminated with 1%-by-weight toluene and the kiln was charged with 4.5 kg of particles and rotated at 1 rpm. The flow rates of toluene in Figure 3 were normalized by division by the initial weight of toluene present.

Toluene flow rates were measured at three average kiln gas temperatures, 770, 840, and 1140 K as determined by suction pyrometry. At 1140 K, virtually no toluene entered the flue. The fraction of toluene burned in the kiln increases with increased temperature. These results show the sensitivity of the desorption and desorption/combustion process to temperature. This is not surprising and it serves to emphasize the importance of understanding the heat transport phenomena occurring within a rotary kiln.

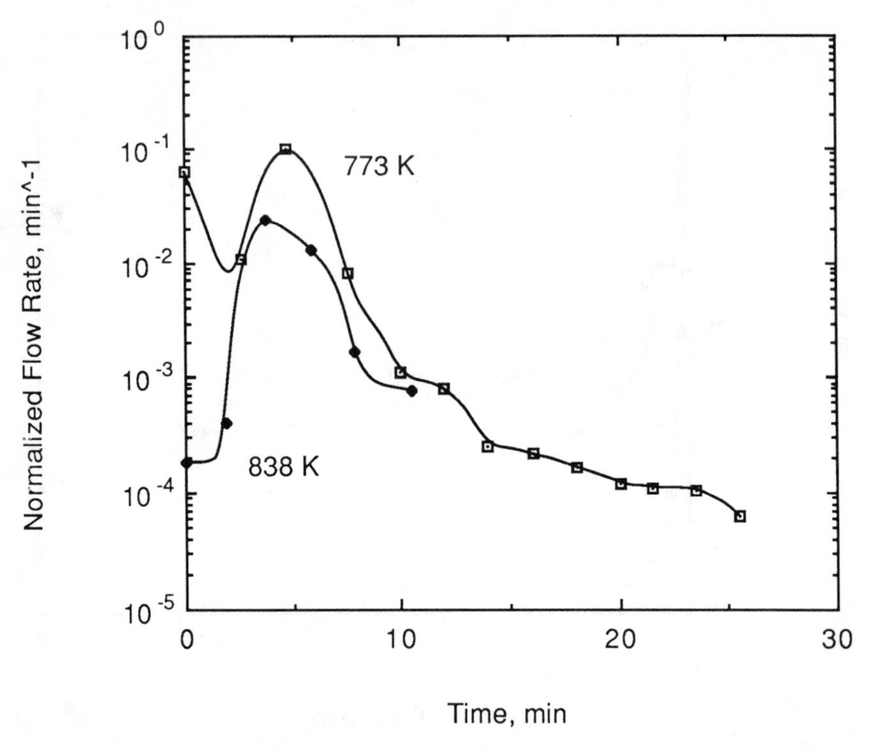


**Figure 3.** Normalized mass flow rates of toluene from the rotary kiln simulator as a function of temperature.

## MATHEMATICAL MODELING OF HEAT TRANSFER

### Introduction

Due to the widespread use of rotary kilns in industry, numerous mathematical models of their heat transfer characteristics have been developed. The models are generally of the zone type and include correlations for the estimation of fluid flow characteristics and entrainment, e.g., Jenkins and Moles,[4] and Gorog et al.[1-3] The model of Jenkins and Moles predicts gas and refractory temperatures which compare well with data they obtained from a full-scale cement kiln. However, their approach neglects the presence of the solids burden and does not provide estimates of solids heating rates. Gorog and coworkers modeled the flame and nonflame zones of rotary kilns. Their studies include the presence of the solids burden, but in their published work, they neglect to make the following extensions. Their flame zone and nonflame zone models are not combined to give a realistic representation of the whole kiln. They also assign a uniform temperature to the bed solids along the entire length of the kiln and do not consider the possibility of coflowing gas and solids streams. Furthermore, radiation between adjacent zones is not allowed.

## Model Description

The heat transfer model described here is an extension of the work of Gorog et al.[3] A schematic of the model is shown in Figure 4. The kiln is divided axially into zones and the afterburner completely surrounds the kiln exhaust zone. The only parameters used to describe it are its temperature and emissivity. The different surfaces within a zone are characterized by their individual, isothermal temperatures. There are no radial temperature gradients in the gas or flame, although the flame and the gas surrounding it are at different temperatures. Radiation is permitted between the surfaces of immediately adjacent zones. The flame is also allowed to radiate to immediately adjacent zones. The surface radiation portion of the model is three-dimensional to account for the asymmetry introduced by the bed solids which rest across the bottom of a horizontal cylinder. The thickness of the kiln wall in the burner zone can be adjusted to any desired value to achieve a realistic heat loss.

The flame is considered a cylinder of constant diameter and length. The combustion rate of the fuel is specified as a function of distance from the burner. The flame is completely surrounded by an annular shell of secondary air, part of which may be attributed to leakage through the rotary seals on the stationary burner wall.

Figure 4 indicates one possible means of feeding solid wastes to the kiln. The waste is packaged in barrels, which are pushed into the kiln at regular intervals. For the purposes of making heat transfer calculations, this discontinuous feeding process is approximated by a continuous process. The model also assumes that the solids burden is well stirred by kiln rotation and that it is isothermal in the radial direction.

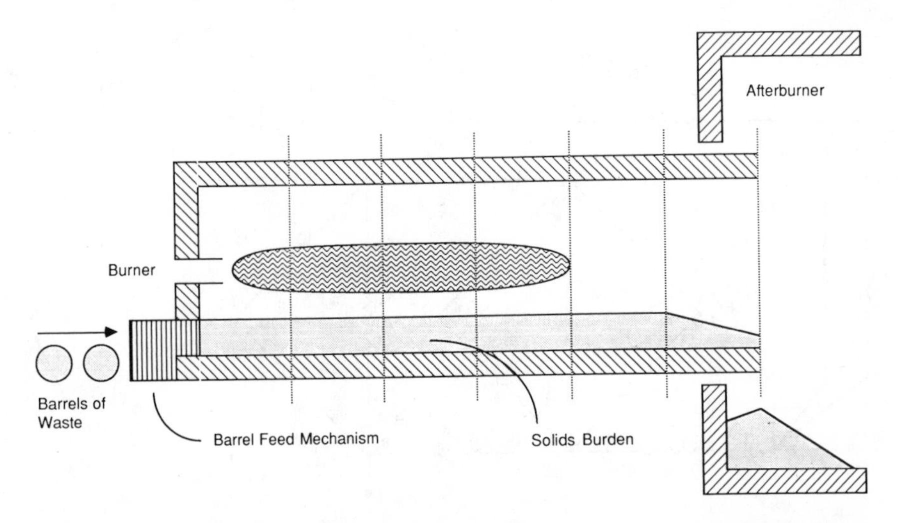


**Figure 4.** Schematic of a rotary kiln showing the division into zones for radiation heat transfer calculations.

The following approach is taken to obtain the governing equations for the model. Energy balances are performed on all materials flowing through the kiln, that is, the gas, flame, and bed solids. The energy balances are made about each zone. The direction of the solids flow can either be with or against the gas flow. For application to contaminated soils, the energy balances include the heat of vaporization of water from the soil.

Separate from the energy balances described earlier, heat transfer calculations are performed to account for all convective, conductive, and radiative energy exchanges. The energy flows are systematically accounted for using electrical network analogs. Figure 5 shows the analog for the burner zone. Radiation to adjacent zones is indicated by superscript plus or minus signs on the appropriate potentials. There are no minus signs in Figure 5 because there is no zone to the left of the burner zone. To obtain the unknown potentials in the analog, (that is, emissive powers, E, or radiosities, J), current balances are performed about each node with an unknown potential. This process is repeated for the entire kiln with appropriate analogs for the burner, flame, nonflame, and exhaust zones.

For example, a current balance about $J_w$ in Figure 5 yields:

$$\frac{E_f^+-J_w}{R_1}+\frac{E_f-J_w}{R_2}+\frac{J_s-J_w}{R_3}+\frac{J_s^+-J_w}{R_4}+\frac{J_w^+-J_w}{R_5}+\frac{E_w-J_w}{R_6}=0$$

where the $R_i$ are appropriate resistances. The set of nonlinear equations that results from the balances is solved using the Newton-Raphson method. Once the nodal balances have been solved, then the energy balances on the gas,

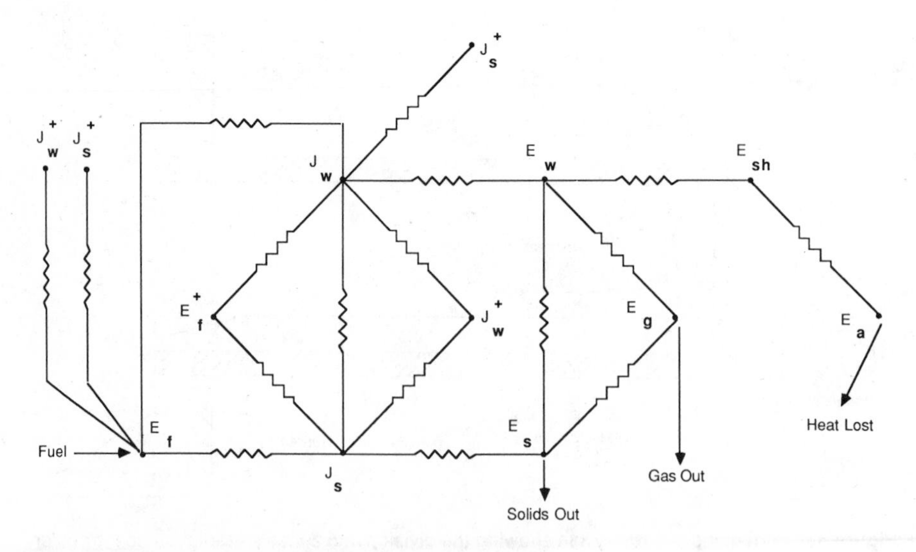


**Figure 5.** Electric cirucit analog for the burner zone.

flame, and bed solids are repeated. An energy balance on the burning jet for the burner zone in Figure 5 yields:

$$T_{jet,2} = 298 - \frac{1}{C_p m_{jet,2}}\left(Q_{s,1} + Q_{w,1} - Q_{g,1} - Q_{in,1}\right) \quad (1)$$

where $T_{jet,2}$ is the temperature of the jet at the exit of the burner zone. Similarly, an energy balance on the solids stream yields:

$$T_{s,2} = T_{s,1} + \frac{Q_s}{C_s m_s} \quad (2)$$

The Q's in Equations 1 and 2 are energy transfer rates due to radiation, convection, and bulk flow.

The iterative solution procedure described here is repeated until converged temperatures are obtained for the bed solids. The entire solution procedure is summarized in Figure 6. A separate program is used to calculate the view factors. Guesses for all unknown temperatures and radiosities are required to

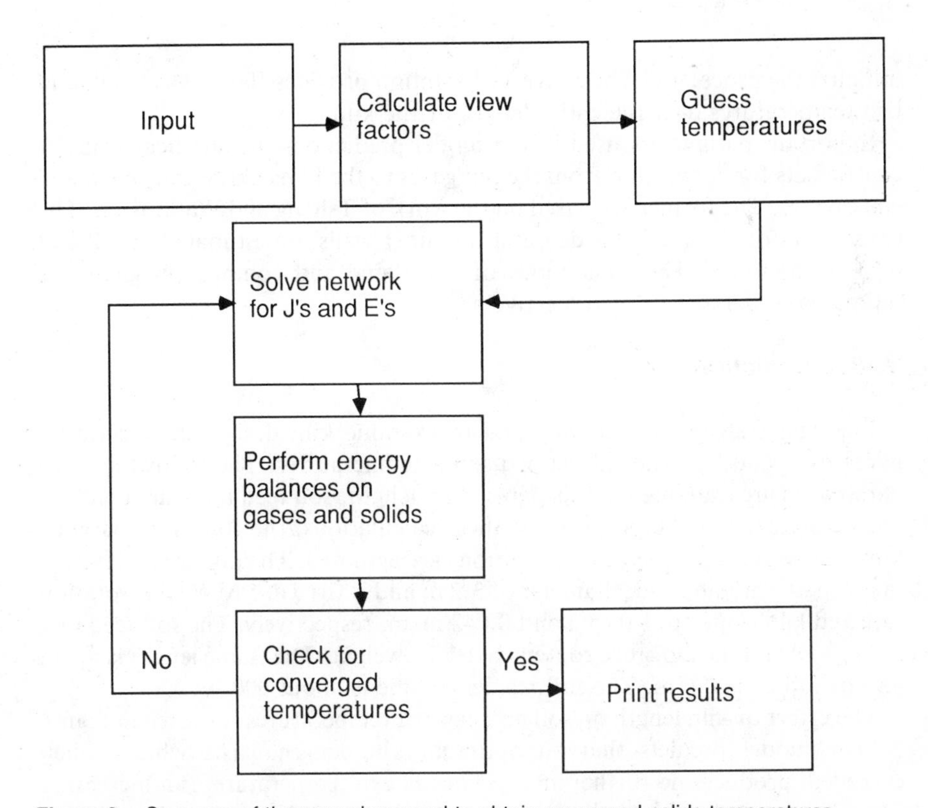


**Figure 6.** Summary of the procedure used to obtain converged solids temperatures.

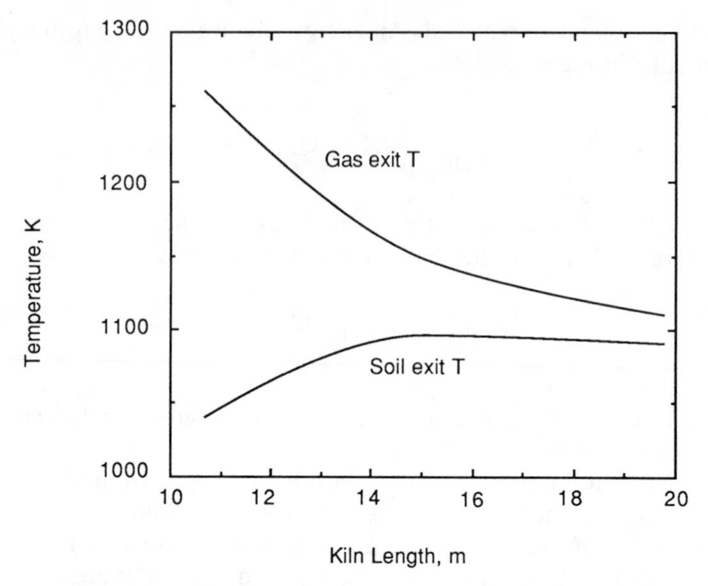


**Figure 7.** Model predictions showing the effect of kiln length on mean gas and soil exit temperatures.

initialize the procedure. The converged solution provides flame, gas, wall, and bed temperatures over the entire length of the kiln.

Important parameters used in the model predictions include heat transfer coefficients for convection from the hot gases to the kiln charge and walls, and emissivities. The former are based on the work of Tsheng and Watkinson.[5] The emissivities of the solids burden and the inner walls are estimated at 0.8 and 0.75, respectively. The emissivities of the flame and combustion gases are estimated at 0.3 and 0.15, respectively.

## Model Predictions

The heat transfer model was used to examine kiln design and operation parameters and how they affect bed solids temperatures. The following three parameters are examined in this paper: kiln length, soil feed moisture content, and feed rate. For the purposes of these calculations, the following baseline kiln dimensions and operating conditions are assumed. The kiln measures 10.7 m in length with an inside diameter of 3.2 m and is fired at 5 MW. The rotation rate and kiln slope are 1.0 rpm and 0.042 m/m, respectively. The soil feed rate is 4 kg/sec and the moisture content is 0% by weight. The flame length is 8.8 m and its radius is 0.6 m. The soil residence time is about 900 sec.

The effect of kiln length on soil and gas exit temperatures is shown in Figure 7. The model predicts that an optimum kiln length exists which, when exceeded, produces no further increase in soil exit temperature. But increasing kiln length does increase soil residence time at elevated temperatures. Hence,

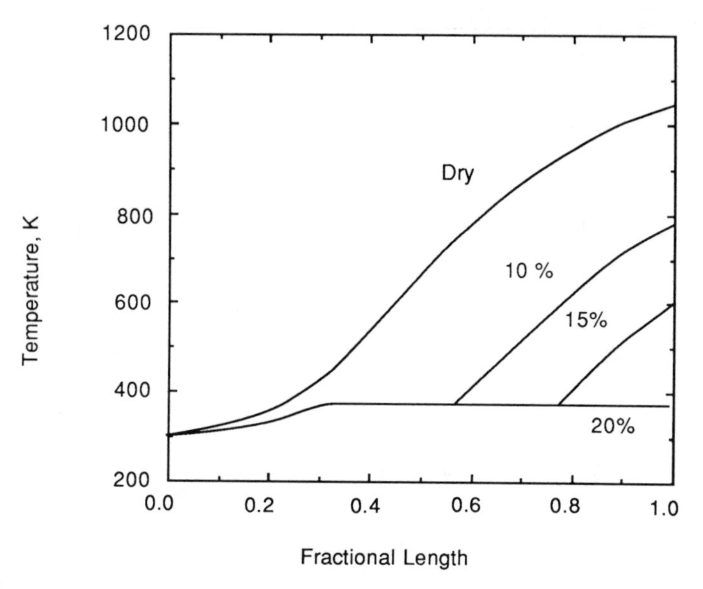


**Figure 8.** Predicted effects of feed soil moisture content on soil temperature history.

for a particular contaminated soil, an optimum kiln length may exist, all other parameters being equal. It is understood that by simply changing kiln rotation rates and feed rates, the soil residence time and exit temperature can be changed. The kiln length parameter only becomes important to kiln design if the soil feed rate is fixed by throughput kiln requirements.

The predicted effect of soil feed moisture weight fraction is illustrated in Figure 8 for weight fractions ranging from 0 to 20%. Moisture level is tremendously important due to the high heat of vaporization of water and evaporative cooling effects are clearly seen in Figure 8. High moisture content requires modification of kiln operation to lower feed rates and lengthen solids residence time.

The effects of soil feed rate on soil exit temperatures are shown in Figure 9. The feed is assumed dry. Increasing the feed rate from 2 to 8 kg/sec decreases bed exit temperatures from 1350 to 750 K, however, the time that the solids spend above a certain temperature is just as important as the final temperature. The effect of feed rate on the time that the bed spends above an arbitrarily chosen reference temperature of 850 K is shown in Figure 10. For example, to achieve a temperature of at least 850 K for 5 min requires that the feed rate be less than 4 kg/sec. This is illustrated in Figure 10 at conditions identical to those used in obtaining Figure 9.

## CONCLUSIONS

The principle conclusion to be drawn from the experimental studies is that time and temperature are important factors affecting soil decontamination

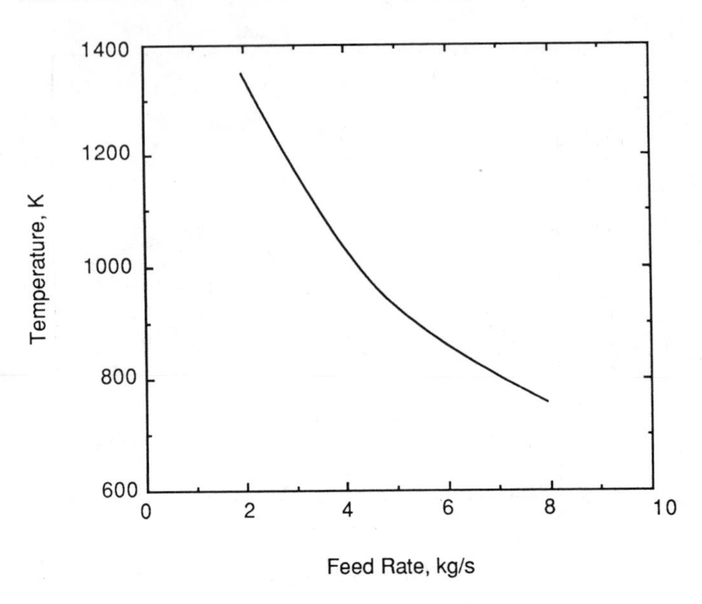


**Figure 9.** Model predictions showing the effect of soil feed rate on soil exit temperature.

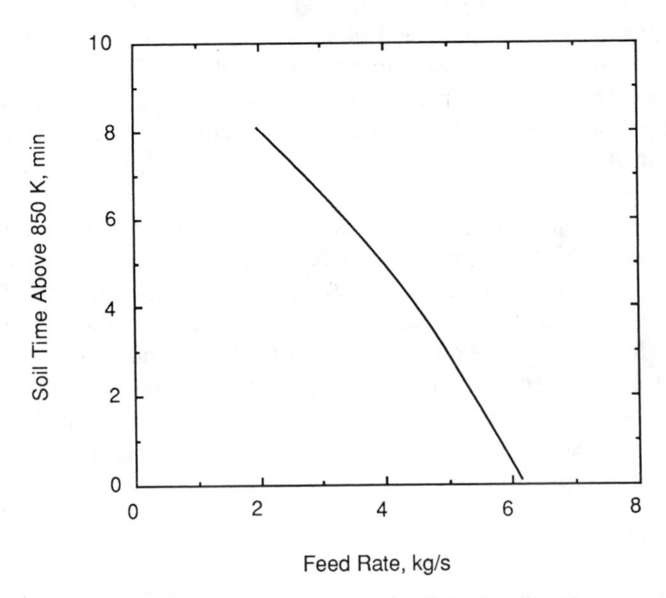


**Figure 10.** The predicted effect of feed rate on time of exposure of bed solids to temperatures in excess of 850 K.

rates. This is not surprising. The critical problem is how to design and operate rotary kilns to provide the required residence times and temperatures sufficient to achieve desired levels of decontamination. Model predictions of the thermal environment and history of soil beds in rotary kilns indicate that an optimum kiln length exists for providing peak bed temperatures. Furthermore, the predictions clearly show the extreme sensitivity of bed temperature to soil feed moisture content. This suggests that kiln operating conditions must be carefully chosen to match soil moisture levels. Finally, the predictions indicate that by adjusting soil feed rate, the desired temperature window needed to assure decontamination can be achieved.

## ACKNOWLEDGMENT

This work was sponsored by the Advanced Combustion Engineering Research Center. Funds for this center are received from the National Science Foundation, the state of Utah, 25 industrial participants, and the U.S. Department of Energy.

## REFERENCES

1. Gorog, J. P., J. K. Brimacombe, and T. N. Adams. "Radiative Heat Transfer in Rotary Kilns," *Met. Trans. B.*, 12B: 55–70 (1981).
2. Gorog, J. P., T. N. Adams, and J. K. Brimacombe. "Regenerative Heat Transfer in Rotary Kilns," *Met. Trans. B.*, 13B: 153–163 (1982).
3. Gorog, J. P., T. N. Adams, and J. K. Brimacombe. "Heat Transfer from Flames in a Rotary Kiln," *Met. Trans. B.*, 14B: 411–424 (1983).
4. Jenkins, B. G., and F. D. Moles. "Modeling of Heat Transfer From a Large Enclosed Flame in a Rotary Kiln," *Trans. Inst. Chem. Eng.*, 59: 17–25 (1981).
5. Tscheng, S. H., and A. P. Watkinson. "Convective Heat Transfer in a Rotary Kiln," *Can. J. Chem. Eng.*, 57: 433–442 (1979).

## CHAPTER 24

# Incineration of a Chemically Contaminated Synthetic Soil Matrix Using a Pilot-Scale Rotary Kiln System

**M. P. Esposito, M. L. Taylor, C. L. Bruffey and R. C. Thurnau**

## INTRODUCTION

The Resource Conservation and Recovery Act (RCRA) Hazardous and Solid Waste Amendments of 1984 (HSWA) prohibit the continued land disposal of untreated hazardous wastes beyond specified dates. The statute requires the U.S. Environmental Protection Agency (EPA) to set "levels or methods of treatment, if any, which substantially diminish the toxicity of the waste or substantially reduce the likelihood of migration of hazardous constituents from the waste so that short-term and long-term threats to human health and the environment are minimized." In addition to addressing future land disposal of specific listed wastes, the RCRA land disposal restrictions also address the disposal of soil and debris from Comprehensive Environmental Response, Compensation, and Liability Act (CERCLA) site response actions. Sections 3004(d) (3) and (e) (3) of RCRA state that the soil/debris waste material resulting from a Superfund-financed response action or an enforcement authority response action implemented under Sections 104 and 106 of CERCLA, respectively, will become subject to the land ban on November 8, 1988.

Because Superfund soil/debris waste often differs significantly from other types of hazardous waste, the EPA is developing specific RCRA Section 3004 (m) standards or levels applying to the treatment of these wastes. These standards will be developed through the evaluation of best demonstrated and available technologies (BDAT). In the future, Superfund wastes, in compliance with these regulations, may be deposited in land disposal units; wastes not meeting the standards will be banned from land disposal unless a variance is issued.

In early 1987, the EPA's Hazardous Waste Engineering Research Laboratory (now known as the Risk Reduction Engineering Laboratory), at the request of the Office of Emergency and Remedial Response, initiated a

**Table 1. Clean Soil Formula**

| Percent (by volume) (%) | Component |
|---|---|
| 30 | Clay[a] |
| 25 | Silt |
| 20 | Sand |
| 20 | Topsoil |
| 5 | Gravel |
| — | |
| 100 | |

[a]Bentonite and Montmorillonite.

research program to evaluate various treatment technologies for contaminated soil and debris from Superfund sites. Under Phase 1 of this research program, which was conducted from April to November 1987, a synthetic soil matrix (SSM) was spiked with chemical contaminants typical of those found at Superfund sites. Portions of the chemically spiked SSM were implemented in bench-scale and pilot-scale performance evaluations of five available treatment technologies including (1) soil washing, (2) chemical treatment (KPEG), (3) thermal desorption, (4) incineration, and (5) stabilization/fixation. This paper covers those segments of Phase 1 related to preparation and spiking of the SSM and evaluation of incineration as a treatment technology. As part of this technology evaluation, a complete series of pilot-scale test burns was conducted and a battery of process and emission samples were collected and analyzed. Results of these tests are described herein.

## METHODS AND EXPERIMENTAL

### Preparation and Spiking of the SSM

The soil matrix components of the SSM (see Table 1) were selected on the basis of an extensive review of soil composition data from nearly 100 Superfund sites from various regional locations. The soil was prepared by first air drying the components to minimize moisture, then mixing the components together in a standard truck-mounted, 6-$yd^3$ cement mixer to obtain two identical 15,000-lb batches. This operation was completed at an Ohio sand and gravel quarry using standard process equipment. The SSM was then dispensed in 500-lb quantities into epoxy-lined, 55-gal steel drums, loaded on a flatbed truck, and transported to the EPA's Center Hill Research Facility in Cincinnati, OH where the chemicals (or analytes) were added. Analysis of the SSM revealed that it had a pH of 8.5 and a cation exchange capacity of 133 meq/100ml. It was also checked for contaminants appearing on the EPA's Hazardous Substance List and found to be contaminant free.[1]

The chemicals which were subsequently added to the SSM were selected to be representative of specific compounds or classes of chemical compounds typically found in contaminated soils at Superfund sites. The organic chemicals added included ethylbenzene, acetone, 1,2-dichloroethane, tetra-

**Table 2. Target Contaminant Levels for Synthetic Soils Used as Waste Feed**

| | Contaminant | Soil I (mg/kg) | Soil II (mg/kg) |
|---|---|---|---|
| Volatiles | | | |
| | Ethylbenzene | 3,200 | 320 |
| | Xylene | 8,200 | 820 |
| | Tetrachloroethylene | 600 | 60 |
| | Chlorobenzene | 400 | 40 |
| | Acetone | 6,800 | 680 |
| | 1,2-Dichloroethane | 600 | 60 |
| | Styrene | 1,000 | 100 |
| | | 20,800 | 2,080 |
| Semivolatiles | | | |
| | Anthracene | 6,500 | 650 |
| | Bis(2-ethylhexyl)phthalate | 2,500 | 250 |
| | Pentachlorophenol | 1,000 | 100 |
| | | 10,000 | 1,000 |
| Inorganics | | | |
| | Lead | 280 | 280 |
| | Zinc | 450 | 450 |
| | Cadmium | 20 | 20 |
| | Arsenic | 10 | 10 |
| | Copper | 180 | 180 |
| | Nickel | 30 | 30 |
| | Chromium | 30 | 30 |
| | | 1,000 | 1,000 |

chloroethylene, chlorobenzene, styrene, xylene, anthracene, pentachlorophenol, and bis(2-ethylhexyl)phthalate. Salts or oxides of the following inorganic were also added: lead, zinc, cadmium, arsenic, copper, chromium, and nickel. In order to challenge the incineration system with both a high and a low range of contaminant concentrations, such as occur at typical Superfund sites, two different spiked soil formulations were prepared in which the concentration of the targeted organic contaminants/analytes differed by a factor of 10X, i.e., total organic contaminants were set at approximately 30,000 mg/kg in one formulation of spiked SSM and 3000 mg/kg in the other formulation. The level for the inorganic contaminants was set at 1000 mg/kg in both formulations. Table 2 provides a complete listing of the contaminant concentrations targeted for each formulation. Nearly 14,000 lb of each of the two formulations of spiked SSM were prepared in July and August of 1987 for the test burns which were subsequently conducted in September of 1987.

## Incineration Test Facility

The series of test burns was conducted at the John Zink Company's RCRA-permitted testing facility located in Tulsa, OK. The test facility consists of several pilot-scale modular components which can be assembled in a variety of configurations.

For this test series, a rotary kiln system was used as shown schematically in Figure 1. The system consisted of a refractory-lined rotary kiln 15 ft in length and 5 ft in outside diameter, fitted with a natural gas fuel burner, a screw conveyor for feeding bulk solids, and a continuous bottom ash removal and quench system. Gases from the kiln were ducted to a horizontal secondary combustion chamber (10 ft in length, 4.5 ft in outside diameter, and fitted with a natural gas burner), a cyclone separator for removal of large particulates in the gas stream, a gas quench section, an adjustable venturi scrubber, and an induced-draft fan. For this test, the rotary kiln was positioned for co-current operation, i.e., the solids/ash traveled in the same direction as the combustion gases. The entire system operated under negative pressure to avoid fugitive losses to the atmosphere. In addition to this equipment, a tertiary afterburner (required by the Oklahoma State Department of Health for final thermal treatment of the secondary combustion system's flue gas) followed the air pollution control equipment . All sampling for this test program was conducted upstream of the tertiary unit.

The test design called for the completion of three 4-hr test burn runs for each SSM formulation. The feed rate to the incinerator was maintained at approximately 1000 lb/hr of contaminated soil. The kiln rotation speed was maintained at 0.35 rpm throughout all tests to provide a solids retention time in the kiln of 45 min. The temperature of the kiln was maintained at 1800°F during all tests and the secondary combustion chamber was maintained at

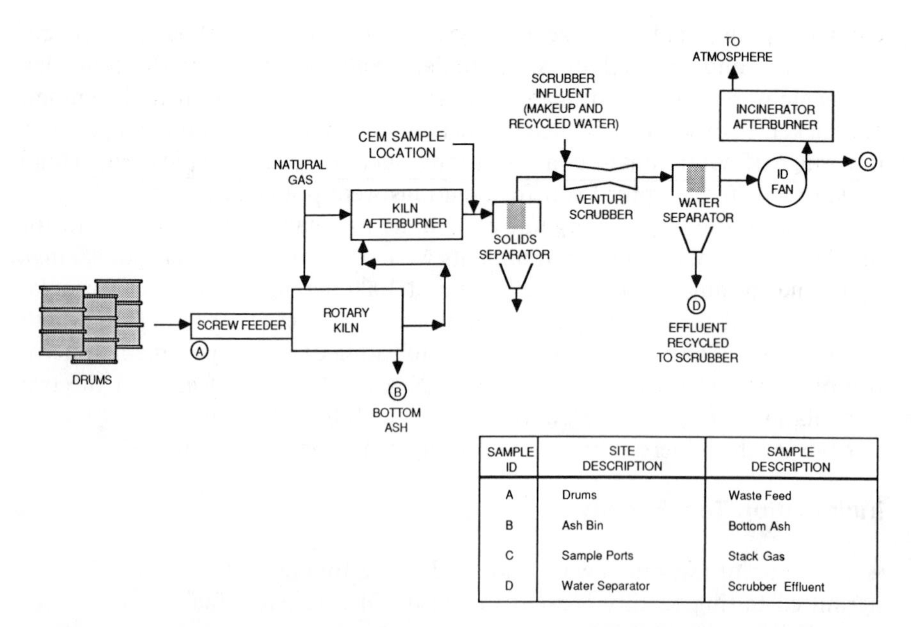


| SAMPLE ID | SITE DESCRIPTION | SAMPLE DESCRIPTION |
|---|---|---|
| A | Drums | Waste Feed |
| B | Ash Bin | Bottom Ash |
| C | Sample Ports | Stack Gas |
| D | Water Separator | Scrubber Effluent |

**Figure 1.** Schematic diagram of the John Zinc Company's pilot-scale rotary kiln incineration system showing feed and residuals sampling sites (points A, B, C, and D).

2000°F. Excess air was fed to both combustion chambers at all times, and the scrubber water was recycled completely, with no blowdown.

The test plan also called for a complete process and stack emission monitoring program. The volatile compounds ethylbenzene and xylene and the semivolatile compounds anthracene and bis(2-ethylhexyl)phthalate were designated as Principal Organic Hazardous Constituents (POHCs) because they were present in the waste feed in the largest amounts. These two compounds as well as each of the other organic and inorganic compounds which were added to the test soils were quantitated in the incinerator influent and effluent samples. Other parameters monitored during the test series included chlorinated dibenzo-p-dioxins and dibenzofurans in the waste feed, bottom ash, and scrubber water, and HCl, particulate, $O_2$, $CO_2$, and CO in the gas stream emerging from the air pollution system. All of the sampling and analysis methods utilized for the monitoring program were EPA-approved methods.

John Zink personnel operated the incineration system throughout the entire test program. They also provided standard system performance monitoring consisting of continuous temperature, gas flow, and $O_2$ monitoring through the kiln and secondary chamber, plus continuous monitoring of CO, $CO_2$, and $NO_x$ in the stack gases exiting the tertiary burner. PEI Associates, Inc. personnel collected all process samples of soil feed, bottom ash, and scrubber water throughout each test run. PEI personnel also collected continuous stack gas samples from sample ports following the venturi for volatiles (VOST), semivolatiles (MM5), metals (EPA Method 12), HCl (EPA Method 5), CO, $CO_2$, and $O_2$ analyses throughout each test run. A chart showing the overall monitoring and sampling plan is shown in Figure 2. The sample ports used by PEI for stack testing purposes were located in the flue between the venturi scrubber

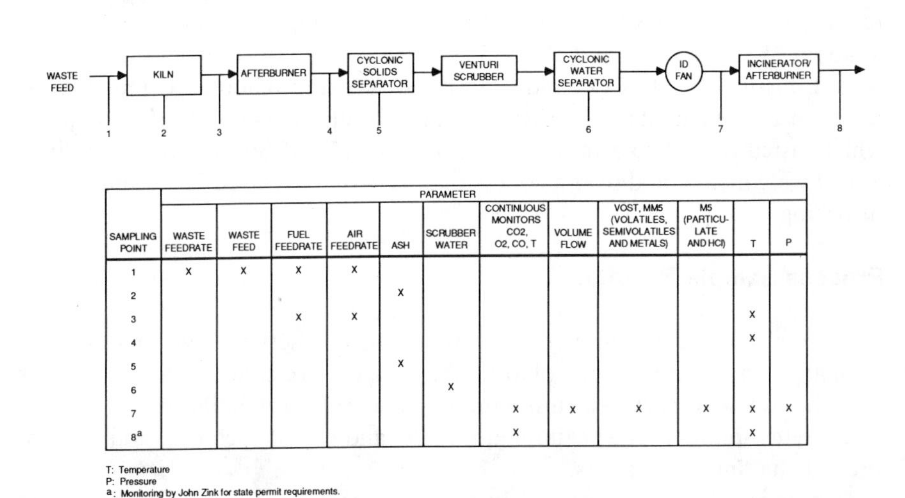


| SAMPLING POINT | PARAMETER | | | | | | | | | | | |
|---|---|---|---|---|---|---|---|---|---|---|---|---|
| | WASTE FEEDRATE | WASTE FEED | FUEL FEEDRATE | AIR FEEDRATE | ASH | SCRUBBER WATER | CONTINUOUS MONITORS CO2, O2, CO, T | VOLUME FLOW | VOST, MM5 (VOLATILES, SEMIVOLATILES AND METALS) | M5 (PARTICULATE AND HCl) | T | P |
| 1 | X | X | X | X | | | | | | | | |
| 2 | | | | | X | | | | | | | |
| 3 | | | X | X | | | | | | | X | |
| 4 | | | | | | | | | | | X | |
| 5 | | | | | X | | | | | | | |
| 6 | | | | | | X | | | | | | |
| 7 | | | | | | | X | X | X | X | X | X |
| 8[a] | | | | | | | X | | | | X | |

T: Temperature
P: Pressure
a: Monitoring by John Zink for state permit requirements.

**Figure 2.** Operating parameters monitored during test burns.

and the tertiary afterburner. The ports were 4 in. in diameter, located on a vertical, 12-in. i.d. section of flue about 15 ft above ground.

Prior to the start of each test, the combustion system was brought up to operating temperature by firing natural gas in the system for several hours. The process of feeding soil to the kiln was started only after the test system was operating satisfactorily at steady-state condition. The auxiliary fuel firing rate was adjusted automatically by system controls to maintain a steady operating temperature throughout each run.

The 55-gal drums of soil were staged on the hazardous waste storage pad located immediately adjacent to the test equipment. Drums were transported to the screw conveyor using a hydraulic drum lifter; once the drums has been emptied of feed material, they were used to containerize the bottom ash, which ultimately was sent for disposal to U.S. Pollution Control, Inc.'s Lone Mountain RCRA landfill in Oklahoma.

## RESULTS

### Operating Conditions

Two runs per day were conducted over the 3-day period of September 16 to 18, 1987. Runs 1, 2, and 3 were conducted with Soil Formulation I (high organics) and runs 4, 5, and 6 were conducted with Soil Formulation II (low organics). No significant problems with the incinerator system were encountered during these tests.

The process operating data collected during each test (see Table 3) indicate the temperatures and feed rates achieved were reasonably close to the test plan goals (i.e., 1800°F in the kiln, 2000°F in the secondary, and a nominal feed rate of 1000 lb/hr). Excess air (measured as percentage of $O_2$) was maintained at about 3% in the kiln and about 5% in the secondary chamber during both tests. Emissions of $O_2$, $CO_2$, and CO were steady throughout; CO remained at less than 10 ppm at all times except for a few brief excursions of 45 to 90 ppm, which lasted from 1 to 5 min. A total of 13,932 lb of Soil I and 13,460 lb of Soil II were incinerated over a course of 3 days that involved 29 hr and 22 min of testing.

### Process Sample Results

Process samples of waste feed, bottom ash, and scrubber water that were ultimately analyzed for semivolatiles and metals were collected as composites over the course of each test run. Samples analyzed for volatiles were collected as discrete samples at the beginning, middle, and end of each run, and composited at the time of analysis.

Two approaches were utilized in analyzing the process samples that were collected during each test run. First of all, the samples were analyzed for the

**Table 3. Summary of Process Operating Conditions**

| | Soil I | | | | Soil II | | | |
|---|---|---|---|---|---|---|---|---|
| **Location/parameter** | **Run 1** | **Run 2** | **Run 3** | **Average** | **Run 4** | **Run 5** | **Run 6** | **Average** |
| **Feed** | | | | | | | | |
| Time elasped, hr:min | 4:23 | 3:45 | 4:54 | 4:21 | 4:06 | 4:12 | 4:02 | 4:07 |
| Amount fed, lb | 4640 | 3939 | 5353 | 4644 | 4392 | 4561 | 4507 | 4487 |
| Feed rate, lb/hr | 1059 | 1050 | 1092 | 1067 | 1071 | 1086 | 1118 | 1092 |
| **Kiln** | | | | | | | | |
| Temperature range, °F | 1764–1824 | 1745–1809 | 1725–1770 | 1745–1801 | 1706–1777 | 1754–1804 | 1764–1799 | 1741–1793 |
| Average temperature, °F | 1795 | 1776 | 1760 | 1777 | 1749 | 1784 | 1780 | 1771 |
| $O_2$,% | 3.3 | 4.7 | 2.5 | 3.5 | 2.6 | 3.3 | 2.4 | 2.8 |
| **Secondary chamber** | | | | | | | | |
| Temperature range, °F | 1942–2038 | 1928–2008 | 1960–2008 | 1961–2018 | 1959–2004 | 1983–2014 | 1999–2010 | 1980–2009 |
| Average temperature, °F | 2004 | 1997 | 1995 | 1999 | 1985 | 2004 | 2005 | 1998 |
| $O_2$,% | 4.1 | 6 | 4.6 | 4.9 | 4.1 | 5.4 | 5.2 | 4.9 |
| **Scrubber** | | | | | | | | |
| Flow, % maximum | 72 | 70 | 76 | 73 | 72 | 72 | 70 | 71 |
| Blowdown, % maximum | 0 | 0 | 0 | 0 | 0 | 0 | 0 | 0 |
| Pressure drop (venturi), in. w.g. | 23 | 25 | 24 | 24 | 26 | 23 | 23 | 24 |
| **Stack** | | | | | | | | |
| $O_2$, % | 5.8 | 6.3 | 6.1 | 6.1 | 5.1 | 5.9 | 5.3 | 5.4 |
| $CO_2$, % | 10.1 | 9.6 | 10.2 | 10 | 10.9 | 10.4 | 11.1 | 10.8 |
| CO, ppm | $<10$ | $<10$ | $<10$ | $<10$ | $<10$[a] | $<10$[b] | $<10$ | $<10$ |

[a]On three occasions, CO spiked to 90, 69, and 69 ppm for about 1 min. each occurrence.
[b]On two occasions, CO spiked to 45 ppm briefly (less than 5 min. each).

total content of each target analyte using EPA SW-846 methods. Second, TCLP procedures were used to develop leachability data for the waste feed and the bottom ash samples from each test run. The TCLP and total analyte analyses have both been used to develop conclusions about the treatment efficiency of incineration for contaminated soils, relative to the pending land ban provisions of HSWA.

### *Waste Feed Analyses*

Table 4 lists the data obtained through total analysis of the waste feed/soil samples collected during the test program; the concentrations found for most of the analytes were reasonably close to the desired targets, although some of the organics were on the low side of the targeted values.

One notable exception is the group of results for pentachlorophenol (PCP). Five of the six feed samples failed to show any detectable concentration of the compound, even though it was known that PCP was added to each batch either at the 1000 mg/kg level or at the 100 mg/kg level. One possible explanation for this may be related to the current SW-846 sample preparation method which calls for an acid digestion of the soil sample prior to analysis; reportedly, this digestion procedure is very ineffective in liberating PCP from the soil into the extract which is ultimately analyzed.[2]

Table 5 compares the results of the average total analysis of the waste feed soils to the results of the TCLP analysis for each soil (both values developed on samples of the untreated soil taken simultaneously from the feedstock during each test run). The table also lists the regulatory limits that have been proposed for several of the analytes. The data in Table 5 indicate that both untreated soils fail the TCLP test for volatiles, but pass for all of the inorganic contaminants.

### *Bottom Ash Analyses*

Table 6 lists the results from analyses of the bottom ash samples collected during each of the test runs. The organic compounds chlorobenzene, styrene, tetrachloroethylene, anthracene, and pentachlorophenol were not detected in any of the ash samples, indicating that they were completely removed and/or destroyed. Measurable quantities of ethylbenzene and xylene were sometimes found in the ash of both soils and 1,2-dichloroethane was found in the ash produced during two of the three tests of Soil II; the amounts reported were very small (in the low parts-per-billion range) and within two to three times the method detection limit. Acetone was found in the ash samples of all runs for both soils at levels ranging from 190 to 790 $\mu$g/kg; these levels are 24 to 99 times higher than the method detection level (8$\mu$g/kg). Similarly, the phthalate analyte was found in the ash of all six runs at levels many times the method detection limit.

The reported findings of ethylbenzene, xylene, 1,2-dichloroethane, acetone,

**Table 4. Total Analysis of Soil Feed[a]**

| Analyte | Method Detection Limit | Soil I | | | | Soil II | | | |
|---|---|---|---|---|---|---|---|---|---|
| | | Run 1 | Run 2 | Run 3 | Average | Run 4 | Run 5 | Run 6 | Average |
| **Volatiles, mg/kg** | | | | | | | | | |
| Ethylbenzene | 7 | 3600 | 2400 | 4000 | 3333 | 240 | 84 | 330 | 218 |
| Xylene | 5 | 5800 | 4000 | 6000 | 5267 | 120 | 150 | 520 | 253 |
| Tetrachloroethylene | 4 | (b) | 260 | 350 | 305 | 29 | 8.5 | 36 | 25 |
| Chlorobenzene | 6 | 340 | 240 | 360 | 313 | 22 | 6.9 | 30 | 20 |
| Acetone | 8 | 3300 | 6000 | 2700 | 4000 | 680 | 570 | 270 | 507 |
| 1,2 Dichloroethane | 3 | 450 | 140 | 340 | 310 | 13 | 3.5 | 28 | 15 |
| Styrene | 3 | 770 | 580 | 810 | 720 | 51 | 16 | 67 | 45 |
| **Semivolatiles, mg/kg** | | | | | | | | | |
| Anthracene | 6 | 6200 | 8500 | 5300 | 6667 | 480 | 420 | 440 | 447 |
| Bis(2-ethylhexyl)phthalate | 44 | 2800 | 3300 | 2200 | 2767 | 290 | 270 | ND | 187 |
| Pentachlorophenol | 3 | ND | 630 | ND | – | ND | ND | ND | – |
| **Metals, mg/kg** | | | | | | | | | |
| Lead | 4.2 | 261 | 296 | 292 | 283 | 328 | 301 | 302 | 310 |
| Zinc | 0.12 | 451 | 551 | 526 | 509 | 548 | 508 | 158 | 405 |
| Cadmium | 0.12 | 26 | 25 | 26 | 26 | 26 | 26 | 26 | 26 |
| Arsenic | 0.04 | 17 | 17 | 20 | 18 | 19 | 19 | 18 | 19 |
| Copper | 0.42 | 244 | 267 | 261 | 257 | 282 | 250 | 255 | 262 |
| Nickel | 0.3 | 28 | 30 | 27 | 28 | 30 | 28 | 28 | 29 |
| Chromium | 0.3 | 24 | 33 | 39 | 32 | 30 | 27 | 27 | 28 |
| **Average feed rate** | | | | | | | | | |
| lb/hr | | 1060 | 1062 | 1092 | 1071 | 1071 | 1086 | 1118 | 1092 |
| kg/hr | | 482 | 483 | 496 | 487 | 487 | 497 | 508 | 497 |
| **Moisture content (Dean Stark Distillation)** | | | | | | | | | |
| %$H_2O$ | | | 19.6 | | | | 22.9 | | |

*Note:* ND = None detected.

[a]See Table 2 for target concentrations.
[b]Analytical result rejected for QA reasons.

Table 5. Analytical Results for Untreated Soils[a]

| Analyte | Soil I | | Soil II | | Proposed TCLP Regulatory Limit (mg/L)[d] |
|---|---|---|---|---|---|
| | Average Total Analysis (mg/kg[b]) | TCLP (mg/L) | Average Total Analysis (mg/kg)[c] | TCLP (mg/L) | |
| **Volatiles** | | | | | |
| Ethylbenzene | 3333 | 49.7 | 218 | 7.31 | – |
| Xylene | 5267 | 84.6 | 253 | 15.8 | – |
| Tetrachloroethylene | 305 | 3.59 | 25 | 0.68 | 0.1 |
| Chlorobenzene | 313 | 6.47 | 20 | 0.79 | 1.4 |
| Acetone | 4000 | 282 | 507 | 26.1 | – |
| 1,2 Dichloroethane | 310 | 18.9 | 15 | 0.48 | 0.4 |
| Styrene | 720 | 2.1 | 45 | 0.58 | – |
| **Semivolatiles** | | | | | |
| Anthracene | 6667 | <0.01 | 447 | <0.02 | – |
| Bis(2-ethylhexyl)phthalate | 2767 | 0.06 | 187 | <0.06 | – |
| Pentachlorophenol | 210[e] | 0.98 | ND | 0.09 | 3.6 |
| **Metals** | | | | | |
| Lead | 283 | 0.46 | 310 | 1.74 | 5.0 |
| Zinc | 509 | 7.96 | 405 | 13.5 | – |
| Cadmium | 26 | 0.62 | 26 | 0.77 | 1.0 |
| Arsenic | 18 | ND (0.15) | 19 | ND (0.15) | 5.0 |
| Copper | 257 | 1.31 | 262 | 4.12 | – |
| Nickel | 28 | 0.34 | 29 | 0.65 | – |
| Chromium | 32 | ND (0.01) | 28 | 0.07 | 5.0 |

[a]See Table 4 for individual analyte values obtained during each run.
[b]Average of runs 1, 2 and 3.
[c]Average of runs 4, 5 and 6.
[d]Limits are those proposed in Federal Register Volume 51, Number 114, June 13, 1986. Several analytes have no proposed regulatory limit; these are indicated by "–".
[e]ND in runs 1 and 3; 630 mg/kg in run 2.

**Table 6. Total Analysis of Bottom Ash**

| Analyte | Method Detection Limit | Soil I | | | Soil II | | |
|---|---|---|---|---|---|---|---|
| | | Run 1 | Run 2 | Run 3 | Run 4 | Run 5 | Run 6 |
| **Volatiles, μg/kg** | | | | | | | |
| Ethylbenzene | 7 | ND[a] | 19 | ND | 8 | ND | 13 |
| Xylene | 5 | ND | 34 | ND | 11 | 6 | 20 |
| Tetrachloroethylene | 4 | ND | ND | ND | ND | ND | ND |
| Chlorobenzene | 6 | ND | ND | ND | ND | ND | ND |
| Acetone | 8 | 440 | 420 | 630 | 190 | 210 | 790 |
| 1,2 Dichloroethane | 3 | ND | ND | ND | ND | 5 | 10 |
| Styrene | 3 | ND | ND | ND | ND | ND | ND |
| **Semivolatiles, mg/kg** | | | | | | | |
| Anthracene | 37 | ND | ND | ND | ND | ND | ND |
| Bis(2-ethylhexyl)phthalate | 63 | 1600 | 540 | 740 | 950 | 710 | 1300 |
| Pentachlorophenol | 370 | ND | ND | ND | ND | ND | ND |
| **Metals, mg/kg** | | | | | | | |
| Lead | 4.2 | 56 | 98 | 107 | 146 | 75 | 88 |
| Zinc | 0.12 | 217 | 227 | 250 | 252 | 199 | 237 |
| Cadmium | 0.12 | <0.2 | <0.2 | <0.2 | <0.2 | <0.2 | <0.2 |
| Arsenic | 0.04 | 38 | 36 | 44 | 46 | 39 | 37 |
| Copper | 0.42 | 111 | 132 | 159 | 125 | 106 | 162 |
| Nickel | 0.3 | 12 | 15 | 11 | 12 | 9 | 12 |
| Chromium | 0.3 | 10 | 14 | 12 | 12 | 7 | 10 |
| **Volatile PICs, μg/kg** | | | | | | | |
| 2-Butanone | 25 | 35 | ND | ND | 14[b] | ND | ND |
| Methylene chloride | 2.8 | 2.9 | 5.4 | 4.2 | ND | ND | ND |
| 2-Chloroethylvinylether | 5 | 70 | ND | ND | ND | ND | ND |

[a]ND = None detected.
[b]Estimated value less than detection limit.

and the phthalate in the ash should be viewed with caution. Because the first three compounds were found intermittently and typically at levels very close to the method detection level (MDL), the values reported could be considered noise or analytical error, and should not necessarily be viewed as positive evidence of the presence of these materials. The values reported for the latter two compounds cannot be dismissed as analytical noise because the concentrations reported were many times higher than the MDL; however, inspection of the numbers shows that the values reported for the ash of Soil I are very similar to those reported for the ash of Soil II, even though the input waste feed level for these compounds (acetone and phthalate) was roughly 10 times higher in Soil I than in Soil II. This is suggestive of sample contamination somewhere in the system, either in the field during sampling or in the lab during analysis.* It is also noteworthy that significant quantities of phthalate

*Ash samples were recovered directly from the kiln prior to the water quench, at the beginning, middle, and end of each run, and placed in 5-gal steel pails to cool prior to compositing and placement in glass laboratory sample jars for shipment to the laboratory for chemical analysis. It is possible that the metal can was the source of the phthalate contamination, but this has not been verified.

**Table 7. Soil I Bottom Ash: Total Waste Analysis Results for Metals**

| Analyte | Average amount found in feed runs 1 to 3 (mg/kg) | Average amount found in ash runs 1 to 3 (mg/kg) | Percent remaining in ash (%) |
|---|---|---|---|
| Arsenic | 18 | 39 | 217 |
| Cadmium | 26 | <0.2 | <0.1 |
| Chromium | 32 | 12 | 33 |
| Copper | 257 | 134 | 52 |
| Lead | 283 | 87 | 31 |
| Nickel | 28 | 13 | 46 |
| Zinc | 509 | 231 | 45 |

were found during analysis in several of the method blanks, and both acetone and phthalates are well-known contaminants that inadvertantly show up in chemical analysis.

The total metals data for the ash samples are perhaps the most interesting. Prior to testing, it was anticipated that most of the metal concentrations in the ash would tend to be elevated compared to the waste feed, due to the combined effects of retention of metals in the ash and losses of water and organics from the feed during the incineration process. Cadmium levels in the ash, however, were anticipated to be lower due to volatilization of the metal in the kiln at the high operating temperature of 1800°F. As expected, cadmium levels in the ash were low, at least 99.9% lower than the waste feed levels. Surprisingly, all of the other metal levels except for arsenic were also considerably lower in the ash (on the order of 50 to 80 percent lower) than in the waste feed. Arsenic levels, on the other hand, were more than double those found in the feed. Table 7 provides a comparison of the metals data in the ash relative to the original feed for Soil I; similar results were obtained for Soil II, which had essentially the same levels of metals in the feed and ash.

Table 8 presents the results of TCLP analyses of the ash samples taken during the first run of each soil; the table also lists the previously mentioned TCLP values for the untreated soils (waste feed) and the proposed regulatory limits for various contaminants. The data clearly show that the ash produced by incinerating both soils meets the proposed regulatory TCLP limits for all organics as well as all metals, whereas both of the untreated soils failed the TCLP test for organics.

### *Scrubber Water Analyses*

Table 9 presents the results of analyses of the scrubber water samples collected prior to and during each test run for total organic and inorganic analytes. The analyses indicate that the scrubber water was essentially free of all organics from the Soil I and Soil II feeds, except for acetone, which appeared in runs 2 and 3 at low parts-per-billion levels, and phthalate, which appeared in four of the six runs, also at low parts-per-billion levels. In both cases, the amounts detected were only two to three times the MDLs.Pentachlorophenol was also detected in the scrubber water from two of the three runs on Soil I, at

**Table 8. TCLP Values (mg/L)[a]**

| Analyte | Soil I Untreated (waste feed) run 1 | Soil I Treated (bottom ash) run 1 | Soil II Untreated (waste feed) run 4 | Soil II Treated (bottom ash) run 4 | Proposed regulatory limit (mg/L)[a] |
|---|---|---|---|---|---|
| **Volatiles** | | | | | |
| Ethylbenzene | 49.7 | 0.65 | 7.31 | 0.19 | – |
| Xylene | 84.6 | 1.05 | 15.8 | 0.23 | – |
| Tetrachloroethylene | 3.59 | 0.04[b] | 0.68 | ND (0.10) | 0.1 |
| Chlorobenzene | 6.47 | 0.04[b] | 0.79 | ND (0.10) | 1.4 |
| Acetone | 282 | 0.74 | 26.1 | 0.14 | – |
| 1,2-Dichloroethane | 18.9 | ND (0.10) | 0.48 | ND (0.10) | 0.4 |
| Styrene | 2.1 | 0.11 | 0.58 | ND (0.10) | – |
| **Semivolatiles** | | | | | |
| Anthracene | <0.01 | <0.01 | <0.02 | <0.01 | – |
| Bis(2-ethylhexyl)phthalate | 0.06 | <0.01 | <0.06 | <0.01 | – |
| Pentachlorophenol | 0.98 | <0.02 | 0.09 | <0.02 | 3.6 |
| **Metals** | | | | | |
| Lead | 0.46 | ND (0.15) | 1.74 | ND (0.15) | 5 |
| Zinc | 7.96 | ND (0.03) | 13.5 | 0.49 | – |
| Cadmium | 0.62 | ND (0.01) | 0.77 | ND (0.01) | 1 |
| Arsenic | ND (0.15) | ND (0.15) | ND (0.15) | ND (0.15) | 5 |
| Copper | 1.31 | ND (0.02) | 4.12 | 0.26 | – |
| Nickel | 0.34 | ND (0.04) | 0.65 | 0.05 | – |
| Chromium | 0.01 | ND (0.01) | 0.07 | 0.04 | 5 |

*Note:* ND = Not detected; numbers in parentheses indicate detection limit.

[a]Federal Resister, Vol. 51, No. 114 (June 13, 1986).
[b]Less than detection limit; estimated value.

4 and 8 $\mu$g/L, levels which are 10 to 20 times higher than the MDL. Two volatile compounds which were not added to the waste feed, chloroform and bromodichloromethane, were detected at low parts-per-billion levels both in the influent scrubber water prior to the start of the tests and throughout the test runs.

The metals data for the scrubber water have been partially evaluated. The total volume of water that recirculated through the scrubber system has been estimated at 400 gal, with evaporative losses made up at the rate of about 4 gal/min. The water was recirculated continuously through runs 1 to 6 without any discharge or blowdown. Assuming all of the metals scrubbed from the stack gases were retained in the scrubber water (i.e., that there was no carryover with the evaporative losses), the values reported from analysis of the scrubber water sample taken during run 6 should be indicative of the total metals capture by the scrubber water over the course of the six runs. Calculations utilizing the average metals content of the waste feed, the total amount of soil fed, the values reported in the scrubber water from run 6, and the total estimated volume of scrubber water (400 gal) have produced the following

**Table 9. Total Analysis of Scrubber Water**

| Analyte | Method Detection Limit | Soil I | | | Soil II | | | |
|---|---|---|---|---|---|---|---|---|
| | | Run 1 | Run 2 | Run 3 | Run 4 | Run 5 | Run 6 | Influent[a] |
| **Volatiles, μg/L** | | | | | | | | |
| Ethylbenzene | 7 | ND[b] | ND | ND | ND | ND | ND | ND |
| Xylene | 5 | ND | ND | ND | ND | ND | ND | ND |
| Tetrachloroethylene | 4 | ND | ND | ND | ND | ND | ND | ND |
| Chlorobenzene | 6 | ND | ND | ND | ND | ND | ND | ND |
| Acetone | 8 | ND | 17 | 12 | ND | ND | ND | ND |
| 1,2 Dichloroethane | 3 | ND | ND | ND | ND | ND | ND | ND |
| Styrene | 3 | ND | ND | ND | ND | ND | ND | ND |
| **Semivolatiles, μg/L** | | | | | | | | |
| Anthracene | ND | ND | ND | ND | ND | ND | ND | ND |
| Bis(2-ethylhexyl)phthalate | 3 | ND | 5 | 2.3 | 5 | ND | 9 | ND |
| Pentachlorophenol | 0.4 | 8 | 4 | ND | ND | ND | ND | ND |
| **Metals, μg/L** | | | | | | | | |
| Lead | 0.105 | 1.8 | 1.5 | 2.3 | 2.2 | 2 | 4.8 | NA[c] |
| Zinc | 0.003 | 2.1 | 4.4 | 1.8 | 2.9 | 1.7 | 3.3 | NA |
| Cadmium | 0.003 | 2.4 | 4.1 | 1.9 | 3.6 | 2 | 5.8 | NA |
| Arsenic | 0.001 | 0.15 | 0.26 | 0.15 | 0.27 | 0.18 | 0.41 | NA |
| Copper | 0.011 | 0.6 | 0.53 | 0.48 | 0.56 | 0.59 | 1.1 | NA |
| Nickel | 0.008 | 0.32 | 0.76 | 0.27 | 0.36 | <0.04 | 0.27 | NA |
| Chromium | 0.008 | 2.4 | 4.1 | 1.9 | 3.6 | 2 | 5.8 | NA |
| **Other Compounds, μg/L** | | | | | | | | |
| Chloroform | 1.6 | 5.7 | 5.6 | 4.8 | 4.2 | 5.9 | 8.6 | 7.1 |
| Bromodichloromethane | 2.2 | ND | ND | ND | ND | ND | 2.4 | 2.7 |

[a]Sample of scrubber water collected prior to start of testing and analyzed for organic content only.
[b]ND = None detected.
[c]NA = Not analyzed.

estimates of the percent metals retained by the scrubber water system during runs 1 to 6:

| Metal | Percent (%) |
|---|---|
| Cadmium | 8.2 |
| Chromium | 6.7 |
| Arsenic | 1.3 |
| Nickel | <0.8 |
| Lead | 0.6 |
| Zinc | 0.2 |
| Copper | 0.2 |

These results are being further studied in an attempt to fully understand the material balance around metallic contaminants that are incinerated with soils under these conditions.

**Table 10. Summary of Organic DRE Data (%)[a]**

| Analyte | Soil I Test 1 | Soil I Test 2 | Soil I Test 3 | Soil II Test 4 | Soil II Test 5 | Soil II Test 6 |
|---|---|---|---|---|---|---|
| **Volatiles** | | | | | | |
| Ethylbenzene | 99.999 | 99.999 | 99.998 | 99.998 | 99.995 | 99.999 |
| Xylene | 99.999 | 99.999 | 99.998 | 99.992 | 99.99 | 99.998 |
| Tetrachloroethylene | ND[b] | 99.999 | 99.997 | 99.998 | 99.989 | 99.991 |
| Chlorobenzene | 99.998 | 99.999 | 99.995 | 99.996 | 99.988 | 99.996 |
| Acetone | 99.999 | 99.999 | 99.999 | >99.999 | >99.999 | 99.999 |
| 1,2 Dichloroethane | >99.99 | 99.999 | 99.998 | >99.99 | >99.99 | 99.998 |
| Styrene | 99.998 | 99.998 | 99.995 | 99.958 | 99.993 | 99.997 |
| **Semivolatiles** | | | | | | |
| Anthracene | >99.99 | >99.99 | >99.99 | >99.99 | >99.99 | >99.99 |
| Pentachlorophenol | ND[c] | >99.99 | ND[c] | ND[c] | ND[c] | ND[c] |
| Bis(2-ethylhexyl)phthalate | 99.99 | 99.999 | 99.967 | 99.992 | 99.6 | ND[d] |

*Note*: ND = Not determined.

[a]%DRE = {(mass in — mass out)/mass in} × 100.
[b]Analytical result for total tetrachloroethylene in waste feed rejected for QA reasons.
[c]None detected in the waste feed (detection limit 3.3 mg/kg) or in stack sample (detection limit 0.5 mg total).
[d]None detected in the waste feed (detection limit 44 mg/kg).

### *Results of Dioxin and Furan Analyses*

Waste feed, scrubber water, and ash samples taken during runs 2 and 5 were analyzed for the presence of chlorinated dibenzo-p-dioxins and dibenzofurans (total tetra-, penta-, hexa-, and octa-chlorinated isomers) using EPA Method 8280 (SW-846, Third Edition). The analyses failed to detect the presence of dioxins or furans in any of the samples. All surrogate compound recoveries were within the limits specified by the method.

## Air Emission Test Results

RCRA regulations require that hazardous waste incinerators meet certain requirements for particulate, HCl, and POHC emissions. The results of analyses of the stack samples collected during the Soil I and Soil II test burns indicate the particulate concentrations (corrected to 7% $O_2$) were below the RCRA allowable limit of 0.08 g/dscf for each soil. In addition, the HCl emission rates (in pounds per hour) were considerably less than the RCRA allowable rate of 4.0 lb/hr for each soil. Also, as noted previously, ethylbenzene, xylene, anthracene and bis(2-ethylhexyl)phthalate were designated POHCs for all test runs. As seen in Table 10, the destruction and removal efficiency (DRE) performance standard of 99.99% was achieved for three of the four designated POHCs (ethylbenzene, xylene, and anthracene) in all six runs. A DRE of 99.99% was achieved for the phthalate compound in three of the six runs; sample contamination (background level) problems are believed to be responsible for the poor DRE results in two of the other three runs. In one run, phthalate was not detected in the waste feed, preventing calculation of a DRE for that run.

The DRE of 99.99% or greater was achieved for the organic compounds (in addition to those discussed earlier) in Soil I in all three runs; for Soil II, which had the lower levels of organics, the 99.99% DRE was demonstrated in all three runs for acetone and dichloroethane, and in two out of three runs for chlorobenzene, styrene, and tetrachloroethylene. In several instances, acetone and dichloroethane could not be found in the stack gases; in these cases, the MDLs of 5 ng for acetone and 1 ng for dichloroethane were used to calculate the minimum DRE achieved. DREs could not be calculated for tetrachloroethylene in Run 1, phthalate in Run 6, and pentachlorophenol in Runs 1, 3, 4, 5, and 6 because of failure to detect the analytes in the waste feed samples; in the one run where pentachlorophenol was detected in the waste feed, a calculation was possible and the DRE exceeded 99.99%. The problems encountered in detecting the pentachlorophenol that was added to the waste feed are believed to be related to the analytical sample preparation procedures outlined in the SW-846 method that was used, as previously discussed.

Table 11 shows the calculated percent loss of metals to the atmosphere from the Soil I feedstock used in Runs 1 to 3. These values were derived from total mass emission data and the total mass feed data for each metal. Cadmium proved to be the most volatile of the metals tested, with losses to the atmosphere of about 39%. Losses of the other metals, on average, were less than three % each. Zinc, copper, and arsenic were found in the least amounts. Though not shown here, the metals emission results for Soil II were very similar to those for Soil I.

Combustion gases in the stack emissions were continuously monitored during each test run for CO, $CO_2$, and $O_2$. Average values for each run are included in Table 3. Values for each analyte remained relatively steady throughout each test with only a few short-term excursions of 1 to 5 min. CO remained at less than 10 ppm at all times, except during these excursions when it rose to 45 to 90 ppm. During the CO excursions, the $O_2$ content of the stack gas typically dropped to near 0%. These events are believed to correspond to brief fluxes or surges in the feed rate of soil to the kiln.

## CONCLUSIONS

These studies have shown that incineration can be a highly effective treatment method for soils containing a broad range of organic contaminants. The residual ash produced by incineration is likely to be considerably less hazardous than the untreated soil, as measured by the TCLP test, because the contaminants are either destroyed, removed, or otherwise altered so as to make them less likely to migrate from the treated residual when landfilled. The data from these tests indicate that under well-controlled conditions, and given sufficient exposure to high temperatures, the organic contaminants likely to be

**Table 11. Percent Loss of Metals to the Atmosphere for Soil I**

| Metal | Percent (%) |
|---|---|
| Arsenic | 0.52 |
| Cadmium | 39 |
| Chromium | 1.3 |
| Copper | 0.48 |
| Lead | 1.7 |
| Nickel | 2.9 |
| Zinc | 0.40 |

encountered in contaminated soils resulting from response actions at Superfund sites can be destroyed and/or removed to levels meeting or exceeding 99.99%. However, the contaminants must be present in levels sufficiently high to allow calculation of the 99.99% DRE and the analytical methods utilized must be highly sensitive to the analytes of concern.

Soils containing a mix of organic and metallic contaminants should be carefully evaluated before they are incinerated. These and other studies have produced evidence that some metals, such as cadmium, can volatilize in the process and be emitted in large part with the stack gases, even when efficient water-based scrubbing systems for particulate removal are employed. The inclusion of a more efficient gas cooling system prior to the scrubber or the addition of an electrostatic precipitator to the air pollution control system may improve the capture of volatile metals. Alternatively, lowering the incineration temperature should reduce the rate of metals volatilization, but the effect on DRE must not be compromised. Further studies in this area are needed.

## ACKNOWLEDGMENTS

This work was funded in its entirety by the U. S. Environmental Protection Agency under Contract No. 68-03-3389, Work Assignment No. 7, R. C. Thurnau, Project Officer. PEI Associates, Inc., Cincinnati, OH, was the prime contractor with subcontractor support from Radian Corporation (sample analysis) and John Zink Company (test facility). M. P. Esposito, who is now with Bruck, Hartman, & Esposito, Inc., Cincinnati, OH, was the overall Technical Project Manager and C. L. Bruffey was the Stack Test Manager for PEI during the project.

The authors wish to acknowledge the assistance provided by Jake Campbell of John Zink Company throughout the planning and testing stages of the project, as well as during the preparation of this paper. His cooperation and guidance have been very valuable and are sincerely appreciated. The technical assistance provided by Ric Traver of the EPA's Hazardous Waste Engineering Research Laboratory in Edison, NJ, during the preparation of the test soils is also gratefully acknowledged.

## REFERENCES

1. PEI Associates, Inc. "CERCLA BDAT SARM Preparation and Results of Physical Soil Washing Experiments," Final Report, Vol. 1 and 2, EPA Contract No. 68-03-341 (January 26, 1988).
2. Jackson, D. Radian Corporation, Austin, TX. Personal communication (April 7, 1988).

# CHAPTER 25

# The Impact on the Environment of Airborne Particulate Matter from the Eruption of Mount Saint Helens in May 1980

**L. L. Lamparski, T. J. Nestrick, and S. S. Cutie′**

## ABSTRACT

The eruption of Mount Saint Helens in the state of Washington on May 18, 1980 introduced large amounts of particulate matter into the atmosphere. The unique circumstances surrounding this geological event provided the impetus for Dow Chemical to initiate a preliminary study concerning atmospheric interactions of airborne particles. Results from this investigation conducted in 1980, involving the examination of groundfall volcanic "ash" for low-volatility organic species content, suggest that the direct interaction of particles suspended in the atmosphere can influence their trace organic composition under typical environmental exposure conditions. One consequence of these findings is that trace organic species of low volatility found on airborne particles are not always related to the originating particulate emission source.

## INTRODUCTION

The information presented in this paper is based upon results from a preliminary investigation conducted by The Dow Chemical Company in 1980. At that time, the recent publication of the "Trace Chemistries of Fire" (TCOF) hypothesis[1] aroused scientific controversy concerning the association of chlorinated dibenzo-p-dioxins (CDDs) on airborne particulate matter with emissions from combustion sources. In an attempt to provide evidence supporting the existence of natural emission of CDDs, we set about to determine if CDDs might be present in the particulate effluent from a volcanic eruption. Interestingly, the findings of this study indicated volcanic "ash" to be free of CDDs at or nearby the emission source. However, groundfall "ash" samples collected at various distances and locations away from the source were found to contain measurable amounts of CDDs and other organic species of presumed anthropogenic origin. Because the analytes selected for determination in this study have extremely low volitility under typical environmental conditions (e.g., CDDs, CBs = chlorinated biphenyls, and PNAs = polynuclear aromatic hydrocarbons), their association with volcanic "ash" traveling through the atmosphere is one indication that direct interaction of particles could be a significant phenomenon. The potential intermixing of trace organic species of low volatility via particle agglomeration may have important analytical consequences.

From an environmental study perspective, the circumstances surrounding particulate emissions from the May 18, 1980 eruption of Mount Saint Helens were nearly ideal for two reasons. First, groundfall particulates collected 23 days after the eruption, at a distance of ~3 to 5 mi from the crater, were found to be essentially free of low volatility organic compounds. These data, as well as subsequent information from analysis of material taken directly from the crater, indicate that relatively "clean" particulates were emitted from the volcano which is located in a national forest area ~40 mi from the nearest major highway or urban-industrialized zone. Second, the area exposed to these particulate emissions where groundfall accumulation was significant extends beyond 300 mi from the source. This area included rural, urban, metropolitan, industrialized, and uninhabited regions.

## EXPERIMENTAL

### Volcanic Particulate Samples

Groundfall volcanic particulate effluent (incorrectly, but often referred to as "ash") samples were collected during the period of 11 to 24 days after the May 18, 1980, eruption of Mount Saint Helens. At each collection site, the particulates were transferred to clean glass bottles equipped with Poly-Seal caps using a stainless-steel spatula. All samples were obtained from surfaces where "ash" accumulation was $\geq$0.5 cm in depth to avoid contamination by local sediments and also only from areas where "ash" deposition visually appeared to have occurred by natural means. The samples were considered to be representative only when they were collected from areas suffering widespread deposition and the "ash" was present as a dry powder. Sample identification information is presented in Table 1.

### Analytical Methodology

The analytical procedures used to identify and quantitate CDDs and CBs in particulate matrices have been described.[2-7] The methodology consists of six steps: (1) analyte removal from the matrix via benzene Soxhlet extraction, (2) gross impurities removal from the crude extract via chemically modified adsorbant treatments, (3) specialized adsorbent treatment of the extract to separate

**Table 1. Volcanic Particulate Effluent Samples**

| Sample identification | Distance and direction from volcano | Exposure time in days following May 18, 1980, eruption |
|---|---|---|
| **Blast zone** | | |
| #53 Blast debris (Unit A) | ~3 mi NW, Studebaker Ridge | 23 |
| #50 Hot ash (Unit C) | ~5 mi N, west of Spirit Lake | 23 |
| **Rural zones** | | |
| #58 Gifford Pinchot National Forest (~3 mi south of Packwood, WA) | ~40 mi NE | 17 |
| #42 Yakima Canyon, WA (~17 mi south of Ellensburg, WA) | ~85 mi NE | 17 |
| #86 Wastucna, WA (~2 mi west of village) | ~190 mi E | 24 |
| #22 Moscow ID (~20 mi north of city) | ~250 mi E | 24 |
| **Interstate highway zones** | | |
| #27 Castle Rock, WA; ~2 mi east (~50 ft west of I-5) | ~50 mi W | 10 |
| #29 Ritzville, WA; ~1 mi east (~500 ft northwest of I-90) | ~190 mi ENE | 14 |
| #31 Spokane, WA; ~10 mi west (~100 ft northwest of I-90) | ~240 mi ENE | 14 |
| **Urban zones** | | |
| #40 Yakima, WA (near US-97 in city) | ~90 mi NE | 15 |
| #38 Ellenburg, WA (central portion of city) | ~100 mi ENE | 14 |
| #26 Portland, OR (northern portion of city) | ~45 mi SSW | 11 |

CBs from CDDs, (4) reverse phase-high performance liquid chromatography (RP-HPLC) fractionation of the extract to isolate and purify CDD analytes, (5) normal phase-HPLC refractionation of the RP-HPLC CDD analyte fractions, and (6) identification and quantitation of CDDs and CBs in their appropriate fractions by gas chromatography/mass spectrometry (GC/MS).

The determination of PNAs was accomplished according to the procedure of Wise et al.[8] which employs RP-HPLC separation with fluorescence emission spectroscopic detection. Samples observed to contain co-eluting interferences via this method were subjected to additional normal phase-HPLC purification prior to RP-HPLC. Sixteen individual PNAs as described in the EPA list of 129 priority pollutants[9] were monitored in this portion of the study.

Capillary column-high resolution gas chromatography with flame ionization detection (HRGC-FID) was accomplished on selected sample extracts to provide qualitative and semiquantitative data regarding detectable organic compounds present. These extracts were prepared by subjecting an aliquot of particulates (30 g) to methylene chloride Soxhlet extraction followed by Kuderna-Danish concentration of the crude extract to a prescribed volume (1.0 mL). HRGC-FID analyses were conducted on a Carlo-Erba Model 4160 gas chromatograph equipped with a 0.50 mm i.d. × 20 m glass capillary column coated with OV-73 silicone ($d_f$ = 0.20 $\mu$m) under temperature-programed conditions using hydrogen as a carrier gas. The chromatograms shown in Figure 1

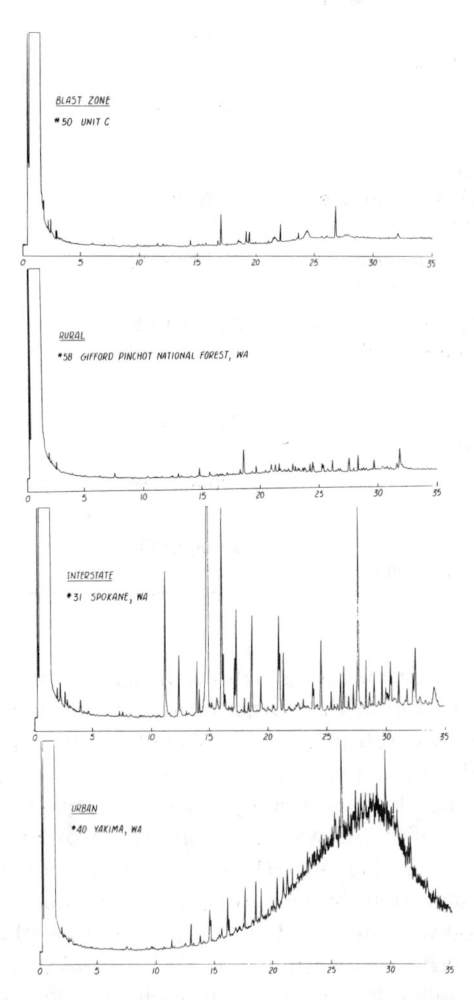


**Figure 1.** HRGC-FID chromatograms of chromatographable organics extracted from the Mount Saint Helens groundfall ash (see text for conditions and sample identifications).

represent a 2.0 $\mu$L splitless injection of each sample extract at a full-scale attenuation of 128×.

## RESULTS

Approximately 50 groundfall "ash" samples were collected during the period of June 1 to 12, 1980, from which the 12 presented in Table 1 were selected for examination. As indicated, we have subjectively grouped these samples into four different categories based upon the environmental conditions observed at their point of collection. The blast zone samples, collected for us by U.S. Geological Survey personnel, were characteristic of particulate matter deposited in the area of complete devastation on the northern side of Mount Saint Helens within line-of-sight of the crater. The Unit A material was classified as blast-related fallout and the Unit C sample was exemplary of particulates related to pyroclastic flow that were deposited while hot. Rural zone samples were obtained from locations reasonably distant from permanent habitations and major highways. Particulates collected immediately adjacent to interstate highways bearing heavy motor traffic on a regular basis were defined as interstate highway zones. As indicated in Table 1, none of the interstate samples were collected from areas where contamination sources other than the highway traffic were evident. The urban samples were collected from areas within a city where habitations, motor traffic, and industrial activity were prevalent. Ellensburg and Yakima, WA, are primarily associated with agricultural activity, whereas Portland, OR, is the closest metropolitan area to Mount Saint Helens. Also included in Table 1 are the estimated environmental exposure times for each sample. Because two eruptions occurred prior to sample collection, one on May 18 and the other on May 25, it is impossible to accurately define the age of these particulates; however, samples #26 and #27 were the only ones anticipated to contain significant amounts of material from the May 25 eruption as indicated by their reduced exposure times.[10]

The HRGC-FID chromatograms in Figure 1 typify detectable organics associated with groundfall "ash" from each environmental sampling category. As indicated, the blast zone particulates have the lowest amount of organic species present. The amount of organics present in the rural sample is also low, but an increase in the number of detectable species is apparent. Since this sample was collected from the Gifford Pinchot National Forest, its increased component diversity could be attributable to natural products associated with flora of the heavily forested area.

Chromatograms for the interstate highway and urban sampling zones show higher concentrations of organic species present on the "ash" relative to the rural and blast areas. Although the interstate sample has a lower concentration of detectable organics than the urban material, its separable component diversity appears to be greater. Examination of these residues by GC/MS in the scanning mode indicates that the interstate "ash" contains many oxygenated

**Table 2. PNA Content of Volcanic Particulate Effluent Samples**

| Sample[a] | Total of monitored PNAs (ppb)[b] |
|---|---|
| **Blast zone** | |
| #53 blast debris | ND (0.1—90)[c] |
| #50 Hot ash | ND (0.1—90) |
| **Rural zones** | |
| #58 Gifford Pinchot National Forest | 160 |
| #86 Washtucna, WA | 64 |
| #22 Moscow, ID | 190 |
| **Interstate highway zones** | |
| #27 Castle Rock, WA | 0.5 |
| #29 Ritzville, WA | 670 |
| #31 Spokane, WA | 20 |
| **Urban zones** | |
| #40 Yakima, WA | 27 |
| #38 Ellensburg, WA | 97 |
| #26 Portland, OR | 640,000 |

[a]Refer to Table 1 for description of sample locations.
[b]Polynuclear aromatics on the EPA list of 129 priority pollutants were determined in this analysis.
[c]Indicated that component was not detected at the limit of detection (LoD), defined as 2.5 times noise, given in parenthesis.

aromatic compounds as well as PNAs, whereas aliphatic hydrocarbons predominate in the urban "ash." Although it is impossible to accurately define exposure time or route of transport for these groundfall "ash" samples, the HRGC-FID evidence presented here suggests that the original volcanic effluent particulates are relatively free of detectable organic contamination. Upon transport through the environment, they become associated with organic compounds characteristic of their location. The estimated component concentrations observed by this HRGC-FID technique range from low part-per-billion (ppb = ng/g) levels in the blast zone to part-per-million (ppm = $\mu$g/g) levels in the urban zone.

The organics determined by HRGC-FID represent what we shall term the "upper-tier" contaminants which have become associated with the volcanic particulates. The observed concentration of these diverse species might indicate their predominance in the atmosphere to which the "ash" particles were primarily exposed. Because the upper-tier contaminants have concentrations in the part-per-million range and appear to display a pattern of increasing concentration when the collection point was moved from rural to urban areas, we opted to specifically analyze several of these "ash" samples for priority pollutant PNAs content. Because many of these PNAs have low volatility but can be readily detected in the atmosphere at concentrations anticipated to be lower than those of upper-tier components, we shall define them as "middle-tier" contaminants. As indicated by the results presented in Table 2, observed PNAs concentrations were typically in the part-per-billion range. Since samples from

**Table 3. PCB Content of Selected Volcanic Particulate Effluent Samples**

| Sample[a] | Total of monitored PCBs (ppt)[b] |
|---|---|
| **Blast zone** | |
| #53 Blast debris | ND (20—170) |
| #50 Hot ash | ND (20—180) |
| **Rural zones** | |
| #58 Gifford Pinchot National Forest | 50 |
| **Interstate highway zones** | |
| #27 Castle Rock, WA | 40 |
| #31 Spokane, WA | 140 |
| **Urban zones** | |
| #40 Yakima, WA | 1200 |

[a]Refer to Table 1 for description of sample locations.
[b]Penta-, hexa-, and heptachlorobiphenyls were determined.

the blast zone contained nondetectable PNA levels, it appears that the original volcanic emissions were initially free of these compounds. However, the effects of aging and transport in the surrounding environment permitted association of PNAs with the "ash." It is interesting to note that the definitive contamination pattern exhibited by upper-tier organic compounds is no longer well defined when very specific organic species, present on the particulates in the part-per-billion concentration range, are examined.

The analysis of "ash" samples for CBs and CDDs was expected to provide information concerning the association of particulate matter with organic species from what we shall term the "lower tier." The determination of penta-, hexa-, and heptachlorobiphenyls in selected "ash" samples from each collection zone showed these CBs to be present at part-per-trillion (ppt = pg/g) levels as shown in Table 3. From these data, it appears that volcanic emissions were free of detectable CBs upon exiting the source and remained so throughout their period of deposition in the blast zone. Particulate/organic association for these CBs demonstrate an increasing concentration pattern as the sampling location is moved from rural to urban environments. The data in Table 4 indicate that CDDs and CBs are similar with respect to their potential for association with "ash" particles, to include the near absence of detectable CDDs in original volcanic emissions and enhanced concentrations in urban samples.

## DISCUSSION

The eruption of Mount Saint Helens has provided an opportunity to study atmospheric interactions with transient, suspended particulate matter under "real world" conditions. The observation that massive amounts of "clean" particulates were released from a source in a nearly uninhabited area and then transported through the atmosphere into a variety of environmentally differ-

**Table 4. CDD Content of Selected Volcanic Particulate Effluent Samples**

| Sample | CDD concentration (ppt) | | | | |
|---|---|---|---|---|---|
| | TCDD | HCDD | $H_7CDD$ | OCDD | Total |
| **Blast zone** | | | | | |
| #53 Blast debris | ND(0.3) | ND(0.7) | 0.8(0.3) | ND(1.6) | 0.8 |
| **Rural zones** | | | | | |
| #58 Gifford Pinchot | ND(1.3) | ND(2.0) | ND(1.0) | ND(2.0) | ND |
| #42 Yakima Canyon | ND(0.3) | ND(0.8) | 1.0(0.4) | 3.2(1.6) | 4.2 |
| #86 Washtucna, WA | —[a] | ND(0.9) | ND(0.5) | ND(1.6) | ND |
| #22 Moscow ID | — | ND(0.8) | ND(0.4) | ND(1.6) | ND |
| **Interstate zones** | | | | | |
| #27 Castle Rock, WA | ND(0.9) | ND(2.0) | 2.0(0.7) | 6.0(1.6) | 8.0 |
| #29 Ritzville, WA | — | ND(0.8) | ND(0.5) | ND(1.8) | ND |
| #31 Spokane, WA | ND(1.6) | ND(2.0) | ND(1.0) | 7.0(2.0) | 7.0 |
| **Urban zones** | | | | | |
| #40 Yakima, WA | ND(0.2) | ND(0.4) | 4.7 | 23 | 27.7 |
| #38 Ellensburg, WA | ND(0.3) | ND(0.9) | 2.6(0.4) | 7.8 | 10.4 |
| #26 Portland, OR | 0.8(0.3) | 22(0.9) | 58(7.0) | 170 | 250 |

[a]Indicates that the sample was not analyzed for the specified component.

ent zones is perhaps unique. Even though the limited data presented appear to confirm the obvious, "clean" particles passing through "dirty" air become contaminated, little is known concerning the mechanism or significance of these interactions under actual environmental conditions. This situation is especially true for low-volatility organic compounds that are known to exist in the atmosphere.

Based upon the analytical data presented, particulate matter that is suspended in the atmosphere can become associated with organic compounds which are not related to the particulate emission source. Many of the higher-concentration (upper-tier) organics observed on "ash" collected away from the source may possess sufficient volatility to permit their adsorption from the gas phase directly onto the particle surface. However, the observation of trace concentrations of many low-volatility compounds such as PNAs, CBs, and CDDs on most of the "ash" samples that were collected away from the source suggests the possible existence of an atmospheric association mechanism other than gas-phase adsorption. Since PNAs and CDDs are known to enter the environment on particulate emissions arising from combustion,[1] it is plausible to hypothesize that particulate agglomeration is another such association mechanism. The extremely low volatility of higher-chlorinated dioxins and many of the five-ring, or larger, PNAs provide evidence in support of this concept.

Examination of several "ash" samples by light microscopy indicated that exogenous particulate matter, if present, must be much smaller than the primary "ash" particles. Assuming from this observation that microparticulates are a means of atmospheric transport for low-volatility organic compounds and that particulate agglomerations can occur at rates consistent with the exposure times estimated for "ash" samples in Table 1, then certain analytical consequences should be considered. First, primary particulate effluent from a point source located in an urban industrialized area will ultimately become contaminated with organic components that are not necessarily related to the original source. Regarding low-volatility organic compounds, the extent of such contamination will be a function of exposure time and concentration of contaminated microparticulates present in the local atmosphere. Subsequent collection of what would appear to be original primary particulates, as is commonly done for many atmospheric monitoring tests, could then result in the detection of many organic compounds which may not be exiting the particulate emission source. Second, the trace organic components present on particulate matter that has been exposed to urban-industrialized atmosphere do not appear to be a reliable means for identifying the source of the original particulates.

## CONCLUSIONS

Fundamentally, two observations have been presented. First, airborne particulate matter can become associated with exogenous organic compounds under typical environmental exposure conditions. These association interactions can occur within a short time frame that does not exceed ~11 days. Second, simple gas-solid adsorption phenomena do not appear to adequately explain particulate association with low-volatility organic compounds. The suggested presence of microparticulate matter, and its subsequent agglomeration potential, have been hypothesized to explain such particulate/organic interactions in the atmosphere. These observations suggest that whenever analytical determinations of trace organic species are conducted upon airborne particulate matter, identification of the particulate emission source via trace organics content may not be reliable. This situation appears to have maximum impact in the atmospheric conditions existent in urban industrialized areas.

## ACKNOWLEDGMENTS

The authors extend their appreciation to R. A. Rasmussen of the Oregon Graduate Center for his many hours of consultation and to his students who provided several of the ash samples. We also appreciate the efforts of T. Casadevall and S. Kieffer of the U.S. Geological Survey, and T. LaDoux of the U.S. Forest Service for their assistance in collecting samples from restricted areas of Mount Saint Helens. The mass spectrometric identifications were provided by T. L. Peters and L. A. Shadoff of The Dow Chemical Company. The light microscopy information was provided by H. L. Garrett, also of The Dow Chemical Company. A special debt of gratitude is extended to R. R. Bumb for his timely initiative and support of this study.

## REFERENCES

1. Bumb, R. R. et al. *Science* 210:385–390 (1980).
2. Lamparski, L. L., and T. J. Nestrick. *Anal. Chem.* 52:2045–2054 (1980).
3. Lamparski, L. L., T. J. Nestrick, and R. H. Stehl. *Anal. Chem.* 51:1453–1458 (1979).
4. Nestrick, T. J., L. L. Lamparski, and R. H. Stehl. *Anal. Chem.* 51:2273–2281 (1979).
5. Nestrick, T. J., L. L. Lamparski, and D. I. Townsend. *Anal. Chem.* 52:1865–1974 (1980).
6. Lamparski, L. L., and T. J. Nestrick. *Chemosphere* 10:3–18 (1981).
7. Lamparski, L. L, and T. J. Nestrick. *Anal. Chem.* 54:402–406 (1982).
8. Wise, S. A., S. N. Chesler, H. S. Hertz, L. R. Hilpert, and W. E. May, *Anal. Chem.* 49:2306–2310 (1977).
9. Keith, L. H., and W. A. Telliard. *Environ. Sci. Technol.* 13:416–423 (1979).
10. Findley, R. *Nat. Geographic* 159:3–65 (1981).

## CHAPTER 26

# Modeling of Chemical Movement in Soil Matrices

**Raymond A. Freeman and Jerry M. Schroy**

## BACKGROUND

Many models have been proposed to describe the movement of pesticides in the environment. Jury et al.[1] developed an analytical model to describe the movement of chemicals by vaporization. Their model assumes constant conditions of soil temperature and density and was developed to allow for environmental screening. Many other investigators such as Oddson et al.[2], Leistra[3], Davidson and McDougal[4], and Lindstrom et al.[5] have modeled the movement of chemicals by convective transport in water.

Previously, we have developed a model that allows for variation in soil properties (thermal conductivity, density, porosity) with depth and accounts for the daily and seasonal variation in soil column temperature.[6,7] This model has been used to study the results of field experiments on the movement of 2,3,7,8-tetrachlorodibenzo-p-dioxin (2,3,7,8-TCDD), Lindane, and Dieldrin[8-10] in soil columns. The model has one adjustable parameter which is used to fit the experimental data. The purpose of this paper is to propose a tentative method for the estimation of the adjustable parameter of the model using only the physical properties of the chemical under study.

## THEORETICAL

### Material Balance Equations

Previously, we[6,7] have presented an unsaturated zone TCDD transport model based on a solid and a vapor phase being in contact with each other. Transport through the vapor phase is assumed to be hindered by the long complex path of the soil air voids. The material balance model is essentially the same as that used by Jury et al. and is given as:

$$\frac{\partial C_d}{\partial t} = \frac{2\,\rho_{molar}\,D_{ab}}{\Phi}\left[\frac{C_a}{M_w\,\rho_{molar}} - \frac{K\,C_d\,P^o}{P_t}\right]\frac{M_w\,a}{\rho\ soil} = R \tag{1}$$

$$\frac{\partial C_a}{\partial t} = h\,\frac{\partial}{\partial Z}\left(D_{ab}\frac{\partial C_a}{\partial Z}\right) - \frac{R}{\varepsilon} \tag{2}$$

These two material balance partial differential equations must be solved simultaneously. Numerical and theoretical studies of the material balance equations found that the vapor phase is always in equilibrium with the solid soil phase. An empirical equilibrium partitioning coefficient, K, is defined by the relation between the gas space partial pressure of TCDD and the vapor pressure as:

$$K = \frac{P_1}{C_d\,P^o} \tag{3}$$

Substituting the vapor-phase equilibrium expression and collecting terms allows the Freeman-Schroy material balance equations to be rewritten as:

$$\frac{\partial C_a}{\partial t} = \frac{(\varepsilon^2 / \tau)}{\varepsilon + P_t / (K\,P^o\,M_w\,\rho_{molar})}\;\frac{\partial}{\partial Z}\left(D_{ab}\frac{\partial C_a}{\partial Z}\right) \tag{4}$$

The material balance, given as Equation 4, is solved using boundary conditions 1 and 2:

B.C. 1 – Air-soil interface concentration
at $Z = 0$; $C_a = 0$ for all t (5)

B.C. 2 – Constant concentration some depth in ground
at $Z = L$; $C_d = 0$ for all t (6)

The diffusivity, $D_{ab}$, the molar density, $\rho_{molar}$, and the vapor pressure, $P^o$, are all functions of temperature. Daily high soil surface temperatures of 40°C have been measured during the day with corresponding lows of 20°C during the night.[11] Since the soil temperature will also vary from season to season, an energy balance model is required to estimate correctly the impact of temperature variations on the mass transport process.

## Energy Balance Equation

The energy balance equation is the same as previously presented by Tung, et al.[11] A brief description is presented below. The one-dimensional transient energy balance equation may be written as:

$$\rho_{soil}\,C_p\,\frac{\partial T}{\partial t} = k\,\frac{\partial^2 T}{\partial Z^2} \tag{7}$$

To solve this equation for the temperature waves that pass through a soil column two boundary conditions are required:

B.C. 3—Surface energy flux

$$\text{at } Z = 0;\ k\frac{\partial T}{\partial Z} = q_r + q_c + q_b \text{ for all } t > 0 \tag{8}$$

B.C. 4—Constant temperature at some depth in ground

$$\text{at } Z = L;\ T = T_g \text{ for all } t > 0 \tag{9}$$

The term $q_r$ represents the radiative solar input into the soil column. The term $q_c$ is convective heat transfer between the soil and the air. The term $q_b$ represents the black body radiative loss of energy from the soil surface. These soil surface energy fluxes are complex functions of soil temperature, weather conditions, and site location. Schroy and Weiss[12] have previously presented methods for the computation of $q_r$, $q_c$, and $q_b$. For details of the soil temperature model and the numerical solution, see Tung et al.[11]

## PHYSICAL INTERPRETATION OF THE GAS-PHASE PARTITION COEFFICIENT, K

As formulated, the gas-phase partition coefficient, K, relates the vapor-phase partial pressure with the soil concentration of a chemical. Soil has an extremely complex structure. Figure 1 presents an idealized structure for a section of soil. The gas phase is actually in contact with a water layer that surrounds each clay particle. A fraction of the chemical will be dissolved in the water, a second fraction will be either bound or dissolved in the soil, and a third fraction will be adsorbed to the surface of the clay particle. Only the chemical dissolved in the water layer is available for transport into the vapor phase. The rate of transport of a chemical into the vapor phase will depend on (1) the water solubility and (2) the vapor pressure. The gas-phase partition coefficient, K, expresses the equilibrium relationship between the gas phase and the bulk soil phase. Thus, K should also depend on the water solubility and the vapor pressure for the chemical.

The Henry's Law constant, H, is the ratio of the vapor pressure of a chemical and the water solubility of a chemical at a specified temperature. Since the vapor pressure and the water solubility are temperature dependent, the Henry's Law constant will also be temperature dependent. A relationship between the gas-phase partition coefficient and the Henry's Law constant, H, should exist if the argument given earlier is correct. To examine this relationship requires experimental data for chemicals with different Henry's law constants.

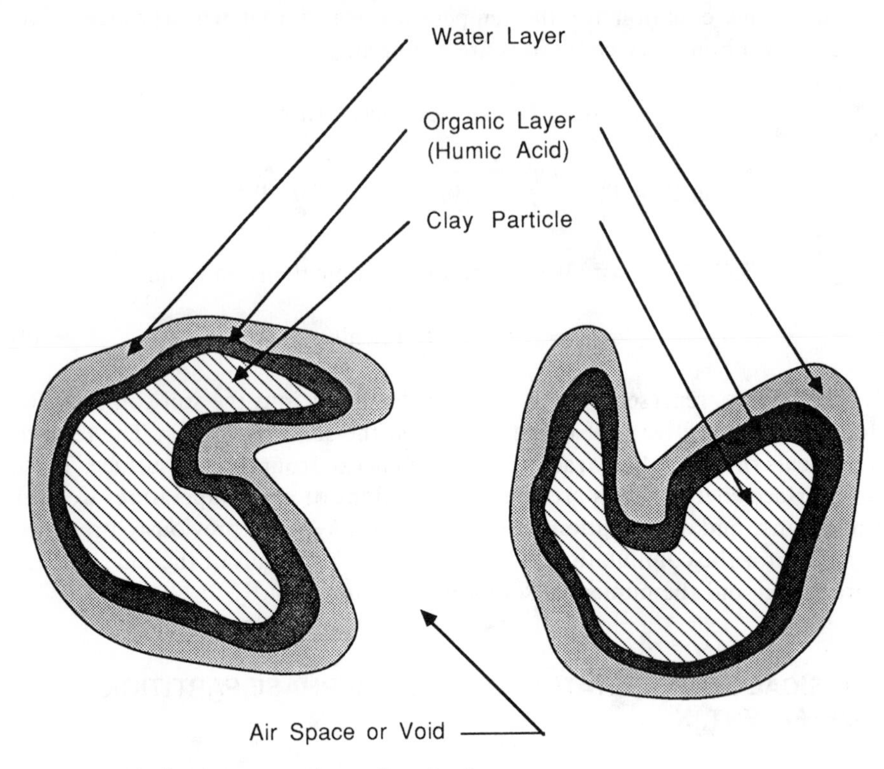


**Figure 1.** Idealized structure of a section of soil.

Fortunately, there exists a body of experimental data on the rate of soil migration of various chemicals. Previously, Freeman et al.[8-10] have presented the results of modeling studies on the movement of 2,3,7,8-TCDD, Lindane, and Dieldrin in soils. The Lindane and Dieldrin models are based upon 25 years of experimental data accumulated by Dr. Ralph Nash[13] at the U.S. Department of Agriculture. The 2,3,7,8-TCDD model is based on 15 years of field experimental data at Eglin AFB,[6], 2 years of data at Times Beach,[9] and laboratory experiments by Yanders et al.[14]

Table 1 presents the physical properties of 2,3,7,8-TCDD, Lindane, and Dieldrin as well as the gas-phase partition coefficient determined for each chemical from the available experimental data. Theoretical studies by Taghavi et al.[15] have demonstrated that average values of the transport and physical properties may often be used to represent the movement of a chemical in soil over long time periods. Thus, a constant temperature of 25°C was used to compute the Henry's Law constant, H. The relationship between the gas-phase partition coefficient, K, and the Henry's Law constant (H at 25°C) is presented graphically in Figure 2. The gas-phase partition coefficient, K, is a linear function of the Henry's Law constant. The linear function is described by:

**Table 1. Physical Properties and Gas-Phase Partition Coefficient for 2,3,7,8-TCDD, Lindane, and Dieldrin (at 25°C)**

| Chemical name | Molecular weight[a,b] | Vapor pressure (mmHg)[a,b] | Water solubility (mg/L)[a,b] | Partition coefficient | Henry's Law Constant [mm Hg/(mg/L)] |
|---|---|---|---|---|---|
| 2,3,7,8-TCDD | 321.974 | $1.50 \times 10^{-9}$ | $7.91 \times 10^{-3}$ | $6.0 \times 10^{4}$ | $1.9 \times 10^{-7}$ |
| Lindane | 290.832 | $6.54 \times 10^{-5}$ | 6.6 | $9.0 \times 10^{5}$ | $9.9 \times 10^{-6}$ |
| Dieldrin | 380.913 | $5.38 \times 10^{-6}$ | 0.186 | $2.1 \times 10^{6}$ | $2.9 \times 10^{-5}$ |

[a]Physical properties of 2,3,7,8-TCDD from Reference 17.
[b]Physical properties of Lindane and Dieldrin from Reference 10.

$$K = 1.12 \times 10^{5} + (6.97 \times 10^{10}) \times H \qquad (10)$$

The coefficient of determination, $\rho^2$, for this linear function is 0.993, which is statistically significant at the $\alpha = 0.10$ level (correlation coefficient test with one degree of freedom).[16] Given the linear relationship between the gas-phase partition coefficient and the Henry's Law constant, the Freeman-Schroy transport model can be used to make first estimates on the soil migration of other

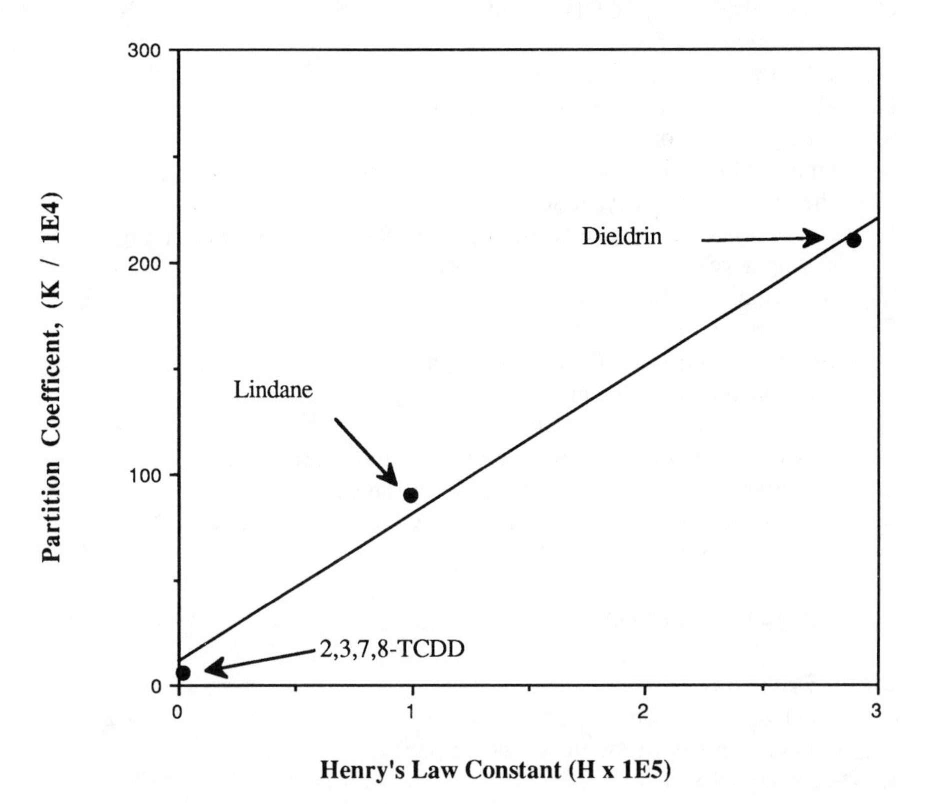


**Figure 2.** Gas-phase partition coefficient as a function of Henry's Law Constant for 2,3,7,8-TCDD, Lindane, and Dieldrin at 25°C.

chemicals without the need for extensive field data. This predictive ability rests on the relationship presented in Equation 10. Obviously, more data on other chemicals are needed to verify the relationship presented in Equation 10.

## CONCLUSIONS

1. For 2,3,7,8-TCDD, Lindane, and Dieldrin, the gas-phase partition coefficient, K, is a linear function of the Henry's Law constant, H. This relationship is presented as Equation 9.
2. The linear relationship of Equation 9 should be tested by the analysis of the rate of soil migration for other chemicals with different Henry's Law constants.

## NOMENCLATURE

$a$ —air-soil interfacial area per unit soil volume
$C_a$ —concentration of TCDD in air
$C_d$ —concentration of TCDD in soil
$C_p$ —heat capacity of soil
$D_{ab}$ —diffusivity of TCDD in air
$H$ —Henry's Law Constant, mmHg/(mg/L)
$h$ —hindrance factor
$K$ —empirical equilibrium partitioning coefficient between soil and air
$k$ —thermal conductivity of soil
$L$ —soil depth where temperature and concentration does not change during a year
$M_w$ —molecular weight of TCDD
$P^o$ —vapor pressure of TCDD
$P_l$ —partial pressure of TCDD in gas space
$P_t$ —total barometric pressure
$q_b$ —black-body radiation loss to the sky
$q_c$ —convective energy exchange between soil and atmosphere
$q_r$ —radiative energy received by soil from the sun
$R$ —volumetric rate of volatilization of TCDD into air voids
$T_g$ —soil temperature at a depth L
$t$ —time
$Z$ —depth into the ground

### Greek Symbols

$\varepsilon$ —soil void fraction
$\rho_{molar}$ —molar density of air in soil void space
$\rho_{soil}$ —density of soil
$\tau$ —tortuosity factor, $\tau = 2$ for an average soil
$\phi$ —average diameter of soil particle

## REFERENCES


1. Jury, W. A., W. F. Spencer, and W. J. Farmer. "Behavior Assessment Model for Trace Organics in Soil I. Model Description," *J. Environ. Qual.* 12:558–564 (1983).
2. Oddson, J. K., J. Letey, and L. V. Weeks. "Predicted Distribution of Organic Chemicals in Solution and Adsorbed as a Function of Position and Time for Various Chemicals and Soil Properties," *Soil Sci. Soc. Am. Proc.* 34:412–417 (1970).
3. Leistra, M. "Computing the Movement of Ethoprophos in Soil After Application in the Spring," *Soil Sci.*, 128:303–311 (1979).
4. Davidson, J. M. and J. R. McDougal. "Experimental and Predicted Movement of Three Herbicides in a Water-Saturated Soil," *J. Environ. Qual.* 2:428–433 (1973).
5. Lindstrom, F. T., L. Boersma, and H. Gardiner. "2,4-D Diffusion in Saturated Soils: A Mathematical Theory," *Soil Sci.* 106:107–113 (1968).
6. Freeman, R. A., and J. M. Schroy. "Modeling the Transport of 2,3,7,8-TCDD and Other Low Volatility Chemicals in Soils," *Environ.Prog.* 28–33 (1986).
7. Freeman, R. A. and J. M. Schroy. "Environmental Mobility of Dioxins," in *Aquatic Toxicology and Hazard Assessment: 8th Symposium*, R. C. Bahner and D. J. Hansen, Eds. (Philadelphia: American Society of Testing Materials, 1985).
8. Freeman, R. A., J. M. Schroy, F. D. Hileman, and R. W. Noble. "Environmental Mobility of 2,3,7,8-TCDD and Companion Chemicals in a Roadway Soil Matrix, in *Chlorinated Dioxins and Dibenzofurans in Perspective* (Chelsea, MI: Lewis Publishers, 1986), Chapter 12.
9. Freeman, R. A., F. D. Hileman, R. W. Noble, and J. M. Schroy. "Experiments on the Mobility of 2,3,7,8-Tetrachlordibenzo-p-dioxin at Times Beach, Missouri," in *Solving Hazardous Waste Problems*, ACS Symposium Series Number 338 (1987), Chapter 9.
10. Freeman, R. A., and J. M. Schroy. "Modeling the Movement of Low Volatility Chemicals," presented at the 1985 Summer National Meeting of the American Institute of Chemical Engineers, Seattle, WA, August 25 to 28, 1985.
11. Tung, L. S., R. A. Freeman, and J. M. Schroy. "Prediction of Soil Temperature Profiles: A Concern in the Assessment of Transport of Low Volatility Chemicals in the Soil Column," presented at the AICHE National Meeting in Seattle, WA, August 25 to 29, 1985.
12. Schroy, J. M., and J. S. Weiss. "Prediction of Wastewater Basin Temperature, A Design and Operating Concern," presented at the AICHE Symposium, Washington, D.C., November 4, 1983.
13. Nash, R. G., and E. A. Woolson. "Distribution of Chlorinated Insecticides in Cultivated Soil," *Soil Sci. Soc. Am. Proc.* 32:525–527 (1968).
14. Palausky, J., J. J. Harwood, T. E. Clevenger, S. Kapila, and A. F. Yanders. "Disposition of Tetrachlorodibenzo-p-Dioxin in Soil," in *Chlorinated Dioxins and Dibenzofurans in Perspective* (Chelsea, MI: Lewis Publishers, 1986), Chapter 15.
15. Taghavi, H., P. Ryan, and Y. Cohen. "Contaminant Transport Under Non-Isothermal Conditions in the Top-soil," *Abst. Pap. Am. Chem. Soc.* 192 (September:70 (1986).

16. Rickmers, A. D., and H. N. Todd. *Statistics: An Introduction* (New York: McGraw-Hill Book Company, 1967), p. 563.
17. Schroy, J. M., F. D. Hileman, and S. C. Cheng. "Physical/Chemical Properties of 2,3,7,8-Tetrachlordibenzo-p-Dioxin," in *Aquatic Toxicology and Hazard Assessment: 8th Symposium*, R. C. Bahner and D. J. Hansen, Eds. (Philadelphia: American Society of Testing Materials, 1985).

CHAPTER 27

# Atmospheric Transport of Combustion-Generated Organic Compounds

**O. Hutzinger and A. Reischl**

This paper summarizes aspects of organic pollutant generation and emission during combustion processes and discusses the influence of vapor/particle distribution on atmospheric transport. Special emphasis will be given to chlorinated dioxins and furans and their deposition onto plant surfaces.

## INTRODUCTION

The most important source of manmade direct atmospheric pollution are combustion processes.[1] Acid rain, which is mainly caused by inorganic nitrogen and sulfur species has highlighted the importance of combustion emissions. In addition, the content of combustion-emitted carbon dioxide in the atmosphere grows with the world's population and may contribute to climatic change.

In addition to the few inorganic products, many organic chemicals, resulting from incomplete combustion and thermal synthesis, are emitted through combustion processes (Table 1). The following discussion will focus on organic compounds of low volatility ($<10^{-4}$ atm) where long-chain aliphatic hydrocarbons, polycyclic aromatic hydrocarbons (PAHs), and halogenated hydrocarbons are important due to amounts produced or their toxic potential. Especially, some aspects of the atmospheric distribution behavior of polychlorinated dibenzo-p-dioxins (PCDDs) and dibenzofurans (PCDFs) and their interaction with terrestrial ecosystems will be discussed, mainly with respect to stationary sources.

## SUBSTANCE GENERATION AND CONTRIBUTION TO ENVIRONMENTAL LEVELS

In most combustion processes, a compound-mixture undergoes thermal reactions and ambient air used as an oxygen source provides additional reac-

**Table 1. Estimated Emission Factors Without Control (Pounds per Ton on Refuse Burned) of Different Incineration Sources[21,65]**

| Pollutant | Municipal Multiple chamber | Industrial and Commercial Single chamber | Industrial and Commercial Multiple chamber |
|---|---|---|---|
| Aldehydes | 1.1 | 5—64 | 0.3 |
| CO | 0.7 | 20—200 | 0.5 |
| Hydrocarbons | 1.4 | 20—50 | 0.3 |
| $NO_x$ | 2.1 | 1.6 | 2.0 |
| $SO_x$ | 1.9 | n.a. | 1.8 |
| Ammonia | 0.3 | n.a. | n.a. |
| Organic acids | 0.6 | n.a. | n.a. |
| Particulate | 6—12[a] | 20—50 | 4.0 |

*Note:* n.a. = Not available.
[a]6 with, 12 without spray chamber.

tants.[2] Measurements in a coal-fired utility boiler plant showed that PCB concentrations in the flue gas were similar to values in the background air.[3] This indicates that at least in this case the emission of PCBs is due to intake air and not to generation during combustion.

In all practical applications, thermal decomposition of carbonaceous materials is incomplete. PAHs are typical reaction products of incomplete combustion. Their formation occurs by free radical mechanisms, combining carbon atoms in a chemically reducing atmosphere at high temperatures (>700°C), resulting in complex PAH molecules (Figure 1).[4] Incomplete combustion will also release original, undecomposed compounds by volatilization from the combustion chamber. The emission of aliphatic hydrocarbons from car engines may serve as an important example.[5]

PAHs and aliphatic hydrocarbons in the environment originate not only from human activity, but occur also in natural cycles. Therefore the contribution of natural sources must be estimated.[6,7] For n-alkanes, a Carbon Preference Index (CPI = odd/even n-alkanes, for C > 20) to determine the contri-

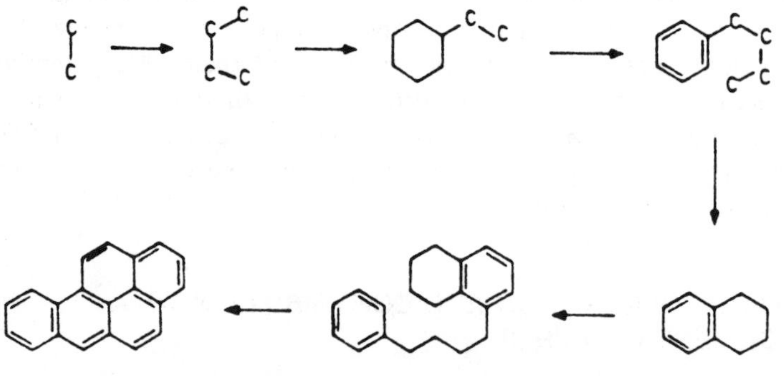

**Figure 1.** Formation of PAH compounds (here benzo[a]pyrene) during combustion processes. From Zander, M. *The Handbook of Environmental Chemistry, Vol. 3, Part A, Anthropogenic Compounds*, 1980, with permission.

**Table 2. Some Chlorinated Compounds in Stack Gas Emissions from a Two-Stage Refuse Incinerator[13]**

| | Stack emissions per ton feed (mg/ton) |
|---|---|
| PCDD | 428 |
| PCDF | 570 |
| PCB | 3,413 |
| Chlorophenol | 18,403 |
| Chlorobenzene | 18,014 |

butions of natural and manmade emissions to ambient air concentrations has been proposed.

The CPI ranges from about five (terrestrial source: plant waxes, soil emissions) to about one (anthropogenic sources) and thus gives significantly different values for remote and urban areas.[8] PAH compounds, occurring naturally in the environment, may be synthesized by plants or microorganisms or produced in natural fires. However, man strongly contributes to environmental PAH levels. Burning of fossil fuels, refuse burning, and agricultural burning contributes more than 90% to the overall environmental PAH levels.[6] The predominance of some PAH types may signify the type of emission source, as it is the case for the alkylphenanthrene series, the dibenzothiophene series, or the benzo(b)naphtho(1,2-d)thiophene.[9] The alkyl homologue distribution of the pyrene/fluoranthene series has also been found useful for the indication of combustion-produced PAH in sediments.

Significant production and emission of halogenated organics depends on the occurrence of halogenated precursors or active chlorine species during the combustion process or the emission of unchanged substance contents in the combusted material. Rather intensive studies have been undertaken on the generation of chlorinated dioxins and furans, especially in municipal incinerators.[11] Formation of PCDDs/PCDFs and their typical isomer and congener distribution pattern resulting from combustion processes has been mainly attributed to condensation reactions (e.g., chlorophenols, chlorinated aromatic ethers, or chlorinated biphenyls) or catalytic chlorination or dechlorination reactions.[12]

Table 2 shows the amounts of several chlorinated species emitted from some Canadian municipal incinerators in relation to combusted material[13] and in terms of flue gas concentrations (Table 3).[14] PCDD/PCDF formation during combustion shows a strong dependence from temperature, with a maximum at about 300°C, and therefore only small amounts could be detected in the high-temperature areas (furnace) of municipal incinerators. Especially, deposits of the inner boiler surface near the boiler off gas exit or even dust filters may be the preferred locations of dioxin/furan generation due to ideal thermal conditions.[15-17]

Reconstructing the historical inputs of PCDDs and PCDFs into the Great Lakes and Siskiwit Lake, a strong increase has been found for sediment concentrations since 1940 (Figure 2).[18] Although some PCDDs/PCDFs are by-

**Table 3. Chlorinated Dioxins and Furans in Flue Gas Samples from a Municipal Incinerator[14]**

| Compound | Mean concentration (ng/m³) |
|---|---|
| Dibenzo-p-dioxin | |
| Monochloro- | 7.4 ± 5.5 |
| Dichloro- | 39 ± 54 |
| Trichloro- | 45 ± 58 |
| Tetrachloro- | 230 ± 180 |
| Pentachloro- | 1200 ± 1000 |
| Hexachloro- | 510 ± 280 |
| Heptachloro- | 160 ± 92 |
| Octachloro- | 41 ± 31 |
| Dibenzofuran | |
| Monochloro- | 360 ± 54 |
| Dichloro- | 510 ± 110 |
| Trichloro- | 2000 ± 830 |
| Tetrachloro- | 1100 ± 670 |
| Pentachloro- | 6200 ± 5700 |
| Hexachloro- | 700 ± 710 |
| Heptachloro- | 200 ± 120 |
| Octachloro- | 14 ± 7.2 |

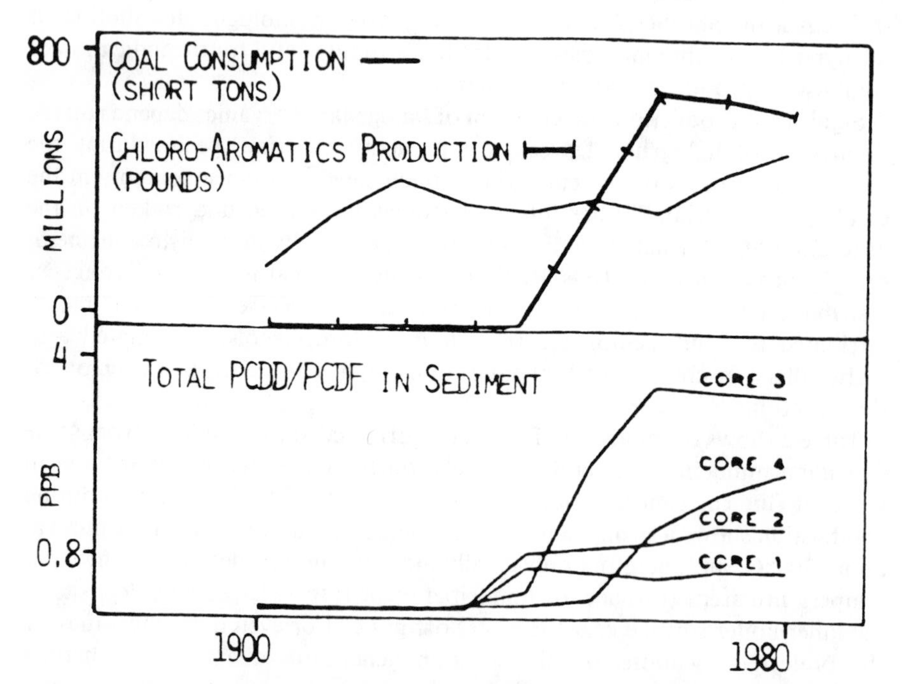


**Figure 2.** Total PCDDs and PCDFs in Lake Huron sediment cores compared to production of chloroaromatics and coal consumption.[18] From Czuczuma, J. M., and R. A. Hites. *Environ. Sci. Technol.* 18:444 (1984) with permission.

Table 4. Particle-Size Distribution of Combustion Gas from a Sewage Incinerator Before and After the Cleaning Stage[21,66]

| Particulate loading (pounds per wet ton)— size distribution (μm) | Percent by weight less than indicated size | |
|---|---|---|
| | Location 1 52.08 | Location 2 0.36 |
| 18.7 | 37.9 | 100.0 |
| 11.7 | 30.6 | 98.0 |
| 8.0 | 16.4 | 94.9 |
| 5.4 | 6.6 | 93.4 |
| 3.5 | 2.6 | 92.8 |
| 1.8 | 1.6 | 83.8 |
| 1.1 | 0.9 | 67.7 |
| 0.76 | 0.1 | 54.6 |

products of industrial chemical synthesis and may therefore be directly emitted, the largest amount occurring now in the environment seems to originate from combustion emissions and atmospheric deposition.[19,20]

## THE EMISSION PROCESS

Quality and pollutant loading of the flue gas varies widely with the type of combustion and cleaning stages.[21] Organics may be emitted in the vapor state or associated with particulate matter. Although particulates may be retained nearly quantitatively (>99.9%), the removal efficiency shows a strong decrease with the particle size and therefore small-size particles are preferentially emitted (Table 4).[21]

The influence of different anthropogenic activities on the volume distributions of background aerosol is shown in Figure 3.[22] where additional differences between combustion sources are pointed out.

Investigations on the size distribution of organic chemicals in municipal incinerator flyash showed the contents of benzene-soluble organics to differ with particle size.[23] As shown in Table 5, the smaller particles did not always contain the highest amounts of the compounds. Smaller particles contained preferentially PAH molecules of higher molecular weight and vice versa. A more nonuniform distribution has also been found for PCDDs/PCDFs (Table 5). However, the quality of flyash and the distribution of organics strongly varies between different incinerators (Table 6).[24]

The benzene extraction of flyash may not necessarily reflect elemental carbon content,[25] and the size fraction far below 30 μm, which is the most important for environmental release, has not been characterized.

In addition to existing flyash particles, submicron particles form due to nucleation and condensation of hot vapors.[26-28] This process includes part of organic compounds in emitted particles, which undergo further coagulation and may not be readily exchangeable in the atmosphere.[29] Therefore, the estimation and the analytical differentiation of the gas/particle-phase distribu-

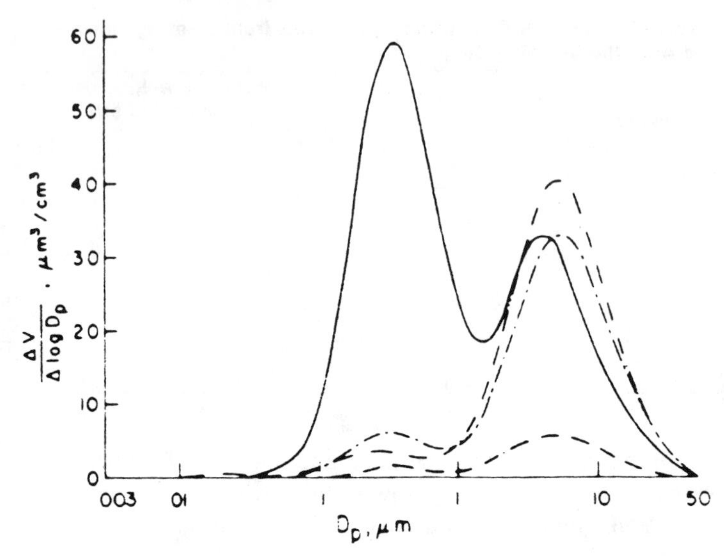


**Figure 3.** Volume distribution of four aerosol background distributions: —urban plume influenced; – – –background average; – ·· –auto influenced; – · –clean background. From Whitby, K. T., and G. M. Sverdrup. *Adv. Environ. Sci. Technol.* 10:477 (1980) with permission.

tion is quite difficult. Flue gas measurements have been made with a distinction between filter-, impinger-, and adsorbent-retained substance content[14,30] using modified EPA Method 5 sampling trains. In these studies, however, blow-off or blow-on effects are not considered.

Table 7 shows the levels of PCDDs and PCDFs in different phases of the stack emission from a municipal incinerator. The filter retained fraction depends on the compound's vapor pressure, however high stack gas temperatures strongly influence the vapor/particle distribution and this may change as the gas leaves the stack. Lützke[31] found that changing the filter temperature of

**Table 5. Distribution of Organic Compounds on Size-Fractionated Flyash from an Incinerator. Concentrations in Nanograms per Grams[23]**

| Compound | Average particle size (μm) | | | | | |
|---|---|---|---|---|---|---|
| | 30 | 80 | 125 | 200 | 550 | >850 |
| Biphenyl | 43 | 35 | 33 | 160 | 240 | 700 |
| Fluorene | 24 | 3.8 | 3.5 | –.– | 69 | –.– |
| Fluoranthene | 76 | 51 | 68 | 97 | 120 | 150 |
| Pyrene | 44 | 16 | 13 | 18 | 9.1 | –.– |
| Alkanes | 14,00 | 7,900 | 5,200 | 5,400 | 11,000 | 5,500 |
| $T_4CDD$ | 2.2 | 2 | 3.9 | 7 | 12 | 8.4 |
| PCDD | 3 | 2 | 4 | 8 | 10 | 5 |
| $H_6CDD$ | 3 | 3 | 6 | 10 | 8 | 2 |
| $H_7CDD$ | 3 | 3 | 4 | 6 | 3 | 0.3 |
| OCDD | 9.5 | 4.5 | 6.9 | 8.2 | 3.1 | 0.2 |

**Table 6. Distribution of PCDDs/PCDFs on Size-Fractionated Flyash from Different Incinerators[23,24,38]**

| Size ($\mu$m) | $T_4CDD/OCDD$ | PCDD/OCDD | $H_6CDD/OCDD$ | $H_7CDD/OCDD$ |
|---|---|---|---|---|
| | | **Canadian sample** | | |
| 30 | 0.23 | 0.60 | 0.70 | 0.67 |
| 80 | 0.44 | 0.84 | 1.52 | 1.46 |
| 125 | 0.57 | 1.12 | 1.98 | 1.26 |
| 200 | 0.85 | 1.88 | 2.79 | 1.59 |
| 550 | 3.87 | 6.21 | 5.86 | 2.11 |
| >850 | 42 | 48 | 23 | 3.26 |
| | | **French sample** | | |
| 30 | –.– | 0.08 | 0.16 | 0.54 |
| 80 | –.– | 0.04 | 0.48 | 1.11 |
| 125 | –.– | 0.15 | 0.75 | 1.46 |
| 200 | –.– | 0.27 | 0.55 | –.– |
| 550 | –.– | 0.33 | 0.80 | 0.37 |
| >850 | –.– | –.– | –.– | 1.04 |

**Table 7. Distribution of PCDDs and PCDFs in Different Stages of the Modified EPA Method 5 Train Sampling the Stack Gas from a Municipal Incinerator[30] The Data Reported are Total Amounts in Nanograms from 24-Hr. Sampling**

| Congener | Filter | Impinger | Florisil | Total |
|---|---|---|---|---|
| $T_4CDD$ | 11 | 450 | 1.0 | 460 |
| PCDD | 44 | 540 | 1.0 | 580 |
| $H_6CDD$ | 77 | 1900 | 18.0 | 2000 |
| $H_7CDD$ | 81 | 1000 | 16.0 | 1100 |
| OCDD | 76 | 200 | 6.0 | 280 |
| $T_4CDF$ | 24 | 870 | 1.0 | 900 |
| PCDF | 44 | 1100 | 3.0 | 1100 |
| $H_6CDF$ | 53 | 2500 | 11.0 | 2600 |
| $H_7CDF$ | 29 | 1200 | 5.0 | 1200 |
| OCDF | 12 | 200 | 3.0 | 220 |

**Table 8. Influence of the Filter Temperature of Stack Gas Sampling Device on Filter-Retained Amounts of Dioxin and Furan Homologs[31]**

| | Sampling procedure—filter–condenser–impinger—filter–temperature | | | | | |
|---|---|---|---|---|---|---|
| | 140°C | | | 90°C | | |
| $T_4CDD$ | 2.0 | 8.2 | 2.0 | 8.8 | 2.0 | 0.6 |
| PCDD | 6.5 | 9.6 | 2.8 | 15.8 | 2.8 | 0.4 |
| $H_6CDD$ | 12.1 | 7.8 | 2.3 | 11.9 | 1.7 | 0.3 |
| $H_7CDD$ | 13.3 | 4.0 | 1.2 | 9.7 | 0.5 | 0.1 |
| OCDD | 6.0 | 1.4 | – | 5.2 | 0.3 | 0.02 |

*Note*: Concentrations in nanograms per cubic meter.

his sampling device strongly influenced the amount of filter-retained dioxins and furans (Table 8).

## COMBUSTION-EMITTED ORGANICS IN THE AMBIENT ATMOSPHERE

Of central interest for atmospheric transport behavior of organics in the atmosphere is the vapor/particle distribution as well as the concentrations in different particle-size ranges.

Particles with a size $>2$ $\mu$m diameter (coarse particles) are rapidly deposited after emission into the atmosphere and Aitken nuclei ($<0.08$ $\mu$m diameter) show fast coagulation. The particulate matter with highest residence time consists of midsized particles ($0.08$ $\mu m < d < 2$ $\mu$m, accumulation mode range).[32] In addition, this size classification of atmospheric particulates may be simplified according to a biomodal distribution, where only coarse ($>2$ $\mu$m diameter) and fine particles ($<2$ $\mu$m diameter) are distinguished.

It has been pointed out that at this size border, there are not only two different mass and volume distributions observed, but also there is a substantial change in chemical composition and generation.[26] Fine particles mainly result from condensation processes (as they preferentially occur during combustion), whereas coarse particles result from mechanical processes such as soil erosion.

The exact analytical differentiation between these two phases is difficult, since sampling methods may influence the results.[29] In addition, combustion-emitted compounds may be embedded within a particle matrix and may thus not readily exchange. Therefore the model proposed by Junge[33] may not be applicable in this case, since it only considers pure physical adsorption and not particle properties and inclusion:

$$\phi = c * \theta/(p_o + c * \theta) \qquad (1)$$

where $\phi$ = adsorbed compound fraction
$\theta$ = particle surface/volume air
$p_o$ = vapor pressure
c = nearly constant for many organics, depends on the heat of condensation and the molecular weight of compound [33]

The following regression, which results from 1-year measurements of the vapor/phase particle distribution of several PAHs has been proposed:[34]

$$D = \lg (PAH_{vap}) (TSP)/(PAH_{par}) = -A/T + B \qquad (2)$$

where $(PAH_{vap})$ = Concentration of PAH in vapor phase
$(PAH_{par})$ = Concentration of PAH on particles
(TSP) = Total Suspended Particulate ($\mu g/m^3$)
T = Temperature (K)

A,B = Constants

Bidleman[29] applied the Junge equation to several compounds (Figure 4) and states—besides a generally good agreement between measured and calculated results—differences between PAHs and organochlorine compounds. Organochlorines show lower particular sorption than predicted, whereas PAHs behave inversely. This indicates the portion of nonexchangeable PAH contents in combustion-generated particulates.

Calculations on the influence of nonexchangeable compound fractions on $\phi$ for different atmospheric particle loadings are shown in Figure 5.[35] It is evident that the effects of nonexchangable substance amounts diminish with the vapor

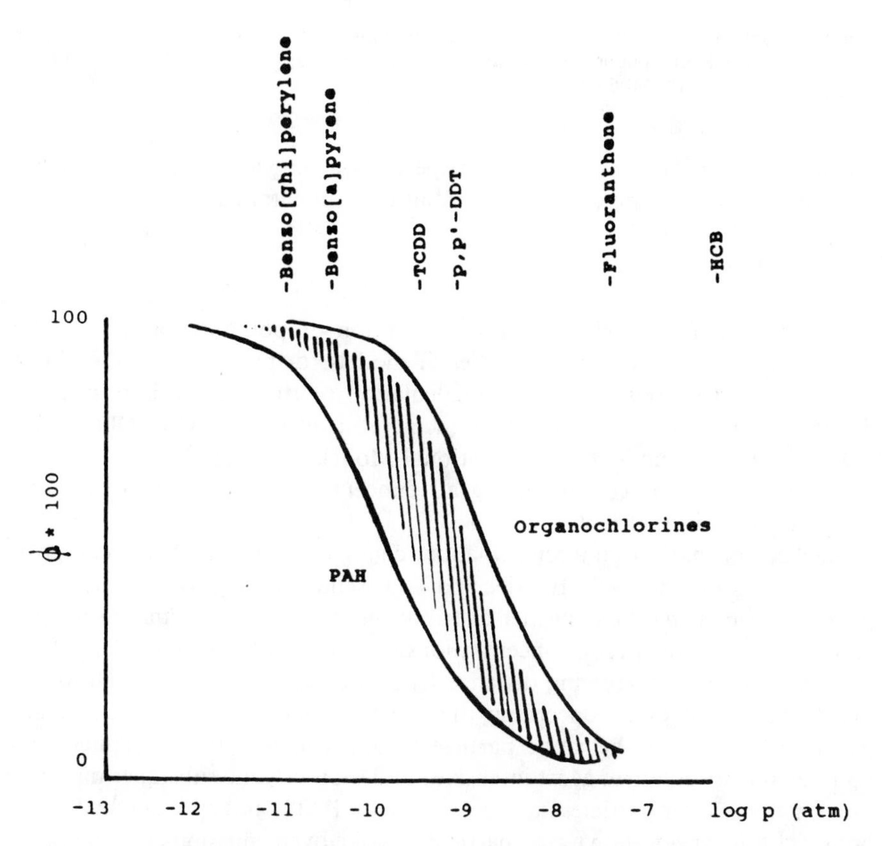


**Figure 4.** Particulate-bound organics as a function of liquid-phase vapor-pressure. The shaded range represents particle-bound compound fractions calculated according to Equation 1, with $1.5 * 10^{-6} — 1.1 * 10^{-5}$ cm$^2$/cm$^3$ as lower- and upper-particle loading, respectively. The solid lines result from field measurements for PAHs and organochlorines in urban air (TSP = 100 $\mu$g/m$^3$). From Bidleman, T. F. *Environ. Sci. Technol.* 22:361 (1988) with permission.

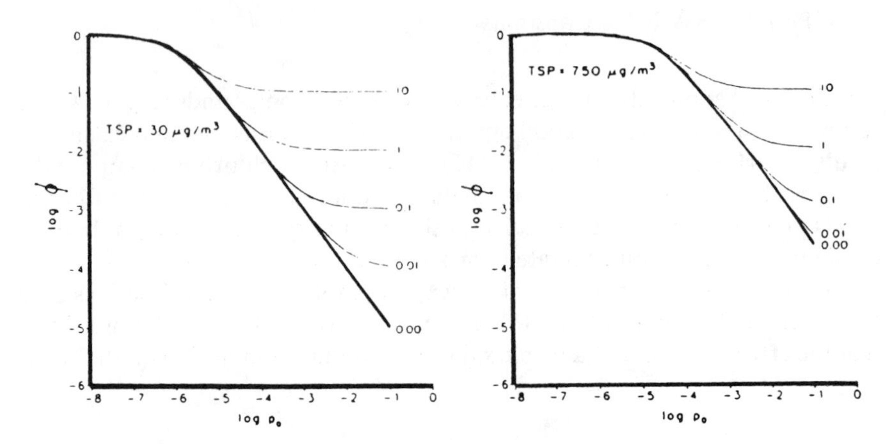


**Figure 5.** Influence of nonexchangable substance parts on the particle-bound fraction of organic compounds with different vapor pressures. From Pankow, J. F. *Atmos. Environ.* 22:1405 (1988) with permission.

pressure. The influence of particle properties on the vapor/particle distribution however cannot be estimated without experimental data.

## Short-distance transport

Extensive literature exists on the description of pollutant behavior after release from the source and a number of models for prediction of the plume rise have been proposed.[36] For the calculation of the average immission of dioxins and furans from a municipal incinerator during one year, Olie et al.[37] used a dispersion model ("Dutch National Model") (Figure 6).

Table 9 shows the corresponding air concentrations at the two sampling stations shown in Figure 4.

Mechanisms have been discussed about changes of PCDD/PCDF composition after leaving the stack. Eduljee and Townsend[38] summarized mechanisms, possibly influencing the congener distribution of combustion-emitted dioxins and furans since dioxin soil concentrations by Crummet[39] showed a shift of the congener ratios with growing distance from the source (Table 10). However, due to a lack of data concerning the distribution of dioxins on small-size particle fractions and the vapor/particle ratio, discussion about this phenomenon remains speculative. Most intensive studies on organic substance distribution on ambient air particles have been done on PAHs and aliphatic hydrocarbons.[40,41] It has been shown that particulate coke oven emissions contain more than 90% of several PAH compounds on sizes smaller than 3 μm;[40] airborne particulate matter from a suburban area showed above 70% of PAH to occur on particles smaller than 1 μm.[41] However, this may be different for dioxins and furans since chlorination of these compounds in the stack or plume seems possible.[42]

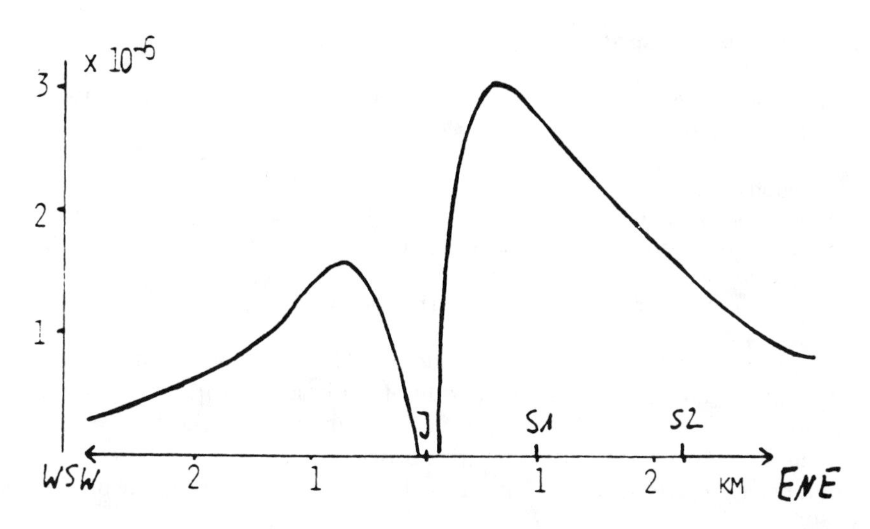


**Figure 6.** Calculated dispersion factors for maximum (ENE-direction) and minimum (WSW-direction) immission of PCDDs and PCDFs from a municipal incinerator (I) and air sampling locations at two distances (S1 and S2). From Olie, K., M. v.d. Berg, and O. Hutzinger. *Chemosphere* 12:627 (1983) with permission.

**Table 9. Air Concentrations of PCDDs and PCDFs in the Vicinity of a Municipal Incinerator[37]**

| Congener | Location S1 ($pg/m^3$) | Location S2 ($pg/m^3$) |
|---|---|---|
| $T_4CDD$ | 1.5 | 0.1 |
| PCDD | 3.4 | 0.5 |
| $H_6CDD$ | 4.0 | 2.0 |
| $H_7CDD$ | 5.2 | 2.1 |
| OCDD | 13.3 | 0.9 |
| $T_4CDF$ | 5.0 | 0.4 |
| PCDF | 4.8 | 0.6 |
| $H_6CDF$ | 5.2 | 2.1 |
| OCDF | 4.8 | 0.4 |

**Table 10. Shift of PCDD Congener Ratios in Soil with Growing Distance to the Stack Emission Source[39]**

| Distance from stack (ft) | PCDD in soil (ppt) | $H_7CDD/OCDD$ |
|---|---|---|
| 100 | 60 | 0.34 |
| 200 | 1,280 | 0.24 |
| 400 | 25,640 | 0.15 |
| 1,000 | 10,040 | 0.17 |

## Long-Range Transport

Principally, the atmospheric residence time of organic compounds is determined by four processes:

- degradation
- vapor dissolution in aqueous atmospheric phases, followed by precipitation
- adsorption of organics in the gaseous state to terrestrial surfaces or solution in aqueous phases
- removal of particles, containing adsorbed or included organics

Atmospheric degradation of the compounds under discussion is mainly attributed to photolytic action and will not be discussed further. It should be mentioned, however, that degradation rates observed in the laboratory are influenced by sorption on atmospheric particles.[43] Behymer and Hites,[44] found a number of PAHs to be stabilized in the sorbed state on flyash or carbon black.

The wet removal of organics from either the vapor state or particle bound may be described by:

$$W = W_{vap}*(1-\phi) + W_{par}*\phi \quad (3)$$

where $W_{vap}$ = washout ratio of vapor = $R*T/H$[46]
$W_{par}$ = washout ratio of particles (substance concentration in rain/substance concentration on particles)

Therefore, compounds with high Henry constants will be mainly removed by wet particle deposition. Ligocki et al.[47] found that for most PAH compounds investigated, gas scavenging mechanisms were more important than particle scavenging (Table 11). However, it may be expected that $W_{par}$ strongly varies with meteorological conditions, season, and location,[32,48,49] with a typical range for $W_{par}$ of $1 \times 10^5 - 1 \times 10^6$.[29] For PCBs, Eisenreich et al.[50] estimated the relation of wet to dry deposition according to:

$$R = J * t_w * W/v_d * t \quad (4)$$

where $J * t_w$ = wet deposition during one year
$W$ = overall washout ratio
$v_d$ = deposition velocity
$t$ = 1 year

and proposed a dry-to-wet deposition ratio of 1.5 to 5 for PCBs into the Great Lakes.

The deposition velocity, as used in Equation 4, is a measure for particle flux to a surface. It is defined as $v_d = F/C$, where F is the flux (mass/area × time) and C is concentration (mass/volume). The deposition velocity of particles is strongly dependent on the particle size[51] and, due to their low deposition

**Table 11. Scavenging Ratios for a Number of Neutral Organic Compounds[47]**

| Compound | Mean $\Phi$ | Mean $W_p$ | Mean $W_g$ | Mean W | DM |
|---|---|---|---|---|---|
| alpha-HCH | 0.0 | n.a. | 31,000 | 31,000 | g |
| Dibenzofuran | 0.008 | 11,000 | 930 | 1,000 | g |
| Fluorene | 0.009 | 15,000 | 1,500 | 1,600 | g |
| Fluoranthene | 0.053 | 11,000 | 6,300 | 6,600 | g |
| Chrysene | 0.71 | 2,600 | 18,000 | 7,000 | g |
| Benz[a]anthracene | 0.75 | 1,300 | 12,000 | 4,000 | g |
| Benzo[a]pyrene | 0.98 | 1,700 | n.a. | 1,700 | p |
| Perylene | 1.0 | 1,800 | n.a. | 1,800 | p |
| Benzo[ghi]perylene | 1.0 | 3,100 | n.a. | 3,100 | p |

*Notes*: $\Phi$—Particle-bound fraction.
$W_p$—Washout ratio for particles.
$W_g$—Washout ratio for gas.
$W—W_p{}^*\Phi + W_g(1 - \Phi)$.
n.a.—Not available
DM—Dominant scavenging mechanism.
g—Gas.
p—Particle.

velocities, particles in the size range between 0.08 and 2 μm diameter are favored for long-range transport.

## ACCUMULATION OF AIRBORNE PCDDs/PCDFs ON PLANT SURFACES

Plant surfaces in nonarid areas represent an important part of the overall terrestrial surface (Table 12)[52] and may therefore function as an important temporary sink of airborne pollutants. This becomes particularly important since the deposition velocity of particles onto vegetation is generally higher than onto many other natural surfaces, e.g., water.[53]

In addition, due to their lipophilic properties,[54] the leaves of higher plants

**Table 12. Leaf Area Index and Deposition Velocity of Particles on Different Trees in Relation to Grass.[52] The Corresponding Values for Soil and Water Surfaces are Listed Below[53]**

| Species | Leaf area index | $v_g$ tree/$v_g$ grass |
|---|---|---|
| Beech | 6.5 | 15.84 |
| European Hornbeam | 8.0 | 9.76 |
| Red Oak | 4.3 | 3.06 |
| Norway Maple | 5.02 | 2.80 |
| European White Birch | 5.3 | 23.52 |
| Spruce | 11.0 | 39.25 |
| Pine | 3.9 | 10.23 |
| Japanese Larch | 4.6 | 26.11 |
| Fir | 8.5 | 18.02 |
| | | $v_g$ s,w/$v_g$ grass |
| Soil | | 0.4 |
| Water | | 0.2 |

**Table 13. Concentrations of PCDDs and PCDFs on One- and Two-Years-Old Needles of Spruce with Increasing Distance from a Municipal Incinerator. Concentrations in Picogram per Gram Needle Dry Weight**

| Distance | 0.5 km | | 2 km | | 16 km | |
|---|---|---|---|---|---|---|
| **Needle age (years)** | **1** | **2** | **1** | **2** | **1** | **2** |
| $T_4CDF$ | 38 | 28 | 34.5 | 39 | 19 | 21 |
| PCDF | 50 | 86 | 34.5 | 39 | 15 | 17 |
| $H_6CDF$ | 82 | 46 | 15.5 | 18 | 9 | 8 |
| $H_7CDF$ | 35.5 | 33 | 6.5 | 9 | 4 | 5 |
| OCDF | 4.5 | 5 | 1 | 2 | – | – |
| **Σ PCDF** | **210** | **198** | **92** | **107** | **47** | **51** |
| $T_4CDD$ | 5.5 | 4 | 4.5 | 4 | 2 | 3 |
| PCDD | 25.5 | 18 | 11.5 | 13 | 8 | 9 |
| $H_6CDD$ | 28 | 34 | 17.5 | 17.5 | 8 | 9 |
| $H_7CDD$ | 32 | 42 | 12.5 | 17 | 16 | 17 |
| OCDD | 54 | 42 | 18 | 22 | 28 | 32 |
| **Σ PCDD** | **145** | **145** | **64** | **73.5** | **62** | **70** |

*Note:*—Below detection limit.

act as accumulators of gaseous and dissolved deposited organics. This function has already been described for different plants and compounds (e.g., organochlorines, PAHs).[55–59]

Even for PCDDs/PCDFs, which occur at the picogram per cubic meter level in the ambient atmosphere,[59] significant accumulation occurred on plant leaves.[60,61] The congener and isomer patterns were found similar to combustion emissions and a clear dependence of leaf concentrations from the distance to emission sources was found (Table 13). However, the relative importance of different deposition mechanisms on plant leaves could not yet be estimated. The findings of Eisenreich et al.[50] and Jonas and Heinemann[53] suggest that for dioxins and furans, as well as the high-molecular PAHs, wet deposition is of minor importance.

Dry deposition fluxes to leave surfaces can be described as follows:

$$\text{Flux}_{dry} = v_g * C_g + \Sigma_i v_{pi} * C_{pi} \tag{5}$$

where $v_g$ = deposition velocity of vapor
$C_g$ = vapor concentration
$v_{pi}$ = deposition velocity of a particle size fraction
$C_{pi}$ = compound concentration on one particle size fraction
$_i$ = particle size fraction-index

However, since the deposition velocities for the deposited particle-size spectrum and deposited gasses varies within a few orders of magnitude, even for compounds, which mainly exist in the sorbed or vapor state alone, no clear decision about the dominant way of accumulation can be made.

The accumulation kinetics of low-volatile vapor organics on plant leaves has not been properly determined.[62] Based on the bioconcentration factors (BCFs) of several chlorinated compounds (two isomers of hexachlorocyclo-hexane p,p′-DDT, p,p′-DDE) and a PCB mixture, calculated from Bacci and Gaggi,[62] Travis and Hattemer-Frey[63] proposed two equations to predict BCFs leaf/vapor:

$$BCF_{l,v} = 9.1 * (K_{ow})^{0.71} \quad (6)$$

where $BCF_{l,v}$ = bioconcentration factor leaf/vapor [ng/g dry weight/ng/g air]

$K_{ow}$ = partition coefficient octanol/water

and

$$BCF_{l,v} = 5 * 10^{-8} * (K_{ow})^{1.5}/H \quad (7)$$

where H = Henry's law constant

Reischl et al.[61] stated that Equation 6 is strongly limited by physicochemical properties and proposed the following two equations, based on accumulation measurements using nine compounds with widely different properties:

$$lgBCF_{l,v} = -0.2946 * lg_{ps} + 3.3917 \quad (8)$$

where $p_s$ = vapor pressure

and

$$lgBCF_{l,v} = 0.4254 * lgK_{ow} - 0.4244 * lgH + 2.2951 \quad (9)$$

Equations 8 and 9 allowed the calculation of PCDD and PCDF vapor concentrations necessary to generate the corresponding concentrations in spruce needles from areas of different pollution. These calculated vapor concentrations were compared to data from ambient air measurements at the same locations.[59]

Although the air measurements were neither continuous, nor were they done at the same time as needle sampling, generally a good agreement was found between calculated and measured air concentration data (Table 14). However, with increasing degrees of chlorination and decreasing vapor pressure,[64] the measured air concentrations clearly exceed the calculated values. This discrepancy may be due to measured concentrations being the sum of vapor and particulate fraction contents. Therefore, the data presented in Table 14 might indicate, that even in the case of dioxins and furans, mainly ambient vapor concentrations are responsible for accumulated levels in higher plants.

## CONCLUSIONS

Particle binding and distribution on different particle-size fractions is of great importance for the environmental distribution of combustion-emitted compounds. This is especially the case for particles in the micrometer and

**Table 14. Comparison of Measured $C_{meas}$ (Vapor + Particle Phase) and Calculated Vapor Concentrations $C_{calc}$ (According to Equation 9). The Needle Samples Used for Calculation were Taken from Three Locations of Different Immission Situation[59]**

| Compound | $C_{calc}/C_{meas}$ I | $C_{calc}/C_{meas}$ II | $C_{calc}/C_{meas}$ III |
|---|---|---|---|
| $T_4CDF$ | 0.058/– | 0.078/– | 0.13/– |
| PCDF | 0.088/– | 0.104/0.05 | 0.156/– |
| $H_6CDF$ | 0.091/0.05 | 0.082/0.59 | 0.0103/0.71 |
| $H_7CDF$ | 0.051/0.21 | 0.049/1.87 | 0.041/1.1 |
| OCDF | 0.033/0.44 | 0.029/1.58 | 0.026/1.01 |
| $T_4CDD$ | 0.29/0.11 | 0.803/1.12 | 1.54/1.84 |
| PCDD | 0.26/0.16 | 0.401/0.87 | 0.701/1.79 |
| $H_6CDD$ | 0.077/0.19 | 0.12/0.61 | 0.181/0.64 |
| $H_7CDD$ | 0.0235/0.28 | 0.033/0.93 | 0.033/1.15 |
| OCDD | 0.00448/0.18 | 0.005/0.30 | 0.006/0.26 |

*Note:*—Below detection limit, concentrations in picogram per cubic meter.

submicrometer range and their generation and behavior during the emission process. In addition, this may correspond with PCDD/PCDF generation and affect the intensity in which dioxins and furans bind to particulate surfaces. However, only few investigations treat these topics. It may be hypothesized that compared to PAHs, PCDDs/PCDFs show, due to different location of synthesis, also a different particle-binding type and size distribution.

The vapor/particle distribution also affects the deposition mechanisms onto terrestrial surfaces. For polychlorinated dioxins and furans, dry deposition of vapors seems to mainly contribute to the concentrations of these compounds on plant leaves.

## REFERENCES

1. Stief-Tauch, H. P. “Energy Production and Air Pollution with Particular Emphasis to the Medium Term within the European Community,” in *Pollutants and their Ecotoxicological Significance*, H. W. Nürnberg, Ed. (Chinchester: John Wiley & Sons, Inc., 1985).
2. Richard, J. J., and G. A. Junk. “Polychlorinated Biphenyls in Effluents from Combustion of Coal/Refuse,” *Environ. Sci. Technol.* 15:1095 (1981).
3. Haile, C. L., J. S. Stanley, T. Walker, G. R. Cobb, and B. A. Boomer. “Comprehensive Assessment of the Specific Compounds Present in Combustion Processes, Vol. 3, National Survey of Organic Emissions from Coal-Fired Utility Boiler Plants,” Report No. 560/5-83-006 (1983).
4. Zander, M. “Polycyclic Aromatic and Heteroaromatic Hydrocarbons,” in *The Handbook of Environmental Chemistry, Vol. 3, Part A, Anthropogenic Compounds*, O. Hutzinger, Ed. (Berlin: Springer, 1980).
5. Graedel, T. E. *Chemical Compounds in the Atmosphere*, (London: Academic Press, Inc., 1978).
6. Edwards, N. T., “Polycyclic Aromatic Hydrocarbons (PAH’s) in the Terrestrial Environment—A Review, *J. Environ. Qual.* 12:427 (1983).
7. Schneider, J. K., R. B. Gagosian, J. K. Cochran, and T. W. Trull. “Particle Size

Distributions of n-alkanes and $^{210}Pb$ in Aerosols off the Coast of Peru," *Nature* 304:429 (1983).

8. Kawamura, K., and I. R. Kaplan. "Biogenic and Anthropogenic Organic Compounds in Rain and Snow Samples Collected in Southern California," *Atmos. Environ.* 20:115 (1986).
9. Sicre, M.-A., J.-C. Marty, A. Saliot, X. Aparaicio, J. Grimalt, and J. Albaiges. "Aliphatic and Aromatic Hydrocarbons in the Mediterranean Aerosol," *Int. J. Environ. Anal. Chem.* 29:73 (1987).
10. Sporstol, S., N. Gjos, R. G. Lichtenthaler, K. O. Gustavsen, K. Urdal, and F. Oreld. "Source Identification of Aromatic Hydrocarbons in Sediments Using GC/MS," *Environ. Sci. Technol.* 17:282 (1983).
11. Olie, K., W. A. Lustenhouwer, and O. Hutzinger. "Polychlorinated Dibenzo-p-Dioxins and Related Compounds in Incinerator Effluents," in *Dioxins and Related Compounds—Impact on the Environment*, O. Hutzinger, R. W. Frei, E. Merian, and F. Pocchiari, Eds. (Oxford: Pergamon Press, 1982).
12. Choudry, G. G., K. Olie, and O. Hutzinger. "Mechanisms in the Thermal Formation of Chlorinated Compounds Including Polychlorinated Dibenzo-p-dioxines." in *Chlorinated Dioxins & Related Compounds in the Environment*, O. Hutzinger, R. W. Frei, E. Merian, and F. Pocchiari, Eds. (Oxford: Pergamon Press, Inc., 1982).
13. Hay, D. J., A. Finkelstein, and R. Klicius. "The National Incinerator Testing and Evaluation Program Two-Stage Incinerator Combustion Test," *Chemosphere* 15:1201 (1986).
14. Haile, C. L., R. B. Blair, J. S. Stanley, D. P. Redford, D. Heggem, and R. M. Lucas. "Emissions of Polychlorinated Dibenzo-p-dioxines and Polychlorinated Dibenzofurans from a Resource Recovery Municipal Incinerator," in *Dioxins and Dibenzofurans in the Environment II*, L. H. Keith, C. Rappe, and G. Choudary, Eds. (Stoneham, MA: Butterworth Publishers, Inc., 1985.
15. Nottrodt, A., and K. Ballschmiter. "Causes for and Reduction Strategies against Emissions of PCDD/PCDF from Waste Incineration Plants—Interpretations of Recent Measurements," *Chemosphere* 15:1225 (1986).
16. Vogg, H., and L. Stieglitz. "Thermal Behavior of PCDD/PCDF in Fly Ash from Municipal Incinerators," *Chemosphere* 15:1373 (1986).
17. Hagenmaier, H., H. Brunner, R. Haag, and M. Kraft. "Die Bedeutung katalytischer Effekte bei der Bildung und Zerstörung von polychlorierten Dibenzodioxinen und polychlorierten Dibenzofuranen," in *VDI-Berichte 634* (Mannheim: VDI-Verlag 1987).
18. Czuczuwa, J. M., and R. A. Hites. "Environmental Fate of Combustion-Generated Dioxins and Furans," *Environ. Sci. Technol.* 18:444 (1984).
19. Bumb, R. R., W. B. Crummett, S. S., Cutie, J. R. Gledhill, R. H. Hummel, R. O. Kagel, L. L. Lamparski, E. V. Luoma, R. H. Miller, T. J. Nestrick, L. A. Shadoff, R. H. Stehl, and J. S. Woods. "Trace Chemistries of Fire: A Source of Chlorinated Dioxins," *Science* 210:385 (1980).
20. Czuczuwa, J. M., and R. A. Hites. "Environmental Sources and Fate of PCDD and PCDF," *Chemosphere* 15:1417 (1986).
21. Brunner, C. R. *Hazardous Air Emissions from Incineration* (New York: Chapman and Hall, 1985).
22. Whitby, K. T., and G. M. Sverdrup. "California Aerosols: Their Physical and Chemical Characteristics," *Adv. Environ. Sci. Technol.* 10:477 (1980).

23. Clement, R. E., and F. W. Karasek. "Distribution of Organic Compounds Adsorbed on Size-Fractionated Municipal Incinerator Fly-Ash Particles." *J. Chromatog.* 234:395 (1982).
24. Karasek, F. W., R. E. Clement, and A. C. Viau. "Distribution of PCDDs and other Toxic Compounds Generated on Fly Ash Particulates in Municipal Incinerators," *J. Chromatog.* 239:173 (1982).
25. Appel, B. R., P. Colodny, and J. J. Wesolowski. "Analysis of Carbonaceous Materials in Southern California Atmospheric Aerosols," in *The Character and Origins of Smog Aerosols*, G. M. Hidy, P. K. Mueller, D. Grosjean, B. R. Appel, and J. J. Wesolowski, Eds. (New York: John Wiley & Sons, Inc. 1980).
26. Whitby, K. T. The Physical Characteristics of Sulfur Aerosols, *Atmos. Environ.* 12:135 (1978).
27. Smith, R. D., J. A. Campbell, and W. D. Felix. "Atmospheric Trace Element Pollutants from Coal Combustion," Proceedings of American Institute Mining Engineering Meeting, New Orleans, LA (1979).
28. Abel, K. H., J. A. Young, and L. A. Rancitelli. "Atmospheric Chemistry of Inorganic Emissions," Atmospheric Science and Power Production, D. Randerson, Ed., Technical Information Center, Office of Scientific and Technical Information, U.S. Department of Energy, Springfield (1984).
29. Bidleman, T. F., "Atmospheric Processes," *Environ. Sci. Technol.* 22:361 (1988).
30. Clement, R. E., H. M. Tosine, J. Osborne, V. Ozvacic, and G. Wong. "Levels of Chlorinated Organics in a Municipal Incinerator," in *Chlorinated Dioxins and Dibenzofurans in the Environment II*, L. H. Keith, C. Rappe, and G. Choudary, Eds. (Butterworth Publishers, Inc., 1985).
31. Lützke, K. "Emissionsmessungen von PCDD und PCDF," in *VDI-Berichte 634* (Düsseldorf: VDI-Verlag, 1987).
32. Müller, J. "Atmospheric Residence Time of Carbonaceous Particles and Particulate PAH-Compounds," *Sci. Total Environ.* 36:339 (1984).
33. Junge, C. E. "Basic Considerations About Trace Constituents in the Atmosphere as Related to the Fate of Global Pollutants," in *Fate of Pollutants in the Air and Water Environments Part 1, Mechanisms of Interaction Between Environments and Mathematical Modeling and the Physical Fate of Pollutants*, I. H. Suffet, Ed. (New York: John Wiley & Sons, Inc. 1977).
34. Yamasaki, H., K. Kuwata, and H. Miyamoto. "Effects of Ambient Temperature on Aspects of Airborne Polycyclic Aromatic Hydrocarbons," *Environ. Sci. Technol.* 16:189 (1982).
35. Pankow, J. F. "The Calculated Effects of Non-Exchangeable Material on the Gas-Particle Distributions of Organic Compounds," *Atmos. Environ.* 22:1405 (1988).
36. Briggs, G. A. "Plume Rise and Buoyancy Effects," Atmospheric Science and Power Production, D. Randerson, Ed., Technical Information Center, Office of Scientific and Technical Information, U.S. Department of Energy, Springfield (1984).
37. Olie, K., M. v.d. Berg, and O. Hutzinger. "Formation and Fate of PCDD and PCDF from Combustion Processes," *Chemosphere* 12:627 (1983).
38. Eduljee, G. H., and D. I. Townsend. "Evaluation of Potential Mechanisms Governing Dioxin Congener Profiles in Soils Near Combustion Sources," *Chemosphere* 16:1096 (1987).
39. Crummet, W. B. "Letters to Science: Origin of Dioxines," *Science* 213:1060 (1981).

40. Broddin, G., L. van Vaeck, and K. van Cauwenberghe. "On the Size Distribution of Polycyclic Aromatic Hydrocarbon Containing Particles from a Coke Oven Emission Source, *Atmos. Environ.* 11:1061 (1977).
41. van Vaeck, L., and K. van Cauwenberghe. "Cascade Impactor Measurements of the Size Distribution of the Major Classes of Organic Air Pollutants in Atmospheric Particulate Matter," *Atmos. Environ.*:2229 (1978).
42. Eiceman, G. A., and H. O. Rghei. "Chlorination Reactions of 1,2,3,4-Tetrachlorodibenzo-p-dioxin on Fly Ash with HCl in the Air," *Chemosphere* 11:833 (1982).
43. Miller, G. C., V. R. Hebert, and R. G. Zepp. "Chemistry and Photochemistry of Low-Volatility Organic Chemicals on Environmental Surfaces," *Environ. Sci. Technol.* 21:1164 (1987).
44. Behymer, T. D., and R. A. Hites. "Photolysis of Polycyclic Aromatic Hydrocarbons Adsorbed on Simulated Atmospheric Particulates," *Environ. Sci. Technol.* 19:1004 (1985).
45. Pankow, J. F., L. M. Isabelle, and W. E. Asher. "Trace Organic Compounds in Rain. I. Sampler Design and Analysis by Adsorption/Thermal Desorption (ATD)," *Environ. Sci. Technol.* 18:310 (1984).
46. Slinn, W. G. N., L. Hasse, B. B. Hicks, A. W. Hogan, D. Lai, P. Liss, K. O. Munnich, G. A. Sehmel, and O. Vittori. "Some Aspects of the Transfer of Atmospheric Trace Constituents Past the Air-Sea-Interface," *Atmos. Environ.* 12:2055 (1978).
47. Ligocki, M. P., C. Leuenberger, and J. F. Pankow. "Trace Organic Compounds in Rain. III. Particle Scavenging of Neutral Organic Compounds," *Atmos. Environ.* 19:1619 (1985).
48. van Vaeck, L., and K. A. van Cauwenberghe. "Characteristic Parameters of Particle Size Distributions of Primary Organic Constituents of Ambient Aerosols," *Environ. Sci. Technol.* 19:707 (1985).
49. Leuenberger, C., J. Czuczuwa, E. Heyerdahl, and W. Giger. "Aliphatic and Polycyclic Aromatic Hydrocarbons in Urban Rain, Snow and Fog," *Atmos. Environ.* 22:695 (1988).
50. Eisenreich, S. J., B. B. Looney, and J. D. Thornton. "Airborne Organic Contaminants in the Great Lakes Ecosystem," *Environ. Sci. Technol.* 15:30 (1981).
51. Sehmel, G. A. "Particle and Gas Dry Deposition: A Review," *Atmos. Environ.* 14:983 (1980).
52. Jonas, R., M. Horbert, and W. Pflug. "Die Filterwirkung von Wäldern gegenüber staubbelasteter Luft," *Forstwiss. Cbl.* 104:289 (1985).
53. Jonas, R., and K. Heinemann. "Schädigung von Pflanzen durch abgelagerte Schadstoffe," *Staub Reinhalt. Luft* 45:112 (1985).
54. Martin, J. T., and B. E. Juniper. *The Cuticles of Plants* (London: Edward Arnold, 1970).
55. Buckley, E. H. "Accumulation of Airborne Polychlorinated Biphenyls in Plant Foliage," *Science* 216:520 (1982).
56. Thomas, W., A. Rühling, and H. Simon. "Accumulation of Airborne Pollutants (PAH, Chlorinated Hydrocarbons, Heavy Metals) in Various Plant Species and Humus," *Environ. Poll.* (Ser. *A*) 36:295 (1984).
57. Gaggi, C., E. Bacci, D. Calamari, and R. Fanelli. "Chlorinated Hydrocarbons in Plant Foliage: An Indication of the Tropospheric Contamination Level," *Chemosphere* 14:1673 (1985).

58. Reischl, A., M. Reissinger, and O. Hutzinger. "Occurrence and Distribution of Atmospheric Organic Micropollutants in Conifer Needles," *Chemosphere* 16:2647 (1987a).
59. Buck, M., and P. Kirschmer. "Immissionsmessungen polychlorierter Dibenzo-p-Dioxine und Dibenzofurane in Nordrhein-Westfalen," *LIS-Berichte* 62 (1986).
60. Reischl, A., H. Thoma, M. Reissinger, and O. Hutzinger. "PCDD und PCDF in Koniferennadeln," *Naturwissenschaften* 74:88 (1987b).
61. Reischl, A., M. Reissinger, H. Thoma, and O. Hutzinger. "Uptake and Accumulation of PCDD/PCDF in Terrestrial Plants: Basic Considerations," paper presented at the 8th International Symposium on Chlorinated Dioxins and Related Compounds, Umea, Sweden, 1988.
62. Bacci, E., and C. Gaggi. "Chlorinated Hydrocarbon Vapours and Plant Foliage: Kinetics and Applications," *Chemosphere*, 16:2515 (1987).
63. Travis, C. C., and H. A. Hattemer-Frey. "Uptake or Organics by Aerial Plant Parts: A Call for Research," *Chemosphere* 17:277 (1988).
64. Rordorf, B. F. "Thermal Properties of Dioxins, Furans and Related Compounds," *Chemosphere* 15:1325 (1986).
65. American Industrial Hygiene Association. Air Pollution Manual, Part 1 (Evaluation), American Industrial Hygiene Association, Westmont (1972).
66. Brunner, C. R. *Incineration Systems: Selection and Design* (New York: D. Van Nostrand Reinhold Company, 1984).

# CHAPTER 28

# Dioxins from Combustion Processes: Environmental Fate and Exposure

**S. Paterson, W. Y. Shiu, D. Mackay, and J. D. Phyper**

## INTRODUCTION

Combustion is now recognized as an important source of dioxin emissions. Once released into the environment, dioxins can become widely dispersed in air and water, sorb strongly onto sediments and soil, and bioaccumulate in the food chain. The purpose of this study is to identify the major pathways of human exposure to dioxins from combustion processes. In order to assess the reliability of the predictions, the estimated concentrations of dioxins in various environmental compartments are compared with measured values. Because of the differences in the physicochemical properties of the individual dioxin homologs, and hence their environmental fate, tetra-, hexa-, and octa-dibenzo-p-dioxin are evaluated.

A steady-state mathematical model, the Level-III Fugacity Model, is used to evaluate environmental partitioning and transport. The model is a multimedia transport model which has been used in several previous studies.[1,2] Estimated emissions from several combustion processes in Southern Ontario are used as input to the air compartment of the model: municipal and sewage sludge incineration, industrial incineration, wood burning, hospital incineration, and automobiles. Emissions from forest fires are not included. Advection terms in air and water are also employed in the model to quantify the inflow of dioxins to the region from the atmosphere and the Niagara River.

Inevitably there are considerable uncertainties about emission rates, partitioning, transport, and transformation processes, thus the quantities calculated here have substantial error margins. The objective of this paper is to demonstrate the concept of multimedia modeling as a tool in exposure assessment, rather than provide a detailed, accurate simulation.

**Table 1. Physical-Chemical Properties of Selected Chlorinated Dibenzo-p-Dioxins and Reaction Rate Constants in Various Media**

| Physical chemical properties | 2,3,7,8-$T_4CDD$ | 1,2,3,4,7,8-$H_6CDD$ | $O_8CDD$ |
|---|---|---|---|
| Molecular weight (g/mol)[4] | 322.0 | 391.0 | 460.0 |
| Solubility ($mmol/m^3$)[4] | 0.00006 | 0.000113 | 0.0000016 |
| log $K_{ow}$[4] | 6.80 | 7.80 | 8.20 |
| Vapor pressure (Pa)[4] | $2 \times 10^{-7}$ | $5.1 \times 10^{-9}$ | $1.1 \times 10^{-10}$ |
| **Reaction rate constants** | **$T_4CDD$** | **$H_6CDD$** | **$O_8CDD$** |
| Air ($hr^{-1}$)[5] | 0.02 | 0.005 | 0.0015 |
| Water ($hr^{-1}$)[5] | 0.008 | 0.002 | 0.0005 |
| Soil ($hr^{-1}$)[5] | $1.1 \times 10^{-5}$ | $2.8 \times 10^{-6}$ | $7.0 \times 10^{-7}$ |
| Sediment ($hr^{-1}$)[6] | $1.5 \times 10^{-5}$ | $4.0 \times 10^{-6}$ | $1.0 \times 10^{-6}$ |

## PHYSICAL CHEMICAL PROPERTIES

An initial difficulty in applying models of this type to chemicals such as "dioxins" is the existence of 75 congeners. Ideally, the model should be applied on a congener-specific basis, but this requires congener-specific data on emission rates, physical chemical properties, and concentrations for model validation. Recently, the merits of this detailed approach have been demonstrated by Oliver and Niimi[3] who have described the detailed behavior of some 60 PCB congeners in the Lake Ontario ecosystem. Because of the present limited availability of data, isomers, e.g., the mixture of all tetrachlorinated congeners, are assigned average properties using reported values for a specific congener (e.g., 2,3,7,8 $T_4CDD$). There is an inherent error introduced by using this approach but it is presently unavoidable. The isomers examined are tetra-, hexa-, and octachlorinated dibenzo-p-dioxins, designated $T_4CDD$, $H_6CDD$ and $O_8CDD$.

Table 1 shows the relevant, assumed physical chemical properties of molecular weight, solubility in water, vapor pressure, and octanol/water partition coefficient. These data were obtained from a recent review by Shiu et al.[4] The reaction half-lives are based on data taken from a variety of sources as indicated in Table 1. In several cases, the reported rate constants were arbitrarily adjusted to give what are believed to be more realistic values. For example, photolytic rate constants have been reported in surface water,[5] but these have been reduced to reflect a substantial volume of deep water in the Southern Ontario environment which is not subject to appreciable photolysis.

## EMISSION SOURCES IN SOUTHERN ONTARIO

Estimating dioxin emissions is complex because of the uncertainty regarding the formation of these compounds during combustion and lack of emission data, especially in terms of the individual congeners or isomers. Table 2 presents the estimated annual emission rates for the three selected isomers. Several assumptions (noted in Table 2) are employed in the estimation of the

**Table 2. Estimated Annual Emissions in Southern Ontario (Grams per Year)**

| | $T_4CDD$ | $H_6CDD$ | $O_8CDD$ |
|---|---|---|---|
| Municipal incinerators | | | |
| Commissioner St.[7] | 96.5 | 587 | 1627 |
| Ashbridge's Bay[8] | 33.1 | 56.8 | 111 |
| SWARU[8] | 1883 | 102 | 26 |
| Hospital incineration[6,9] | 0.6 | 16 | 27 |
| Industrial incineration | 17.2 | 47.8 | |
| Residential wood fuel[6] | 107 | 617 | 537 |
| Automobiles[6] | 66.7 | NA | NA |
| Total | 2204 | 1427 | 2328 |

*Notes:* NA—not available.
Assumptions employed in estimation of emissions:
- No contribution from forest fires and apartment building incinerators.
- Distribution of dioxins from wood burning is based on emissions from untreated wood.
- Distribution of dioxins from hospital incinerators is based on Royal Jubilee Hospital, Victoria, B.C.
- Distribution of dioxins from auto emissions is assumed to be equal to a coal-fired source.
- Distribution of dioxins from industrial emissions is representative of those from major Ontario sources.

emission rates, hence a large degree of uncertainty is associated with the presented values.

## ADVECTION

The concentrations of dioxins in air and water which flow into Southern Ontario are shown in Table 3. A background atmospheric concentration of 0.05 pg/m$^3$ flowing into the region was assumed for $T_4CDD$ and the values for $H_6CDD$ and $O_8CDD$ were calculated on the basis of their respective ratios to total emissions.

The Niagara River is considered to be the primary source of water-borne dioxins with a reported inflow of approximately 1 g/hr of total dioxins and

**Table 3. Advective Flows in Air and Water in Southern Ontario**

| | Inflow concentration | Influx (g/hr) |
|---|---|---|
| **Air** — Inflow rate $3.3 \times 10^{12}$ m$^3$/hr | (pg/m$^3$) | |
| $T_4CDD$ | 0.048 | 0.16 |
| $H_6CDD$ | 0.035 | 0.11 |
| $O_8CDD$ | 0.054 | 0.18 |
| **Water** — Inflow rate $3.3 \times 10^{8}$ m$^3$/hr | (ng/m$^3$) | |
| $T_4CDD$ | 0.6 | 0.2 |
| $H_6CDD$ | 0.6 | 0.2 |
| $O_8CDD$ | 0.6 | 0.2 |

furans.[6] An arbitrary value of 0.2 g/hr was assumed for each of the dioxin isomer groups considered. It is important to estimate and document these inflow rates to obtain an appreciation of the contribution to local contamination of local sources versus more distant sources. The air and water flow rates are also given from which the influx in grams per hour is deduced as shown.

## THE ENVIRONMENTAL MODEL

The model which is described in detail elsewhere[2] is based on the fugacity concept and is similar to that described by Mackay et al.[1] Instead of being evaluative in nature, it is applied to the region of Southern Ontario, which is shown in Figure 1. The model describes partitioning and intermedia transport between four primary compartments of air, water, soil, and sediment as depicted in Figure 2. One prevailing fugacity is deduced for each compartment, thus equilibrium is assumed to exist (1) between gas and aerosol; (2) between water and all suspended matter in the water, including fish; (3) the soil matrix consisting of organic mineral, air, and water material; and (4) sediment solids and pore water.

Intermedia transfer processes are described by a series of appropriate equa-

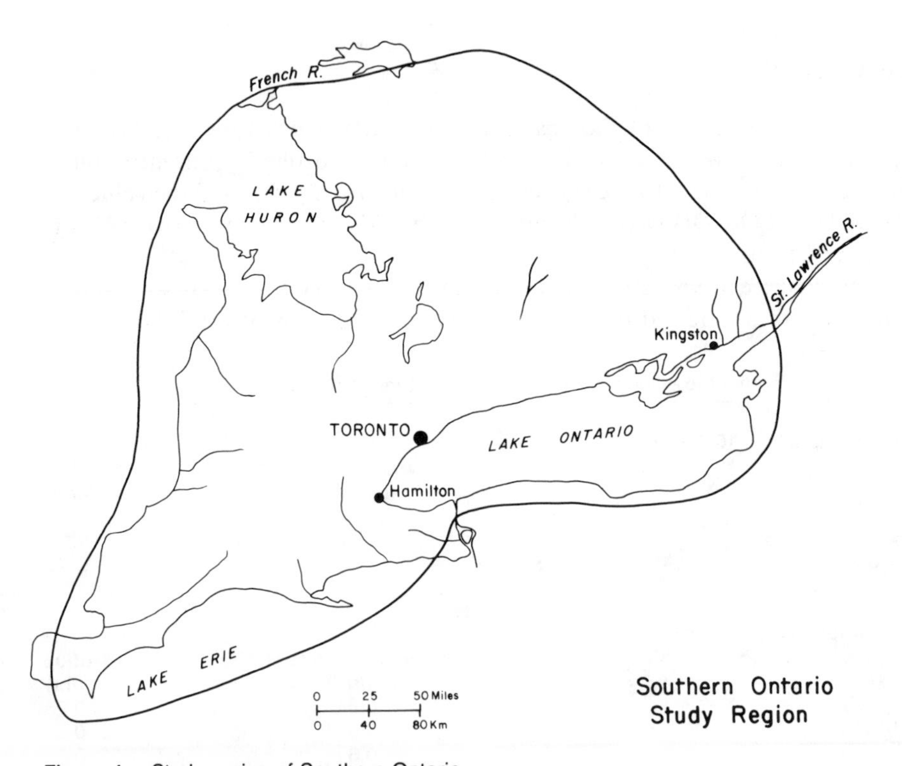


**Figure 1.** Study region of Southern Ontario.

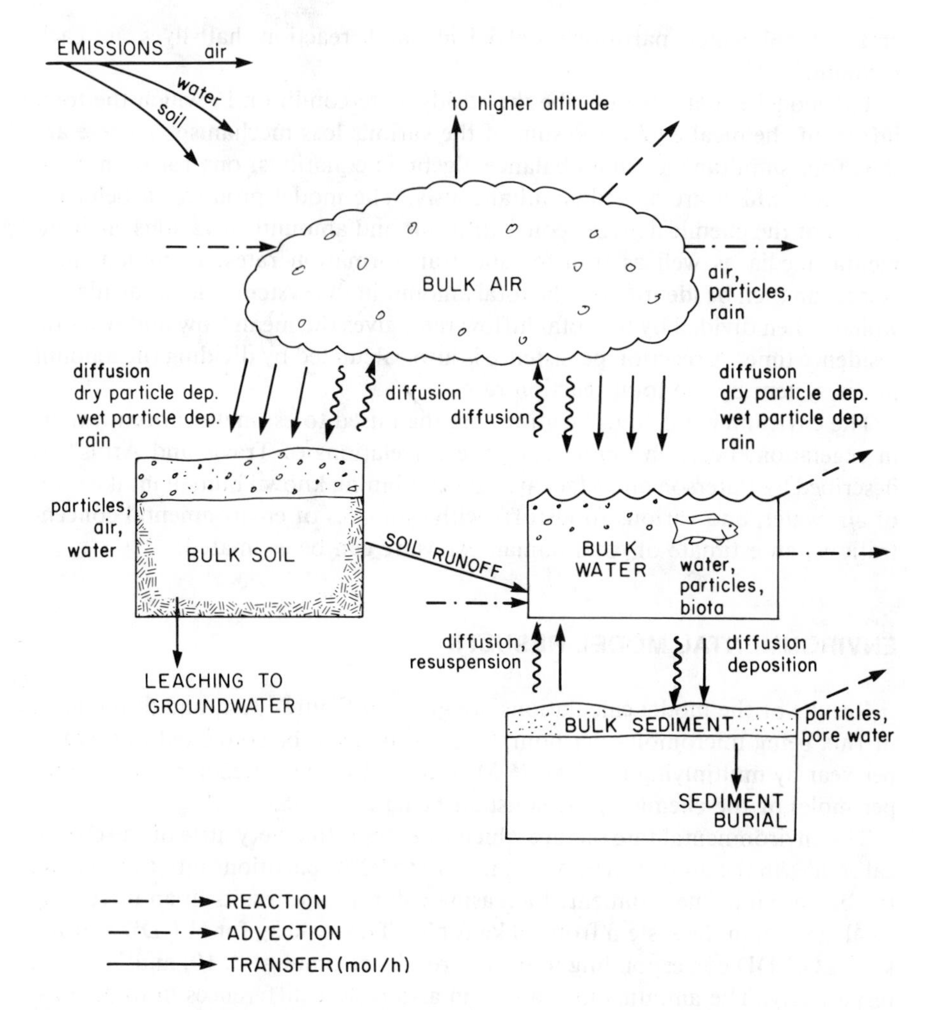


**Figure 2.** Environmental model illustrating four primary compartments and transport and reaction processes.

tions. These include wet and dry deposition from the atmosphere and adsorption from air to soil and air to water; volatilization from soil and water to the atmosphere; exchange between sediment and water by deposition, resuspension, and diffusion; and soil to water transfer by runoff. In addition to the loss mechanisms by degrading reactions in all media and advective flows in air and water, expressions are included for loss of chemical from the system by sediment burial, transfer to the stratosphere, and leaching from soil to ground water.

Input to the model consists of emission rates into each compartment, the physical chemical properties of molecular weight, solubility, vapor pressure, and octanol/water partition coefficient, and reaction half-lives in each medium.

The model equations describe the steady-state condition in which the total inflow of chemical equals the sum of the various loss mechanisms. There are thus four simultaneous mass balance algebraic equations, one for each compartment, which are solved simultaneously. The model produces a behavior profile of the chemical giving concentrations and amounts in various environmental media as well as transfer and transformation rates. Dominant processes can then be identified. The total amount in the system can be calculated, which, when divided by the total inflow rate, gives the mean flow and reaction residence time. A reaction persistence is also calculated by dividing the amount in the system by the total reaction rate.

The environmental concentrations are then used to estimate concentrations in vegetation, beef, and milk using the correlations of Travis and Arms[10] as described by Paterson and Mackay.[11] By combining known human intake rates of air, water, and various foodstuffs with estimates of environmental concentrations, an estimate of total human exposure can be estimated.

## ENVIRONMENTAL MODEL RESULTS

Results of the model calculations are given in Figures 3, 4, and 5, the units of flux being micromoles per hour. These fluxes can be converted into grams per year by multiplying by 0.00876 M, where M is the molecular mass (grams per mole) of the chemical, the constant being $24 \times 365 \times 10^{-6}$.

The environmental fate picture which emerges is that very little of the chemical resides in the air or water. Most, i.e., 70 to 90%, partitions into the soil and the balance into the sediment. Increasing chlorination results in an increasing total amount in the system from 10 kg for $T_4CDD$ to 40 kg for $H_6CDD$ and 160 kg for $O_8CDD$ corresponding to overall residence times of 2, 10, and 31 years, respectively. The amounts in the system also reflect differences in total emission rates which as shown in Table 2 are in the ratio 1.6:1:1.7.

The important input processes are emissions or local point sources and air and water advective inflow. These rates appear to be similar in magnitude, i.e., within a factor of two. The important removal processes are reaction and advection in air and water. Essentially, the amounts which are emitted into the atmosphere and which fail to be reacted or advected from the atmosphere migrate to soil or water compartments primarily by wet deposition and to a lesser extent by dry deposition. The rates of these processes control the atmospheric concentration which is in the range $2 \times 10^{-5}$ to $4 \times 10^{-5}$ ng/m$^3$. Most of the chemical entering the soil reacts there with relatively little being subject to runoff to water, except in the case of octa, which is very persistent in soil and thus survives long enough to be subject to runoff. The lower chlorinated congeners entering the water column are primarily subject to reaction, but as chlorination increases, the octanol/water partition coefficient increases,

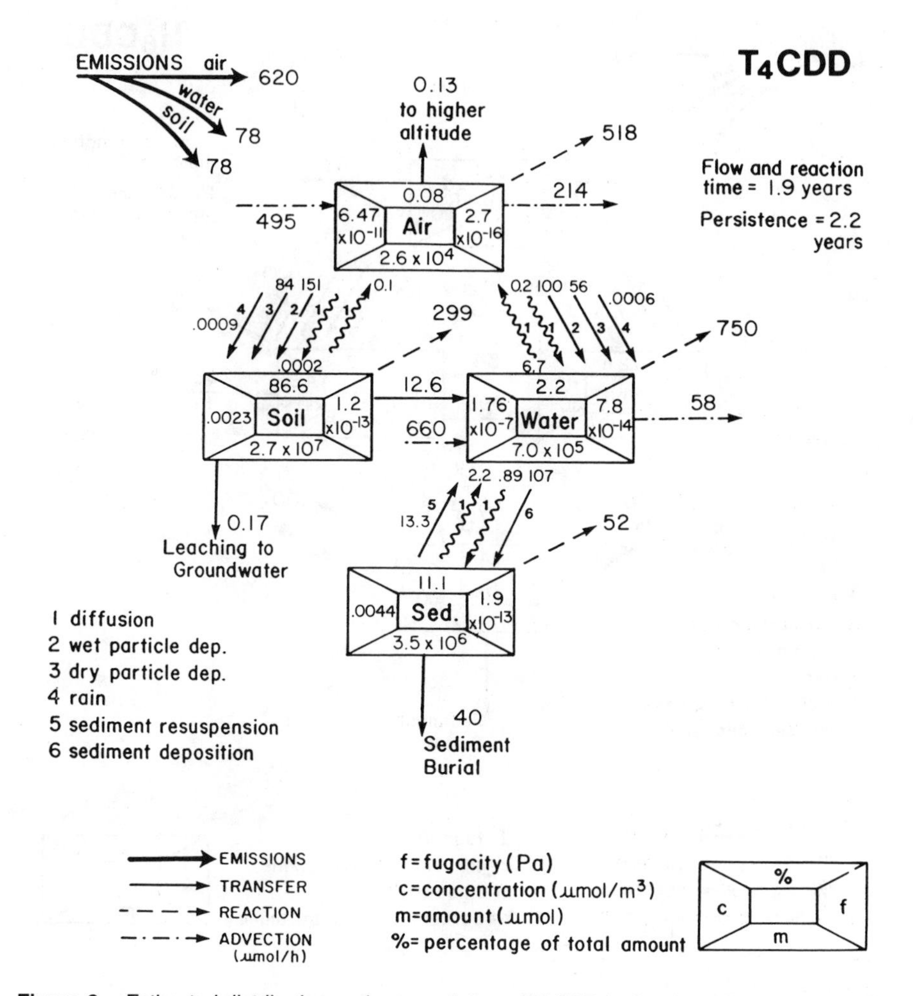


**Figure 3.** Estimated distribution and accumulation of $T_4CDD$ in Southern Ontario.

attachment to sediment and particles increases, and an increased amount becomes subject to deposition to the sediment. In view of the very slow degradation rates in sediments, most of the material sedimented is ultimately buried, especially in the case of $O_8CDD$.

In view of the considerable uncertainty about emission rates, reaction rates, and certain intermedia transfer rates, these fluxes should be regarded as only tentative order-of-magnitude estimates. However, it is believed that the general pattern of appreciable atmospheric and water column reaction, wet and dry deposition, and sediment deposition as generated by the model is probably correct.

The amounts partitioning into fish are negligible, but the concentrations in fish may be appreciable multiples (bioconcentration factors or BCFs) of the

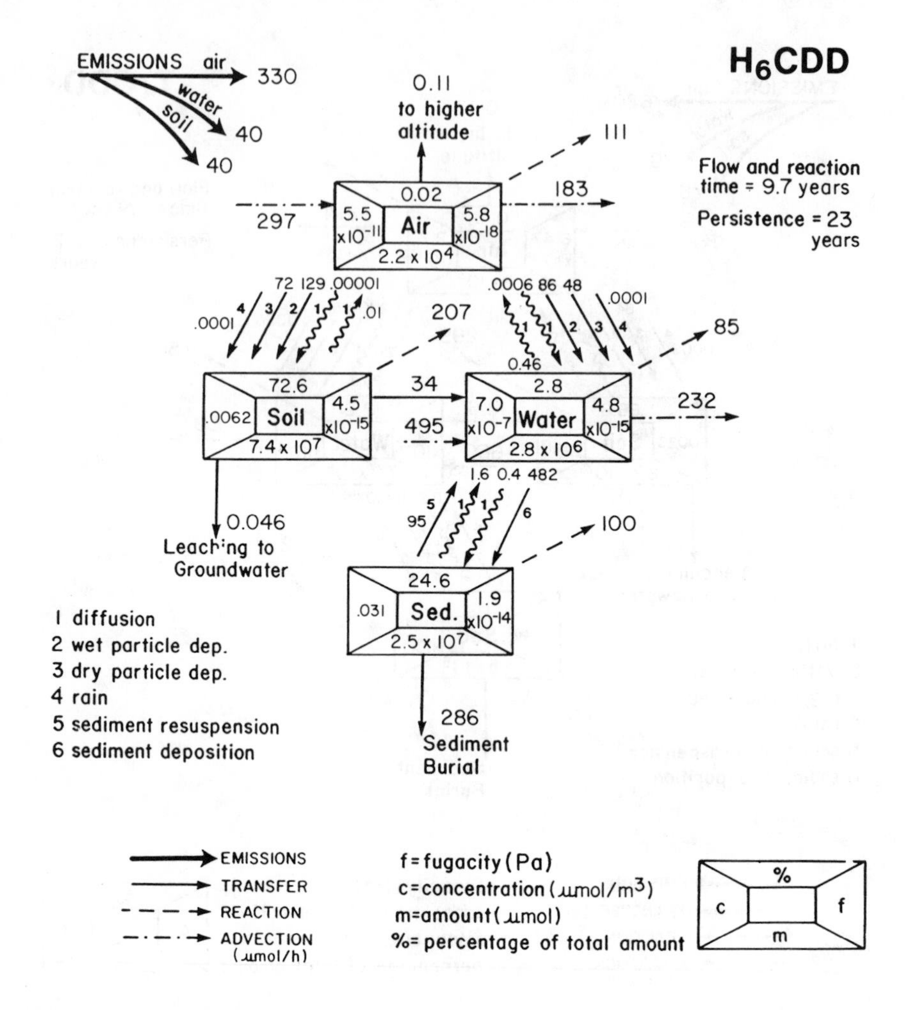


**Figure 4.** Estimated distribution and accumulation of $H_6CDD$ in Southern Ontario.

water concentrations. The magnitude of these BCFs are speculative and subject to some controversy because it appears that dioxins are not always readily absorbed by fish, some congeners are subject to metabolism and there may be specific binding between certain dioxins and protein.[12]

An implication of this behavior profile is that there may be more merit in monitoring for dioxins in the soil, sediment, and selected biotic environments than in air and water. The small amounts in air and water suggest that concentrations will be small and may be quite variable in time and space as they respond to changes in point source emission rates and atmospheric distribution patterns.

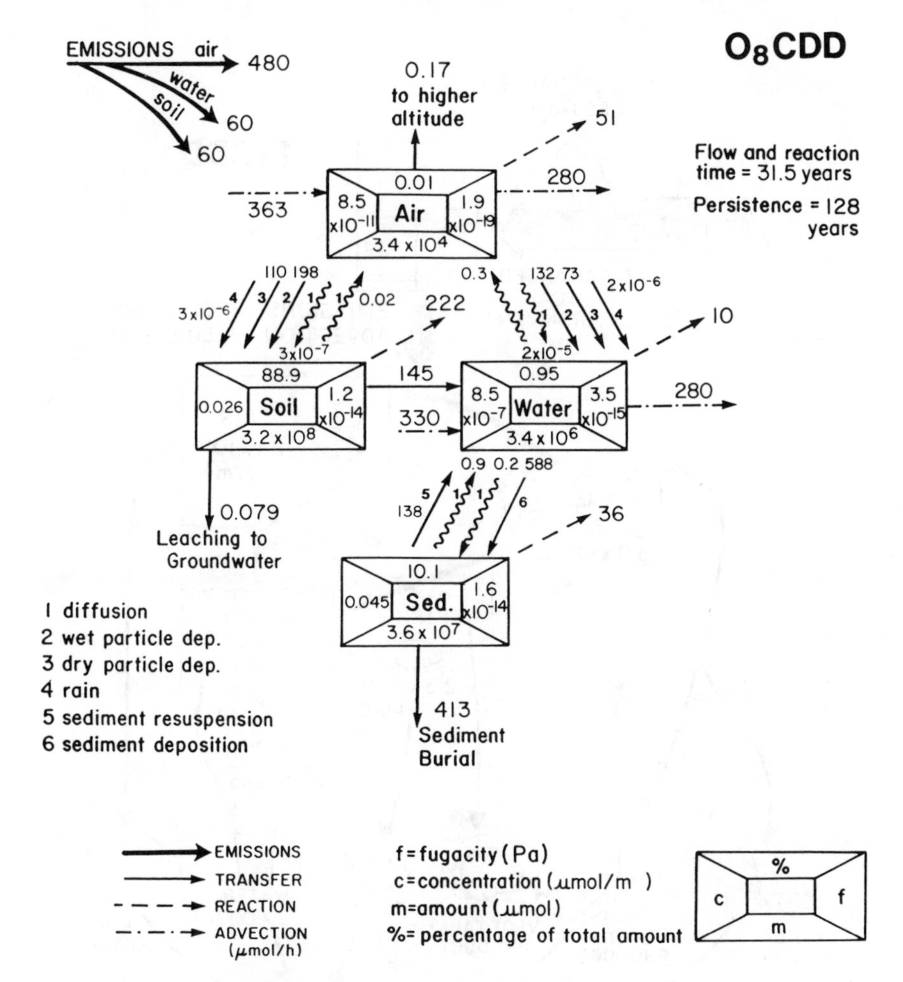


**Figure 5.** Estimated distribution and accumulation of $O_8CDD$ in Southern Ontario.

## HUMAN EXPOSURE MODEL

Figures 6, 7, and 8 show the results of the environmental exposure model. In this model, the air, soil, water, sediment, and fish concentrations are used to estimate concentrations in inhaled air, ingested water, consumed fish, vegetable, dairy, and meat products. It is assumed that the water is untreated and the fish consumed are entirely from the region of Southern Ontario. The principal task, and source of error, in the human exposure model lies in translating the estimated prevailing air and soil concentrations into those in vegetation, in domestic animals which eat this vegetation, and their meat and milk. The correlations used are based largely on the work of Travis and Arms[10] who developed correlations between octanol/water partition coefficient and bioconcentration of organic chemicals in vegetation, beef, and cow milk.

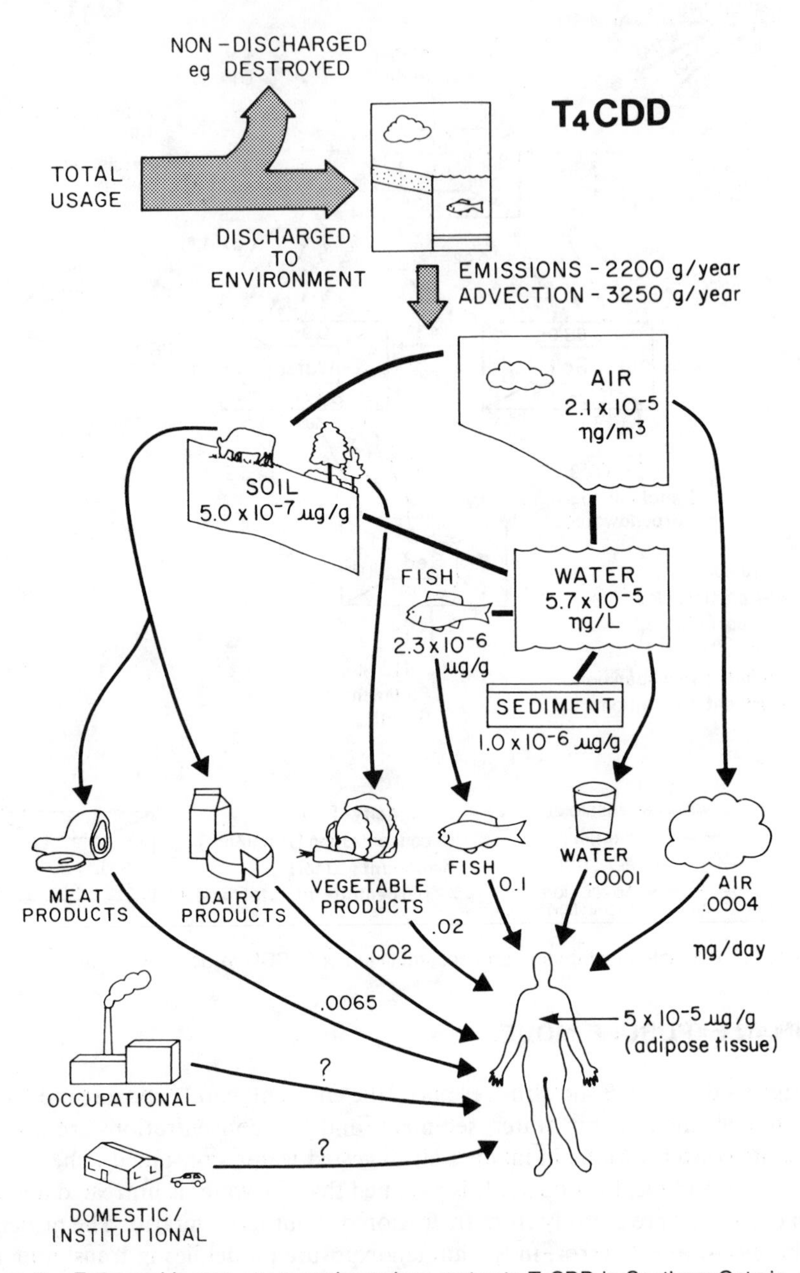


**Figure 6.** Estimated human exposure by various routes to $T_4CDD$ in Southern Ontario.

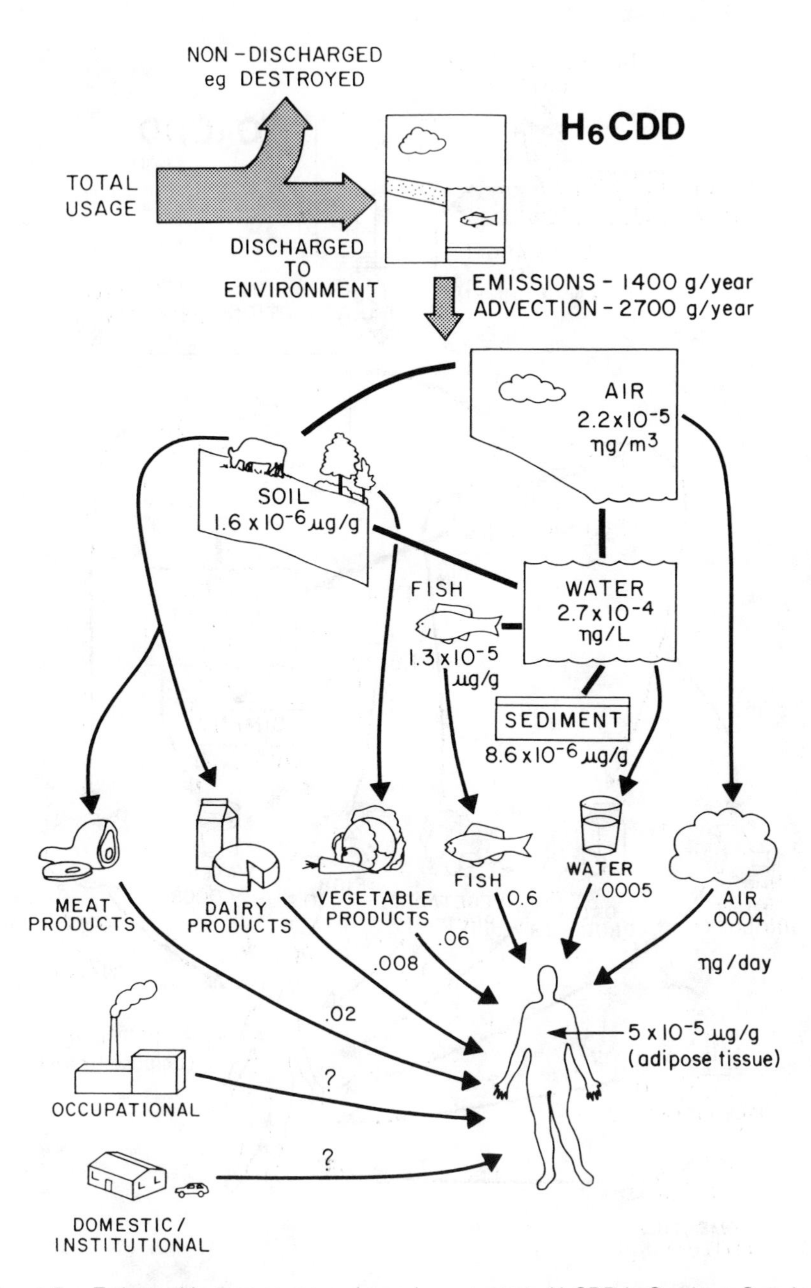


**Figure 7.** Estimated human exposure by various routes to $H_6CDD$ in Southern Ontario.

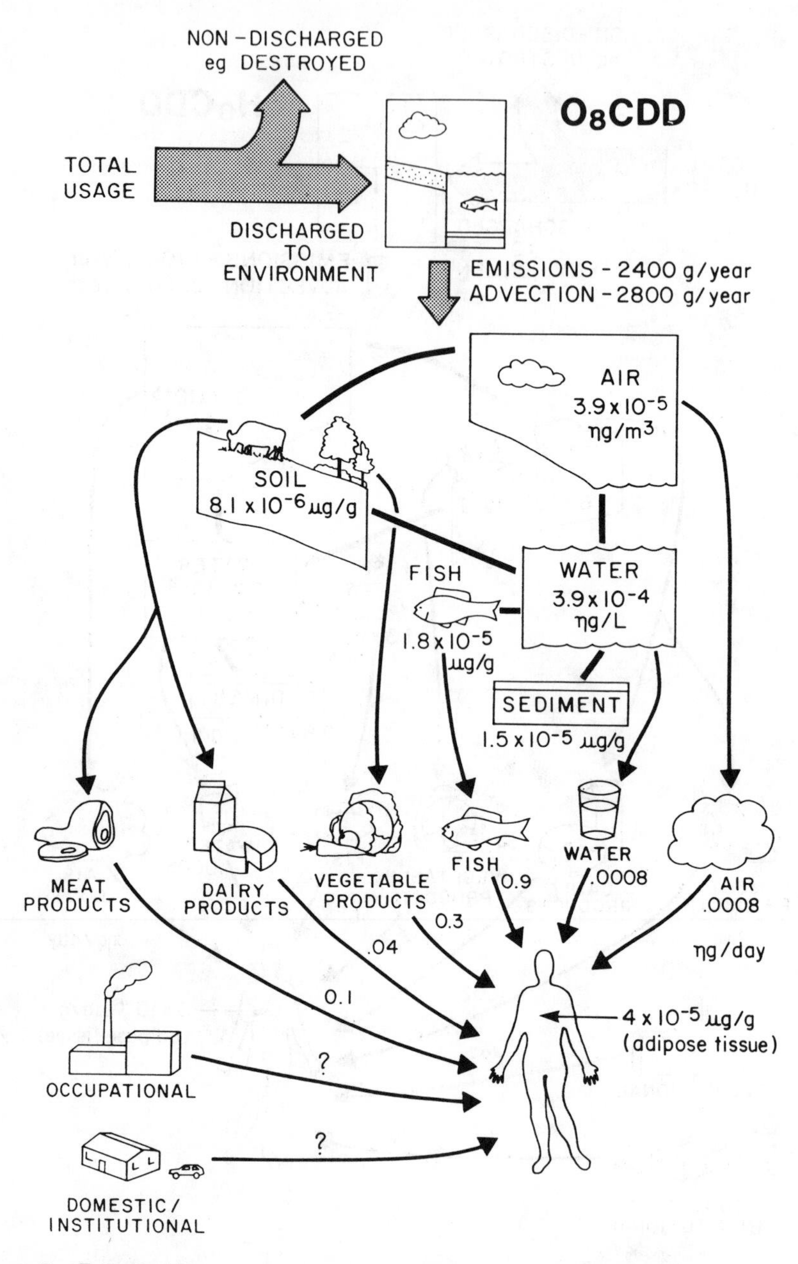


**Figure 8.** Estimated human exposure by various routes to $O_8CDD$ in Southern Ontario.

Uptake of chemical by plants from soil is defined by a BCF which is also a vegetation/soil partition coefficient, $K_{vs}$, where:

$$K_{vs} = \frac{\text{concentration in above-ground plant parts (mg of compound/kg of dry plant)}}{\text{concentration in soil (mg of compound/kg of dry soil)}}$$

Subsequent uptake by cows through ingestion of this vegetation is described by biotransfer factors (BTFs) to beef, $B_b$, and milk, $B_m$, where:

$$B_b = \frac{\text{concentration in beef (mg/kg)}}{\text{daily intake of chemical (mg/d)}}$$

$$B_m = \frac{\text{concentration in milk (mg/kg)}}{\text{daily intake of chemical (mg/d)}}$$

As discussed previously,[11] the data of Travis and Arms[10] indicate that very hydrophobic, persistent organic compounds have a similar BCF in vegetation of approximately 0.1, a similar BTF in beef of approximately 0.056 d/kg, and in milk of 0.01 d/kg. These values were assumed to be representative of the dioxins considered here.

By combining the BTFs for beef and milk with feed rates of 8 and 16 kg/day, respectively, for nonlactating and lactating cows, BCFs for beef, $K_{bv}$, and milk, $K_{mv}$, from vegetation (i.e., ratios of concentration in beef to that in vegetation) can be obtained as:

$$K_{bv} = 0.056 \times 8 = 0.45$$

$$K_{mv} = 0.01 \times 16 = 0.16$$

Beef/soil, $K_{bs}$, and milk/soil, $K_{ms}$, partition coefficients are determined as:

$$K_{bs} = K_{bv} \times K_{vs} = 0.45 \times 0.1 = 0.045$$

$$K_{ms} = K_{mv} \times K_{vs} = 0.16 \times 0.1 = 0.016$$

Concentrations in vegetation, beef, and milk are calculated from these partition coefficients and the estimated soil concentrations, and subsequently combined with appropriate consumption rates to determine daily human exposure to the various homolog groups.

It is emphasized that at the present state of knowledge, these correlations must be viewed as being tentative. Clearly more accurate, detailed, and comprehensive correlations are necessary to describe the migration of dioxins from the atmosphere and soil into plant material. For example, the model does not adequately treat wet and dry deposition from the atmosphere to plant surfaces. Further, it is assumed that during the process of food preparation, these materials are not decontaminated by washing or cooking and that the entire market basket is derived from the area of interest, i.e., the consumer eats no imported food from other regions which may be more or less contaminated.

It is believed that for many chemicals, occupational, domestic, and institutional exposures may be appreciable. However, these exposures are not included in this model due to lack of data.

A final stage of the model, which is treated only tentatively here and is thus included primarily for illustrative purposes, is translation of the total human intake rate in nanograms per day into a concentration prevailing in human adipose tissue. This task can be treated by one of several approaches ranging from simple correlations of body burden or adipose tissue concentration as a function of input rate to more complex pharmocokinetic models. The latter method describes intake by food and water ingestion, air inhalation, and possible dermal contact, and resulting concentrations prevailing in the blood stream and various tissue groups or organs. An example is the fugacity-based pharmocokinetic model of Paterson and Mackay.[13] In this study, we illustrate the former concept by calculating adipose tissue concentrations using the correlation of Geyer et al.[14] who derived the relationship:

$$BCF_L = 2.54 \log K_{ow} - 0.22 (\log K_{ow})^{-2} - 4.56$$

$BCF_L$ is the ratio (on a lipid basis) between concentration of chemical in human fat tissue after prolonged exposure and the concentration in diet, and $K_{ow}$ is the octanol/water partition coefficient. Concentration in human adipose tissue can be calculated as the product of $BCF_L$ and an average food concentration. By assuming that these compounds will be stored mainly in the fat and adopting a fat volume of 0.0147 $m^3$, a total body burden can be calculated. A residence time in humans can be calculated by dividing the body burden by the intake rate.

## EXPOSURE MODEL RESULTS

The results of the exposure analysis show that fish consumption is probably a major potential route of dioxin uptake. There is a tendency for fish concentrations to increase from tetra to octa, reflecting the longer persistence and the higher bioconcentration factor. In reality it appears that the octa congener is not bioconcentrated to the extent which is predicted from simple bioconcentration factor octanol/water partition coefficient correlations. This may be due to an unusually large resistance to uptake through the fish gut or to reduced bioavailability. This uptake route is important because the fish concentrations of these chemicals are primarily a reflection of the concentration in food consumed by the fish. This "discrepancy" illustrates the dangers of using models of this type based on general correlations without a full understanding of the assumptions underlying the correlations.

It appears that vegetable, dairy product, and meat consumption are fairly comparable as routes of exposure to these chemicals. A typical rate is in the range 0.01 to 0.1 ng/day. In contrast, consumption of water is not an important exposure route especially in view of the probable removal of a significant fraction of these chemicals during conventional water treatment processes. Similarly, inhalation of ambient air results in an exposure of only 0.0004 to 0.0008 ng/day, which is also negligible in comparison. However, this exposure is variable and it is probable that individuals living close to incinerator facilities may receive appreciably higher exposures by this route, possibly by a factor of about 10. But even in these cases, it is likely that exposure to dioxins by food consumption is more important.

The total exposure of approximately 0.1 ng/day is derived approximately one half from emissions and one half from advective inflow. It is possible to use the model to explore how reductions in local emissions could result in reductions in exposure and to test how local increases in air concentration attributable to proximity to incinerators would result in increases in human exposure.

## COMPARISON OF REPORTED AND MODEL CONCENTRATIONS

Table 4 gives concentrations which have been reported in the literature and those calculated by the model in a variety of units for air, soil, water, sediment, vegetation, dairy products, fish, and human adipose tissue. Comparison of the reported and "predicted" concentrations gives an impression of the accuracy of the model.

At the outset it must be emphasized that the model gives only an approximate characterization of the fluxes and concentrations of these chemicals. There are considerable uncertainties in the emission rates, in reactivities, and in intermedia transfer process rates. As important, the prevailing concentrations in the various media of the environment are believed to vary considerably because of varying proximity to sources. There is thus no single "correct" concentration against which the model output can be compared. Concentrations are most frequently measured by regulatory agencies in areas which are believed to be subject to severe contamination. Thus reported data may not be for average conditions. We believe that ultimately the best reconciliation between reported and predicted data is not to compare two single numbers as in Table 4, but is to compare concentration distribution functions. Unfortunately, this requires acquisition of a considerable amount of concentration data from a variety of locations and media throughout the study area. Instead of using single values of the various parameters as input to the model, values can be supplied which are allowed to vary in a random manner between defined limits. It is possible to run the model to develop a concentration

**Table 4. Comparison of Predicted and Reported Concentrations**

| Medium | $T_4CDD$ | $H_6CDD$ | $O_8CDD$ |
|---|---|---|---|
| Air (ng/m³) | | | |
| Estimated | $2 \times 10^{-5}$ | $2.2 \times 10^{-5}$ | $3.8 \times 10^{-5}$ |
| Reported | $1 \times 10^{-4}$ [16] | $6 \times 10^{-4}$ [16] | $5 \times 10^{-4}$ [16] |
| Water (ng/L) | | | |
| Estimated | $5.7 \times 10^{-5}$ | $2.7 \times 10^{-4}$ | $3.8 \times 10^{-4}$ |
| Reported | — | — | — |
| Soil (μg/g) | | | |
| Estimated | $4.8 \times 10^{-7}$ | $1.6 \times 10^{-6}$ | $7.9 \times 10^{-6}$ |
| Reported | $(<1–9) \times 10^{-3}$ [17] | $1.7 \times 10^{-4}$ [16] | $3.5 \times 10^{-3}$ [16] |
| Sediment (μg/g) | | | |
| Estimated | $1.0 \times 10^{-6}$ | $8.7 \times 10^{-7}$ | $1.4 \times 10^{-5}$ |
| Reported | $2.6 \times 10^{-5}$ [18] | $1 \times 10^{-5}$ [18] | $5.6 \times 10^{-4}$ [18] |
| Vegetation (mg/kg) | | | |
| Estimated | $4.8 \times 10^{-8}$ | $1.6 \times 10^{-7}$ | $7.9 \times 10^{-7}$ |
| Reported | $4 \times 10^{-7}$ [17] | $<1 \times 10^{-6}$ [19] [a] | $<1 \times 10^{-6}$ [19] [a] |
| Reported | $<2 \times 10^{-6}$ [20] | $<3 \times 10^{-7}$ [20] | $1 \times 10^{-6}$ [20] |
| Meat (mg/kg) | | | |
| Estimated | $2.2 \times 10^{-8}$ | $7.4 \times 10^{-8}$ | $3.5 \times 10^{-7}$ |
| Reported | $<1 \times 10^{-6}$ [17] | $<1 \times 10^{-6}$ [19] | $1 \times 10^{-6}$ [19] |
| Reported | $<2 \times 10^{-6}$ [20] | $<1 \times 10^{-6}$ [20] | $(3–24) \times 10^{-6}$ [20] |
| Dairy (mg/kg) | | | |
| Estimated | $7.7 \times 10^{-9}$ | $2.6 \times 10^{-8}$ | $1.3 \times 10^{-7}$ |
| Reported | $2.5 \times 10^{-6}$ [19] | $<1 \times 10^{-6}$ [19] [a] | $3.6 \times 10^{-5}$ [19] |
| Reported | $<1 \times 10^{-7}$ [20] | $<1 \times 10^{-7}$ [20] | $1 \times 10^{-6}$ [20] |
| Fish (mg/kg) | | | |
| Estimated | $2.3 \times 10^{-6}$ | $1.3 \times 10^{-5}$ | $1.8 \times 10^{-5}$ |
| Reported | $<1 \times 10^{-6}$ [17] | | |
| Human adipose tissue (mg/kg) | | | |
| Estimated | $4.8 \times 10^{-5}$ | $5.3 \times 10^{-5}$ | $3.9 \times 10^{-5}$ |
| Reported | $5 \times 10^{-6}$ [15] | $7.2 \times 10^{-5}$ [15] | $5.6 \times 10^{-4}$ [15] |

[a]Detection limit is $1 \times 10^{-6}$ mg/kg.[19]

distribution function in the various media. One can then match this distribution function with monitoring data. This can lead to statements of the type that 5% of the region is subject to concentrations which exceed x ng/m³. This conveys an impression of the regional heterogeneity of exposure and identifies the fraction of the population which may experience unusually high exposures.

Table 4 shows that air concentrations are generally within an order of magnitude of those reported. The soil concentrations tend to be uniformly low by approximately two to three orders of magnitude. The reason for this may be an overestimation of the rate of degradation in surface soils or it could be the result of other soil sources such as pesticide use. The model uses a soil depth of 10 cm, whereas dioxins deposited on soil surface may not be mixed to this depth. Reducing the soil depth would result in an increased estimated concentration at the soil surface. Sediment concentrations are generally within an order of magnitude of those predicted, as are fish concentrations, except for the octa congener, which is known not to bioconcentrate appreciably. Vegetation, dairy, and meat concentrations appear to be of the correct general order of magnitude. Estimated concentrations for the three isomers of $4 \times 10^{-5}$ to $5 \times 10^{-5}$ $\mu$g/g calculated for human adipose tissue are in good agreement with those reported by Stanley et al.[15]

It is thus concluded that the model is able to predict environmental concentrations with an accuracy of about a factor of 10 in most cases, but occasionally within a factor of 100. It is clearly desirable to examine the nature of these descrepancies and to explain them in terms of erroneous or atypical analysis or erroneous input data to the model or to structural defects in the model.

It is believed that by examining data from all media using the model as a linking mechanism, it will become possible to establish an approximate mass balance for these chemicals as they migrate between the various media of the Southern Ontario environment, its foodstuffs, and biota. The model helps to elucidate the need for better emission data, better inflow advection data, and more information on properties, and their temperature dependence. It is clearly desirable to be able to describe more reliably transfer rates between fish and water, air, soil, and vegetation, and into animals and humans.

Finally, it is believed that ultimately it will be essential to model on a congener-specific basis rather than using isomers or toxic equivalents or similar characterizations of total amounts of dioxin. This has been accomplished recently with notable success for PCBs, the quality of the resulting information being demonstrably superior to that obtained by treating the mixture as if it consisted of a group of chemicals of identical properties.

## CONCLUSIONS

Tentative environmental transport, transformation, and human exposure models have been developed and applied to tetra-, hexa-, and octa congeners of dioxins in the region of Southern Ontario. There is generally order of magnitude agreement between model and measured concentrations. The model is believed to give an adequate description of how these chemicals migrate through the environment and ultimately reach humans.

## ACKNOWLEDGMENTS

The authors are grateful to the Hazardous Contaminants Coordination Branch of the Ontario Ministry of the Environment for their financial support and scientific contributions.

## REFERENCES

1. Mackay, D., S. Paterson, B. Cheung, and W. B. Neely. "Evaluating the Environmental Behavior of Chemicals with a Level III Fugacity Model," *Chemosphere* 14:335–374 (1985).
2. Mackay, D., and S. Paterson. "Regional and Evaluative Level III Fugacity Models Describing the Multimedia Environmental Partitioning, Reaction, and Transport of Organic Chemicals," Ontario Ministry of the Environment (1988).
3. Oliver, B. G., and A. J. Niimi. "Trophodynamic Analysis of Polychlorinated Biphenyl Congeners and other Chlorinated Hydrocarbons in the Lake Ontario Ecosystem," *Environ. Sci. Technol.* 22:388–397 (1988).
4. Shiu, W. Y., W. Doucette, F. A. P. C. Gobas, A. Andren, and D. M. Mackay. "Physical Chemical Properties of Chlorinated Dibenzo-p-dioxins," *Environ. Sci. Technol.* 22:651–658 (1988).
5. Mill, T. "Prediction of the Environmental Fate of Tetrachlorodibenzodioxin," in *Dioxins in the Environment*, M. A. Kamrin, and P. W. Rodgers, Eds. (Hemisphere Publishers, 1985) pp. 173–194.
6. Ontario Ministry of the Environment. "Polychlorinated Dibenzo-p-Dioxins (PCDDs) and Polychlorinated Dibenzofurans (PCDFs)," Scientific Criteria Document for Standard Development No. 4-84, Intergovernmental Relations and Hazardous Contaminants Coordination Branch, Ontario Ministry of the Environment, Toronto (1985).
7. SENES Consultants Ltd. "Assessment of Risks from Solid Waste Incineration in the Commissioners/Cherry Street Area of Toronto," Metropolitan Toronto Works Department (1988).
8. Ontario Ministry of the Environment. "Dioxins in the Environment: The Inside Story," in *Proceedings of Seminar on Dioxins and Furans* (Toronto: Ontario Ministry of the Environment, 1986).
9. MacPherson, A. F. "Environmental Health Effects of Waste Incineration in the City of Toronto, 1987," City of Toronto, Department of Public Health, Toronto (October 1987).
10. Travis, C. C., and A. D. Arms. "Bioconcentration of Organics in Beef, Milk and Vegetation," *Environ. Sci. Technol.* 22:271–274 (1988).
11. Paterson, S., and D. Mackay. "A Model Illustrating the Environmental Fate, Exposure and Human Uptake of Persistent Organic Chemicals," *Ecol. Model.* 47:85–114 (1989).
12. Gobas, F. A. P. C., W. Y. Shiu, and D. Mackay. "Factors Determining Partitioning of Hydrophobic Organic Chemicals in Aquatic Organisms," in *QSAR in Environmental Toxicology–II*, K. L. E. Kaiser, Ed. (Dordrect, Holland: D. Reidel Publishers, 1987) pp. 107–124.
13. Paterson, S., and D. Mackay. "A Pharmacokinetic Model of Styrene Inhalation using the Fugacity Approach," *Toxicol. Appl. Pharmacol.* 82:444–453 (1986).
14. Geyer, H. J., I. Scheunert, and F. Korte. "Correlation between the Bioconcentration Potential of Organic Environmental Chemicals in Humans and their n-Octanol/Water Partition Coefficients," *Chemosphere* 16:239–252 (1987).
15. Stanley, J. S., K. E. Boggers, J. Onstot, T. M. Sack, J. C. Remmers, J. Breen, F. W. Kutz, J. Carra, P. Robinson, and G. A. Mack. "PCDDs and PCDFs in Human Adipose Tissue from the EPA FY82 NHATS Repository," *Chemosphere* 15:1605–1612 (1986).

16. Cantox Inc. "Health Hazard Evaluation of Specific PCDD, PCDF and PAH in Emissions from the Proposed Petrosun/SNC Resource Recovery Incineration and From Ambient Background Sources," Petrosun/SNC Operations Ltd., Oakville, Ontario (February 1988).
17. Travis, C. C., and H. A. Hattemer-Frey. "Human Exposure to 2,3,7,8-TCDD," *Chemosphere* 16:2331–2342 (1987).
18. Astle, J. W., F. A. P. C. Gobas, W. Y. Shiu, and D. Mackay. "Lake Sediments as Historic Records of Atmospheric Contamination by Organic Chemicals," in *Sources and Fates of Aquatic Pollutants*, R. A. Hites, and S. J. Eisenreich, Eds. (Washington, D.C.: American Chemistry Society, 1987), pp. 57–77.
19. Davies, K. "Concentrations and Dietary Intake of Selected Organochlorines, including PCBs, PCDDs and PCDFs in Fresh Food Composites Grown in Ontario, Canada," *Chemosphere*, 17:263–276 (1988).
20. Ontario Ministry of Agriculture and Food, Ontario Ministry of the Environment, Toxics in Food Steering Committee. "Polychlorinated Dibenzo-p-dioxins and Polychlorinated Dibenzofurans and other Organochlorine Contaminants in Food," Toronto (1988).

CHAPTER 29

# Practical Application of Risk Assessment in the Permitting of a Hazardous Waste Incinerator

**C. N. Park**

## INTRODUCTION

This paper will discuss the practical problems of applying risk assessment methodologies to evaluate the safety of a proposed incinerator. The risk assessment methods to be discussed are not unique, but their real world application presents unique problems and decision points to be addressed.

The particular incinerator evaluation to be described is for a proposed facility in Pittsburg, CA. The proposed unit is a relatively small (7000 tons per year capacity) hazardous waste incinerator, designed to be part of the total waste management program for a chemical production facility owned by Dow Chemical. As with all modern chemical manufacturing operations, the emphasis at the Dow site is on minimizing waste through source reduction and recycle/reuse. Some waste is inevitable, however, and a well-run state-of-the-art incinerator is felt to be an integral part of a waste management program. Commercial waste is not being considered for incineration in the proposed facility.

The risk assessment can be broken down into four major phases:

- identification of compounds and quantities emitted
- dispersion modeling of the emitted compounds
- identification of appropriate standards for ambient air concentrations (and concentrations in other media)
- comparison of predicted levels to appropriate standards to determine the potential for human hazard.

## IDENTIFICATION OF EMISSIONS

In theory, this phase of the assessment is quite straight forward, but in practice it can be complex. The potential input streams were analyzed for their major components and impurities. The *total* yearly pounds and *maximum*

**Table 1. Incinerator Input Streams**

| Component | Maximum feed rate (lb/hr) | Annual average feed rate (lb/hr) |
|---|---|---|
| Hexachlorobenzene | 2125 | 8 |
| Hexachloroethane | 625 | 2 |
| Hexachlorobutadiene | 625 | 2 |
| Substituted pyridines | 3500 | 980 |
| Carbon Tetrachloride | 5000 | 52 |
| Perchloroethylene | 5000 | 67 |
| Methylene chloride | 5000 | 37 |
| 1,1,1-Trichloroethane | 5000 | 37 |

hourly rates for the input streams are listed in Table 1. Inevitably, a small fraction of these input wastes will be released as air emissions. A practical way of estimating these emissions is by assuming a certain destruction and removal efficiency (DRE). The DRE will vary from compound to compound and will depend to some extent upon the other compounds which will be present. As a requirement of the permit, however, the DREs for each compound must be demonstrated to exceed 99.99% in actual trial burns. Experience has shown that this condition is met with relative ease in a well-managed facility. Therefore as *maximum* emission levels from the stack, 0.01% of the compounds present in the input streams will be listed as emissions, as shown in Table 2. There will also be products of incomplete combustion (PICs) formed at low levels in the incineration process. Specifically, benzene, chloroform, dioxin, and furan emission levels are estimated from data collected at comparable facilities. These data, which will be verified in trial burns, are shown in Table 3. In addition to the stack emissions, there will also be fugitive emissions from

**Table 2. Incinerator Stack Emissions**

| Component | Maximum emissions (lb/hr) | Annual average emissions (lb/hr) |
|---|---|---|
| Hexachlorobenzene | 0.2125 | 0.000571 |
| Hexachloroethane | 0.0625 | 0.000114 |
| Hexachlorobutadiene | 0.0625 | 0.000114 |
| Substituted pyridines | 0.3500 | 0.0696 |
| Carbon tetrachloride | 0.5000 | 0.00365 |
| Perchlorethylene | 0.5000 | 0.00479 |
| Methylene chloride | 0.5000 | 0.00263 |
| 1,1,1-Trichloroethane | 0.5000 | 0.00263 |
| Hydrogen chloride | 0.1146 | 0.0811 |
| Chlorine | 0.1712 | 0.126 |
| Sulfur dioxide | 3.0927 | 2.17 |
| Nitrogen oxides | 55.2000 | 39.2 |
| Carbon monoxide | 13.5223 | 9.60 |
| Ferric oxide | 0.2800 | 0.0924 |

[a]Annualized for 6220 hr operations per year.

**Table 3. Incinerator Dibenzodioxin and Dibenzofuran Stack Emissions**

| | Quantity (lb/hr) |
|---|---|
| Tetrachlorodibenzodioxin | |
| 2,3,7,8 Substituted | $9.3 \times 10^{-9}$ |
| Not 2,3,7,8 substituted | $1.6 \times 10^{-6}$ |
| Pentachlorodibenzodioxin | |
| 2,3,7,8 Substituted | $5.7 \times 10^{-9}$ |
| Not 2,3,7,8 substituted | $8.6 \times 10^{-8}$ |
| Hexachlorodibenzodioxin | |
| 2,3,7,8 Substituted | — |
| Not 2,3,7,8 substituted | — |
| Heptachlorodibenzodioxin | |
| 2,3,7,8 Substituted | $1.6 \times 10^{-8}$ |
| Not 2,3,7,8 substituted | $2.6 \times 10^{-8}$ |
| Tetrachlorodibenzofuran | |
| 2,3,7,8 Substituted | $3.3 \times 10^{-9}$ |
| Not 2,3,7,8 substituted | $2.9 \times 10^{-5}$ |
| Pentachlorodibenzofuran | |
| 2,3,7,8 Substituted | $1.6 \times 10^{-8}$ |
| Not 2,3,7,8 substituted | $5.3 \times 10^{-7}$ |
| Hexachlorodibenzofuran | |
| 2,3,7,8 Substituted | $1.1 \times 10^{-8}$ |
| Not 2,3,7,8 substituted | $7.9 \times 10^{-8}$ |
| Heptachlorodibenzofuran | |
| 2,3,7,8 Substituted | $3.0 \times 10^{-8}$ |
| Not 2,3,7,8 substituted | $4.7 \times 10^{-8}$ |
| Annualized total in 2378 equivalents (TEFs) | $2.54 \times 10^{-8}$ |
| Benzene | 0.00031 |
| Chloroform | 0.00016 |

the project. These include emissions from seals and flanges in the kiln and emissions from loading, unloading, and storage of the compounds. Table 4 summarizes these emission levels.

As can be seen from Tables 2 and 3, emission levels down to 1 lb/yr (and less for dioxins/furans) have been estimated. In theory, it is possible that other PICs, for example other hydrocarbons, could be formed and reach a formation/destruction equilibrium which would depend upon the particular waste streams being incinerated. The concentration of these compounds would be very low, but they could exist. To ensure that the cumulative hazard from unknown PICs would not impact this assessment, a worst-case scenario for unknown PICs was requested by one of the permitting agencies. Under this scenario, it was assumed that a DRE of 99.99% was met, but that all of the

**Table 4. Fugitive Emission Summary**

| Component | Total annualized (lb/hr) |
|---|---|
| Gas | |
| HCl | 0.589 |
| $Cl_2$ | 0.000685 |
| Light liquids | |
| Carbon tetrachloride | 0.0525 |
| Perchloroethylene | 0.0178 |
| Methylene chloride | 0.00639 |
| 1,1,1-Trichloroethane | 0.00502 |
| Methanol | 0.00360 |
| Ethylene glycol | 0.498 |
| Acetone | 0.000685 |
| Heavy liquids | |
| Substituted pyridines | 0.185 |
| Lubricating oil | 0.00852 |
| Compressor oil | 0.00284 |
| Hydraulic oil | 0.00065 |

remaining 0.01%, by weight, was a mixture of carcinogens with an average potency equal to that of benzene.*

## AIR DISPERSION MODELING

### Meteorology Data

The proposed site for this facility leads to unique site-specific meteorological conditions. The location is near Pittsburg, CA, on the Sacramento River approximately halfway between San Francisco and Sacramento. The river location substantially modifies wind direction and frequency due to the channeling effect up the river valley. Local meteorology was characterized as a combination of wind speed and direction data collected at the proposed site and other relevant data from nearby meteorological stations.

Wind speed, wind direction, and temperature data collected by Dow at the Pittsburg plant site were used as input to the RAMMET program. The meteorological data were processed to create sequential meteorological files for the years 1978 through 1982. These sequential meteorological files were used as input into the ISCST and COMPLEX 1 models.

Hourly cloud cover and ceiling height data from Travis Air Force Base were used along with the Dow wind speed data to determine atmospheric stability categories. Twice daily mixing heights determined from the upper air soundings taken at Oakland, CA were also used by RAMMET. The mixing heights

*California uses a potency (unit risk) for benzene which is approximately eightfold higher than that used for the federal EPA. Thus a risk of $5.3 \times 10^{-5}$ $(\mu g/m^3)^{-1}$ was applied to an assumed emission level of 0.01% of all input streams, as a worst-case scenario for unknown emissions.

were calculated using the aloft data from Oakland with surface temperature from Travis Air Force Base.

### Study Area

In order to define a practical limit to the area under study, the California Source Assessment Manual (1986) defines the following methodology:

> In upper bound on carcinogenic risk is calculated for receptors in a complete polar grid. This upper bound risk is calculated by summing the upper bound risk for each compound and for fugitive and stack emissions.*

The $10^{-7}$ contour is then estimated from the grid, and the smallest circle centered at the source and encompassing *all* $10^{-7}$ risk points is drawn. This circle defines the outer perimeter of the study area and all census tracts with population inside the circle are included in the evaluation. Census tracts fully outside the circle are not considered further.

### Modeled Dispersion Results

Two sets of computer runs form the basis for the estimation of possible ambient levels. One set calculates annual average as well as maximum 1-, 3-, 8-, and 24-hr average concentrations for each of 96 possible discrete receptors in the study area. These receptors included census tract population centroids and worst-case locations for each census tract. Other discrete receptors include all nearby industrial locations and "sensitive receptors," schools, hospitals, and elder care facilities.

It was felt that these sensitive receptors, even though they would not have the highest modeled concentrations, should be separately evaluated due to the potentially more sensitive populations which could require different risk management evaluation criteria.

The second set of computer runs predict ambient concentrations over a polar grid. These results later proved to be very useful for answering "what if" types of questions, such as what if the proposed site were moved to a slightly different location? The results from the polar grid were also used to evaluate potential ambient levels at future population sites.

## COMPARISON STANDARDS AND RESULTS

### Long-term Criteria for Noncarcinogens

In general, separate risk assessment criteria should be derived for acute, chronic, and subchronic exposure scenarios. In practice, however, subchronic criteria are unnecessary since for air emissions which are relatively constant and which involve compounds that are regulated as possible carcinogens,

*Note that the sum of 95% upper confidence bounds results in an upper bound with a confidence level exceeding the nominal individual levels of 95%.

**Table 5. Annual Ambient Concentration at the Maximum Receptor Site—Noncarcinogens**

| Compound | Maximum annual concentration ($\mu g/m^3$) | Relevant criteria ($\mu g/m^3$) | Maximum as a percent of criteria |
|---|---|---|---|
| Chlorine | 0.016 | 30 | <1% |
| Ferric oxide | 0.012 | 50 | <1% |
| Hydrogen chloride | 1.08 | 70 | 1.5% |
| Substituted pyridines | 0.34 | 11[a] | 3.1% |
| 1,1,1-Trichloroethane | 0.0095 | 19,000 | <1% |
| Acetone | 0.0012 | 17,800 | <1% |
| Ethylene glycol | 0.91 | 1,250 | <1% |
| Methanol | 0.0065 | 2,600 | <1% |

[a]NOEL/100.

chronic exposure criteria for carcinogens are much more restrictive than the comparable subchronic criteria.

For compounds which are regulated as potential carcinogens, the exposure evaluation criteria is the cumulative upper bound on lifetime risk summed across all the compounds. For noncarcinogens, chronic evaluation criteria were not generally available when this assessment was done. Since then, the EPA has developed reference doses for many of the compounds under consideration, but these values have only been recently available and are still far from complete. The American Conference for Governmental Industrial Hygienists has, however, generated TLVs for a large number of chemicals. TLVs should not be used *directly* for regulatory purposes, outside of the workplace, but *fractions* of the TLV are often used as bench mark comparisons for air emissions. The appropriate fraction to use depends upon the toxicological properties of the individual compounds and also involves societal risk management criteria. For air permits, many states use fractions of between TLV/24 and TLV/420 with 100 being the most commonly used value. Table 5 shows TLV/100 and the predicted annual average concentration at the maximum receptor site for the noncarcinogenic compounds. The use of TLV/100 does not endorse this criterion as a "safe" value, instead it provides a bench mark for the permitting agencies to evaluate the potential emissions from the proposed facility.

The inclusion of fugitive emissions in this analysis adds a level of complexity to the calculation of maximum predicted ambient concentrations. For fugitive emissions, this maximum will occur close to the proposed facility since for low release heights, concentration will decrease with distance.

For stack emissions, on the other hand, the expected ambient concentrations will be almost zero at the base of the stack, then will increase over distance to a maximum, then fall off again. Therefore, the *location* of the maximum predicted levels in Table 5 and other subsequent tables is not constant. The location will depend upon whether the emissions for a particular compound are stack emissions, fugitive emissions, or a combination of the two.

**Table 6. One-Hour Ambient Concentrations at the Maximum Receptor Site**

| Compound | Maximum annual concentration ($\mu g/m^3$)($\mu g/m^3$) | Relevant criteria of criteria | Maximum as a percent |
|---|---|---|---|
| Chlorine | 1.6 | 3,000 | <1% |
| Ferric oxide | 2.1 | 5,000 | <1% |
| Hydrogen chloride | 294.0 | 7,400 | 4% |
| Substituted pyridines | 96.0 | 1,100 | 9% |
| Perchloroethylene | 13.0 | 335,000 | <1% |
| Carbon tetrachloride | 30.0 | 30,000 | <1% |
| Methylene chloride | 6.9 | 350,000 | <1% |
| 1,1,1-Trichloroethane | 6.1 | 1,900,000 | <1% |
| Hexachlorobenzene | 1.5 | 1,300 | <1% |
| Benzene | 0.0032 | 30,000 | <1% |
| Chloroform | 0.0016 | 50,000 | <1% |
| Hexachloroethane | 0.46 | 100,000 | <1% |
| Hexachlorobutadiene | 0.46 | 240 | <1% |
| 2,3,7,8-$T_4CDD$ (Tox equivalent) | 0.44 $pg/m^3$ | 330 $pg/m^3$ | <1% |
| Acetone | 1.6 | 1,780,000 | <1% |
| Methanol | 3.6 | 260,000 | <1% |
| Ethylene glycol | 258 | 125,000 | <1% |

## Acute Exposures

Predicted annual average ambient levels represent concentrations averaged across many different meteorological conditions. For any given receptor site, the predicted ambient level is zero for a majority of the time since the wind does not consistently blow in any one direction. For conditions of low wind speed and high atmospheric stability, however, the predicted short-term levels will exceed the average by a large amount. To ensure that emissions do not pose an *acute* hazard to nearby residents, the maximum predicted 1-hr concentrations across all receptor sites are calculated for each of the emitted compounds and compared to the TLV as shown in Table 6. The TLV may not always be the most appropriate criteria for comparison, but other regulatory acute criteria do not generally exist. Since the TLV is derived on the basis of being protective as an 8-hr average for a healthy working population, it at least represents a relevant bench mark for the *maximum* 1-hr ambient concentration which would be predicted to occur once in a year. For the substituted pyridines, a 2-week animal NOEL was used, and for dioxins, an EPA subchronic criteria is reported.

## Carcinogenic Risk

Methodology has been developed by regulatory agencies to estimate an upper bound on lifetime risk resulting from exposure to possible carcinogens.[1] The procedures are conservative in that they consistently use worst-case assumptions, but they do provide a consistent framework for demonstrating that some risks are negligible.

Conservative factors include biological assumptions, such as the assumption

**Table 7. Average Annual Ambient Concentrations at Nearest Residential Location in Census Tract 3050**

| Compound | Fugitive emission rate (lb/hr) | Resulting ambient concentration[a] (μg/m³) | Stack emission rate (lb/hr) | Resulting ambient concentration (μg/m³) | Total concentration (μg/m³) | Upper bound on risk × $10^{-6}$ |
|---|---|---|---|---|---|---|
| Perchloroethylene | 0.0178 | 0.00213 | 0.00479 | 0.000250 | 0.00238 | 0.0014 |
| Carbon tetrachloride | 0.0525 | 0.00627 | 0.00365 | 0.000191 | 0.00646 | 0.097 |
| Methylene chloride | 0.00639 | 0.000764 | 0.00263 | 0.000137 | 0.000901 | 0.0037 |
| Hexachlorobenzene | 0 | | 0.000571 | 0.0000298 | 0.0000298 | 0.015 |
| Benzene | 0 | | 0.000308 | 0.0000161 | 0.0000161 | 0.00085 |
| Chloroform | 0 | | 0.000160 | $8.35 \times 10^{-6}$ | $8.35 \times 10^{-6}$ | 0.00019 |
| Hexachlorobutadiene | 0 | | 0.000114 | $5.95 \times 10^{-6}$ | $5.95 \times 10^{-6}$ | 0.00013 |
| Hexachloroethane | 0 | | 0.000114 | $5.95 \times 10^{-6}$ | $5.95 \times 10^{-6}$ | 0.000024 |
| TCDD-TEF | | | | | | |
| California method | 0 | | $2.54 \times 10^{-8}$ | $1.33 \times 10^{-9}$ | $1.33 \times 10^{-9}$ | 0.051 |
| EPA method | | | $4.27 \times 10^{-8}$ | $2.23 \times 10^{-9}$ | $2.23 \times 10^{-9}$ | 0.085 |
| Total | | | | | | |
| California | | | | | | 0.169 |
| EPA | | | | | | 0.203 |

[a]For the maximum receptor site, the predicted fugitive and stack ambient concentrations are 0.1195 μg/m³ and 0.0522 μg/m³/lb/hr of emission.

that some carcinogenic risk exists at any dose, and that man is more sensitive than laboratory rodents on a weight basis. Exposure calculations assume that no degradation or fallout of the compounds occur between the source and the receptors and that individuals live at the same receptor site for 70 years during which time the incinerator operates at the same level and emissions are the same.

Table 7 shows a sample risk calculation for a specific site and Table 8 shows upper bounds on individual lifetime risks calculated at each of the population receptor sites. The risks are calculated from the cumulative exposures to stack and fugitive emissions for all of the compounds. Similarly, Table 9 shows upper bound risks calculated from ambient exposures at nearby industrial locations (46-year exposure) and schools (16-year exposure). It can be seen from the tables that none of the risks approach the regulatory criteria of $10^{-6}$ to $10^{-4}$, which are commonly used to evaluate hazard potential.

Table 10 recalculates the upper bound risk estimates on the basis of the total *population* cancer burden. The calculation represents the upper bound on the expected value of the number of excess cases per 70 years for each census tract. This assumes that the census tract grows in population over the 70-year lifetime of the facility and that no one in the census tract moves away. The last column in the table is the same calculation, expressed as the reciprocal: the *lower* bound on expected number of years between cases.

For the scenario under which it was assumed that 0.01% of all input waste

**Table 8. Accumulated Hypothetical Risk Estimates for Census Tracts in the Study Area**

| Census tract | Receptor | Ambient concentration per lb/hr of stack emissions | Ambient concentration per lb/hr of fugitive emissions | Upper limit on lifetime risk ($\times 10^{-6}$) |
|---|---|---|---|---|
| 3050 | Centroid | 0.0373 | 0.0651 | 0.104 |
| | Maximum | 0.0522 | 0.1195 | 0.169 |
| 3071 | Centroid | 0.0229 | 0.0437 | 0.067 |
| | Maximum | 0.0370 | 0.0760 | 0.112 |
| 3072.01 | Centroid | 0.0110 | 0.0202 | 0.031 |
| | Maximum | 0.0120 | 0.0322 | 0.043 |
| 3072.02 | Centroid | 0.0305 | 0.0659 | 0.095 |
| | Maximum | 0.0198 | 0.0711 | 0.085 |
| 3072.03 | Centroid | 0.0140 | 0.0238 | 0.038 |
| | Maximum | 0.0363 | 0.0724 | 0.108 |
| 3090 | Centroid | 0.0209 | 0.0541 | 0.073 |
| | Maximum | 0.0185 | 0.0653 | 0.079 |
| 3100 | Centroid | 0.0126 | 0.0336 | 0.045 |
| | Maximum | 0.0185 | 0.0653 | 0.079 |
| 3110 | Centroid | 0.00962 | 0.0181 | 0.028 |
| | Maximum | 0.0134 | 0.0287 | 0.042 |
| 3120 | Centroid | 0.0151 | 0.0373 | 0.051 |
| | Maximum | 0.0198 | 0.0711 | 0.085 |
| 3131.01 | Centroid | 0.00425 | 0.00774 | 0.012 |
| | Maximum | 0.00906 | 0.0261 | 0.034 |
| 3131.02 | Centroid | 0.00752 | 0.0153 | 0.023 |
| | Maximum | 0.0120 | 0.0322 | 0.043 |
| 3131.03 | Centroid | 0.00425 | 0.00740 | 0.012 |
| | Maximum | 0.00802 | 0.0151 | 0.023 |
| 3132.01 | Centroid | 0.00681 | 0.0116 | 0.019 |
| | Maximum | 0.0111 | 0.0199 | 0.031 |
| 3132.02 | Centroid | 0.00316 | 0.00549 | 0.009 |
| | Maximum | 0.00423 | 0.00671 | 0.011 |

streams were emitted from the stack as carcinogens with an average potency equal to benzene, annual average ambient concentrations, and upper bounds on risk, were calculated at the maximum census tract location. From Table 1, the annual average feed rate to the incinerator for all compounds is 1185 lb/hr. Assuming 0.01% emissions, the emission rate will be 0.1185 lb/hr, resulting in an annual average concentration at the maximum receptor site of 0.00619 $\mu g/m^3$ (0.1185 * 0.0522). For a unit risk of $5.3 \times 10^{-5}$, this results in an upper bound risk of $0.33 \times 10^{-6}$. Thus, even under this extreme worst-case scenario, risk is negligible.

## Other Routes of Exposure

The quantification of risk from inhalation exposures has a considerable degree of uncertainty associated with the end results. Generally, these uncertainties are resolved in favor of overestimating, rather than underestimating the true value of the risk. Even though the uncertainties in these assessments may seem large, they are much less than the uncertainties associated with exposures which could result through other media. These other media gener-

**Table 9. Summed Hypothetical Risk Estimates for Other Receptors**

| Site | Census tract | Upper limit on lifetime risk × $10^{-6}$ |
|---|---|---|
| Lifetime residence | | |
| Low Medanos Hospital | 3131.02 | 0.0292 |
| Pittsburg Manor Hospital | 3120 | 0.0439 |
| Elders Wind | 3072.03 | 0.0670 |
| 16-years residence, 1200 hr per year | | |
| Turner School | 3072.01 | 0.000980 |
| Central Jr. High School | 3101.01 | 0.000696 |
| Delta High School | 3100 | 0.00142 |
| Martin Luther King School | 3120 | 0.00141 |
| Pittsburg High School | 3110 | 0.00131 |
| Village School | 3131.02 | 0.00131 |
| Low Medanos College | 3072.02 | 0.00111 |
| Kindercare | 3072.03 | 0.00372 |
| 40-years residence, 1960 hr per year | | |
| Continental Can | 3090 | 0.1073 |
| Hersey | 3090 | 0.1180 |
| Imperial West | 3090 | 0.0402 |
| McConnell Pallets | 3090 | 0.0832 |
| POSCO Administration | 3090 | 0.2085 |
| POSCO Manufacturing | 3090 | 0.0748 |
| Sewage Treatment Distributors | 3050 | 0.0656 |
| Stauffer | 3090 | 0.0817 |
| USS Realty—Loveridge Road | 3090 | 0.0692 |
| USS Realty—Old Pittsburg Antioch Hwy. | 3090 | 0.0339 |

**Table 10. Hypothetical Cancer Burden for Census Tracts**

| Census tract | Hypothetical risk × $10^{-6}$ | Projected population (2022) | Expected value of number of cases per 70 years | Expected value of years until first case at 2022 population size |
|---|---|---|---|---|
| 3050 | 0.104 | 38,800 | 0.0040 | 17,000 |
| 3071 | 0.067 | 12,400 | 0.00083 | 84,000 |
| 3072.01 | 0.031 | 7,900 | 0.00024 | 290,000 |
| 3072.02 | 0.095 | 7,300 | 0.00069 | 100,000 |
| 3072.03 | 0.038 | 17,700 | 0.00067 | 104,000 |
| 3090 | 0.073 | 10,200 | 0.00074 | 94,000 |
| 3100 | 0.045 | 6,700 | 0.00030 | 230,000 |
| 3110 | 0.028 | 8,100 | 0.00023 | 310,000 |
| 3120 | 0.051 | 4,000 | 0.00020 | 340,000 |
| 3131.01 | 0.012 | 12,600 | 0.00015 | 460,000 |
| 3131.02 | 0.023 | 5,100 | 0.00012 | 600,000 |
| 3131.03 | 0.012 | 21,900 | 0.00026 | 270,000 |
| 3132.01 | 0.019 | 9,400 | 0.00018 | 390,000 |
| 3132.02 | 0.009 | 9,800 | 0.000088 | 790,000 |
| Total | | | 0.0088 | |
| Expected value of years until first case for total study area | | | | 8,000 |

ally involve deposition and subsequent exposure through contact. One of the largest variable factors in determining the potential for exposures from these media is the deposition velocity of the incinerated materials. Deposition velocity is defined as the rate of gravitational settling of the compound onto the surface in units of grams per second per square meter. The potential for this deposition depends upon a number of factors, including the physical state of the compound being deposited (e.g., gaseous or adsorbed on particulate) and the characteristics of the receiving surface. These characteristics include the type of surface (plant, soil, concrete, etc.) as well as the moisture content of the surface and the water or lipid solubility of the compound being deposited. In some cases, empirical studies have indicated the appropriate range for the relevant parameters, but these values tend to be very specific to the local situation. In addition to dry deposition, wet deposition can occur through in-cloud and below-cloud scavenging.[3] For different applications, various modelers have used deposition velocity estimates which differ by orders of magnitude.[2,3] The different estimates of this key parameter lead to widely different conclusions. It should also be remembered that all the *inhalation* estimates were calculated under the conservative assumption that there is *no* deposition velocity, i.e., that all of the compound is airborne with no removal, either by degradation or by deposition.

To avoid the problem of selecting the most appropriate parameter estimates, in the face of considerable uncertainty, and the further problem of interpreting what the parameter estimates mean as far as actual deposition, a simplifying but conservative approach was used: It was assumed, for the purpose of calculating multimedia exposure, that *all* of the relevant compounds were deposited uniformly in the study area. The compounds which would be of concern for multimedia exposure are those which have low degradation rates in the environment and which have higher bioconcentration potentials in fish. Dioxins are generally discussed in this context and are assumed to be adsorbed onto the flyash particulate. By analogy, hexachlorobenzene was also evaluated.

The assumption of uniform deposition is highly conservative in that it predicts complete deposition of small particulates (less than 1 $\mu$m) within a 2.5-mi radius from the source. If all the particulate *were* deposited in this area, there would be patterns of higher and lower concentrations reflecting different wind patterns and decreasing concentrations as a function of distance. These differences will be small, however, relative to the highly conservative assumption of complete deposition in a small area. This approach has the advantage of being understandably relative to the assumptions being made rather than attempting to interpret deposition velocities.

A number of exposure scenarios were evaluated. These include calculations of soil concentrations, for comparison to the CDC criteria of 1 ppb, as well as exposures through consumption of fish from the river and vegetables grown in home gardens. These scenarios will not be explained in detail, but the same basic approach was used for all cases. It was assumed that all the dioxin (and

**Table 11. Noninhalation Exposure Pathways**

| Media | Exposure | Risk or Criterion |
|---|---|---|
| Soil | 0.019 ppb | 1 ppb (CSC) |
| Lettuce | $7 \times 10^{-13}$ mg/kg/day | $0.09 \times 10^{-6}$ |
| Tomatoes | $3 \times 10^{-13}$ mg/kg/day | $0.03 \times 10^{-6}$ |
| Fish | 0.0005 ppt | 25 ppt (FDA) |
| | $= 3 \times 10^{-14}$ mg/kg/day | $= 0.003 \times 10^{-6}$ |

hexachlorobenzene) was deposited in the study area and that no degradation of the compounds took place in the lifetime of the facility (70 years). This resulted in a final maximum soil concentration of 0.019 ppb for dioxin, well below the CDC criteria. For human consumption, U.S. average consumption patterns were used together with the unlikely scenario that all vegetables consumed were from home gardens with completely accumulated deposition and that there was no removal of the compounds either through washing or cooking. For fish exposure, two scenarios were considered. In one scenario, it was assumed that all of the dioxin deposited in the river remained suspended in the water and was available for uptake by the fish. In the other scenario, it was assumed that for 70 years all of the emitted dioxin was mixed into the sediment on the river bottom with no potential for sediment flushing. It was then assumed that the concentration in the flesh of the fish equilibrated with the sediment concentration. This equilibration has been suggested on the basis of some empirical evidence,[4] although there are also some studies which have shown somewhat higher fish-to-sediment ratios. The latter scenario is more likely to represent the actual values, although 70 years of accumulation in the sediment with no removal, by flushing or added deposition or degradation, is unlikely. Nonetheless, these scenarios set simple upper bounds on the potential for human exposure and, as shown in Table 11, demonstrate that no potential for significant human hazard exists under the scenarios considered.

## CONCLUSIONS

It has been shown in the preceeding risk assessment that a well-run hazardous waste incinerator poses no significant risk to human health, even when evaluated under the most stringent criteria. The methodologies used in this assessment have numerous interesting decision points which can lead to different evaluation criteria. When the most health-protective option is taken at each of these decision points, the results must be carefully presented as worst-case evaluations.

## REFERENCES

1. Anderson, E. L. "*Quantitative Methods in Use in the United States to Assess Risk*," paper presented at the Workshop on Quantitative Estimation of Risk to Humans in Rome, Italy, July 12, 1982.

2. Hosker, R. P. and S. E. Lindberg. "*Review: Atmospheric Deposition and Plant Assimilation of Gases and Particles*," *Atmos. Environ.* 16(5): 899–910 (1982).
3. U.S. Environmental Protection Agency. "*Report of the Environmental Effects, Transport and Fate Committtee*," SAB-EETFC-88-25 (April 1988a).
4. U.S. Environmental Protection Agency. "*Estimating Exposures to 2,3,7,8-TCDD*," EPA/600/6-88/005A (March 1988b).

# CHAPTER 30

# Human Exposure to Dioxin from Combustion Sources

**Holly A. Hattemer-Frey and Curtis C. Travis**

## INTRODUCTION

Because of their extreme toxicity, much concern and debate has arisen about the nature and extent of human exposure to dioxin. Since municipal solid waste (MSW) incinerators are known to emit polychlorinated dibenzo-p-dioxins (PCDDs) and polychlorinated dibenzofurans (PCDFs),[1] many people who live near MSW incinerators fear that they will be exposed to high levels of dioxin and subsequently develop cancer. What is often overlooked in this debate, however, is the fact that the general population is continuously being exposed to trace amounts of dioxin as exemplified by the fact that virtually *all* human adipose tissue samples contain dioxin at levels of 3 parts per trillion (ppt) or greater.[2,3] This paper provides a perspective on MSW incineration as a source of human exposure to dioxin by comparing this exposure source with exposure to background environmental contamination and evaluates some of the potential key sources of PCDD/PCDF input into the environment.

## BACKGROUND EXPOSURE

Background levels of 2,3,7,8-tetrachlorodibenzo-p-dioxin (TCDD) have been measured in air, soil, sediment, suspended sediment, fish, cow milk, human adipose tissue, and human breast milk samples.[2,4–13] Travis and Hattemer-Frey[14] used measured environmental concentrations to quantify the amount of TCDD entering the food chain and the extent of human exposure. Potential pathways of human exposure include (1) inhalation; (2) accumulation in vegetation via direct deposition and root uptake and ingestion of contaminated fruit, vegetables, and grains; (3) ingestion of contaminated soil and forage by terrestrial organisms, bioaccumulation of TCDD in their tissues, and human consumption of contaminated beef and milk products; (4) deposition onto water surfaces, accumulation in aquatic biota, and transferral to humans through the aquatic food chain; and (5) ingestion of contaminated drinking water.

Because TCDD is highly lipophilic, it partitions mainly into soil (85%) and sediment (14%) and readily bioaccumulates in living organisms.[14] Less than 1% of the TCDD released into the environment partitions into air, water, suspended sediment, and biota.[14] Table 1 shows that the food chain, especially meat and dairy products, accounts for more than 98% of human exposure to TCDD and that the long-term, average daily intake by the general population of the United States is 47 pg/day[14] (Table 1). Consumption of contaminated vegetation and fish were minor pathways of human exposure, and inhalation was *not* a major pathway of human exposure.[14]

This environmental pathway estimate was verified using a one-compartment pharmacokinetic model,[15] which was used to estimate daily intake from human body burdens (i.e., adipose tissue levels). Using a half-life of 2120 days for TCDD in the human body[16] and an average human adipose tissue concentration of 6 ng/kg,[2] the long-term, average daily intake of TCDD was estimated to be 28 pg/day. Both the environmental pathway and the body burden estimates agree well with an estimate of 25 pg/day reported by Beck et al.[17]

Beck et al.[17] and Ono et al.[18] analyzed randomly collected food samples of the human diet and estimated the average, background daily intake of all PCDDs and PCDFs as a whole [expressed in Toxic Equivalents Dioxins and Furans (TEDFs)] resulting from ingestion of contaminated food items to be about 90 and 60 pg TEDFs/day, respectively. (When PCDD and PCDF concentrations are expressed in TEDFs, the toxicity of all isomers is weighted relative to the known toxicity of 2,3,7,8-TCDD. Thus, ingesting 1 g of TEDFs is equivalent to ingesting 1 g of TCDD.) The Japanese intake estimate, however, reflects the dietary consumption of an individual weighing 50 kg. If the Japanese estimate is adjusted for an individual weighing 70 kg, then the average daily intake of PCDDs and PCDFs is 84 pg TEDFs/day.

## EXPOSURE TO MSW INCINERATOR EMISSIONS

It is widely believed that MSW incineration is the major source of human exposure to dioxin. To assess the extent of human exposure to facility-emitted PCDDs and PCDFs, Travis and Hattemer-Frey[19] evaluated 11 risk assessment documents prepared for proposed MSW incinerators designed to use modern, efficient pollution control equipment. The geometric mean of the data from those 11 documents was used to represent exposure to a typical, *modern* MSW incinerator. (Hence, the following results may not apply to older incinerators that do not use efficient pollution control equipment.) Results suggest that MSW incineration is *not* a major source of human exposure to dioxins and furans for the following reasons.

First, exposure to background levels of PCDDs and PCDFs accounts for more than 99% of the total daily intake by the *maximally exposed* individual

**Table 1. Predicted Average Background Daily Intake of TCDD by the General Population of the United States**

| Source | Daily intake (pg/day) | Percentage of the total daily intake (%) |
|---|---|---|
| Air | 1 | 2 |
| Water | 0.007 | <0.01 |
| Food (total) | 46 | 98 |
| Produce | 5 | 11 |
| Milk | 13 | 27 |
| Meat | 23 | 50 |
| Fish | 5 | 10 |
| Total intake | 47 | 100 |

living near a typical, modern MSW incinerator. Table 2 shows that the predicted daily intake of PCDDs/PCDFs by the maximally exposed individual is 80 times less than exposure to background levels. These data indicate that human exposure to PCDDs and PCDFs emitted from a typical, state-of-the-art MSW incinerator is *not excessive* relative to exposure to background levels, since the individual lifetime cancer risk associated with exposure to facility-emitted PCDDs and PCDFs is one in a million ($1 \times 10^{-6}$), while the cancer risk associated with exposure to background environmental contamination is two in ten thousand ($2 \times 10^{-4}$).

Secondly, the *maximum* air concentration of PCDDs and PCDFs predicted to occur at the point of *maximum* individual exposure is slightly less than measured background levels of PCDDs and PCDFs in urban air (Table 3). Background concentrations of PCDDs and PCDFs have been measured in

**Table 2. Total Daily Intake of PCDDs/PCDFs by the Maximally Exposed Individual Living Near a Typical, Modern MSW Incinerator**

| Source | Daily intake (pg TEDFs/day) | Percentage of the total daily intake (%) | Reference |
|---|---|---|---|
| Background[a] | 87.0 | 99.3 | 17, 18 |
| Incinerator | 0.6 | 0.7 | 19 |
| Total | 87.6 | 100 | |

[a]Geometric mean of estimates reported by Beck et al.[17] and Ono et al.[18] for a 70-kg individual.

**Table 3. Predicted Concentration of PCDDs and PCDFs Around a Typical, Modern MSW Incinerator Versus Background Urban Air Concentrations**

| Source | Concentration (ng TEDFs/m$^3$) | Reference |
|---|---|---|
| Predicted | $1.2 \times 10^{-2}$ | 19 |
| Measured background[a] | $1.6 \times 10^{-2}$ | 7, 9, 11 |

*Note:* ng = nanogram = $10^{-9}$ g.

[a]Represents the geometric mean of reported measurements for the United States, Germany, and Sweden.[7, 9, 11]

urban air in the United States ($1.2 \times 10^{-5}$ ng TEDFs/m$^3$), West Germany ($3.3 \times 10^{-5}$ ng TEDFs/m$^3$), and Sweden ($9.3 \times 10^{-6}$ ng TEDFs/m$^3$), respectively,[7,9,11] with a geometric mean of $1.6 \times 10^{-5}$ ng TEDFs/m$^3$. Thus, atmospheric concentrations of PCDDs and PCDFs emitted from modern MSW incinerators should be indistinguishable from measured background levels, which suggests that, on the local level, MSW incinerators are not substantially increasing background atmospheric levels of dioxin.

This conclusion is supported by actual measurements of PCDDs and PCDFs in ambient air near operating MSW incinerators. Rappe et al.[9] reported that the concentration of PCDDs and PCDFs measured at 13 sites 1500 m downwind from a MSW incinerator near Hamburg, West Germany, was $5.8 \times 10^{-5}$ ng TEDFs/m$^3$. Similarly, Olie et al.[20] found that concentrations of PCDDs and PCDFs in the air 2 km from a MSW incinerator near Amsterdam was $4.1 \times 10^{-5}$ ng TEDFs/m$^3$. Thus, measured air concentrations around these two operating MSW incinerators are about three to four times higher than the mean background urban air concentration of PCDDs and PCDFs ($1.6 \times 10^{-5}$ ng TEDFs/m$^3$). These data indicate that air concentrations of PCDDs and PCDFs near operating MSW incinerators are *slightly* elevated relative to background urban air levels. The fact that measured air concentrations of PCDDs and PCDFs near operating MSW incinerators are at or near background levels, however, supports modeling conclusions that MSW incinerators are not substantially elevating background levels of PCDDs and PCDFs (model predictions of PCDD/PCDF concentrations around modern MSW incinerators plus the geometric mean background air concentration equals $2.8 \times 10^{-5}$ ng TEDFs/m$^3$ versus $4.1 \times 10^{-5}$ to $5.8 \times 10^{-5}$ ng TEDFs/m$^3$ actually measured around operating MSW incinerators).

One might wonder why, if concentrations of PCDDs and PCDFs around operating MSW incinerators are at or near measured background levels, is the contribution of facility-emitted PCDDs and PCDFs to total daily intake by the maximally exposed individual (Table 2) so small? One reason is that risk assessments for proposed MSW incinerators generally assume that only a small percentage of food consumed by the maximally exposed individual is actually produced in the local area (and therefore affected by incinerator emissions). Since the food chain is the primary pathway of human exposure to a large class of organic compounds, including TCDD, DDT, and most pesticides,[14,21] this assumption can substantially reduce intake estimates. In areas of high agricultural production, for example, the 0.6 pg TEDFs per day daily intake estimate for incinerators could be low.

## SOURCES OF PCDD/PCDF INPUT

If MSW incineration is not the major source of human exposure to PCDDs and PCDFs, what is? Although the magnitude of PCDD/PCDF input into the environment remains unknown, principal sources of PCDDs and PCDFs are suspected to be (1) high-temperature industrial processing facilities, such as metal processing/treatment plants and copper smelting plants;[1,22] (2) motor vehicles;[9,23-25] (3) chemical manufacturing processes;[26] (4) chemical/industrial waste incineration;[26,27] (5) MSW incinerators;[1,9,22,26] and (6) pulp and paper mills.[9,28] The magnitude of emissions from sources for which empirical data are available are discussed here.

### Industrial Sources

Rappe et al.[1] and Marklund et al.[22] reported that emissions of PCDDs and PCDFs from high-temperature industrial processing plants seem to be of the same magnitude as emissions from MSW incinerators. The Swedish EPA[29] estimated that emissions of PCDDs and PCDFs from these types of plants operating in Sweden range from 50 to 150 g TEDFs per year, while normal-sized, modern MSW incinerators (50 to 200,000 tons of waste per year capacity) emit about 1 to 50 g TEDFs per year.[1,19]

Rappe et al.[9] found that levels of PCDDs and PCDFs in air collected from a West German industrial area located 600 m from a copper smelting plant ($3.6 \times 10^{-2}$ ng/m$^3$) were about 12 times higher than measured background levels in suburban air collected about 13 km from Hamburg, West Germany ($3.0 \times 10^{-3}$ ng/m$^3$) and about three times higher than levels measured around a MSW incinerator operating near Hamburg ($1.3 \times 10^{-2}$ ng/m$^3$).

Rappe et al.,[1] moreover, concluded that because there are many more industrial sources, their contribution to total PCDD/PCDF emissions could be much greater than the contribution from MSW incinerators. Marklund et al.[22] observed that "total emissions from industrial incinerators could be of the same magnitude or even higher than emissions from MSW incinerators," while Nakano et al.[30] reported that "PCDDs and PCDFs in the urban air are surmised to be derived from domestic and industrial waste incinerators." Thus, various researchers agree that industrial sources may contribute equal or even larger amounts of PCDDs and PCDFs into the environment than MSW incinerators.

### Motor Vehicles

Studies have shown that motor vehicle emissions may be a larger source of PCDD/PCDF input than emissions from MSW incinerators. Ballschmiter et al.[23] argued that a more prevalent source of PCDD/PCDF input than MSW incineration must exist to account for the widespread, background contamination of PCDDs and PCDFs and suggested that the ubiquitous, nonpoint character of motor vehicle emissions "strongly recommends this source for consideration as a major environmental input."

Marklund et al.[25] found that total emissions of PCDDs and PCDFs from cars in Sweden using unleaded gasoline were 10 to 100 g TEDFs per year, which is equivalent to the amount of PCDDs and PCDFs emitted from 2 to 20

MSW incinerators of normal size and technology. Thus, in Sweden, motor vehicles and MSW incinerators emit about the same amount of PCDDs and PCDFs. Jones[31] reached the same conclusion for the United States and contends that due to the widespread source of motor vehicle emissions, roadside exposures are equal to or greater than exposures from the elevated stacks of MSW incinerators.

Rappe et al.[9] also measured PCDD/PCDF levels in ambient air within a traffic tunnel. PCDD/PCDF levels in the tunnel ($2.8 \times 10^{-2}$ ng/m$^3$) were two times higher than levels downwind from the MSW incinerator ($1.3 \times 10^{-2}$ ng/m$^3$) and nine times higher than the suburban air concentration ($3.0 \times 10^{-3}$ ng/m$^3$).[9] Rappe et al.[9] concluded that "measurements made in the traffic tunnel clearly indicate that motor vehicles are a source of PCDDs/PCDFs in the ambient air." Hence, several researchers confirm that motor vehicle emissions may also be a significant source of PCDDs/PCDFs in the environment.

### Pulp and Paper Mills

The U.S. EPA[13] first reported that pulp and paper mills using a chlorine bleaching process may be another source of PCDD/PCDF input into the environment. Swanson et al.[32] analyzed the effluents from one mill producing bleached pulp in Sweden and found that PCDD/PCDF emissions ranged from 2 to 5.8 g TEDFs per year. Since the mill they sampled used a less efficient processing method than most operating mills in Sweden, they estimated that Swedish pulp and paper mills as a whole discharge about 5 to 15 g of TEDFs per year.[32] These *preliminary* findings suggest that the amount of PCDDs and PCDFs formed in pulp and paper mills in Sweden is small relative to other sources.

Thus, these data indicate that environmental concentrations cannot be linked to any one combustion source. Combustion processes *in general* (not just MSW incinerators) are the dominant source of PCDDs and PCDFs in the environment. It is premature to conclude that MSW incineration is *the* major source. The magnitude of PCDD/PCDF emissions from and PCDD/PCDF concentrations in the ambient air and other environmental media (e.g., soil and cow's milk) around other operating combustion sources known to emit dioxins and furans (i.e., copper smelting plants, steels mills, motor vehicles, and pulp and paper mills) are needed before definitive statements about the major source(s) of PCDD/PCDF input can be made.

## CONCLUSIONS

This paper in not intended to resolve the incineration debate or to criticize or recommend MSW incineration as a waste management tool. It does, however, reevaluate some widely held beliefs concerning human exposure to PCDDs and PCDFs emitted from MSW incinerators. MSW incinerators are one source of PCDD/PCDF input into the environment. Empirical evidence demonstrates, however, that well-operated, modern MSW incinerators may not be the major source of human exposure to PCDDs and PCDFs, since exposure to background levels overwhelms exposure to facility-emitted contaminants. The relatively small contribution by incinerators to total daily intake suggests that some, as of yet unidentified, source(s) (possibly automobiles or industrial sources) are substantially contributing to background levels of dioxins and furans. We recommend that future research efforts focus on characterizing the source(s) of *background* levels of PCDDs and PCDFs, because they may pose far greater threats to human health than PCDD/PCDF emissions from modern MSW incinerators.

## ACKNOWLEDGMENT

Research was sponsored by the U.S. Environmental Protection Agency under Martin Marietta Energy Systems, Inc., Contract No. DE-AC05-840R21400 with the U.S. Department of Energy.

## REFERENCES

1. Rappe, C., R. Andersson, P.-A. Bergquist, C. Brohede, M. Hansson, L.-O. Kjeller, G. Lindstrom, S. Marklund, M. Nygren, S. E. Swanson, M. Tysklind, and K. Wiberg. "Overview on Environmental Fate of Chlorinated Dioxins and Dibenzofurans. Sources, Levels and Isomeric Pattern in Various Matrices," *Chemosphere* 16(8/9):1603–1618. (1987).
2. Patterson, D. G., J. S. Holler, C. R. Lapeza, Jr., L. R. Alexander, D. F. Groce, R. C. O'Connor, S. J. Smith, J. A. Liddle, and L. L. Needham. "High-Resolution Gas Chromatographic/High-Resolution Mass Spectrometric Analysis of Human Adipose Tissue for 2,3,7,8-Tetrachlorodibenzo-p-dioxin," *Anal. Chem.* 58:705–713 (1986).
3. Ryan, J. J., R. L. Lizotte, and B. P. Y. Lau. "Chlorinated Dibenzo-p-Dioxins and Chlorinated Dibenzofurans in Canadian Human Adipose Tissue," *Chemosphere*. 14(6/7):697–706 (1985).
4. Beck, H., K. Eckart, M. Kellert, W. Mathar, Ch-S. Ruhl, and R. Wittowski. "Levels of PCDFs and PCDDs in Samples of Human Origin and Food in the Federal Republic of Germany," *Chemosphere*. 16(8/9):1977–1982 (1987).
5. Crummett, W. B. Dow Chemical Company. Personal correspondence (1987).
6. Czuczwa, J. M., and R. A. Hites. "Dioxins and Dibenzofurans in Air, Soil and Water," in *Dioxins in the Environment*, M. A. Kamrin and P. W. Rodgers, Eds. (Washington, D.C.: Hemisphere Publishing Corp., 1985), pp. 85–99.
7. Eitzer, B. D., and R. A. Hites. "Dioxins and Furans in the Ambient Atmosphere: A Baseline Study." *Chemosphere* 18:593–598 (1989).
8. O'Keefe, P., C. Meyer, D. Hilker, K. Aldous, B. Jelus-Taylor, K. Dillon, R. Donnelly, E. Horn, and R. Sloan. "Analysis of 2,3,7,8-Tetrachlorodibenzo-p-Dioxin in Great Lakes Fish," *Chemosphere* 12(3):325–332 (1983).
9. Rappe, C., L.-O. Kjeller, P. Bruckmann, and K.-H. Hackhe. "Identification and

Quantification of PCDDs and PCDFs in Urban Air," *Chemosphere* 17(1):3–20 (1988).

10. Rappe, C., M. Nygren, G. Lindstrom, H. R. Buser, O. Blaser, and C. Wuthrich. "Polychlorinated Dibenzofurans, Dibenzo-p-Dioxins and Other Chlorinated Contaminants in Cow Milk from Various Locations in Switzerland," *Environ. Sci. Technol.* 21(10):964–970 (1987).
11. Rappe, C., and L.-O. Kjeller. "PCDDs and PCDFs in Environmental Samples Air, Particulates, Sediments and Soil," *Chemosphere* 16(8/9):1775–1780 (1987).
12. van den Berg, M., F. W. M. van der Wielen, K. Olie, and C. J. van Boxtel. The Presence of PCDDs and PCDFs in Human Breast Milk from the Netherlands," *Chemopshere* 15(6):693–706 (1986).
13. U.S. Environmental Protection Agency. "The National Dioxin Study: Tiers 3,4,5, and 7," Office of Water Regulation and Standards, EPA-440/87/003, Washington, D.C. (1987).
14. Travis, C. C., and H. A. Hattemer-Frey. "Human Exposure to 2,3,7,8-TCDD," *Chemosphere* 16(10–12):2331–2342 (1987).
15. Geyer, H. J., I. Scheunert, J. G. Filser, and F. Korte. "Bioconcentration Potential (BCP) of 2,3,7,8-Tetrachlorodibenzo-p-dioxin (2,3,7,8-TCDD) in Terrestrial Organisms Including Humans," *Chemosphere* 15:1494 (1986).
16. Poiger, H., and C. Schlatter. "Pharmacokinetics of 2,3,7,8-TCDD in Man," *Chemosphere* 14:1489–1494 (1986).
17. Beck, H., K. Eckart, W. Mathar, and R. Wittowski. "PCDD and PCDF Body Burden from Food Intake in the Federal Republic of Germany," *Chemosphere* 18:587–592 (1989).
18. Ono, M., Y. Kashima, T. Wakimoto, and R. Tatsukawa. "Daily Intake of PCDDs and PCDFs by Japanese Through Food," *Chemosphere* 16(8/9):1823–1828 (1987).
19. Travis, C. C., and H. A. Hattemer-Frey. "Human Exposure to Dioxin from Municipal Solid Waste Incineration," *Waste Management* 9:151–156 (1989).
20. Olie, K., M. van den Berg, and O. Hutzinger. Formation and Fate of PCDD and PCDF from Combustion Processes," *Chemosphere* 12(4/5):627–637 (1983).
21. Travis, C. C., and A. D. Arms. "The Food Chain as a Source of Toxics Exposure," in *Toxic Chemicals, Health, and the Environment*, L. B. Lave and A. C. Upton, Eds. (New York: Plenum Publishing Corporation, 1987).
22. Marklund, S., L.-O. Kjeller, M. Hansson, M. Tysklind, C. Rappe, C. Ryan, H. Collazo, and R. Dougherty. "Determination of PCDDs and PCDFs in Incineration Samples and Pyrolytic Products," in *Chlorinated Dioxins and Dibenzofurans in Perspective*, C. Rappe, G. Choudhary, and L. Keith, Eds. (Chelsea, MI:Lewis Publishers, 1986).
23. Ballschmiter, K., H. Buchert, R. Niemczyk, A. Munder, and M. Swerev."Automobile Exhausts Versus Municipal Waste Incineration as Sources of the Polychloro-Dibenzodioxins (PCDD) and -Furans (PCDF) Found in the Environment," *Chemosphere* 15:901–915 (1986).
24. Bumb, R. R., W. B. Crummett, S. S. Cutie, J. R. Gledhill, R. H. Hummel, R. O. Kagel, L. L. Lamparski, E. V. Luoma, D. L. Miller, T. J. Nestrick, L. A. Shadoff, R. H. Stehl, and J. S. Woods. "Trace Chemistries of Fire: A Source of Chlorinated Dioxins," *Science* 210:385–390 (1980).
25. Marklund, S., C. Rappe, M. Tysklind, and K.-E. Egeback. "Identification of

Polychlorinated Dibenzofurans and Dioxins in Exhausts from Cars run on Leaded Gasoline," *Chemosphere* 16(1):29–36 (1987).

26. Hutzinger, O., M. J. Blumich, M. van den Berg, and K. Olie. "Sources and Fate of PCDDs and PCDFs: An Overview," *Chemosphere* 14(6/7):581–600 (1985).
27. Weerasinghe, N. C. A., and M. L. Gross. "Origins of Polychlorodibenzo-p-dioxins (PCDD) and Polychlorodibenzofurans (PCDF) in the Environment," in *Dioxins in the Environment*, M. A. Kamrin and P. W. Rodgers, Eds. (Washington, D.C.: Hemisphere Publishing Company, 1985), pp. 133–151.
28. Beck, H., K. Eckart, W. Mathar, and R. Wittowski. "Occurrence of PCDD and PCDF in Different Kinds of Paper." *Chemosphere* 17(1):51–57 (1988a).
29. Swedish Environmental Protection Agency. *Dioxin* (May 1987).
30. Nakano, T., M. Tsuji, and T. Okuno. "Level of Chlorinated Organic Compound in the Atmosphere," *Chemosphere* 16(8/9):1781–1786 (1987).
31. Jones, K. H., J. Walsh, and D. Alston. "The Statistical Properties of Available Worldwide MSW Combustion Dioxin/Furan Emissions Data as They Apply to the Conduct of Risk Assessments," *Chemosphere* 16(8/9):2183–2186 (1987).
32. Swanson, S. E., C. Rappe, J. Malmstrom, and K. P. Kringstad. "Emissions of PCDDs and PCDFs from the Pulp Industry," *Chemosphere* 17(4):681–691 (1988).

CHAPTER 31

# Assessing the Risks of Combustion Emissions in a Multi-media Context

**Brendan Birmingham**

## INTRODUCTION

Combustion emissions have been shown to directly impact on the urban atmosphere ever since fire was invented. These emissions may also impact on soil and water quality within the zone of influence of major combustion sources. Recently, it has been claimed that the quality of food produced near major sources has also been affected by certain components of emissions.[1-4]

Because of the need for landfill sites near most urban centers, incineration of waste (municipal or otherwise) is being actively considered as one of several waste management options. Emissions of incinerators do not exist in isolation and their risk assessment must be integrated with the risks of exposure from other sources. As waste management is a complex social and political issue, the public and public officials making decisions in this matter demand the best scientific analysis possible.

Many incinerator risk assessments have been performed in recent years in North America.[5-7] All contain estimates of human exposure to the proposed emissions. However, most of these risk assessments have not addressed human exposure to all sources from a multimedia perspective.

What is meant by a multimedia approach is an integrated assessment of human exposure to a specific contaminant in the total environment, that is, from air, food, water, soil, and possibly consumer products.[8] The EPA, for instance, calls this process a multiple pathway approach.[6]

This multimedia approach is used to address the problem of long-term exposures to persistent chemicals such as polychlorinated dibenzo-dioxins (PCDDs) and polychlorinated dibenzofurans (PCDFs), which are found in several environmental media. Human beings are considered the most important receptor for multimedia contaminants because of their exposure via a number of routes and because of their place at the top of several food chains. However, other organisms may also be important receptors to multimedia exposure.

What I plan to discuss are some of Ontario's experiences in developing a multimedia approach to persistent chemicals and the example I will take will be our work with PCDDs and PCDFs. I will attempt to compare the results from this multimedia exposure model with some of the estimates derived from other exposure models based on incinerator emissions.

## MULTIMEDIA EXPOSURE FROM INCINERATORS

Modeling the dispersion of emissions from point sources such as incinerators is now a growth industry. Sophisticated mathematical models for all aspects of the dispersion of emissions and their ensuing environmental fates in the air, soil, water, sediment, and biological compartments of the environment are well described.[4-7,11] A brief overview of the various facets of the environmental fate of incinerator emissions are shown in Figures 1 to 3.

Figure 1 is a diagram of the physical extent of the problem and Figure 2 shows the geographical extent of the problem.

Typically, the incinerator stack emissions are released from a fixed point, usually 30 m or higher above ground level. The emissions are transported downwind and dispersed depending on the local topography and the prevailing wind and weather conditions. Locally there will be a region or zone of maximum impact which is delineated with hypothetical isopleths. This zone will occur within 10 km of the source and may impact on urban sites, farm production areas, or streams and lakes. Typically, as you move away from the incinerator source, other combustion sources, e.g., automotive, power generation, industrial, and home heating sources will also have effects on receptors in their



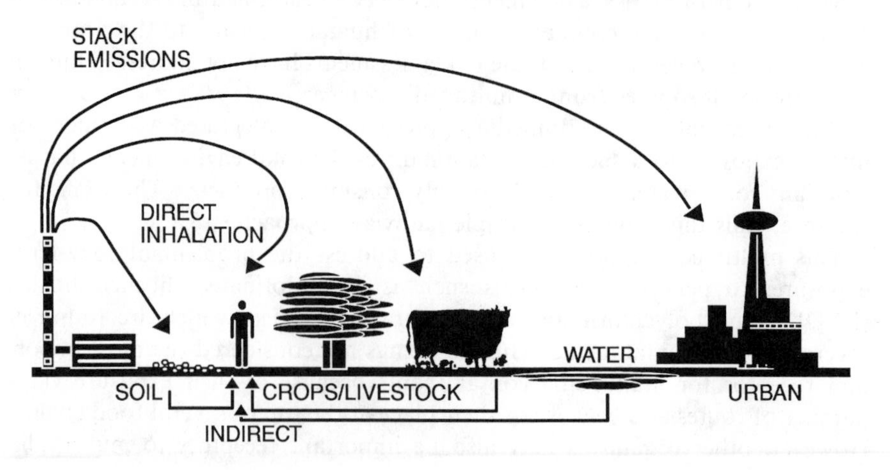


**Figure 1.** Incinerator exposure pathways.

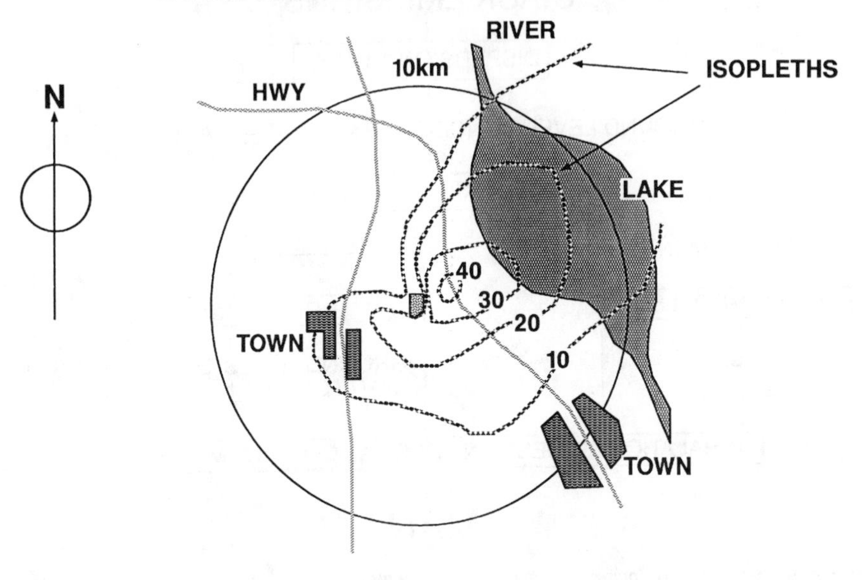


**Figure 2.** Aerial view of incinerator exposure.

vicinity. Outside of the main region of impact, dispersion processes will reduce the emissions to negligible levels.

Inside this zone of impact, which is delineated mathematically, a number of models can be used to evaluate the potential exposure of receptors. Figure 3 gives a simple outline of the pertinent models and exposure pathways.

As can be seen, the potentially significant pathways to humans are

1. direct inhalation of dispersed emissions
2. indirect pathways through ingestion of untreated water and fish, livestock, and crops that may have absorbed PCDDs and PCDFs from the emissions; direct dermal contact with soils or dusts contaminated by emissions is also possible

For persistent multimedia contaminants like PCDDs and PCDFs, we can integrate each of these exposures to assess the potential risks of the incinerator in question. The empirical validation of the output of these models is very limited. This is partially due to the expense of PCDD and PCDF analysis and the need for extremely low levels of sensitivity. In general, we have an extensive set of data on PCDDs and PCDFs in incinerator emissions, however, empirical values for levels of PCDDs and PCDFs in ambient air, soil, produce, livestock, or water that is impacted by specific incinerators of known output are scarce.[1,2] Consequently, after the PCDD and PCDF levels in the emissions are measured

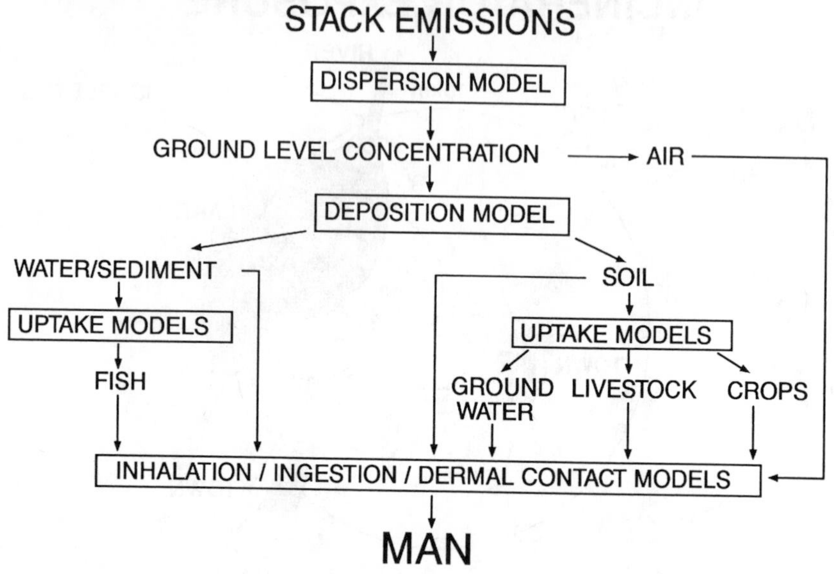


**Figure 3.** Flowchart of incinerator exposure pathways.

or predicted, we are largely at the mercy of mathematical algorithms to predict the exposures and risks involved.

Table 1 shows a comparison of emission rates, calculated air levels, and calculated deposition rates for some U.S. and Canadian incinerators. The focus of these calculations is on measured emission rates, ground level concentrations, and deposition rates, since these are the values driving exposure models of incinerator emissions. It should be pointed out that all values are in 2,3,7,8-TCDD toxic equivalents (TEQ), calculated using the international toxic equivalency factor (TEF) scheme proposed by NATO[12] and the Nordic Council.[13] The values are corrected as far as possible for content of 2,3,7,8-substituted isomers. The U.S. incinerators data are based on the survey published by Travis and Hattemer-Frey. The Canadian data are based on emissions from an Ontario coal-fired power generator,[14] an uncontrolled MSW incinerator of modern design at Charlottetown, Prince Edward Island,[15] and an older MSW incinerator in Toronto, Ontario.[10]

Table 1 shows the daily emission rate in gram TEQ per day, the resulting calculated ground level maximum concentration is expressed *on an annual average basis* to account for wind direction changes. This value, in turn, is used to calculate the resulting deposition rate and the amount of TEQ that might be deposited daily within a 10-km radius around the source. This daily deposition is also expressed in gram TEQ per day and can be used to estimate

**Table 1. Comparison of Emission Rates, Calculated Air Levels, and Deposition Rates of North American Incinerators**

| | | Emission rate (g TEQ/d) | Ground level concentration ($10^{-12}$g TEQ/m$^3$) | Vd (m/sec) | Deposition rate ($10^{-12}$g TEQ/m$^2$/day) | Deposition over 10 km radius (g TEQ/day) |
|---|---|---|---|---|---|---|
| U.S. Incinerators[5] | low | 0.038 | 0.0013 | $2 \times 10^{-2}$ | 2.3 | 0.0007 |
| | high | 0.153 | 0.14 | $1.3 \times 10^{-3}$ | 14.7 | 0.0049 |
| Lambton, Ontario (coal fired) | | 0.024 | 0.001 | $5 \times 10^{-3}$ | 0.46 | 0.0001 |
| Charlottetown, P.E.I., (MSW) | | 0.0026 | 0.047 | $5 \times 10^{-3}$ | 23.5 | 0.0064 |
| Toronto, Ontario (MSW) | | 0.325 | 0.15 | $5 \times 10^{-3}$ | 65 | 0.02 |
| Urban background (Niagara Falls, NY; Bloomington, IN) | | — | 0.03—0.9 | $5 \times 10^{-3}$ | 13—390 | 0.004—0.122 |
| Great lakes Sediments[19,20] | | — | — | — | 3.8—460 | 0.0012—0.14 |

the fraction of the emissions that might contribute to indirect exposure following incorporation into water, soil, livestock, or produce.

It can be seen that the daily amount deposited within 10 km of the source that can contribute to these indirect exposure pathways is only 1 to 6% of the daily emissions. Well over 90% will fall outside this area and while this 90% or more contributes to the total loading of the environment, the fraction available for incorporation into the indirect pathways of local exposure is small.

As a cross check using actual data, Table 1 shows some calculations based on levels of PCDDs and PCDFs found in urban air[16–18] and in a variety of Great Lakes sediments.[19,20] These latter calculations suggest that MSW incinerators are just one of many contributors to levels of PCDDs and PCDFs in ambient air.

The issue of the background exposure of the receptor to other sources of PCDDs and PCDFs is an important component of the assessment of the risk of that receptor.

Combustion sources of PCDDs and PCDFs are very pervasive and are emitted directly to the atmosphere. By comparison, chemical sources are not emitted directly to the atmosphere and their major impacts are localized with impacts on soil quality, ground water quality, and the aquatic food chain to take the best-known examples.[9,10,21,23,41] More widespread impacts on say the human food chain may be possible via contamination of livestock through the presence of wood preservatives in animal bedding[24] or the presence of PCDDs and PCDFs in food packaging derived from bleached pulp products.[25] PCDDs and PCDFs may also be present in a variety of consumer products as well as cigarettes.[26]

To address this problem of background exposure to PCDDs and PCDFs, I will present some background on Ontario's approach. In 1985, Ontario released a comprehensive Scientific Criteria Document on PCDDs and PCDFs, which took a multimedia approach and included a toxicological assessment.[10]

The focal point of Ontario's toxicity assessment of PCDDs and PCDFs was on 2,3,7,8-TCDD and its capacity to adversely affect reproduction and cause cancer in laboratory animals. Animal studies indicate that 2,3,7,8-TCDD is a promoter rather than a complete carcinogen. This means that 2,3,7,8-TCDD will not start or initiate tumors, but may affect development of other tumors if exposure is long enough. 2,3,7,8-TCDD is not mutagenic and does not bind to DNA.

Since studies on humans were negative or inconclusive, it was felt that the maximum allowable daily intake of PCDDs and PCDFs from all sources equivalent to 10 pg 2,3,7,8-TCDD TEQ per kilogram of body weight was not expected to cause adverse effects in humans (Table 2).

Risk analysis leading to development of standards also involves a review of environmental levels and exposure pathways. Using standard inhalation, food, and water ingestion and dermal exposure models, intakes of PCDDs and PCDFs from all major sources can be estimated.[27,28]

**Table 2. Rationale for Ontario PCDD and PCDF Standard**

| | |
|---|---|
| **Toxicological overview** (standard based on 2, 3, 7, 8-$T_4CDD$) | |
| Acute toxicity tests | Extremely toxic (wide range of sensitivity) |
| Mutagenicity tests | Negative (inconclusive) Indirect mode of action |
| Animal reproduction tests | Positive (NOEL indicated) |
| Animal cancer tests | Positive (NOEL indicated) Indirect nongenetic mode of action |

**Toxic Criterion for Standard** (most sensitive effect)
Incidence of liver tumors in female rats (Kociba et al., 1978)[42]
Well-designed, well-validated study
NOEL (0.001 $\mu$g 2,3,7,8-$T_4CDD$/kg body weight/day)

**Assumptions**
Threshold for liver tumors exists (NOEL)
Humans are no more sensitive than rodents
100-fold safety factor will provide adequate margin of safety
Toxicity of other PCDDs and PCDFs can be prorated to that of 2,3,7,8-$T_4CDD$ using appropriate factors

**Recommended maximum allowable daily intake**
10 pg 2,3,7,8-$T_4CDD$ equivalent per kilogram body weight

In Table 3, the left-hand column shows estimated exposures for all major pathways for an average adult.

The ambient air level data for PCDDs and PCDFs are from Table 1.[16-18,22,29] Ambient levels in drinking water are based on Ontario's long-term monitoring of raw and treated tap water.[10,30,40] Soil data were those of Ontario and the U.S. EPA.[31,32] Food data are based on an analysis of Canadian,[24,33,34] Ontario,[26,27,35] U.S. FDA,[36] and German[37] food basket monitoring programs. Consumer products estimates were based on recent data on levels of PCDDs and PCDFs in bleached paper products.[25]

These measured PCDD and PCDF values were converted to 2,3,7,8-TCDD TEQ using the new international TEF scheme. The resulting calculated total

**Table 3. Comparison of Intakes of Average Urban Adult and an Adult Living Near an Incinerator**

| | Average Adult | Adult Near Incinerator |
|---|---|---|
| | (pg TEQ/kg body weight/day) | |
| Air | 0.07 | 0.3 |
| Water | 0.002 | 0.002 |
| Soil | 0.02 | 0.02 |
| Food | 2.28 | 2.28 |
| Consumer products | 0.01 | 0.01 |
| Total | 2.38 | 2.61 |

exposure, on average, is 2.38 pg TEQ/kg body weight/day. These estimates are approximations and not absolute values. The conservative assumptions used should over estimate rather than under estimate exposures.

As a check on how close these estimates are to real life, a calculation based on current levels of PCDDs and PCDFs in the adipose tissue of Canadian adults was made. In this calculation, it was assumed that levels in adult adipose tissue were in steady state equilibrium with the amounts absorbed from all sources and subsequently eliminated. Animal studies suggest that these processes may be adequately described using first order kinetics.[10] Therefore, published reports[38] were converted into a average level of TEQ in adipose tissue.

Using an elimination half-life of 5 years[39] and back calculating, the estimated daily intake of 2,3,7,8-TEQ per kg body weight per day is estimated to be 3 pg. This calculation suggests that the average exposure model in Table 3 is quite reasonable.

Also in Table 3, exposure for an adult living near an incinerator has been estimated by using the top end of the range of measured ambient air levels (Table 1) to estimate the increase in inhaled TEQ. While the overall increase in the TEQ intake is not large when compared to other sources such as the food pathway, the increase is significant.

In closing, I do want to point out that for both cases, total estimated exposure for an adult is still below the Ontario TDI of 10 pg TEQ/kg/body weight. I also want to point out that I have only examined the contribution of PCDD and PCDF in combustion emissions. There are other potentially toxic components of these emissions.

## REFERENCES


1. Rappe, C., M. Nygren, G. Lindstrom, H. R. Buser, O. Blaser, and C. Wuthrich. "Polychlorinated Dibenzofurans and Dibenzo-p-dioxins and Other Chlorinated Contaminants in Cow Milk from Various Locations in Switzerland," *Environ. Sci. Technol.* 21:964-970 (1987).
2. Yasuhara, A., H. Ito, and M. Morita. "Isomer-Specific Determination of Polychlorinated Dibenzo-p-dioxins and Dibenzofurans in Incinerator-Related Environmental Samples," *Environ. Sci. Technol.* 21:971–979 (1987).
3. Davies, K. "Concentrations and Dietary Intake of Selected Organochlorines, including PCBs, PCDDs and PCDFs in Fresh Food Composites Grown in Ontario, Canada," *Chemosphere* 17:263–276 (1988).
4. World Health Organization. "Environmental Health Series. 17. Dioxins and Furans from Municipal Incinerators," World Health Organization, Copenhagen (1987).

4a. Schroeder, W. H., and D. A. Lane. "The Fate of Toxic Airborne Pollutants," *Environ. Sci. Technol.* 22:240–246 (1988).

5. Travis, C., and H. A. Hattemer-Frey. "Human Exposure to Dioxin from Municipal Waste Incineration." Submitted.
6. U.S. EPA. "Methodology for the Assessment of Health Risks Associated with

Multiple Pathway Exposure to Municipal Waste Combustor Emissions," Draft–OAQPS, RTP, North Carolina/ECAO, Cincinnati, OH (1986b).
7. Fries, G. F., and D. J. Paustenbach. "A Critical Evaluation of the Factors Used in Assessing Incinerator Emissions as a Potential Source of TCDD in Foods of Animal Origin," Dioxin '87, Las Vegas, NV.
8. Birmingham, B., R. Clement, D. Harding, R. Pearson, D. Rokosh, W. Smithies, A. Szakolcai, B. Hanna Thorpe, H. Tosine, and D. Wells. "Chlorinated Dioxins and Dibenzofurans in Ontario–Analysing and Controlling the Risks. Development of Scientific Criteria Document Leading to Multi-Media Standards For Polychlorinated Dibenzo-p-dioxins (PCDDs) and Dibenzo Polychlorinated Dibenzofurans (PCDFs)," *Chemosphere* 15:1835–1850 (1986).
9. U.S. EPA "Dow Chemical Wastewater Characterization Study–Tittabawassee River Sediments and Native Fish," U.S. EPA–Region 5, Westlake, OH (1986a).
10. Ontario Ministry of the Environment. "Scientific Criteria Document for Standard Development No. 4-84. Polychlorinated Dibenzo-p-dioxins (PCDDs) and Polychlorinated Dibenzofurans (PCDFs)," Intergovernmental Relations and Hazardous Contaminants Coordination Branch, OME, Toronto, Ontario (1985).
11. Connett, P., and T. Webster. "Critical Factors in the Assessment of Food Chain Contamination by PCDD/PCDF from Incinerators," (1987); *Chemosphere* 18:1123–1129.
12. Boddington, M. "Final Plenary Session–Risk Management Considerations," Dioxin, Las Vegas, NV (1988); *Chemosphere* 18:1337–1339.
13. Nordisk Ministerrad. *Nordisk Dioxinriskbedomning*. Copenhagen (1988).
14. Ontario Research Foundation. "Airborne Trace Organic Emission Program at the Ontario Hydro Lambton Thermal Generating Station," Final Report–P-5177/1C (1986).
15. Environment Canada. "The National Incinerator Testing and Evaluation Program: Two-Stage Combustion (Prince Edward Island)," Report EPS 3/UP/1 (1985).
16. Eitzer, B. D., and R. A. Hites. "Concentrations of Dioxins and Dibenzofurans in the Atmosphere," *Int. J. Environ. Chem. Chem.* 27:215–230 (1986).
17. Eitzer, B. D., and R. A. Hites. "Dioxins and Furans in the Ambient Atmosphere: A Baseline Study," (1987); *Chemosphere* 18:593–598.
18. Smith, R. M., P. W. O'Keefe, D. R. Hilker, K. M. Aldous, S. H. Mo, and R. M. Stelle. "Ambient Air and Incinerator Testing for Chlorinated Dibenzofurans and Dioxins by Low Resolution Mass Spectrometry," (1987); *Chemosphere* 18:585–592.
18a. Smith, R. M., D. Hilker, P. O'Keefe, and K. M. Aldous. "Determination of Chlorinated Dibenzo-p-dioxins and Chlorinated Dibenzofurans: Toxics Ambient Air Monitoring Program," New York State Department of Health, Wadsworth Center for Laboratories and Research (May 21, 1986).
19. Czuczwa, J. M., and R. A. Hites. "Airborne Dioxins and Dibenzofurans: Sources and Fates," *Environ. Sci. Technol.* 20:195–200 (1986).
20. Czuczwa, J. M., and R. A. Hites. "Environmental Fate of Combustion-Generated Polychlorinated Dioxins and Furans," *Environ. Sci. Technol.* 18:444–450 (1984).
21. Kleopfer, R. D., W. W. Bunn, K. T. Yue, and D. J. Harris. "Occurrence of tetrachlorodibenzo-p-dioxin in environmental samples from southwest Missouri," in *Chlorinated Dioxins and Dibenzofurans in the Total Environment*, G.

Choudhary, L. H. Keith, and C. Rappe, Eds. (Boston: Butterworth Publishers, 1983), pp. 193–201.

22. New York State, Department of Environmental Conservation. "Ambient Air Monitoring for Chlorinated Furans and Dioxins at the New York State December Air Monitoring Station, Niagara Falls, New York," Niagara River Toxics Committee Report (1984).
23. Petty, J. D., L. M. Smith, P. A. Bergqvist, J. L. Johnson, D. L. Stalling, and C. Rappe. "Composition of Polychlorinated Dibenzofuran and Dibenzo-p-dioxin Residues in Sediments of the Hudson and Housatonic Rivers," in *Chlorinated Dioxins and Dibenzofurans in the Total Environment*, G. Choudhry, L. H. Keith, and C. Rappe, Eds. (Boston: Butterworth Publishers, 1983), pp. 203–208.
24. Ryan, J. J., R. Lizotte, T. Sakuma, and B. Mori. "Chlorinated Dibenzo-p-dioxins, Chlorinated Dibenzofurans, and Pentachlorophenol in Canadian Chicken and Pork Samples," *J. Agric. Food Chem.* 33:1021–1026 (1985b).
25. National Council of the Paper Industry for Air and Stream Improvement (NCASI). Special Report 87-11 and Technical Bulletin No. 534 (1987).
26. Birmingham, B., R. Clement, H. Tosine, G. Fleming, J. Ashman, J. Wheeler, B. D. Ripley, J. J. Ryan, B. Thorpe, and R. Frank. "Dietary Intake of PCDD and PCDF from Food in Ontario, Canada," *Chemosphere* 19:507–512 (1988).
27. Birmingham, B., A. Gilman, D. Grant, J. Salminen, M. Boddington, B. Thorpe, I. Wile, P. Toft, and V. Armstrong. "PCDD/PCDF Multimedia Exposure Analysis for the Canadian Population: Detailed Exposure Estimation," *Chemosphere* 19:637–642 (1988).
28. Gilman, A., B. Birmingham, D. Grant, J. Salminen, M. Boddington, B. Thorpe, I. Wile, P. Toft, and V. Armstrong. "PCDD/PCDF Multimedia Exposure Analysis for the Canadian Population: Overview," Dioxin '88, Umea, Sweden (1988).
29. Smith, R. M., D. Hilker, P. O'Keefe, K. M. Aldous, D. Spink, S. Connor, H. Valente, L. Wilson, and R. Donnelly. "Determination of Chlorinated Dibenzo-p-dioxins and Chlorinated Dibenzofurans: Toxics Ambient Air Monitoring Program," New York State Department of Health, Wadsworth Center for Laboratories and Research (August 15, 1986).
30. Ontario Ministry of the Environment. "A Survey of Selected Drinking Water Supplies in Ontario for Chlorinated Dibenzo-p-dioxins and Chlorinated Dibenzofurans" (1984).
31. McLaughlin, D. L., R. G. Pearson, and R. E. Clement. "Concentrations of PCDD and PCDF in Soil from the Vicinity of a Large Refuse Incinerator in Hamilton, Ontario," (1987); *Chemosphere* 18:851–854.
32. U.S. Environmental Protection Agency. "Soil Screening Survey at Four Midwestern Sites," EPA 905/4-85-005 (June 1985).
33. Ryan, J. J. Letter to OMAF-OME Toxics in Food Committee concerning Health Protection Branch data on PCDD and PCDF in food (October 28, 1987).
34. Ryan, J. J., P. Y. Lau, J. C. Pilon, D. Lewis, H. A. McLeod, and A. Gervais. "Incidence and Levels of 2,3,7,8-Tetrachlorodibenzo-p-dioxin in Lake Ontario Commercial Fish," *Environ. Sci. Technol.* 18:719–721 (1984).
35. Ontario Ministry of the Environment. "Polychlorinated Dibenzo-p-dioxins and Polychlorinated Dibenzofurans and Other Organochlorine Contaminants in Food" (August 1988).
36. Firestone, D., R. A. Niemann, L. F. Schneider, J. R. Gridley, and D. E. Brown. "Dioxin Residues in Fish and Other Foods," in *Chlorinated Dioxins and Dibenzo-*

*furans in Perspective*, C. Rappe, G. Choudhary, and L. Keith, Eds. (Chelsea MI: Lewis Publishers, 1986).

37. Mathar, W., H. Beck, K. Eckart, and R. Wittkowski. "Body Burden with PCDDs and from Food Intake in Germany," (1987): *Chemosphere* 18:417–424.
38. Ryan, J. J., R. Lizote, and B. P.-Y. Lau. "Chlorinated Dibenzo-p-dioxins and Chlorinated Dibenzofurans in Canadian Human Adipose Tissue, *Chemosphere* 14:697–706 (1985a).
39. Poiger, H., and C. Schlatter. "Pharmacokinetics of 2,3,7,8-TCDD in man," *Chemosphere* 15:1489–1494 (1986).
40. Ontario Ministry of the Environment. "Drinking Water Survey, St. Clair—Detroit River Area," Update (August 1986).
41. The Niagara River Toxics Committee. "Report of the Niagara River Toxics Committee." [Sponsored by the Ontario Ministry of the Environment, Environment Canada, U.S. Environmental Protection Agency and the New York State Department of Environmental Conservation].
42. Kociba, R. J., D. G. Keyes, J. E. Beyer, R. M. Carreson, E. E. Wade, D. A. Dittenber, P. O. Kalmins, L. F. Franson, P. N. Park, S. D. Barnard, P. A. Hummel and C. G. Humiston. "Results of a two year chronic toxicity and oncogenicity study of 2,3,7,8 tetrachlorodibenzo-p-dioxin (TCDD) in rats." *Toxicol. Appl. Pharmacol.* 46:279–303 (1978).

# List of Authors

Sumanta Acharya, Department of Mechanical Engineering, Louisiana State University, Baton Rouge, Louisiana

Elmar R. Altwicker, Department of Chemical Engineering, Rensselaer Polytechnic Institute, Troy, New York 12180-3590

E. J. Anthony, CANMET, Energy, Mines and Resources Canada, Ottawa, Ontario, CANADA K1A 0G1

Bruce A. Benner, Organic Analytical Research Division, National Institute of Standards and Technology, Gaithersburg, Maryland 20899

Brendan Birmingham, Hazardous Contaminants Coordination Branch, Ministry of the Environment, Toronto, Ontario, CANADA

D. Boomer, Spectroscopy Unit, Laboratory Services Branch, Ontario Ministry of the Environment, Rexdale, Ontario, CANADA

C. L. Bruffey, Environmental Technology Department, P.E.I. Associates, Inc., Cincinnati, Ohio

Raymond E. Clement, Dioxin/Mass Spectrometry Research, Laboratory Services Branch, Ontario Ministry of the Environment, Rexdale, Ontario, CANADA

V. A. Cundy, Department of Mechanical Engineering, Louisiana State University, Baton Rouge, Louisiana 70803

Lloyd A. Currie, Gas and Particulate Science Division, National Institute of Standards and Technology, Gaithersburg, Maryland 20899

Sergio S. Cutié, Analytical Sciences, Dow Chemical Company, Michigan Division Research and Development, Midland, Michigan 48667

Barry Dellinger, Environmental Sciences Laboratories, University of Dayton Research Institute, Dayton, Ohio 45469

Barry I. Diamondstone, Center for Analytical Chemistry, National Institute of Standards and Technology, Gaithersburg, Maryland 20899

G. A. Eiceman, Department of Chemistry, New Mexico State University, Las Cruces, New Mexico.

M. P. Esposito, Marketing and Business Development, Bruck, Hartman and Esposito, 4055 Executive Park Drive, Cincinnati, Ohio 45241

A. Finkelstein, Office of Solid Waste Division, Waste Management Branch, Environment Canada, Ottawa, Ontario, CANADA K1A 0H3

Raymond A. Freeman, Monsanto Company, St. Louis, Missouri

John H. Garrett, Toxic Contaminant Research Program, Wright State University, Dayton, Ohio 45435

E. Gibson, Department of Occupational and Environmental Health, D.O.F.A.S.C.O. Inc., P.O. Box 460, Hamilton, Ontario, CANADA L8N 3J5

L. A. Harden, Department of Chemistry, Wright State University, Dayton, Ohio

Holly A. Hattemer-Frey, Advanced Sciences Inc., Oak Ridge, Tennessee 37830

D. J. Hay, Office of Solid Waste Division, Waste Management Branch, Environment Canada, Ottawa, Ontario, CANADA K1A 0H3

R. V. Hoffman, Department of Chemistry, New Mexico State University, Las Cruces, New Mexico

O. Hutzinger, Ecological Chemistry and Geochemistry, University of Bayreuth, Bayreuth, WEST GERMANY D-8580

Ronald O. Kagel, Environmental Quality Department, The Dow Chemical Company, Midland, Michigan 48667

C. Kaiser-Farrell, Environmental Health and Safety, McMaster University, 1280 Main St. West, Hamilton, Ontario, CANADA L8S 4M1

F. W. Karasek, Department of Chemistry, University of Waterloo, Waterloo, Ontario, CANADA N2L 3G1

W. R. Kelly, Center for Analytical Chemistry, National Institute of Standards and Technology, Gaithersburg, Maryland

Ravic Kimar, Department of Chemical Engineering, Rensselaer Polytechnic Institute, Troy, New York

R. Klicius, Office of Solid Waste Division, Waste Management Branch, Environment Canada, Ottawa, Ontario, CANADA K1A 0H3

G. A. Klouda, Gas and Particulate Science Division, National Institute of Standards and Technology, Gaithersburg, Maryland 20899

N. V. Konduri, Department of Chemical Engineering, Renesselaer Polytechnic Institute, Troy, New York

Lester L. Lamparski, Analytical Sciences, Dow Chemical Company, Michigan Division Research and Development, Midland, Michigan 48674

G. K. Lee, CANMET, Energy, Mines and Resources Canada, Ottawa, Ontario, CANADA K1A 0G1

Christopher Leger, Department of Mechanical Engineering, Louisiana State University, Baton Rouge, Louisiana

Thomas W. Lester, Department of Mechanical Engineering, Louisiana State University, Baton Rouge, Louisiana

J. S. Lighty, Department of Chemical Engineering, University of Utah, Salt Lake City, Utah 84112

D. Mackay, Department of Chemical Engineering, University of Toronto, Toronto, Ontario, CANADA M5S 1A4

Nels H. Mahle, Environmental Analysis Research Laboratory, Dow Chemical Company, Midland, Michigan 48667

J. Marson, Approvals Branch, Maclaren Engineering, ATRIA N, Phase II, Willowdale, Ontario, CANADA

Robert A. Martini, Michigan Research and Development Operations, Dow Chemical Company, Midland, Michigan 48667

D. R. McCalla, Department of Biochemistry, McMaster University, Hamilton, Ontario, CANADA L8S 4M1

A. Melanson, Air Resources Branch, Ontario Ministry of the Environment, Toronto, CANADA

Raymond G. Merrill, Air and Energy Engineering Research Laboratory, U.S. Environmental Protection Agency, Research Triangle Park, North Carolina 27711

Alfred N. Montestruc, Department of Mechanical Engineering, Louisiana State University, Baton Rouge, Louisiana

John S. Morse, Department of Mechanical Engineering, Louisiana State University, Baton Rouge, Louisiana

K. P. Naikwadi, Department of Chemistry, University of Waterloo, Waterloo, Ontario, CANADA N2L 3G1

Terry John Nestrick, Analytical Sciences, Dow Chemical Company, Michigan Division Research and Development, Midland, Michigan 48674

John M. Ondov, Department of Chemistry and Biochemistry, University of Maryland, College Park, Maryland 20742

W. D. Owens, Department of Chemical Engineering, University of Utah, Salt Lake City, Utah 84112

V. M. Ozvacic, Air Resources Branch, Ontario Ministry of the Environment, Toronto, CANADA

Dale A. Pahl, Air and Energy Engineering Research Laboratory, U.S. Environmental Protection Agency, Research Triangle Park, North Carolina

C. N. Park, Dow Chemical Company, Midland, Michigan, 48667

Sally Paterson, Department of Chemical Engineering, University of Toronto, Toronto, Ontario, CANADA M5S 1A4

D. W. Pershing, Department of Chemical Engineering, University of Utah, Salt Lake City, Utah 84112

J. D. Phyper, Stelco Inc., Hamilton, Ontario, CANADA L8N 3T2

M. A. Quilliam, Atlantic Research Laboratory, National Research Council of Canada, 1411 Oxford Street, Halifax, Nova Scotia, CANADA B3H 3Z1

A. Reischl, Ecological Chemistry and Geochemistry, University of Bayreuth, Bayreuth, WEST GERMANY D-8580

Ronald Ryan, Alliance Technologies Corporation, Chapel Hill, North Carolina

Jeffrey A. Schonberg, Department of Chemical Engineering, Rensselaer Polytechnic Institute, Troy, New York

Jerry M. Schroy, Monsanto Company, St. Louis, Missouri

Walter M. Shaub, Coalition of Resource Recovery and the Environment (CORRE), U.S. Conference of Mayors, Washington, D.C.

Ann E. Sheffield, Gas and Particulate Science Division, National Institute of Standards and Technology, Gaithersburg, Maryland 20899

Wan Ying Shiu, Department of Chemical Engineering, University of Toronto, Toronto, Ontario, CANADA M5S 1A4

G. D. Silcox, Department of Chemical Engineering, University of Utah, Salt Lake City, Utah 84112

J. G. Solch, Toxic Contaminant Research Program, Wright State University, Dayton, Ohio 45435

Arthur M. Sterling, Department of Chemical Engineering, Louisiana State University, Baton Rouge, Louisiana

Robert K. Stevens, Source Apportionment Research, U.S. Environmental Protection Agency, Research Triangle Park, North Carolina 27711

C. Tashiro, Laboratory Services Branch, Ontario Ministry of the Environment, 125 Resources Road, Rexdale, Ontario, CANADA M9W 5L1

J. K. Taylor, Quality Assurance Consultant, 12816 Tern Drive, Gaithersburg, Maryland 20878

M. L. Taylor, Environmental Technology Department, P.E.I. Associates Inc., Cincinnati, Ohio

Philip H. Taylor, Environmental Sciences Laboratories, University of Dayton Research Institute, Dayton, Ohio 45469

Aaron J. Teller, Corporate, Air and Water Technologies—Research Cottrell Companies, Branchburg, New Jersey

R. C. Thurnau, Risk Reduction Engineering Laboratory, Environmental Protection Agency, Cincinnati, Ohio

Thomas O. Tiernan, Department of Chemistry, Director, Toxic Contaminant Research Program, Wright State University, Dayton, Ohio 45435

Debra A. Tirey, Environmental Sciences Laboratories, University of Dayton Research Institute, Dayton, Ohio 45469

Curtis C. Travis, Office of Risk Analysis, Health and Safety Research Division, Oak Ridge National Laboratory, Oak Ridge, Tennessee

Robert D. Van Dell, Applied Engineering Research and Process Development, Dow Chemical Company, Midland, Michigan 48667

Garrett F. VanNess, Toxic Contaminant Research Program, Wright State University, Dayton, Ohio 45435

Derryl J. von Lehmden, U.S. Environmental Protection Agency, Atmosphere Research and Exposure Assessment Lab, Research Triangle Park, North Carolina

Daniel J. Wagel, Toxic Contaminant Research Program, Wright State University, Dayton, Ohio 454435

H. Whaley, CANMET, Energy, Mines and Resources Canada, Ottawa, Ontario, CANADA K1A 0G1

Stephen A. Wise, Organic Analytical Research Division, National Institute of Standards and Technology, Gaithersburg, Maryland 20899

G. Wong, Air Resources Branch, Ontario Ministry of the Environment, Toronto, CANADA

David Zimmerman, Alliance Technologies Corporation, Chapel Hill, North Carolina

# INDEX